Drug Discovery with Privileged Building Blocks

Drug Discovery with Privileged Building Blocks

Tactics in Medicinal Chemistry

Jie Jack Li and Minmin Yang

CRC Press
Taylor & Francis Group
Boca Raton London New York

CRC Press is an imprint of the
Taylor & Francis Group, an **informa** business

First edition published 2022
by CRC Press
6000 Broken Sound Parkway NW, Suite 300, Boca Raton, FL 33487-2742

and by CRC Press
2 Park Square, Milton Park, Abingdon, Oxon, OX14 4RN

Library of Congress Cataloging-in-Publication Data
Names: Li, Jie Jack, author. | Yang, Minmin, author.
Title: Drug discovery with privileged building blocks : tactics in medicinal chemistry / Jie Jack Li, Minmin Yang.
Description: Boca Raton : CRC Press, 2021. | Includes bibliographical references and index.
Identifiers: LCCN 2021006031 | ISBN 9781032041735 (hardback) | ISBN 9781032041681 (paperback) | ISBN 9781003190806 (ebook)
Subjects: LCSH: Pharmaceutical chemistry. | Drugs—Design. | Drug development.
Classification: LCC RS403 .L52 2021 | DDC 615.1/9—dc23
LC record available at https://lccn.loc.gov/2021006031

ISBN: 978-1-032-04173-5 (hbk)
ISBN: 978-1-032-04168-1 (pbk)
ISBN: 978-1-003-19080-6 (ebk)

Typeset in Times
by codeMantra

Contents

Preface

Who has time to read a book these days?! This book is the answer to the question. Our favorite feature of this book, like the title entails, is a collection of gems in medicinal chemistry. Each chapter may be read and studied without learning the previous chapters. Therefore, the reader can peruse each chapter as a separate mini-book. Even though there are some overlaps, they make each chapter independent of the other.

The genesis of this book traces back to PharmaBlock's founding philosophy of designing privileged building blocks. High-quality building blocks are crucial not only to molecules' biological activities but also to ADMET properties, which eventually will impact the success rate of drug discovery projects. A thorough study of how building blocks perform in drug molecules and regular analysis of new building block structures in the latest researches has proven to be a fruitful strategy to generate novel building blocks. Using this strategy, PharmaBlock has supplied the drug industry with a great number of building blocks, which are increasingly being adopted by drug hunters. We hope this book will be a good starting point for novice medicinal chemists and veteran medicinal chemists who find it useful as well.

We are very much indebted to Dr. Vivek Kumar at Albany Molecular Research Inc. for proofreading the entire manuscript. As always, we welcome your critique. Please email your comments to: lijiejackli@hotmail.com or minminyang@pharmablock.com

Jie Jack Li and Minmin Yang
January 1, 2021

Authors

Dr. Jie Jack Li is an established chemist with over 20 years of experience in both medicinal chemistry and process chemistry. He is also widely published as an author or co-author of 34 peer-reviewed articles, 12 patents, and 29 books. Prior to joining ChemPartner, Dr. Li worked at Pfizer, BMS, and Revolution Medicines in oncology, antivirals, metabolic disease, CNS, anti-inflammatory, and dermatology, targeting enzymes, receptors, or ion channels. Dr. Li was also a professor at the University of San Francisco for four years, teaching organic and medicinal chemistry. He earned his Ph.D. from Indiana University and was a post-doctoral fellow at MIT.

Dr. Minmin Yang received his Ph.D. in organic synthesis from Auburn University, US in 2001 and did post-doctoral training at the same school from 2002 to 2004. Prior to founding PharmaBlock, he worked with increasing responsibilities from a group leader to a section head of medicinal chemistry at Roche Palo Alto and Roche China for 4 years. Dr. Yang has authored 80 publications, including 16 international patents.

1 Alkynes

The alkyne fragment may be found in approximately two dozen drugs on the market in a variety of therapeutic areas. Despite being a structural alert, giving rise to reactive metabolites on occasions, the alkyne motif has a special place in drug discovery, serving as a unique bioisostere for many functional groups such as halogen, nitrile, carboxamide, ethyl, and so on; non-polar rigid and sterically less demanding spacers; and a part of a warhead of targeted covalent inhibitors.

ALKYNE-CONTAINING DRUGS

norethindrone (Norlutin, **1**)
Syntex, 1957
progestin antagonist

mifepristone (RU-486, **2**)
Roussel-Uclaf, 2000
anti-progestin and anticorticoid

Marketed in the late 1950s, the alkyne-containing drug norethindrone (Norlutin, **1**) was revolutionary as the first oral contraceptive pill working as a progesterone receptor antagonist.[1] Afterward, approximately a dozen "me-too" alkyne-containing steroid drugs followed as birth control pills. More successful ones include Roussel Uclaf's mifepristone (Korlym, RU486, **2**),[2] Schering's ethinyl estradiol (Yaz in combination with drospirenone), levonorgestrel (Liletta, Plan B) as an emergency birth control pill, and so on. Mifepristone (**2**) is also a mechanism-based inhibitor (MBI) of CYP 3A4 apoprotein and its alkyne motif is implicated as a ketene precursor during metabolism.

Alkyne-containing drug selegiline [(–)-deprenyl, Eldepryl, **3**], a weak and selective inhibitor of monoamine oxidase-B (MAO-B), was marketed by Knoll in 1971 as an antidepressant.[3] Now it has been explored in the transdermal treatment of neurodegenerative disorders and neurological diseases such as Parkinson's and Alzheimer's diseases. On the other hand, Novartis's terbinafine (Lamisil, **4**) is an orally and topically active hindered alkyne-containing allylamine antifungal agent. It is a squalene epoxidase inhibitor.[4] Terbinafine (**4**) causes rare cases of hepatotoxicity possibly because it undergoes *N*-dealkylation to a reactive iminium and an aldehyde.

selegiline (Eldepryl, **3**)
Knoll, 1971
MAO-B inhibitor

terbinafine (Lamisil, **4**)
Novartis, 1991
squalene epoxidase inhibitor

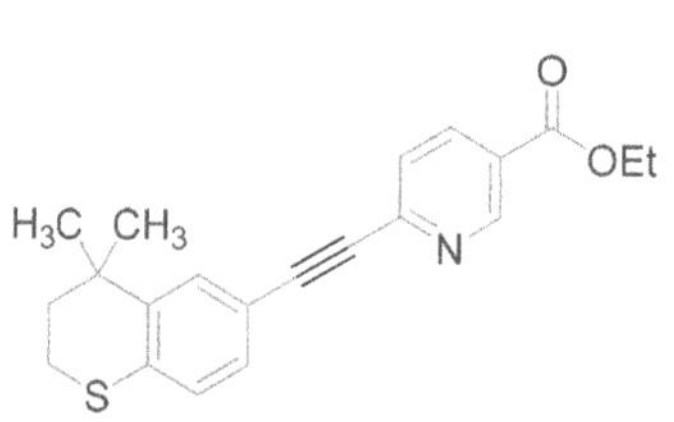

tazarotene (Tazorac, **5**)
Allergan, 1996
retenoid

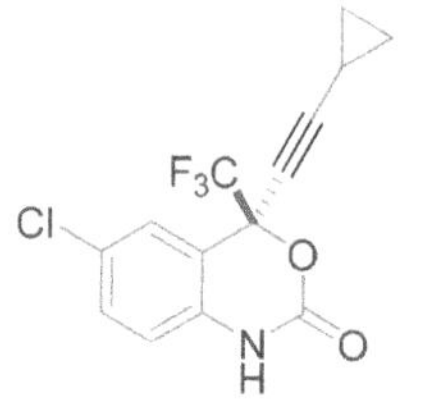

efavirenz (Sustiva, **6**)
BMS/Merck, 1998
NNRTI

Alkyne-containing tazarotene (Tazorac, **5**) is developed by Allergan as a treatment for acne vulgaris. It is a topical receptor-selective retinoid that normalizes the differentiation and proliferation of keratinocytes. Its major metabolite, tazarotenic acid, binds to retinoic acid receptors (RARs) with high affinity.[5] Like most topically administered retinoids, tazarotene (**5**) may avoid some of those drawbacks including teratogenicity by systemically administered retinoids such as isotretinoin (Accutane). On the other hand, efavirenz (Sustiva, **6**) is an alkyne-containing non-nucleoside reverse transcriptase inhibitor (NNRTI) developed by BMS/Merck for the treatment of AIDS.[6] It is an ingredient of the cocktail treatment of HIV along with a protease inhibitor (PI), and a nucleoside reverse transcriptase inhibitor (NRTI).

Some additional alkyne-containing drugs emerged in the 2000s. Schering's iloprost (Ilomedine, **7**) is an analog of prostacyclin (epoprostenol, PGI_2, produced in the vascular endothelium) with improved metabolic and chemical stability. It is a platelet aggregation inhibitor prescribed for the treatment of peripheral vascular diseases.[7] Inhaled iloprost (**7**) has been used for the treatment of primary pulmonary hypertension (PPH). In addition, among more than 60 protein kinase inhibitors on the market today, one of the only two alkyne-containing drugs is erlotinib (Tarceva, **8**), which is an epithelial growth factor receptor (EGFR) kinase inhibitor approved for the treatment of non-small cell lung cancer (NSCLC) and pancreatic cancer.[8] The other alkyne-containing kinase inhibitor is ponatinib (Iclusig, **33**, *vide infra*).

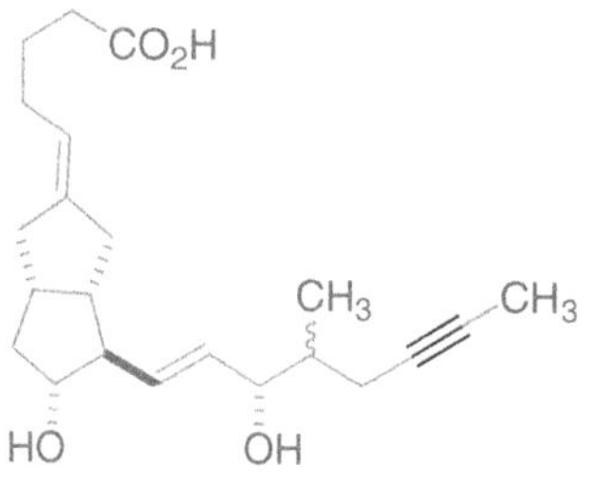

iloprost (Ilomedine, **7**)
Schering, 2004
platelet aggregation inhibitor

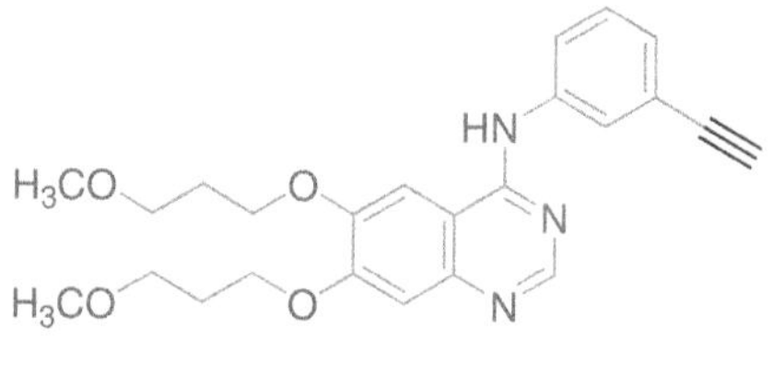

erlotinib (Tarceva, **8**)
OSI, 2004
EGFR kinase inhibitor

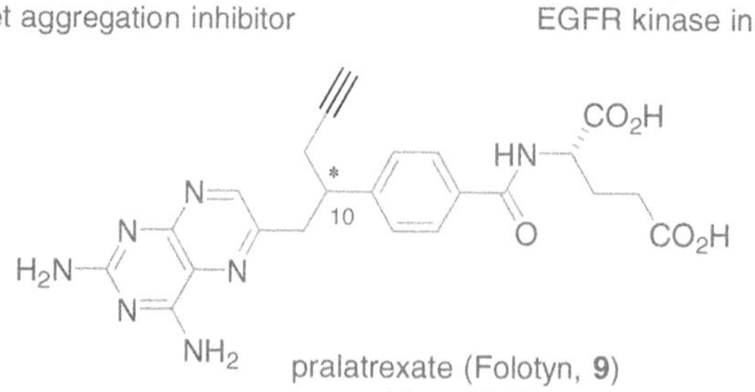

pralatrexate (Folotyn, **9**)
Allos, 2009
dihydrofolate reductase inhibitor
refractory peripheral T-cell lymphoma

Allos's pralatrexate (Folotyn, **9**) is an antifolate antimetabolite approved by the FDA in 2009 for the treatment of relapsed or refractory peripheral T-cell lymphoma (PTCL).[9] In 2020, it received the orphan drug designation in the EU for the treatment of cutaneous T-cell lymphoma (CTCL). Its distinctive ethynylethyl motif is a bioisostere of the methylamine fragment of methotrexate, the prototype of this class of thymidylate inhibitors that block the functions of dihydrofolate reductase (DHFR). Given in its injectable form, pralatrexate (**9**) has a superior potency and toxicity profile to other DHFR inhibitors. For chemistry aficionados, the stereochemistry on C-10 is racemic, and thus pralatrexate (**9**) is a mixture of two diastereomers.

Approximately ten dipeptidyl peptidase-IV (DPP-4) inhibitors known as "gliptins" are currently on the market for the treatment of type II diabetes mellitus (T2DM). Boehringer Ingelheim's orally active DPP-4 inhibitor linagliptin (Tradjenta, **10**) was approved by the FDA in 2011. *In vitro*, linagliptin (**10**) has an IC_{50} value of 1 nM, more potent than sitagliptin (19 nM), alogliptin (24 nM), saxagliptin (50 nM), and vildagliptin (62 nM). Furthermore, linagliptin (**10**) also exhibited prolonged pharmacodynamic activity with long-lasting DDP-4 in several preclinical species.[10]

linagliptin (Tradjenta, **10**)
Boehringer Ingelheim, 2011
DPP-4 inhibitor

acalabrutinib (Calquence, **11**)
AZ/Acerta, 2017
BTK inhibitor

Recently, targeted covalent inhibitors have emerged as an exciting strategy to overcome the perpetual struggle against cancer cells' drug resistance. Bruton's tyrosine kinase (BTK) is a non-receptor tyrosine kinase. Inhibition of BTK activity prevents downstream activation of the B-cell receptor (BCR) pathway and subsequently blocks cell growth, proliferation, and survival of malignant B cells. The second BTK inhibitor acalabrutinib (Calquence, **11**) gained approval from regulatory agencies in 2017.[11] While it works through the same mechanism of action (MOA) as its progenitor ibrutinib (Imbruvica), acalabrutinib (**11**)'s warhead is a but-2-ynamide. Ibrutinib and acalabrutinib (**11**) are known as the first-generation BTK inhibitors. Although they can do wonders to cancer patients afflicted with B-cell malignancies, 75% of them develop resistance to them within 2 years. Close scrutiny revealed that a substitution of serine for cysteine at residue 481 (C481S) took place. Such a mutation led to a less nucleophilic serine so that the first-generation BTK inhibitors are no longer effective. Efforts are underway to discover the second-generation BTK inhibitors.

ALKYNES IN DRUG DISCOVERY

Hench and Kendall's landmark discovery of cortisone (**12**) in the late 1940s heralded the steroid era. Isolation and later synthesis by Marker of progesterone (**13**) promoted research on its pharmacological properties in the treatment of menstrual disorders. Two features of progesterone (**13**) stand out. One, it has poor oral bioavailability and thus has to be given intravenously. Two, almost all initial chemical alterations of the progesterone (**13**) molecule would either diminish or destroy its biological activity. Using today's jargon, the structure–activity relationship (SAR) of progesterone (**13**) is very narrow. Syntex succeeded in synthesizing the alkyne-containing drug norethindrone (**1**).[1] Removal of the C-19 methyl afforded the 19-nor-progesterone that was 4- to

8-times more potent than progesterone (**13**). More importantly, installation of the ethynyl group at the C-17 position not only did not abolish norethindrone (**1**)'s antiprogestin activity, but also effectively blocked CYP450 enzymatic oxidation to estrone (**14**), making the drug more orally bioavailable. Here the ethynyl moiety indeed plays a crucial role in its biological activity and oral bioavailability.

cortisone (**12**)   progesterone (**13**)

norethindrone (Norlutin, **1**)   estrone (**14**)

The unique contributions of the ethynyl moiety are further demonstrated by comparing the bioavailability of estradiol (**15**) and ethinylestradiol (**16**), respectively. The C-17 alcohol of estradiol (**15**) is vulnerable to CYP450 enzymatic oxidation to estrone (**14**), thus it only has a half-life of less than 30 minutes and an ~4% bioavailability. In stark contrast, ethinylestradiol (**16**) has a half-life of 10 hours and 51% bioavailability. More interestingly, even though ethinylestradiol (**16**) is a MBI of CYP3A4 *in vitro*, no impact on CYP3A4 activity was observed in women taking the pill. A small clinical dose was most likely the reason for the absence of drug–drug interactions (DDIs).

estradiol (**15**)   ethinylestradiol (**16**)
$t_{1/2} < 0.5$ h, F 3–5% $t_{1/2}$ 10 h, F 51%

One of the latest oral contraceptives is Roussel Uclaf's mifepristone (**2**), approved in 2000. Its unique structural feature is the installation of a *para*-dimethylaminophenyl substituent on the C-11 position of the steroid skeleton. It has a half-life of 15–48 hours and 40%–70% of oral bioavailability. All in all, mifepristone (**2**) is a perfectly fine medicine. However, its inventors' repeated statement, "The discovery of RU486 [mifepristone (**2**)] and its potent activity as an antiglucocorticoid and antiprogestin brought the long story on steroid hormones and antihormones to its logical conclusion... the armamentarium of the steroid endocrinologist is by now complete... It thus seemed that the last gap in sexual antihormone research has been successfully closed",[12] is hazardous. In science, touting your own achievement as the self-claimed pinnacle is always precarious.

The metabolism of selegiline (**3**) is quite interesting. After it was taken *in vivo*, three major metabolites **17–19** were detected. In addition to the demethylation metabolite, desmethyselegiline (**17**), the propargyl group was metabolically removed to afford methamphetamine (**18**) under the influence of monoamine oxidase (MAO). Methamphetamine (**18**) was further metabolized by CYP-450 to give another metabolite, (*R*)-(–)-amphetamine (**19**).[13] Since both methamphetamine (**18**) and (*R*)-(–)-amphetamine (**19**) are illicit drugs, is it now against the law to take selegiline (**3**)?!

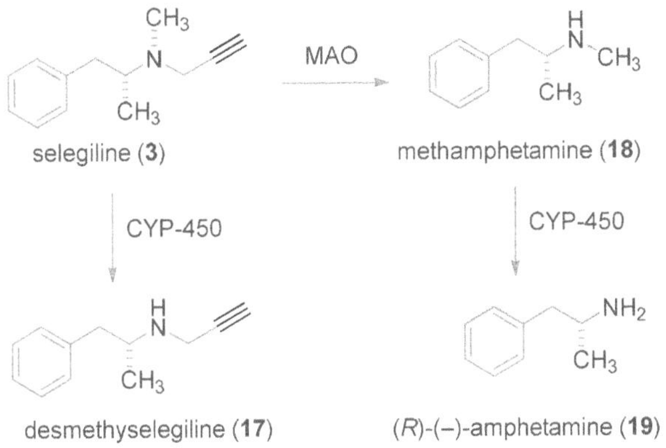

selegiline (**3**)

methamphetamine (**18**)

desmethyselegiline (**17**)

(*R*)-(–)-amphetamine (**19**)

Speaking of metabolism, the alkyne moieties are considered structural alerts because some of the alkyne-containing drugs could potentially be metabolized to reactive metabolites. A general scheme is shown below. Under the action of CYP450, an alkyne is potentially oxidized to reactive intermediates such as a ketocarbene or a ketene via the intermediacy of an oxirene.

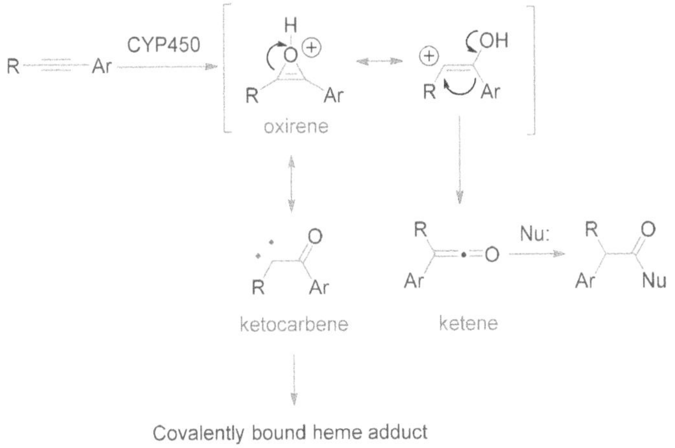

oxirene

ketocarbene   ketene

Covalently bound heme adduct

A case in point is BMS/Merck's alkyne-containing NNRTI efavirenz (**6**).[13] Intensive investigations of the metabolites indicated that efavirenz (**6**) added glutathione (GSH) after α-hydroxylation of cyclopropyl in addition to adducts of some reactive metabolism intermediates with nucleophiles *in vivo*.[14a] In a rat specific metabolic pathway, one of the glutathione-containing reactive metabolites, facilitated by acivicin (**19**), formed a cysteinylglycine adduct **20**, which showed species-specific nephrotoxicity in rats, but not in monkeys and humans. It was also shown that deuterated analog **21** replacing H by D slowed the metabolism and reduced the incidence and severity of nephrotoxicity.

In humans, however, the alkyne functionality in efavirenz (**6**) was latent toward oxidative metabolism and monohydroxylaion was observed on the C-8 position on the aromatic ring and on the cyclopropyl methine.[14b]

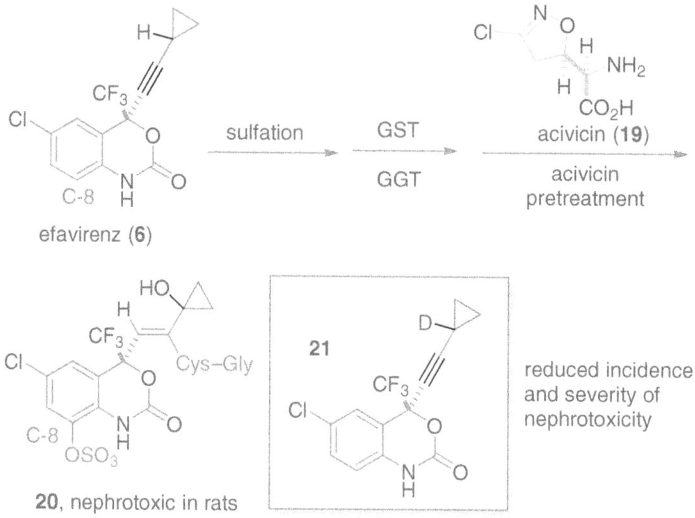

efavirenz (**6**)

sulfation → GST / GGT → acivicin (**19**) / acivicin pretreatment →

20, nephrotoxic in rats

21

reduced incidence and severity of nephrotoxicity

Tactics exist to minimize an alkyne's metabolic instability by replacing it with a metabolically more stable bioisostere. Sazetidine-A (**22**) was discovered as an α4β2 subtype selective partial agonist of neural nicotinic acetylcholine receptors (nAChRs). It showed potent analgesic effects in animal models and robust antidepressant activities. However, the acetylenic bond in sazetidine-A (**22**) may be oxidized to generate a labile, highly reactive oxirene. This metabolic transformation is a potential source of toxicity and discouraged further advancement of the compound. Novel analogs were designed to avoid the acetylene functional group while maintaining the important pharmacophoric element of sazetidine-A (**22**). Kozikowski's group succeeded in obtaining pyridine-isoxazole ether **23**, which was selective, potent, $K_i(\alpha 4\beta 2)=0.7\,\text{nM}$, and efficacious, $EC_{50}(\alpha 4\beta 2)=36\,\text{nM}$, and showed antidepressant-like behavior in the mouse forced swim test at 10 mg/kg following either intraperitoneal (IP) or PO administration. The isoxazole analog **23** also had a good absorption, distribution, metabolism, and excretion–toxicity (ADMET) profile.[15]

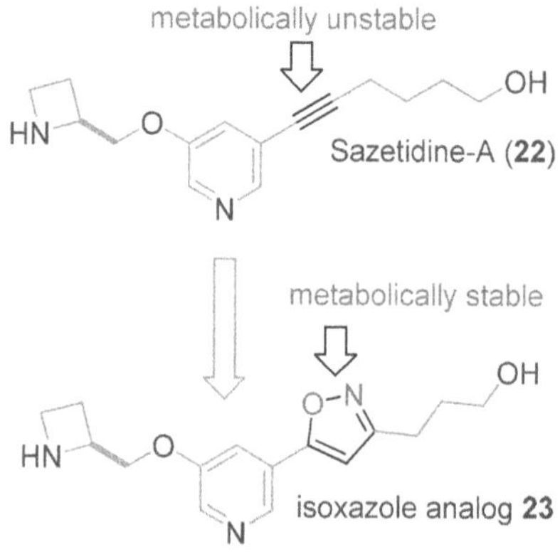

metabolically unstable

Sazetidine-A (**22**)

metabolically stable

isoxazole analog **23**

(S)-alkyne **24**

isoxazole **25**

thiazole **26**

In addition to using oxazole as a metabolically more stable bioisostere of acetylene, thiazole has been employed for the same purpose as well. Gu and coworkers at Abbott discovered an alkyne-containing drug **24** as a selective acetyl-CoA-2 (ACC-2) inhibitor as a potential treatment of type II diabetes mellitus (T2DM).[16a] Regrettably, a preliminary safety evaluation of **24** revealed serious neurological and cardiovascular liabilities of this chemotype. A systemic structure–toxicity relationship (STR) study identified the *alkyne linker* as the key motif responsible for these adverse effects. Replacement of the alkyne linker with alternative linker groups led to new ACC-2 inhibitors such as isoxazole **25** and thiazole **26** with drastically improved cardiovascular and neurological profiles.[16b]

In 2020, Talele published an excellent review titled "Acetylene Group, Friend or Foe in Medicinal Chemistry".[17] He summarized the roles an alkyne group plays in drug activities including potency enhancement by a complimentary fit into a receptor binding pocket, reactive warhead, non-polar linear rigid spacer, bioisostere to a wide range of functionalities, and modulating of the drug metabolism pharmacokinetic (DMPK) profile.

The ethynyl group serves as a non-classical bioisostere because of its versatility. Its π cloud is useful for mimicking aromatic systems, and its polarized –CH moiety is a weak hydrogen bond donor, similar to halogens.[18a]

The EGFR kinase inhibitor erlotinib (**8**) has a distinctive terminal acetylene group, which serves as an isostere of the chlorine atom on gefitinib (Iressa, **27**), the prototype of EGFR inhibitors. The chlorine on gefitinib (**27**) forms a weak halogen bonding interaction with the backbone carbonyl oxygen atom of Leu788 within the active site of the ATP binding pocket of EGFR2.[18b] Similar to the chlorophenyl motif on gefitinib (**27**), the 3-ethynylphenyl group occupies the back hydrophobic pocket and makes hydrophobic contact with the gatekeeper region of the EGFR kinase protein. Indeed, the hydrophobic part of erlotinib (**8**) is surrounded by five leucines in addition to Val702, Ala719, and Lys721. The hydrophobic interaction was assumed to be the significant contribution to optimum lipophilicity of molecules to cross the membrane. In addition, both C–Cl and C≡CH show a positive charge at the tip of these groups.[18a] Since an acetylene fragment also has a polarized –CH group that can serve as a weak hydrogen bond donor, it mimics the chlorine atom on the C-8 position as confirmed by its cocrystal structure with EGFR2.[18c]

The metabolism of erlotinib (**8**) follows the general scheme of that of a terminal alkyne. Therefore, metabolic oxidation of erlotinib (**8**) by CYP450 3A4 gives rise to two reactive metabolites oxirene **28** and ketene **29**, eventually leading to stable metabolite acetic acid **30**. The two reactive metabolites, oxirene **28** and ketene **29**, are response for causing time-dependent inhibition (TDI) of CYP3A4.[19]

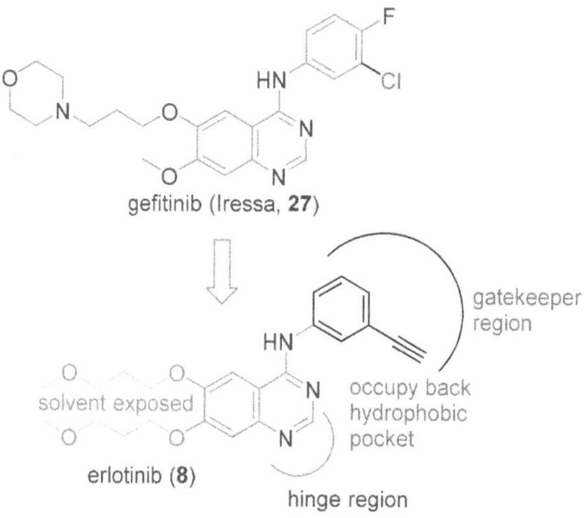

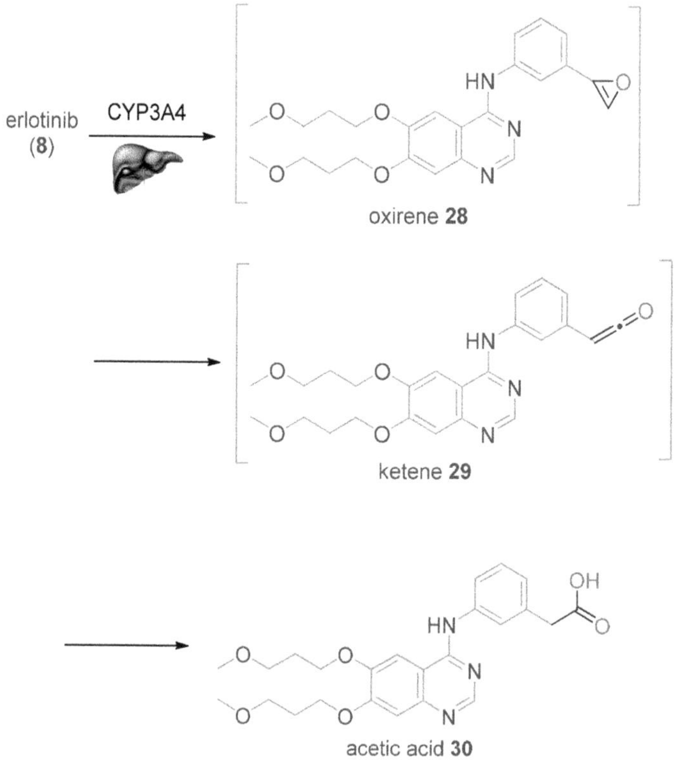

gefitinib (Iressa, **27**)

erlotinib (**8**)

gatekeeper region

occupy back hydrophobic pocket

solvent exposed

hinge region

erlotinib (**8**) CYP3A4

oxirene **28**

ketene **29**

acetic acid **30**

As alluded earlier, the alkyne motif often serves as a spacer of a drug molecule.

Isotretinoin (Accutane, **31**) is a naturally occurring metabolite of vitamin A. It is an orally active drug for the treatment of severe refractory acne that functions by reducing sebaceous gland size and sebum production. Tazarotene (**5**) is a topical receptor-selective retinoid that normalizes the differentiation and proliferation of keratinocytes. Its major metabolite, tazarotenic acid, binds to RARs with high affinity. As far as the internal alkyne on tazarotene (**5**) is concerned, it serves as a spacer to afford the conformationally constrained non-isomerizable analog of the flexible isotretinoin (**31**). Thanks to its rigidity, tazarotene (**5**) is more selective for RARβ and RARγ.[20]

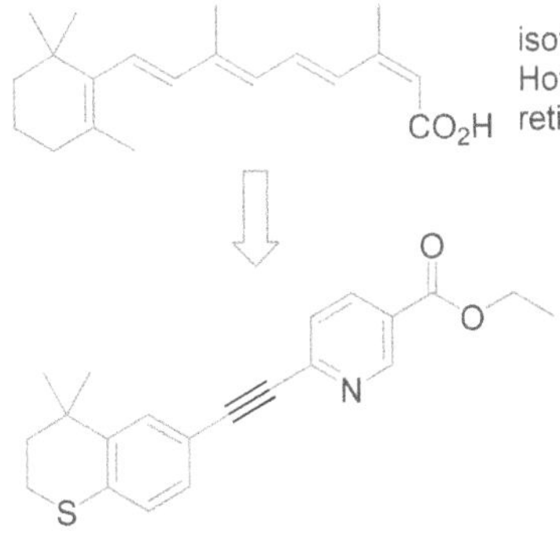

isotretinoin (Accutane, **31**)
Hoffman-La Roche, 1982
retinoic acid analog

tazarotene (**5**)

In the same vein, Ariad's ponatinib (Iclusig, **33**) also has an internal alkyne as a spacer to afford a conformationally more constrained BCR-ABL inhibitor.[21] Novartis's imatinib (Gleevec, **32**) was the first kinase inhibitor on the market. It is a breakpoint cluster region-Abelson (BCR-ABL) tyrosine kinase inhibitor to treat chronic myeloid leukemia (CML). Regrettably, the T315I gatekeeper mutant has emerged as resistant to imatinib (**32**) when amino acid threonine at the 315 position of the protein mutated to isoleucine. The flag methyl group on imatinib (**32**) would then have unfavorable steric hydrophobic interaction with the Ile315 bulky side chain and lost binding potency. Ariad discovered a novel series of potent pan-inhibitors of BCR-ABL, including the T315I mutation as represented by ponatinib (**33**). Here the triple bond linker between the adenine-binding fragment and allosteric pocket-binding fragment skirts the increased bulk of the I315 bulky side chain. The least sterically demanding acetylene linker made more favorable van der Waals's contact with gatekeeper Ile315 and Phe382 of the DFG motif. As a consequence, ponatinib (**33**) inhibited the kinase activity of both native BCR-ABL and the mutant T315I mutant with low nM IC_{50}s.[21] Very interestingly, ponatinib (**33**) suffers some toxicities such as blood clots and narrowing of blood vessels. Are they the direct results of reactive metabolites from the alkyne motif?

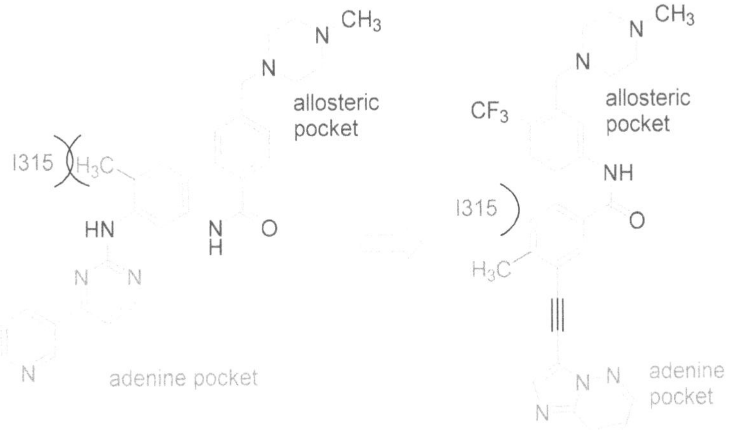

imatinib (Gleevec, **32**)
Novartis, 2008
Bcr-abl inhibitor

ponatinib (Iclusig, **33**)
Ariad, 2012
Bcr-abl/VEGFR inhibitor

DPP-4 inhibitors block the functions of the DPP-4 enzyme and inactivate glucogen-like protein-1 (GLP-1) and incretin, which stimulate insulin release and inhibit glucagon release. The end result of taking DPP-4 inhibitors is lowering of blood glucose.

Boehringer Ingelheim's DPP-4 inhibitor linagliptin (**10**) is more potent than many of its gliptin analogs. Installation of the 2-butynyl substituent eliminated the human ether-a-go-go (hERG) channel and M_1 receptor issues associated with its corresponding *N*-dimethylallyl analog. Scrutiny of its binding to the DPP-4 enzyme revealed that the 2-butynyl fragment at the N-7 position offers a favorable binding to the S_1 hydrophobic pocket on the DPP-4 protein as shown below. In the event, not only did the 2-butynyl group help with potency and selectivity, but also it contributed to the long residence time of linagliptin (**10**).[22]

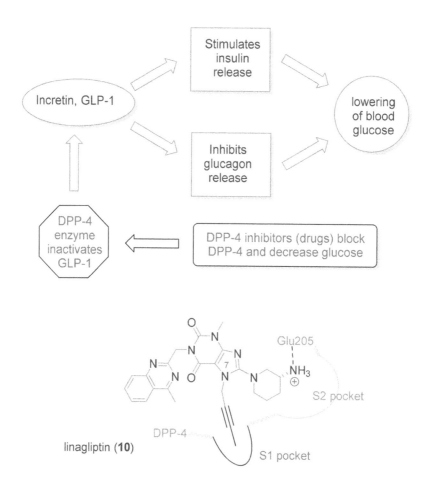

linagliptin (**10**)

Pharmacyclics' ibrutinib (Imbruvica, **34**) is the first-in-class BTK inhibitor for treating mantle cell lymphoma, chronic lymphocytic leukemia, and Waldenstrom's macroglobulinemia. AstraZeneca/Acerta's covalent BTK inhibitor acalabrutinib (**11**) has an ynamide "warhead" in place of ibrutinib (**34**)'s acrylamide warhead. Both the acrylamide and ynamide warheads form a covalent C–S bond with BTK's cysteine-481 amino acid.[23] Regrettably, after approximately 1 year's administration, both ibrutinib (**34**) and acalabrutinib (**11**) begin developing resistance via C481S mutation. Namely, the cysteine-481 amino acid on the BTK protein mutates to serine, which no longer can form a covalent bond with the first-generation BTK inhibitor drugs. Therefore, we need second-generation BTK inhibitors.

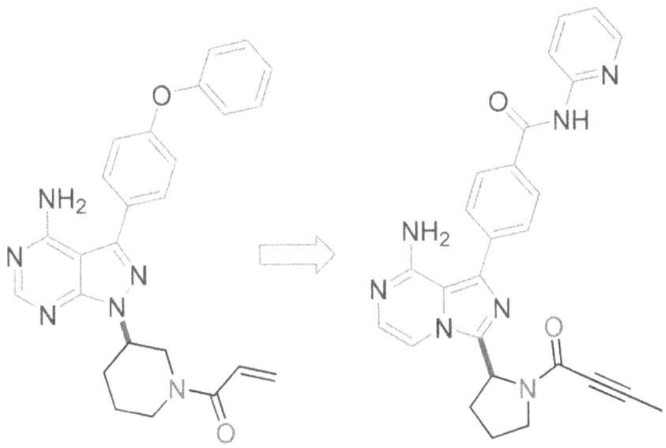

ibrutinib (Imbruvica, **34**)
Pharmacyclics/Jansen, 2013
BTK Inhibitor

acalabrutinib (**11**)

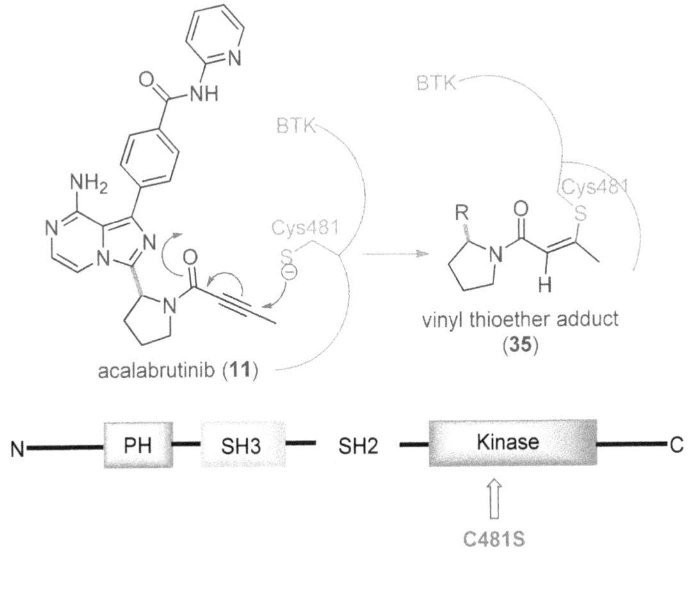

acalabrutinib (**11**)

vinyl thioether adduct
(**35**)

N ▬ PH ▬ SH3 ▬ SH2 ▬ Kinase ▬ C

C481S

SYNTHESIS OF SOME ALKYNE-CONTAINING DRUGS

The key operation of the Allergan's discovery synthesis of tazarotene (**5**) is a Negishi coupling of alkynylzinc chloride **38** with α-chloropyridine **39**. Thus, the transformation of methyl ketone **36** to terminal alkyne **37** was accomplished via the formation of an enol phosphate intermediate. Alkynylzinc chloride **38** was generated *in situ* by treatment of terminal alkyne **37** with *n*-BuLi followed by the addition of $ZnCl_2$. Subsequently, the Negishi coupling of **38** with α-chloropyridine **39** using $Pd(Ph_3P)_4$ as the catalyst delivered tazarotene (**5**).[24]

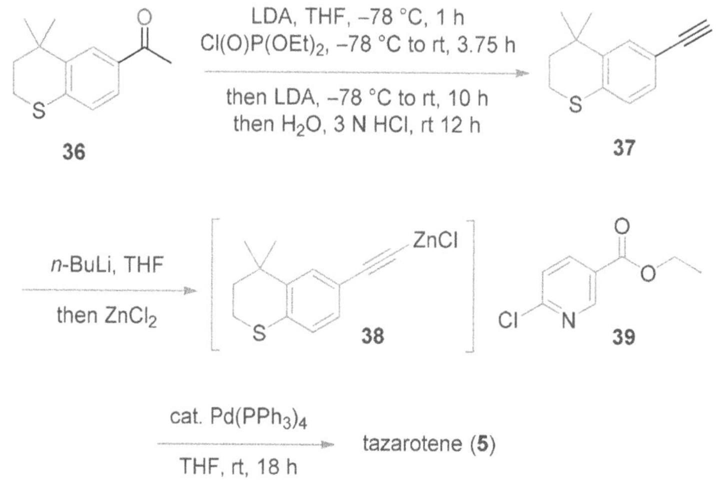

While the aforementioned synthesis of tazarotene (5) was adequate for discovery chemistry, but not for large-scale and manufacturing processes since its use of LDA, *n*-BuLi, and so on, a practical and efficient process for the preparation of tazarotene (5) was reported in 2005.[25] The synthesis used crystalline *S*-oxide 40 as the starting material for the Sonogashira coupling with acetylene 41 to assemble adduct 42. Indeed, 2-methyl-3-buty-2-ol (41) is an inexpensive synthetic equivalent of acetylene. Treatment of internal alkyne 42 with a catalytic amount of base led to terminal alkyne 43 and acetone. Another Sonogashira coupling of 43 with α-chloropyridine 39 then assembled tazarotene *S*-oxide 44. Reduction of the *S*-oxide was carried out using PCl₃ in DMF to deliver tazarotene (5). The entire process avoided expensive starting materials and dangerous reagents. Introduction of the sulfoxide was beneficial for both reaction and purification.

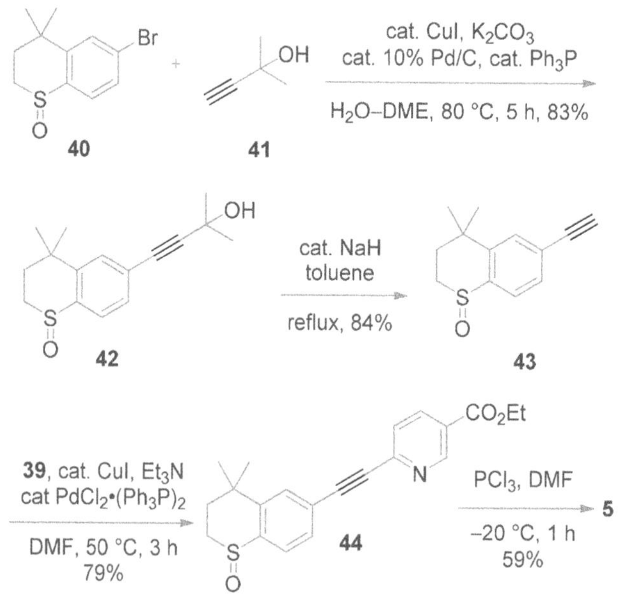

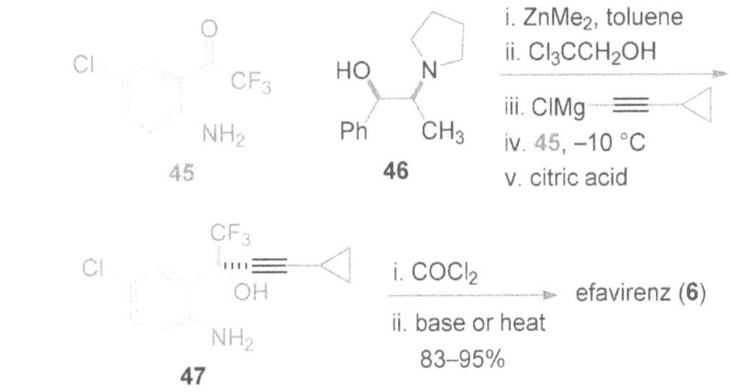

Merck's process synthesis of efavirenz (**6**) is more efficient. In a general procedure, a solution of the chiral ligand **46** is treated with dimethylzinc. This mixture is stirred for 1 hour followed by the addition of 2,2,2-trichloroethanol as an additive. This solution is mixed with a solution of chloromagnesium acetylide. The mixture is stirred for about 30 minutes and then cooled to −10°C. A solution of trifluoroketone **45** is added and the mixture is allowed to stir for 7 hours to afford chiral alcohol **47** after treatment with citric acid. The final step of the synthesis is to complete the benzoxazinone ring of efavirenz (**6**). The most direct and economically desirable route utilizes phosgene (THF–heptane, 0°C–25°C). After aqueous sodium bicarbonate workup, efavirenz (**6**) was crystallized in excellent yields (93%–95%) with excellent chemical and optical purities (>99.5%, >99.5% ee).[26]

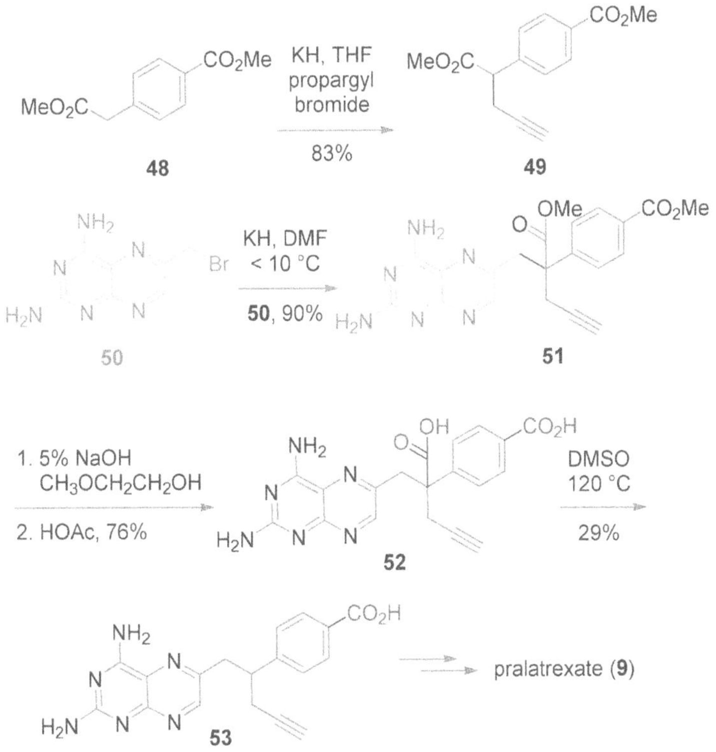

Synthesis of Allos's dihydrofolate reductase (DHFR) inhibitor pralatrexate (**9**) commenced with the installation of the propargyl fragment onto bis-ester **48** to produce **49**. Another S_N2 reaction between **49** and bromide **50** assembled tertiary ester **51**. Hydrolysis of **51** led to bis-acid **52**, which underwent a thermally induced decarboxylation selectively to afford carboxylic acid **53**. Installation of the glutamate moiety then delivered pralatrexate (**9**).[27]

Preparation of Ariad's BCR-ABL inhibitor ponatinib (**33**) employed the Sonogashira coupling twice.[21] The first Sonogashira coupling was between 3-bromoimidazo[1,2-*b*]pyridazine (**54**) and ethynyltrimethylsilane to assemble **55**. After desilylation, the resulting terminal alkyne **56** was coupled with iodide **57** to deliver ponatinib (**33**) after the formation of the HCl salt.[21]

Netherland-based Acerta (now part of AstraZeneca) discovered acalabrutinib (**11**), a targeted covalent BTK inhibitor with an ynamide "warhead". The final step of their discovery synthesis route was the amide formation of pyrrolidine **58** with but-2-ynoic acid to deliver acalabrutinib (**11**).[28]

In summary, despite being a structural alert with potential safety concerns for its reactive metabolites, the alkyne motif has its unique place in drug discovery. Alkyne serves as a unique bioisostere for many functional groups such as halogen, nitrile, carboxamide, ethyl, and so on. As a consequence, the alkyne fragment has been employed to enhance potency via tight binding to the drug target, serving as a non-polar rigid and sterically less demanding spacer, improving bioavailability, and serving as a part of a warhead of a targeted covalent inhibitor.

REFERENCES

1. Djerassi, C. *Steroids* **1992**, *57*, 631–641.
2. Brogden, R. N.; Goa, K. L.; Faulds, D. *Drugs* **1993**, *45*, 384–409.
3. Niklya, I. *Mol. Psychiatry* **2016**, *21*, 1–5.
4. Balfour, J. A.; Faulds, D. *Drugs* **1992**, *43*, 259–284.
5. Foster, R. H.; Brogden, R. N.; Benfield, P. *Drugs* **1998**, *55*, 705–712.
6. (a) Namasivayam, V.; Vanangamudi, M.; Kramer, V. G.; Kurup, S.; Zhan, P.; Liu, X.; Kongsted, J.; Byrareddy, S. N. *J. Med. Chem.* **2019**, *62*, 4851–4883. (b) Bastos, M. M.; Costa, C. C. P.; Bezerra, T. C.; da Silva Fernando, de C; Boechat, N. *Eur. Med. Chem.* **2016**, *108*, 455–465.

7. Grant, S. M.; Goa, K. L. *Drugs* **1992**, *43*, 889–924.

8. Grabe, T.; Lategahn, J.; Rauh, D. *ACS Med. Chem. Lett.* **2018**, *9*, 799–782.

9. Parker, T.; Barbarotta, L.; Foss, F. *Fut. Oncol.* **2013**, *9*, 21–29.

10. Doupis, J. *Drug Res. Devel. Ther.* **2014**, *8*, 431–446.

11. Markham, A.; Dhillon, S. *Drugs* **2018**, *78*, 139–145.

12. Teutsch, G.; Philibert, D. *Human Reprod.* **1994**, *9*, 12–31.

13. Mahmood, I. *Clin. Pharmacokinet.* **1997**, *33*, 91–102.

14. (a) Mutlib, A. E.; Chen, H.; Nemeth, G. A.; Markwalder, J. A.; Seitz, S. P.; Gan, L. S.; Christ, D. D. *Drug Metab. Depos.* **1999**, *27*, 1319–1333. (b) Mutlib, A. E.; Gerson, R. J.; Meunier, P. C.; Haley, P. J.; Chen, H.; Gan, L. S.; Davies, M. H.; Gemzik, B.; Christ, D. D.; Krahn, D. F.; et al. *Tox. Appl. Pharmacol.* **2000**, *169*, 102–113.

15. Yu, L.-F.; Eaton, J. B.; Fedolak, A.; Zhang, H.-K.; Hanania, T.; Brunner, D.; Lukas, R. J.; Kozikowski, A. P. *J. Med. Chem.* **2012**, *55*, 9998–10009.

16. (a) Gu, Y. G.; Weitzberg, M.; Clark, R. F.; Xu, X.; Li, Q.; Zhang, T.; Hansen, T. M.; Liu, G.; Xin, Z.; Wang, X.; et al. *J. Med. Chem.* **2006**, *49*, 3770–3773. (b) Gu, Y. G.; Weitzberg, M.; Clark, R. F.; Xu, X.; Li, Q.; Lubbers, N. L.; Yang, Y.; Beno, D. W. A.; Widomski, D. L.; Zhang, T.; et al. *J. Med. Chem.* **2007**, *50*, 1078–1082.

17. Talele, T. T. *J. Med. Chem.* **2020**, *63*, 5625–5663.

18. (a) Wilcken, R.; Zimmermann, M. O.; Bauer, M. R.; Rutherford, T. J.; Fersht, A. R.; Joerger, A. C.; Boeckler, F. M. *ACS Chem. Biol.* **2015**, *10*, 2725–2732. (b) Nawaz, F.; Alam, O.; Perwez, A.; Rizvi, M. A.; Naim, M. J.; Siddiqui, N.; Pottoo, F. H.; Jha, M. *Archiv. Pharm.* **2020**, *353*, e900262. (c) Park, J. H.; Liu, Y.; Lemmon, M. A.; Radhakrishnan, R. *Biochem. J.* **2012**, *448*, 417–423.

19. (a) Zhao, H.; Li, S.; Yang, Z.; Peng, Y.; Chen, X.; Zheng, J. *Drug Metab. Dispos.* **2018**, *446*, 442–450. (b) Li, X.; Kamenecka, T. M.; Cameron, M. D. *Drug Metab. Dispos.* **2010**, *38*, 1238–1245.

20. (a) Chandraratna, R. A. *J. Am. Acad. Derm.* **1997**, *37*, S12–S17. (b) Talele, T. T. *J. Med. Chem.* **2018**, *61*, 2166–2210.

21. Huang, W.-S.; Metcalf, C. A.; Sundaramoorthi, R.; Wang, Y.; Zou, D.; Thomas, R. M.; Zhu, X.; Cai, L.; Wen, D.; Liu, S.; et al. *J. Med. Chem.* **2010**, *53*, 4701–4719.

22. Eckhardt, M.; Langkopf, E.; Mark, M.; Tadayyon, M.; Thomas, L.; Nar, H.; Pfrengle, W.; Guth, B.; Lotz, R.; Sieger, P.; et al. *J. Med. Chem.* **2007**, *50*, 6450–6453.

23. Barf, T.; Covey, T.; Izumi, R.; van de Kar, B.; Gulrajani, M.; van Lith, B.; van Hoek, M.; de Zwart, E.; Mittag, D.; Demont, D.; et al. *J. Pharmacol. Exp. Ther.* **2017**, *363*, 240–252.

24. Chandraratna, R. A. S. WO96/11686 (1996).

25. Frigoli, S.; Fuganti, C.; Malpezzi, L.; Serra, S. *Org. Process Res. Dev.* **2005**, *9*, 646–650.

26. Chen, C. Y.; Tillyer, R.; Tan, L. WO98/51676 (1998).

27. DeGraw, J. L.; Colwell, W. T.; Piper, J. R.; Sirotnak, F. M. *J. Med. Chem.* **1993**, *36*, 2228–2231.

28. Barf, T. A.; Jans, C. G. J. M.; De Man, A. P. A.; Oubrie, A. A.; Raaijmakers, H. C. A.; Rewinkel, J. B. M.; Sterrenburg, J.-G.; Wijkmans, J. C. H. M. U.S. Patent US20190276456 (2019).

2 Azaindoles

Azaindoles, also known as pyrrolopyridines, are bioisosteres for both indole and purine systems. Depending on the position of the nitrogen atom, they are 4-azaindoles, 5-azaindoles, 6-azaindoles, and 7-azaindoles, respectively. Rare in nature, azaindoles are privileged structures in medicinal chemistry—addition of a nitrogen atom to the indole ring in a prospective drug could potentially modulate its potency and physicochemical properties, as well as creating novel intellectual property space. They have found extensive utility in drug discovery, especially in the field of kinases, a class of particularly fruitful drug targets.

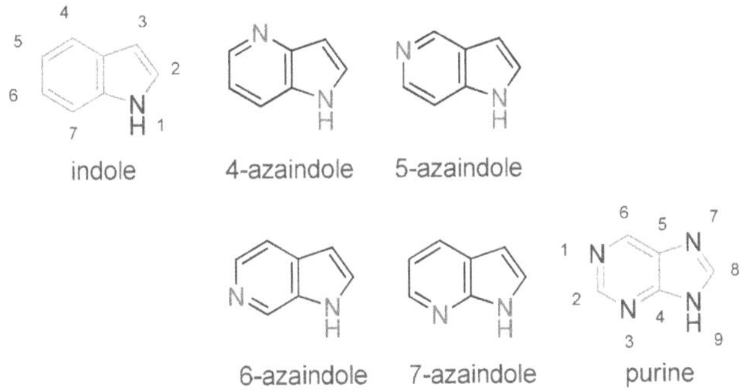

indole   4-azaindole   5-azaindole

6-azaindole   7-azaindole   purine

AZAINDOLE-CONTAINING DRUGS

Two azaindole-containing drugs currently on the market include Plexxikon's BRAF inhibitor vemurafenib (Zelboraf, **1**) and Abbvie's Bcl-2 inhibitor venetoclax (Venclexta, **2**). Both of these 7-azaindole-containing molecules are cancer drugs discovered from the fragment-based drug discovery (FBDD) strategy.

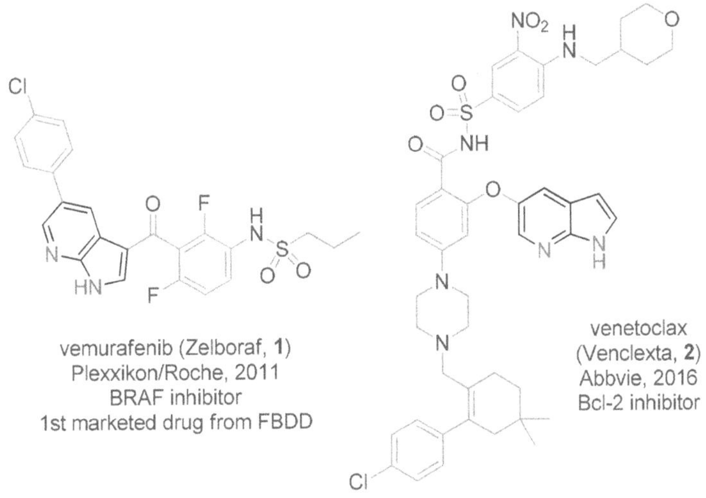

vemurafenib (Zelboraf, **1**)
Plexxikon/Roche, 2011
BRAF inhibitor
1st marketed drug from FBDD

venetoclax
(Venclexta, **2**)
Abbvie, 2016
Bcl-2 inhibitor

Plexxikon's vemurafenib (Zelboraf, **1**) was the first marketed drug discovered employing the FBDD (or scaffold-based drug design) strategy under the guidance of co-crystallography. No sooner than the BRAF^{V600E} mutant allele as a cancer target became known in 2002, Plexxikon began pursuing this target because BRAF^{V600E} is the most frequent oncogenic protein kinase mutation known and exists only in tumors that are dependent on the well-known RAF/MEK/ERK pathway. A library of 20,000 fragment compounds with molecular weights ranging from 150 to 350 (fewer than eight hydrogen bond donors and acceptors and few rotatable bonds) were screened at a concentration of 200 μM. One of the 238 high throughput screening (HTS) hits, binding to the ATP site, 7-azaindole co-crystallized with a kinase called proviral integration site of moloney murine leukemia virus-1 (PIM1) enzyme while 3-anilinyl-7-azaindole **3** also co-crystallized with PIM1 with an IC$_{50}$ value of approximately 100 μM for PIM1. The 7-azaindole scaffold **3** represented a general framework capable of presenting *two hydrogen-bonding interactions with the kinase hinge region*. Minor variation afforded benzyl-7-azaindole **4**, which co-crystallized with another kinase fibroblast growth factor receptor-1 (FGFR1) with an IC$_{50}$ value of 1.9 μM for FGFR1. Structure–activity relationship (SAR) investigations led to PLX4720 (**5**),[1] which was a potent and selective (including wide-type B-Raf and many other kinases) BRAF^{V600E} inhibitor with an IC$_{50}$ value of 13 nM. Installation of a chlorophenyl fragment to replace the 5-chlorine atom on the 7-azaindole core of **5** led to vemurafenib (**1**).[2] Vemurafenib (**1**) displays similar potency for BRAF (31 nM) and c-RAF-1 (48 nM) and selectivity against other kinases, including wide-type B-Raf (100 nM). It was chosen for development over **5** because its pharmacokinetic properties scaled more favorably in beagle dogs and cynomolgus monkeys. The FDA approved vemurafenib (Zelboraf, **1**) for the treatment of BRAF-mutant metastatic melanoma in 2011.[3]

7-azaindole
crystallized with PIM1
PIM1, IC$_{50}$ > 200 μM,
LE < 0.56

anilinyl-7-azaindole **3**
crystallized with PIM1
PIM1, IC$_{50}$, ~100 μM, LE = 0.34

benzyl-7-azaindole **4**
crystallized with FGFR1
FGFR1, IC$_{50}$, 1.9 μM, LE = 0.43

PLX4720 (**5**)
BRAF^{V660E}, IC$_{50}$, 13 nM, LE = 0.40

vemurafenib (Zelboraf, **1**)
BRAF^{V660E}, IC$_{50}$, 31 nM, LE = 0.31

Abbvie's B-cell lymphoma 2 (Bcl-2) inhibitor venetoclax (Venclexta, **2**) was discovered employing the FBDD strategy as well. Instead of the co-crystallography tactic, the "SAR by NMR" method was key to generate their fragment hits. From initial screening of a 10,000-compound library with MW < 215 at 1 mM concentration, *p*-fluorophenyl-benzoic acid emerged as one of the first-site (P1) ligands. Later on, screening a 3,500 compound library with MW ~ 150 at 5 mM concentration identified the second site (P2) ligand 5,6,7,8-tetrahydro-naphthalen-1-ol.[4] A protracted and winding road consisting of identifying the third binding site (P3), designing away from serum deactivation from domain II of human serum albumin (HSA-III) binding, boosting oral bioavailability, and removing a potential nitro structural alert cumulated to the discovery of navitoclax (**6**) as a potent and orally bioavailable Bcl-2 inhibitor.[5] Eventually, the fourth binding site (P4) was replaced with 7-azaindole ether and its N atom captured an additional hydrogen bond with Arg104 on the target, giving rise to venetoclax (Venclexta, **2**) as a potent, selective (against Bcl-x$_L$, Bcl-w, and Bcl-1), and orally bioavailable Bcl-2 inhibitor. In 2016, it was approved by the FDA for treating chronic lymphocytic leukemia (CLL) with the 17p deletion.[6]

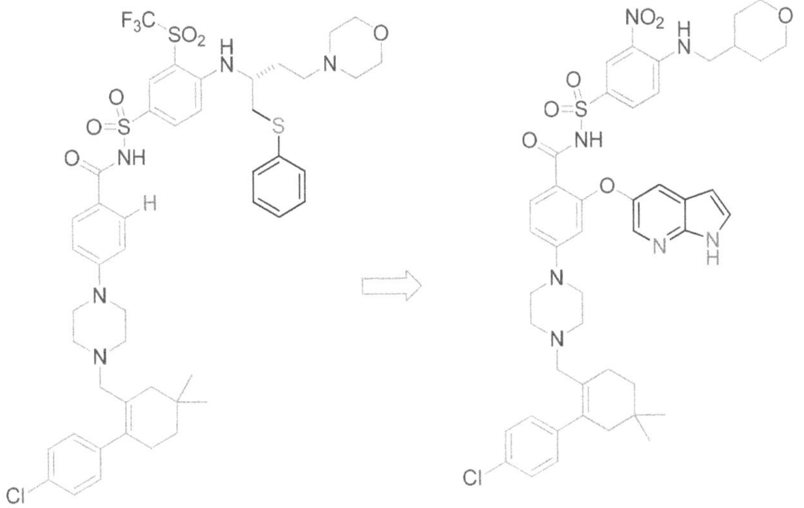

navitoclax (**6**)
K_i = 0.04 nM, F ~ 30%, LE = 0.2

venetoclax (Venclexta, **2**)
K_i = 0.01 nM, F ~ 29%, LE = 0.2

AZAINDOLES IN DRUG DISCOVERY

In terms of utility in drug discovery, 7-azaindoles are the most frequently used, followed by 6-azaindoles, whereas 4- and 5-azaindoles are less frequently encountered in the literature.[7] Azaindoles have found widespread applications in the design of kinase inhibitors. This is probably not surprising considering that azaindoles are structurally similar to the adenine fragment of adenosine triphosphate (ATP), which is critical to the phosphorylation process, the key function of kinases. Kinase inhibitors mimic ATP and bind to the catalytic domain, making azaindoles especially valuable scaffolds in this field.

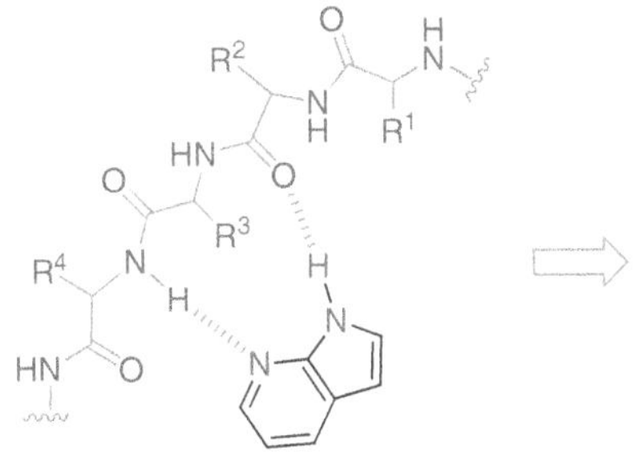

7-azaindole

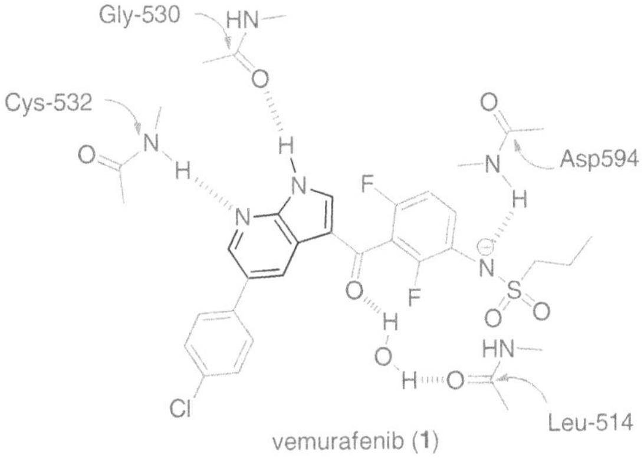

vemurafenib (**1**)

The pyridine N atom and its pyrrole NH of the 7-azaindole ring provide a hydrogen bond acceptor and donor, respectively to make bidentate hydrogen bonds with the kinase's hinge-binding region. Analysis of many 7-azaindole-containing kinase inhibitors revealed that 7-azaindole may bind to kinases in three modes: normal, flipped, and non-hinge binding modes.[8] Vemurafenib (**1**), for instance, has the pyridine N of its 7-azaindole scaffold serving as a hydrogen bond acceptor and forming a hydrogen bond with the NH of BRAF's backbone amide of cysteine-532 (Cys-532) near the hinge region, which overlaps with the ATP-binding site. Meanwhile, its pyrrole NH serves as a hydrogen bond donor, making contact with glycine-530 (Gly-530)'s carbonyl oxygen to form another hydrogen bond. The hydrogen bonding of azaindole, which is tightly confined within the adenine-binding region of the ATP pocket, to the hinge residues anchors the structure.[1]

Plexxikon's encore pexidartinib (**7**), a colony-stimulating factor-1 receptor (CSF-1R) kinase inhibitor for treating tenosynovial giant-cell tumor, emerged from their old hit benzyl-7-azaindole PLX070 (**4**).[9] Pexidartinib (**7**)'s phase III clinical trials completed in 2018. Astallas's peficitinib (**8**), a Janus kinase-3 (JAK3) inhibitor developed for the treatment of rheumatoid arthritis (RA), also contains a 7-azainodle framework.[10] Two 6-azaindole-containing drugs in phase III trials include fevipiprant (**9**), a potent and selective prostaglandin D_2 (DP_2) receptor antagonist for the treatment of asthma,[11] and BMS's fostemsavir (**10**), a human immunodeficiency virus type 1 (HIV-1) attachment inhibitor for treating AIDS.[12]

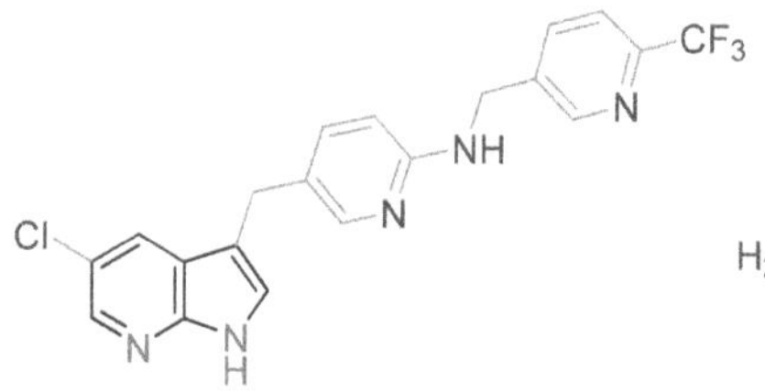

pexidartinib (**7**)
Plexxikon, phase III
CSF-1R inhibitor

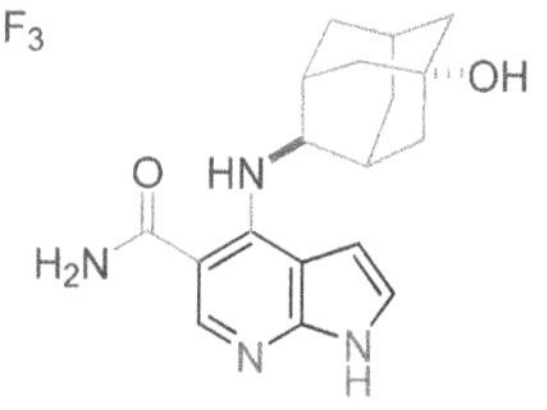

peficitinib (**8**)
Astellas, phase III
JAK3 inhibitor

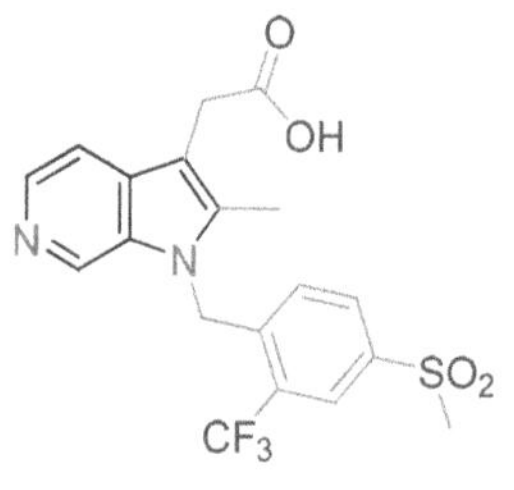

fevipiprant (**9**)

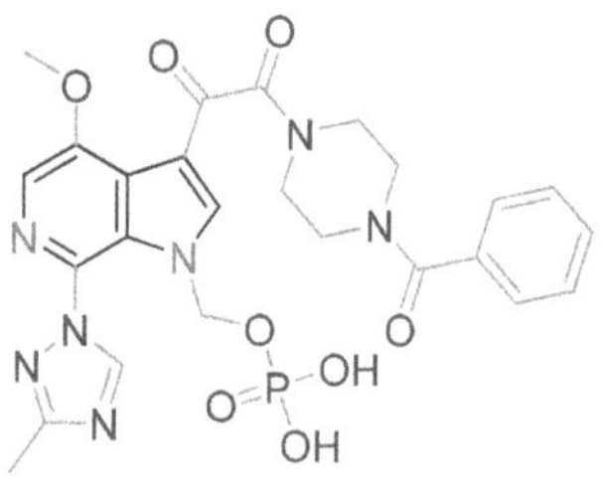

fostemsavir (**10**)

The abundance of 6- and 7-azaindole-containing drugs shown here thus far does not imply that only 6- and 7-azaindoles are useful in medicinal chemistry. A plethora of 4- and 5-azaindole-containing drugs are currently going through the pipelines of phase I and phase II clinical trials.[13] Even more of them are in preclinical investigations.

En route to the discovery of fostemsavir (**10**), systematic replacement of each of the unfused carbon atoms in the phenyl ring of the indole moiety by a nitrogen atom provided four different azaindole derivatives that displayed a clear SAR for antiviral activity and all of which displayed marked improvements in pharmaceutical properties.[14] The prototype indole **11** was a potent, non-cytotoxic inhibitor of HIV-1 in cell culture using a pseudo-virus (LAI strain) assay and displayed high permeability in a Caco-2 assay (Pc 169 nm/s at pH 6.5). But it exhibited a relatively short half-life when incubated in human liver microsomes (HLMs, $t_{1/2} = 16.9$ minutes) and a low crystalline solubility of

16 µg/mL at 25°C and at pH 6.5 in an aqueous buffer solution. For azaindoles, 4-azaindole **12** and 7-azaindole **15** had better efficacy than that of indole **11**; while 5-azaindole **13** and 6-azaindole **14** saw their efficacy reduced in comparison to the parent indole **11**. Conspicuously, all four possible azaindoles **12–15** uniformly displayed enhanced solubility by more than 25-fold (419–936 µg/mL) over that of the prototype indole **11**. They all displayed enhanced metabolic stability as measured by half-life in HLM (38.5 to > 100 minutes).

indole **11** 4-azainole **12** 5-azainole **13** 6-azainole **14** 7-azainole **15**

Compound	EC$_{50}$ (µM)	HLM, t$_{1/2}$ (minutes)	Caco-2 (nm/s)	Solubility (mg/mL)	pK$_a$	Log D at pH 6.5
11	4.85	16.9	169	0.016	10.0	1.9
12	1.56	>100	76	0.932	9.0, 5.8	0.9
13	576.90	>100	19	0.419	6.2, 9.8	1.2
14	21.55	38.5	<15	0.487	6.0, 9.3	1.5
15	1.65	49.5	168	0.936	2.0, 9.7	1.8

This case offers a glimpse of azaindole derivatives' ability to provide superior physiochemical properties to the parent indole compound. Azaindoles can also offer an additional hydrogen bond acceptor, which may translate to higher binding affinity, higher potency, and enhanced efficacy.

SYNTHESIS OF SOME AZAINDOLE-CONTAINING DRUGS

Because pyridine ring is electron-deficient, many classic indole synthesis methods do not work as well to synthesize azaindoles. For instance, the Fischer indole synthesis generally gives poor yields using pyridyl hydrazines and it also requires harsh conditions. There are tactics to circumvent pyridine's electron-deficiency by the addition of electron-pushing groups such as methoxyl and methylsulfide groups. In contrast, the Bartoli reaction and the Batcho–Leimgruber reaction have proven to be productive in preparing azaindoles.

Bartoli Reaction

Although discovered only 20 years ago, the Bartoli reaction has found more and more applications in indole and azaindole synthesis. Applying the classic Bartoli conditions, 3-nitropyridine **16** was converted to 4-azaindole **17** in 17% yield, which was then transformed to 5-HT$_6$ inhibitor **18** as a potential treatment of schizophrenia.[15]

3-nitropyridine **16** 4-azaindole **17** 5-HT$_6$ inhibitor **18**

The Bartoli reaction of 4-nitropyridine **19** with propenylmagnesium bromide produced 5-azaindole **20** in a 35% yield. 5-Azaindole **20**, in turn, was eventually converted to a brain penetrant cannabinoid receptor 2 (CB2) agonist GSK554418A (**21**) as a potential treatment of chronic pain.[16]

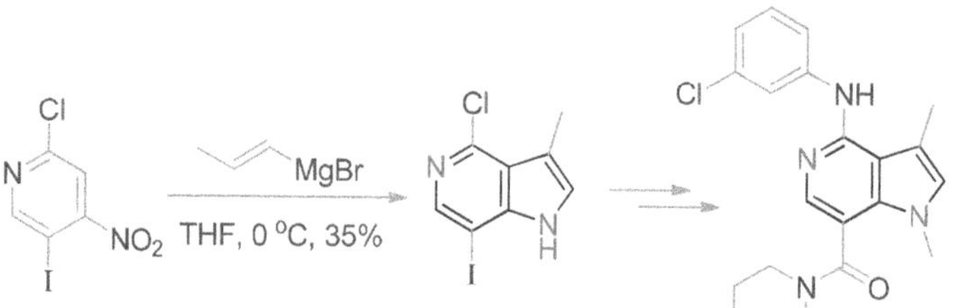

4-nitropyridine **19**   5-azaindole **20**   GSK554418A (**21**)

For the discovery route to prepare fostemsavir (**10**), 4-nitropyridine **22** underwent the Bartoli reaction with vinylmagnesium bromide to assemble the 6-azaindole scaffold **23**.[14]

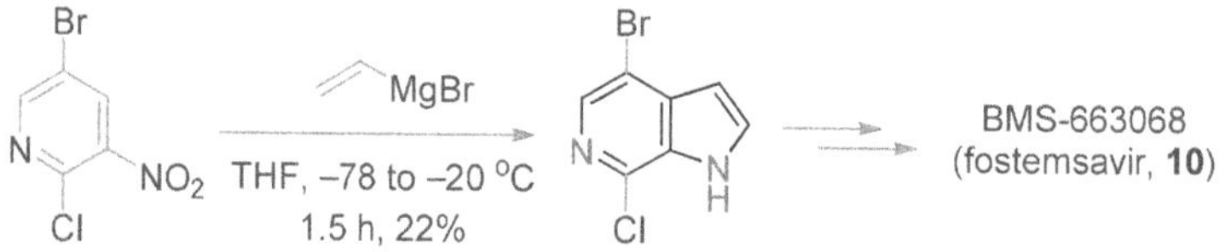

4-nitropyridine **22**   6-azaindole **23**

Notwithstanding its low yields, the Bartoli reaction is adequate for medicinal chemists to synthesize the desired azaindole analogs.

BATCHO–LEIMGRUBER REACTION

Pfizer constructed a 6-azaindole core structure using the Batcho–Leimgruber reaction to prepare a series of azaindole hydroxamic acids **27** as HIV-1 integrase inhibitors. The Batcho–Leimgruber reaction involved treating 4-nitropyridine **24** with N,N-dimethylformamide dimethyl acetal [DMF-DMA, $Me_2NCH(OMe)_2$] to afford enamine intermediate **25**, which was reduced via palladium-catalyzed hydrogenolysis to give 6-azaindole **26**. The 6-azaindole scaffold **26** was then transformed to potent HIV-1 integrase inhibitors **27** after further functional group manipulations.[17]

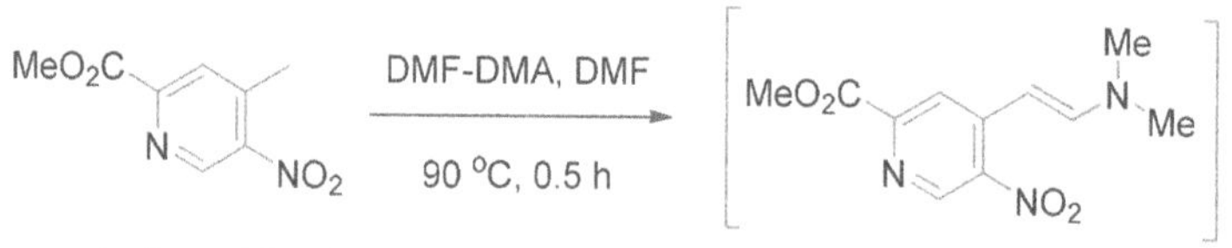

4-nitropyridine **24**   enamine **25**

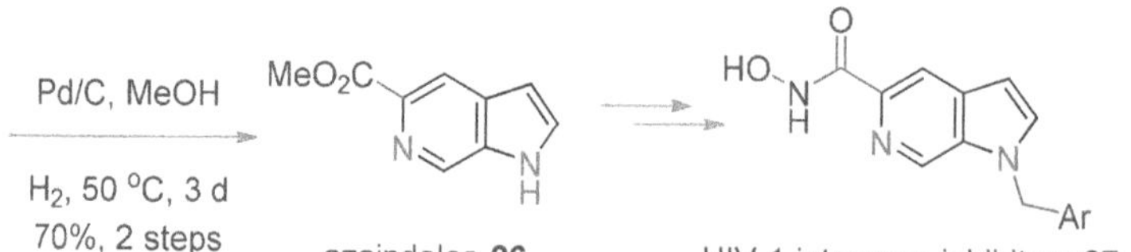

azaindoles **26**   HIV-1 integrase inhibitors **27**

The Process Chemistry at BMS developed an enabling preparation route employing the Batcho–Leimgruber reaction for their first scale-up campaign. Thus, 3-nitropyridine **28** was converted to intermediate enamine **29**, which underwent an Ullman coupling with NaOMe to afford enamine **30**. Palladium-catalyzed hydrogenation reduced the nitro group and led to 6-azaindole **31**, which was further manipulated to deliver fostemsavir (**10**).[18]

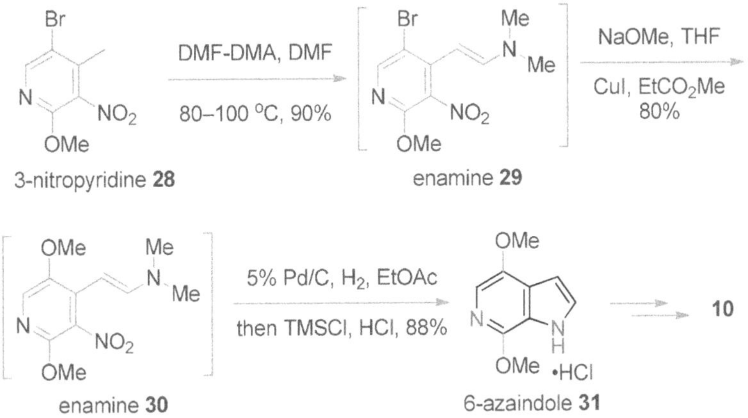

3-nitropyridine **28**   enamine **29**

enamine **30**   6-azaindole **31**

RADICAL AROMATIZATION

The Process Chemistry at BMS made a herculean effort to develop a commercial route to manufacture fostemsavir (**10**). Baran dubbed it an example of the majesty of chemistry.

3-Ketopyrrole **32**, assembled using a Pictet–Spengler cyclization, was exposed to methanesulfonic acid (MSA) and trimethylorthoformate [TMOF, CH(OMe)$_3$] in methanol to prepare methyl enol ether **33**. In the same pot, the addition of cumene hydroperoxide (CHP) initiated a radical aromatization process with the concomitant elimination of the sulfonate group to afford 6-azaindole **34**.[19] Functionalization of the 7-position involved oxidation of 6-azaindole **34** with H$_2$O$_2$ and methyltrioxorhenium (MTO, MeReO$_3$) as the catalyst to make 6-azaindole N-oxide **35**. Treatment of **35** with PyBroP in the presence of K$_3$PO$_4$ in trifluoromethyl toluene (TFT), followed by addition of NaOH in isopropyl alcohol (IPA), generated 7-bromo-6-azaindole **36**, which was converted to fostemsavir (**10**) in due course.

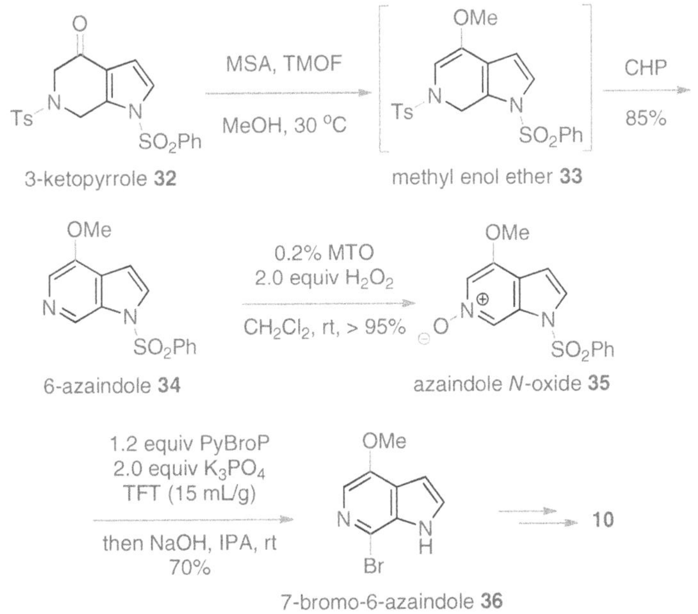

3-ketopyrrole **32**   methyl enol ether **33**

6-azaindole **34**   azaindole N-oxide **35**

7-bromo-6-azaindole **36**

SUZUKI COUPLING

As azaindoles find more and more applications in drug discovery, many advanced azaindole intermediates are now commercially available as building blocks. Here, only azaindole-boron intermediates are highlighted as their widespread utility in Suzuki coupling reactions.

Commercially available 7-azaindole-2-boronic acid **36** coupled with bromide **37** under standard Suzuki coupling conditions to assemble adduct **38**, which was eventually manipulated to a covalent EGFR inhibitor **39**. This particular drug and its analogs with a rigidized hinge binding motif act as single-digit inhibitors of clinically relevant EGFR L858R/T790M and L858R/T790M/C797S mutants.[20]

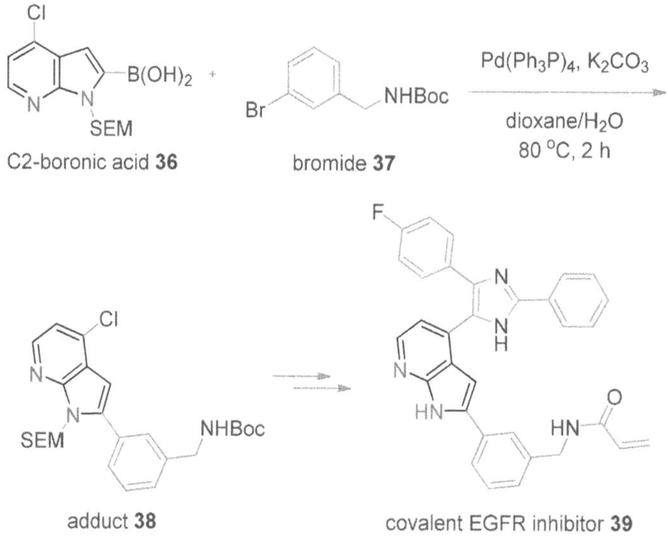

C2-boronic acid **36**   bromide **37**

adduct **38**   covalent EGFR inhibitor **39**

A Suzuki coupling reaction was carried out between commercially available 7-azaindole-3-boronate ester **40** and chloropyrimidine **41** to assemble adduct **42**. Oxidation of the sulfide on **42** gave the corresponding sulfone as a good leaving group, which was then displaced with a primary amine to deliver **43** as a cyclin-dependent kinase-2/9 (CDK2/9) inhibitor.[21]

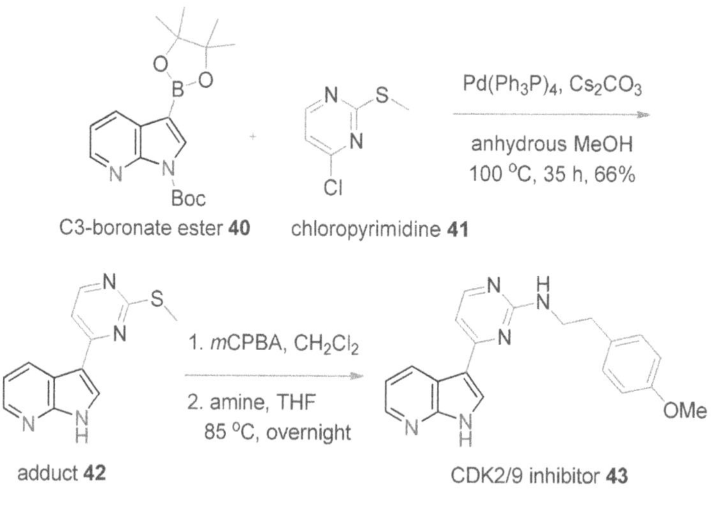

C3-boronate ester **40**   chloropyrimidine **41**

adduct **42**   CDK2/9 inhibitor **43**

A selective protein kinase C iota (PKCι) inhibitor **46** was discovered using the FBDD approach. Interestingly, the fragment expansion employed 4-bromo-7-azaindole as the starting point. In the course of SAR investigations, commercially available 7-azaindole-4-boronate ester **44** was coupled with aryl bromide **45** to deliver **46** after acidic deprotection.[22]

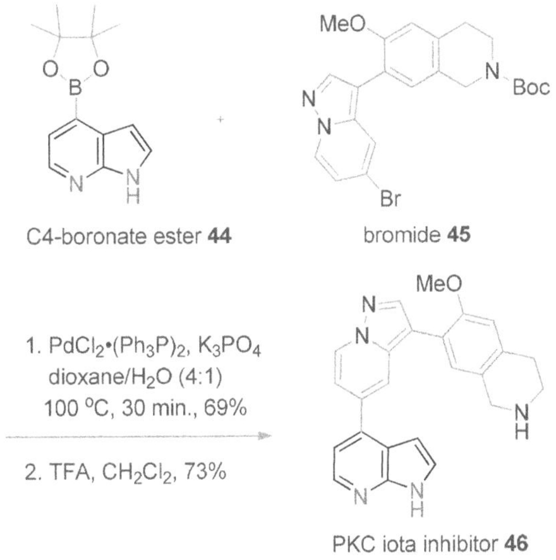

C4-boronate ester **44**

bromide **45**

1. PdCl₂•(Ph₃P)₂, K₃PO₄
   dioxane/H₂O (4:1)
   100 °C, 30 min., 69%

2. TFA, CH₂Cl₂, 73%

PKC iota inhibitor **46**

Glyoxalase I (GLO1) is a zinc enzyme that isomerizes glutathione (GSH) and methylglyoxal to lactic thioester. GLO1 inhibitors have the potential to be used in the treatment of cancer and other diseases. Synthesis of GLO1 inhibitor **49** entails a Suzuki coupling between C4-boronate ester **47** and chloride **48**, followed by two more steps. Although hidden in a ring on **49**, the cyclic hydroxamic acid that chelates with the catalytic zinc cation is a structural alert similar to other linear hydroxamic acids. Here, the nitrogen on the pyridine of the 7-azaindole core forms a hydrogen bond with a water molecule in a hydrogen bond network.[23]

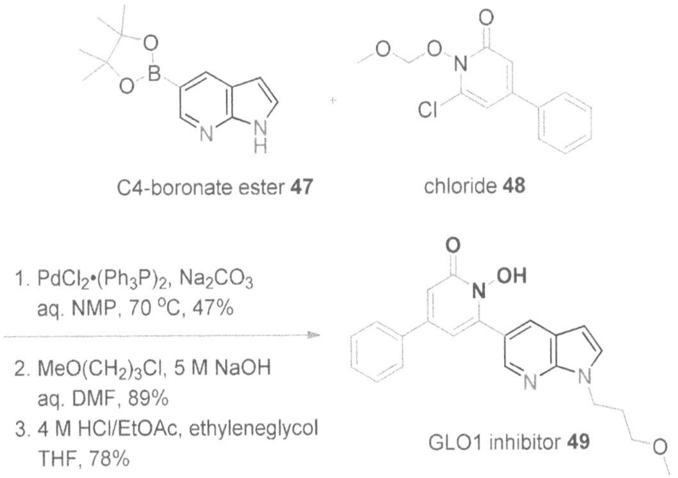

C4-boronate ester **47**

chloride **48**

1. PdCl₂•(Ph₃P)₂, Na₂CO₃
   aq. NMP, 70 °C, 47%

2. MeO(CH₂)₃Cl, 5 M NaOH
   aq. DMF, 89%
3. 4 M HCl/EtOAc, ethyleneglycol
   THF, 78%

GLO1 inhibitor **49**

SONOGASHIRA REACTION

A Sonogashira reaction was key to assemble the 7-azaindole core of Merck's focal adhesion kinase (FAK) inhibitors.[24] Coupling between 3-iodopyridine **50** and *para*-fluorophenylacetylene led to the formation of alkyne **51**, which was exposed to KO*t*-Bu to afford 7-azaindole **52**. Selective C4-chlorination of 7-azaindole **52** required a two-step sequence involving *m*CPBA oxidation to form *N*-oxide **53**, followed by treatment with POCl$_3$ to give rise to 4-chloro-7-azainolde **54**, which was then converted to FAK inhibitor **55** in several additional steps.

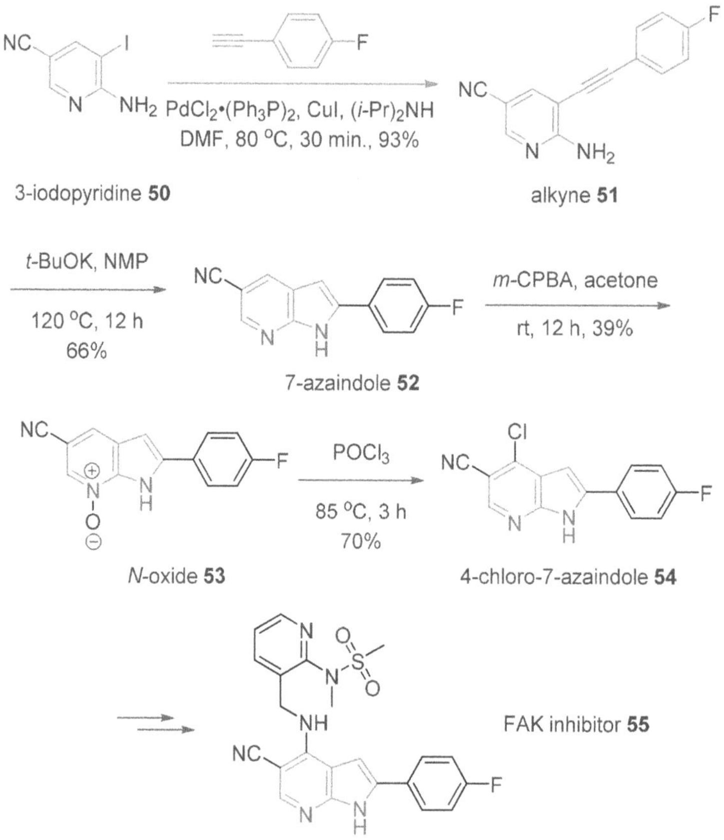

3-iodopyridine **50**   alkyne **51**

7-azaindole **52**

N-oxide **53**   4-chloro-7-azaindole **54**

FAK inhibitor **55**

In addition to what we have discussed here, many additional synthetic methods exist for making all four possible azaindoles.[25]

In summary, azaindoles, fruitful scaffolds in the field of kinase inhibitors and others, may provide higher binding affinity than and superior physiochemical properties to its indole prototypes. Many synthetic methods exist to prepare all four possible azaindoles. Thanks to their widespread applications in medicinal chemistry, many azaindole intermediates are nowadays commercially available, greatly reducing the time and resources required to prepare azaindole-containing compounds.

REFERENCES

1. Tsai, J.; Lee, J. T.; Wang, W.; Zhang, J.; Cho, H.; Mamo, S.; Bremer, R.; Gillette, S.; Kong, J.; Haass, N. K.; et al. *Proc. Natl. Acad. Sci. USA* **2008**, *105*, 3041–3046.
2. Bollag, G.; Hirth, P.; Tsai, J.; Zhang, J.; Ibrahim, P. N.; Cho, H.; Spevak, W.; Zhang, C.; Zhang, Y.; Habets, G.; et al. *Nature* **2010**, *467*, 596–599.
3. Bollag, G.; Tsai, J.; Zhang, J.; Zhang, C.; Ibrahim, P.; Nolop, K.; Hirth, P. *Nat. Rev. Drug Discov.* **2012**, *11*, 873–886.
4. Wendt, M. D.; Shen, W.; Kunzer, A.; McClellan, W. J.; Bruncko, M.; Oost, T. K.; Ding, H.; Joseph, M. K.; Zhang, H.; Nimmer, P. M.; et al. *J. Med. Chem.* **2006**, *49*, 1165–1181.
5. Park, C.-M.; Bruncko, M.; Adickes, J.; Bauch, J.; Ding, H.; Kunzer, A.; Marsh, K. C.; Nimmer, P.; Shoemaker, A. R.; Song, X.; et al. *J. Med. Chem.* **2008**, *51*, 6902–6915.
6. Souers, A. J.; Leverson, J. D.; Boghaert, E. R.; Ackler, S. L.; Catron, N. D.; Chen, J.; Dayton, B. D.; Ding, H.; Enschede, S. H.; Fairbrother, W. J.; et al. *Nat. Med.* **2013**, *19*, 202–208.

7. Mérour, J.-Y.; Buron, F.; Ple, K.; Bonnet, P.; Routier, S. *Molecules* **2014**, *19*, 19935–19979.
8. Irie, T.; Sawa, M. *Chem. Pharm. Bull.* **2018**, *66*, 29–36.
9. Tap, W. D.; Wainberg, Z. A.; Anthony, S. P.; Ibrahim, Prabha N.; Zhang, C.; Healey, J. H.; Chmielowski, B.; Staddon, A. P.; Cohn, A. L.; Shapiro, G. I.; et al. *N. Engl. J. Med.* **2015**, *373*, 428–437.
10. Hamaguchi, H.; Amano, Y.; Moritomo, A.; Shirakami, S.; Nakajima, Y.; Nakai, K.; Nomura, N.; Ito, M.; Higashi, Y.; Inoue, T. *Bioorg. Med. Chem.* **2018**, *26*, 4971–4983.
11. Sandham, D. A.; Barker, L.; Brown, L.; Brown, Z.; Budd, D.; Charlton, S. J.; Chatterjee, D.; Cox, B.; Dubois, G.; Duggan, N.; et al. *ACS Med. Chem. Lett.* **2017**, *8*, 582–586.
12. Wang, T.; Ueda, Y.; Zhang, Z.; Yin, Z.; Matiskella, J.; Pearce, B. C.; Yang, Z.; Zheng, M.; Parker, D. D.; Yamanaka, G. A.; et al. *J. Med. Chem.* **2018**, *61*, 6308–6327.
13. El-Gamal, M. I.; Anbar, H. S. *Expert. Opin. Ther. Pat.* **2017**, *27*, 591–606.
14. Meanwell, N. A.; Krystal, M. R.; Nowicka-Sans, B.; Langley, D. R.; Conlon, D. A.; Eastgate, M. D.; Grasela, D. M.; Timmins, P.; Wang, T.; Kadow, J. F. *J. Med. Chem.* **2018**, *61*, 62–80.
15. Ahmed, M.; Briggs, M. A.; Bromidge, S. M.; Buck, T.; Campbell, L.; Deeks, N. J.; Garner, A.; Gordon, L.; Hamprecht, D. W.; Holland, V.; et al. *Bioorg. Med. Chem. Lett.* **2005**, *15*, 4867–4871.
16. Giblin, G. M. P.; Billinton, A.; Briggs, M.; Brown, A. J.; Chessell, I. P.; Clayton, N. M.; Eatherton, A. J.; Goldsmith, P.; Haslam, C.; Johnson, M. R.; et al. *J. Med. Chem.* **2009**, *52*, 5785–5788.
17. Plewe, M. B.; Butler, S. L.; R. Dress, K.; Hu, Q.; Johnson, T. W.; Kuehler, J. E.; Kuki, A.; Lam, H.; Liu, W.; Nowlin, D.; et al. *J. Med. Chem.* **2009**, *52*, 7211–7219.
18. Fox, R.J.; Tripp, J. C.; Schultz, M. J.; Payack, J. F.; Fanfair, D. D.; Mudryk, B. M.; Murugesan, S.; Chen, C.-P. H.; La Cruz, T. E.; Ivy, S. E.; et al. *Org. Process Res. Dev.* **2017**, *21*, 1095–1109.
19. Bultman, M. S.; Fan, J.; Fanfair, D.; Soltani, M.; Simpson, J.; Murugesan, S.; Soumeillant, M.; Chen, K.; Risatti, C.; La Cruz, T. E.; et al. *Org. Process Res. Dev.* **2017**, *21*, 1131–1136.
20. Juchum, M.; Guenther, M.; Doering, E.; Sievers-Engler, A.; Laemmerhofer, M.; Laufer, S. *J. Med. Chem.* **2017**, *60*, 4636–4656.
21. Singh, U.; Chashoo, G.; Khan, S. U.; Mahajan, P.; Nargotra, A.; Mahajan, G.; Singh, A.; Sharma, A.; Mintoo, M. J.; Guru, S. K.; et al. *J. Med. Chem.* **2017**, *60*, 9470–9489.
22. Kwiatkowski, J.; Liu, B.; Tee, D. H. Y.; Chen, G.; Binte Ahmad, N. H.; Wong, Y. X.; Poh, Z. Y.; Ang, S. H.; Tan, E. S. W.; Ong, E. H. Q.; et al. *J. Med. Chem.* **2018**, *61*, 4386–4396.
23. Chiba, T.; Ohwada, J.; Sakamoto, H.; Kobayashi, T.; Fukami, T. A.; Irie, M.; Miura, T.; Ohara, K.; Koyano, H. *Bioorg. Med. Chem. Lett.* **2012**, *22*, 7486–7489.
24. Heinrich, T.; Seenisamy, J.; Emmanuvel, L.; Kulkarni, S. S.; Bomke, J.; Rohdich, F.; Greiner, H.; Esdar, C.; Krier, M.; Gradler, U.; et al. *J. Med. Chem.* **2013**, *56*, 1160–1170.
25. Mérour, J.-Y.; Routier, S.; Suzenet, F.; Joseph, B. *Tetrahedron* **2013**, *69*, 4767–4834.

3 Azetidines

AZETIDINE-CONTAINING DRUGS

Azetidines are a good compromise between a satisfactory stability and a strong molecular rigidity, allowing an efficient tuning of pharmacological properties displayed by molecules bearing this moiety.[1] Two azetidine-containing drugs are currently on the market. Dihydropyridine azelnidipine (Calblock, **1**) is Sankyo's calcium channel blocker.[2] Exelixis' cobimetinib (Cotellic, **2**), as a targeted cancer therapy, is a mitogen-activated protein kinase-1/2 (MEK1/2) inhibitor.[3] Another azetidine-containing drug ximelagatran (Exanta, **3**) as a direct thrombin inhibitor was discovered by AstraZeneca. Initially sold as an anticoagulant, it was pulled off the market in 2006 due to hepatotoxicity.[4]

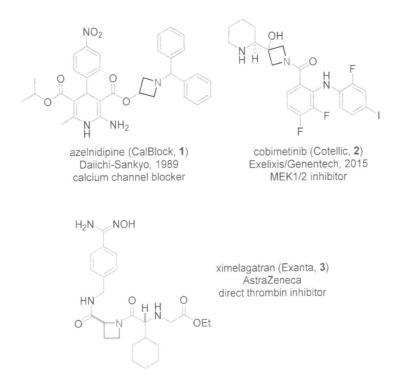

azelnidipine (CalBlock, **1**)
Daiichi-Sankyo, 1989
calcium channel blocker

cobimetinib (Cotellic, **2**)
Exelixis/Genentech, 2015
MEK1/2 inhibitor

ximelagatran (Exanta, **3**)
AstraZeneca
direct thrombin inhibitor

AZETIDINES IN DRUG DISCOVERY

Azetidine carbamate **4** is an efficient, covalent inhibitor of monoacylglycerol lipase (MAGL) discovered by Pfizer.[5] The hexafluoroisopropanol (HFIP) group here serves as the leaving group when attacked by the key serine residue (Ser$_{122}$) at the enzyme's active site. Covalent inhibition is attractive in that it offers the potential for an extended duration of pharmacodynamic modulation relative to the pharmacokinetic profile of the inhibitor.

Fatty acid amide hydrolase (FAAH) inhibitors are a potential treatment for pain. Vernalis discovered a mixture of chiral azetidine-ureas VER-24052 (**5**) as a FAAH inhibitor (rat FAAH, IC$_{50}$ = 188 nM, t = 3 hours). Interestingly, while the isomer with a positive optical rotation was active (rat FAAH, IC$_{50}$ = 78 nM, t = 3 hours), its corresponding enantiomer was completely inactive in the same assay.[6]

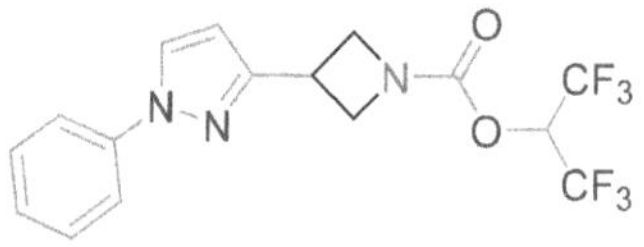

4
MAGL, IC_{50} = 0.18 nM
LipE = 5.5

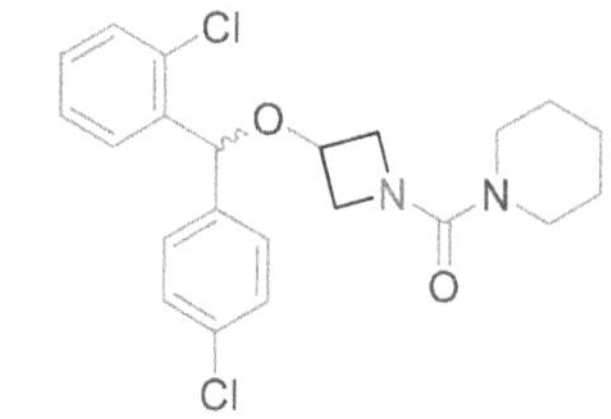

VER-24052 (**5**)
rat FAAH, IC_{50} = 1205 nM

Marketed since 2012, Pfizer's tofacitinib (Xeljanz, **6**) is the first-in-class Janus kinase (JAK) inhibitor for the treatment of rheumatoid arthritis (RA). In 2013, Roche reported an azetidine-containing bis-amide **7** as a selective JAK3 inhibitor (IC_{50}=0.26 nM) with a ten-fold selectivity over JAK1 (IC_{50}=3.2 nM). In addition, the combination of its selectivity over the kinome, good solubility, and reasonable exposure was translated to *in vivo* potency and selectivity in an acute PK/PD mouse model.[7]

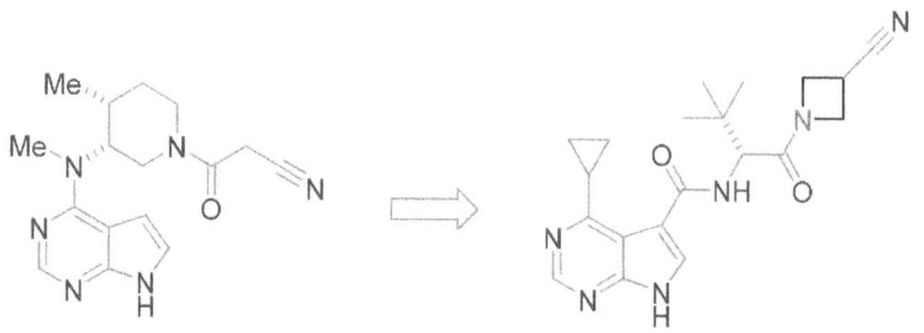

tofacitinib (Xeljanz, **6**)
Pfizer, 2012
JAK Inhibitor for RA

bis-amide **7**
JAK3, IC_{50} = 0.26 nM
JAK1, IC_{50} = 3.2 nM

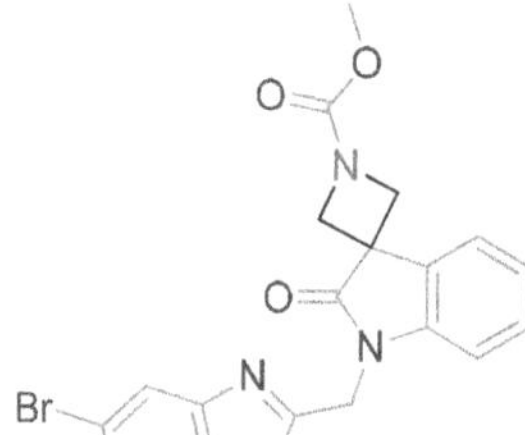

8
RSV EC_{50} = 0.8 nM
CC_{50} > 100 μM
%F = 71%

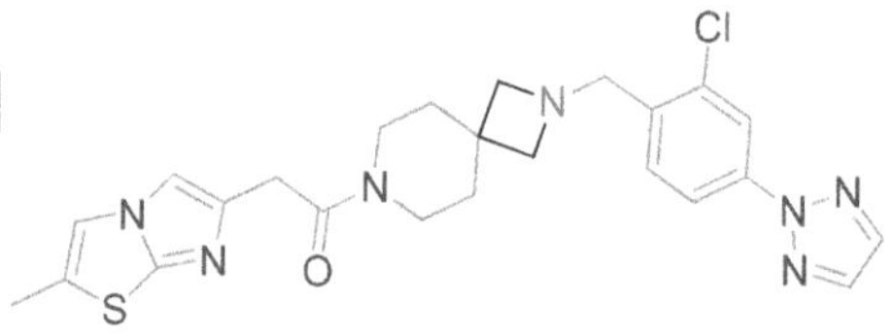

9
HGR IC_{50} = 4.6 nM
HGR K_i = 7.0 nM
LE = 0.33

Spirocyclic azetidines, like all spirocyclic scaffolds, are inherently three dimensional and offer structural novelty. For example, 3,3′-spiro[azetidine]-2-oxo-indoline derivative **8** was discovered as a fusion inhibitor for the treatment of respiratory syncytial virus (RSV).[8] On the other hand, spiro-cyclic piperidine-azetidine **9** is an inverse agonist of the ghrelin receptor (GR), a GPCR target that plays a role in obesity and glucose homeostasis.[9]

SYNTHESIS OF SOME AZETIDINE-CONTAINING DRUGS

Sankyo's synthesis of azelnidipine (Calblock, **3**) began with 1-benzhydryl-3-hydroxyazetidine (**10**), readily assembled from the condensation of benzhydrylamine with epichlorohydrin. Subsequent 1,3-dicyclohexylcarbodiimide (DCC)-mediated esterification of **10** with cyanoacetic acid produced ester **11**, which was converted to amidine **12** in two additional steps. A Hantzsch dihydropyridine synthesis between amidine **12** and enone **13** then delivered azelnidipine (**3**).[10]

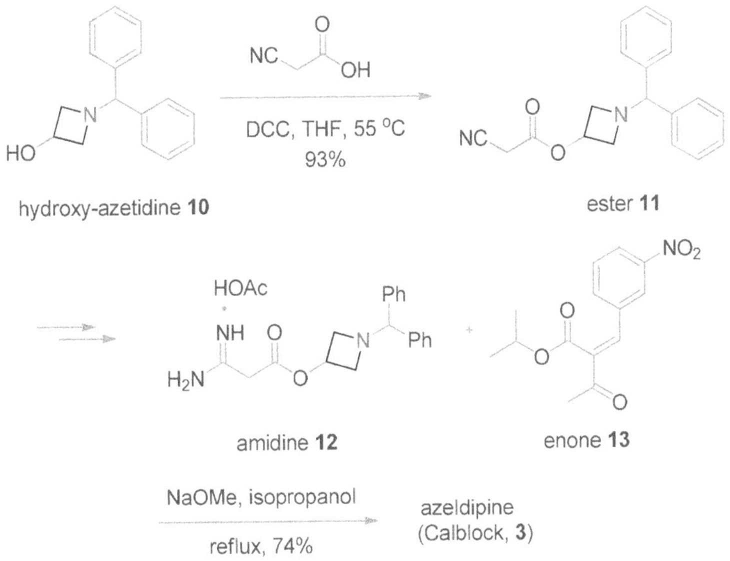

hydroxy-azetidine **10**   ester **11**

amidine **12**   enone **13**

azeldipine
(Calblock, **3**)

Exelixis' synthesis of cobimetinib (Cotellic, **4**) commenced with the addition of piperidine-Grignard reagent **14** to azetidinone **15**. The (*S*)-adduct **16** was secured after chiral resolution employing the Mosher's ester technique. Palladium-catalyzed hydrogenation of **16** removed the Cbz protection to afford the exposed azetidine **17**. Ester formation from the coupling between **17** and acid chloride **18** in the presence of diisopropylethylamine (DIPEA) produced cobimetinib (**4**) after deprotection of the Boc group.[3]

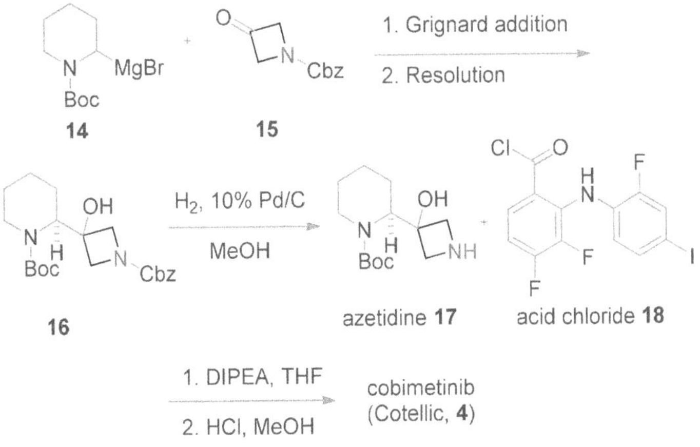

14   **15**

16   azetidine **17**   acid chloride **18**

cobimetinib
(Cotellic, **4**)

AstraZeneca's synthesis of ximelagatran (Exanta, **3**) started with the formation of amide **21** from azetidine ester **19** and chiral amino acid **20**. Subsequently, LiOH-promoted saponification of **21** afforded azetidine acid **22**. An additional amide formation between azetidine acid **22** and benzylamine **23** prepared bis-amide **24**, which was converted to the desired ximelagatran (**3**) after several additional steps.[11]

In summary, azetidines are a good compromise between a satisfactory stability and a strong molecular rigidity, allowing an efficient tuning of pharmacological properties displayed by molecules bearing this moiety. Therefore, azetidine is considered a privileged scaffold in drug discovery.

REFERENCES

1. Brandi, A.; Cicchi, S.; Cordero, F. M. *Chem. Rev.* **2008**, *108*, 3988–4035.
2. Yagil, Y.; Miyamoto, M.; Frasier, L.; Oizumi, K.; Koike, H. *Am. J. Hypertens.* **1994**, *7*, 637–646.
3. Rice, K. D.; Aay, N.; Anand, N. K.; Blazey, C. M.; Bowles, O. J.; Bussenius, J.; Costanzo, S.; Curtis, J. K.; Defina, S. C.; Dubenko, L.; et al. *ACS Med. Chem. Lett.* **2012**, *3*, 416–421.
4. Ericksson, B. I.; Carlsson, S.; Halvarsson, M.; Risberg, B.; Mattsson, C. *Thromb. Haemostasis* **1997**, *78*, 1404–1407.
5. Butler, C. R.; Beck, E. M.; Harris, A.; Huang, Z.; McAllister, L. A.; Am Ende, C. W.; Fennell, K.; Foley, T. L.; Fonseca, K.; Hawrylik, S. J.; et al. *J. Med. Chem.* **2017**, *60*, 9860–9873.
6. Hart, T.; Macias, A. T.; Benwell, K.; Brooks, T.; D'Alessandro, J.; Dokurno, P.; Francis, G.; Gibbons, B.; Haymes, T.; Kennett, G.; et al. *Bioorg. Med. Chem. Lett.* **2009**, *19*, 4241–4244.
7. Soth, M.; Hermann, J. C.; Yee, C.; Alam, M.; Barnett, J. W.; Berry, P.; Browner, M. F.; Frank, K.; Frauchiger, S.; Harris, S.; et al. *J. Med. Chem.* **2013**, *56*, 345–356.
8. Shi, W.; Jiang, Z.; He, H.; Xiao, F.; Lin, F.; Sun, Y.; Hou, L.; Shen, L.; Han, L.; Zeng, M.; et al. *ACS Med. Chem. Lett.* **2018**, *9*, 94–97.
9. Kung, D. W.; Coffey, S. B.; Jones, R. M.; Cabral, S.; Jiao, W.; Fichtner, M.; Carpino, P. A.; Rose, C. R.; Hank, R. F.; Lopaze, M. G.; et al. *Bioorg. Med. Chem. Lett.* **2016**, *24*, 2146–2157.
10. Koike, H.; Nishino, H.; Yoshimoto, M. *Dihydropyridine derivatives, their preparation and their use.* U.S. Patent US4772596 (1988).
11. Lila, C.; Gloanec, P.; Cadet, L.; Hervé, Y.; Fournier, J.; Leborgne, F.; Verbeuren, T. J.; De Nanteuil, G. *Synth. Commun.* **1998**, *28*, 4419–4429.

4 Bicyclic Pyridines Containing Ring-Junction N

There are three classes of bicyclic pyridines that contain a ring-junction nitrogen: imidazo[1,2-*a*]pyridines, imidazo[1,5-*a*]pyridines, and pyrazolo[1,5-*a*]pyridines. Their utility in drug discovery and preparations are reviewed by Larry Yet in a chapter in his excellent book: *Privileged Structures in Drug Discovery, Medicinal Chemistry and Synthesis.*[1]

imidazo[1,2-*a*]pyridine   imidazo[1,5-*a*]pyridine   pyrazolo[1,5-*a*]pyridine

BICYCLIC PYRIDINE-CONTAINING DRUGS

There are at least four imidazo[1,2-*a*]pyridine-containing drugs and one pyrazolo[1,5-*a*]pyridine-containing drug on the market.

Synthélabo's alpidem (Anaxyl, **1**) is a γ-aminobutyric acid (GABA$_A$) agonist specifically used for treating anxiety, approved in France in 1991. Its close analog zolpidem (Ambien, **2**), also a GABA$_A$ agonist, is a blockbuster drug to treat insomnia because, unlike alpidem (**1**), zolpidem (**2**) has a sedative effect. It is highly bioavailable (70%) with a short duration of action (t$_{1/2}$=2 hours). In contrast, alpidem (**1**) has a half-life of 19 hours, a testimony to the fact that its two chlorine atoms are more resistant to CYP450 metabolism in comparison to the two methyl groups on zolpidem (**2**).[2] Two similar imidazo[1,2-*a*]pyridine-based GABA$_A$ agonists saripidem and necopidem were investigated in clinical trials but did not gain government approval for marketing. Olprinone (Coretec, **3**) is a cardiotonic agent only available in Japan. It is a phosphodiesterase-3 (PDE3) inhibitor with positive ionotropic and vasodilator effects.[3] On the other hand, minodronic acid (Recalbon, **4**) is the third-generation bisphosphonate oral drug to treat loss of bone density for diseases such as osteoporosis.

alpidem (Anaxyl, **1**)
Synthelabo, 1991
GABA agonist

zolpidem (Ambien, **2**)
Sanofi-Aventis, 1992
GABA agonist

olprinone (Coretec, **3**)
Eisai, 1996
PDE3 inhibitor

minodronic acid (Recalbon, **4**)
Ono/Astellas, 2009
bisphosphonate

ibudilast (Ketas, **5**)
Kyorin, 1992
PDE inhibitor

Finally, ibudilast (Ketas, **5**) has the pyrazolo[1,5-*a*]pyridine core structure. Only available in Japan for treating asthma and stroke, it is a neuroimmune modulator. It is a pan-PDE inhibitor with activities against PDE-3, PDE-4, PDE-10, and PDE-11. Other more PDE-4 selective inhibitors include roflumilast (Daliresp) for treating chronic obstructive pulmonary disease (COPD) and apremilast (Otezla) for treating plaque psoriasis.[4]

BICYCLIC PYRIDINES IN DRUG DISCOVERY

Receptor interacting protein kinase-2 (RIPK2) is an intracellular serine/threonine/tyrosine kinase, a key signaling partner, and an obligate for nucleotide-binding oligomerization domain-containing protein 2 (NOD2). Employing virtual library screening (VLS), He and colleagues chose pyrazolo[1,5-*a*]pyridine **6** as their starting point among other hits because although it had only micro-molar (1.5 μM) activity, it exhibited attractive ligand efficiency (LE=0.32) and lipophilic efficiency (LiPE=3.5). Guided by structure-based drug design (SBDD) combined with extensive structure–activity relationship (SAR) investigations, they arrived at imidazo[1,2-*a*]pyridine **7**, which was potent and selective with excellent oral bioavailability. In both *in vitro* and *in vivo* assays, imidazo[1,2-*a*]pyridine **7** showed activities in suppressing cytokine secretion upon activation of the NOD2:RIPK2 pathway.[5]

6, RIPK2, IC$_{50}$ = 1.5 μM
CLogP = 2.31
LE = 0.32, LiPE = 3.5

7, RIPK2, IC$_{50}$ = 3 nM

The imidazo[1,2-*a*]pyridine core structure was used as an isostere of imidazo[1,2-*a*]pyrimidine to reduce metabolism mediated by aldehyde oxidase (AO). Pfizer identified imidazo[1,2-*a*]pyrimidine **8** as a full antagonist of the androgen receptor (AR) with excellent *in vivo* tumor growth inhibition (TGI) in castration-resistant prostate cancer (CRPC). Regrettably, compound **8**'s core structure imidazo[1,2-*a*]pyrimidine moiety was rapidly metabolized by AO. Indeed, heteroaryls, such as imidazo[1,2-*a*]pyrimidines, are versatile synthetic building blocks commonly used in medicinal chemistry because they are often capable of binding to diverse biological targets with high affinity and providing useful pharmacological activities. In addition, electron-deficient heteroaryls are often resistant to CYP-450-mediated metabolism. However, an electron-deficient nature may also make the ring carbons susceptible to nucleophilic attack by aldehyde oxidase (AO), particularly when they are adjacent to heterocyclic nitrogen(s). Guided by an AO protein structure-based model, Pfizer chemists discovered that imidazo[1,2-*a*]pyridine core structure on compound **9** (with one nitrogen atom removed from the original core structure on **8**) was clean of AO metabolism although it was more susceptible to CYP450 oxidation. Another tactic was also successful via blocking the AO metabolism by installing a methoxyl group at C29 on the imidazo[1,2-*a*]pyrimidine ring. It was speculated that C29 was the most probable AO oxidation site.[6]

8
metabolism by AO

9
clean of AO metabolism

Fragment-based drug discovery (FBDD) has attracted more and more attention, especially with the FDA approval of Plexxikon's vemurafenib (Zelboraf) in 2011 and Abbvie's venetoclax (Venclexta), both of which started with fragment hits.

10
32% @ 100 μM

11
3.3 nM

Astex obtained fragment **10** as the hit using a protein thermal shift assay (T_M) in their pursuit of selective discoidin domain receptor 1 and 2 (DDR1/2) inhibitors. Fragment **10** placed a chlorophenyl in the back-pocket region and a pyridyl in the selectivity pocket proximal to the small gatekeeper residue (Thr701 in DDR1/2) and lacked a hinge-binding moiety. With the help of crystal structures and computer-aided drug design (CADD) by overlaying with FGFR inhibitor dasatinib, Astex installed an imidazo[1,2-*a*]pyridine in place of the thiazole hinge binder. The resulting compound **11**'s imidazo[1,2-*a*]pyridine fragment indeed formed the anticipated hydrogen bonds with the hinge. More interestingly, it was demonstrated that compound **13**'s sp^3 center in the linker region can be used in conjunction with a variety of linker groups. It is potent, selective and also displays promising pharmacokinetic properties.[7]

Pyrazolo[1,5-*a*]pyridines have been employed as bioisosteres for imidazo[1,2-*a*]pyridines.

A pyrazolo[1,5-*a*]pyridine substituent is a preferred fragment of covalent epidermal growth factor receptor (EGFR) inhibitors. EGFR inhibitors were among the earliest kinase inhibitors on the market. But resistance invariably developed and covalent inhibitors have been invented to combat the L858R and T790M mutations by taking advantage of Cys-797 at EGFR's active site. T790M mutation is also known as the gatekeeper mutation. AstraZeneca chose Dana–Farber's WZ-4002 (**12**, Log $D_{7.4} > 4.3$) as their starting point because it showed activities against EGFR's L858R and T790M double mutation (DM). In an effort to maintain activities against double mutation while reducing lipophilicity, AstraZeneca arrived at covalent inhibitor **13** with the pyrazolo[1,5-*a*]pyridine fragment. While it is less active in the exon 19 deletion activating (AM) using PC9 cell line, inhibitor **13** achieved a remarkable DM/WT margin (WT stands for the wild-type enzyme with a human LoVo cell line). More importantly, it has a Log $D_{7.4}$ value of 3.6 and an LLE (for DM) value of 3.4.

The compound showed encouraging antitumor efficacy in H1975 double mutant and PC9 activating mutant models although it had relatively poor solubility (1.6 µM) and the human ether-a-go-go (hERG) potassium channel activity IC_{50} of 4.2 µM.[8]

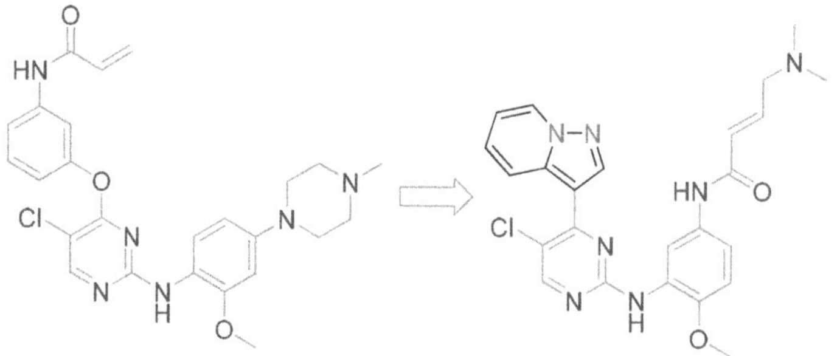

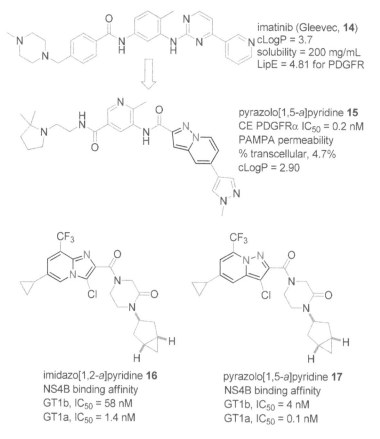

WZ-4002 (12)
DM cell (µM), 0.023
AM cell (µM), 0.044
WT cell (µM), 1.18
DM/WT margin, 51

13
DM cell (µM), 0.096
AM cell (µM), 0.40
WT cell (µM), 23
DM/WT margin, 240

The first approved kinase inhibitor imatinib (Gleevec, 14) inhibits a panel of kinases including bcr-abl, c-kit, and platelet-derived growth factor receptor (PDGFR). A Novartis team chose imatinib (14) as a starting point and employed a novel occupancy assay to directly measure target occupancy. At the end of their SAR, pyrazolo[1,5-a]pyridine-containing compound 15 showed 24 hours occupancy of the PDGFR kinase domain after a single i.t. dose and had efficacy at 0.03 mg/kg in rat monocrotaline model for pulmonary arterial hypertension.[9]

imatinib (Gleevec, 14)
cLogP = 3.7
solubility = 200 mg/mL
LipE = 4.81 for PDGFR

pyrazolo[1,5-a]pyridine 15
CE PDGFRα IC_{50} = 0.2 nM
PAMPA permeability
% transcellular, 4.7%
cLogP = 2.90

imidazo[1,2-a]pyridine 16
NS4B binding affinity
GT1b, IC_{50} = 58 nM
GT1a, IC_{50} = 1.4 nM

pyrazolo[1,5-a]pyridine 17
NS4B binding affinity
GT1b, IC_{50} = 4 nM
GT1a, IC_{50} = 0.1 nM

In GSK's pursuit of hepatitis C replication inhibitors targeting the viral NS4B protein, several iso-steres for imidazopyridine were explored. In comparison to imidazo[1,2-*a*]pyridine **16**, pyrazolo[1,5-*a*]pyridine **17** was tested to be more potent in NS4B binding affinity for both genotype 1b and 1a.[10]

SYNTHESIS OF SOME BICYCLIC PYRIDINES

Several synthetic routes exist for making alpidem (Anaxyl, **1**). One of the earliest and more robust routes began with the condensation of aminopyridine **18** with α-bromoketone **19** to assemble imidazo[1,2-*a*]pyridine **20**. Installation of the dimethylaminomethyl group was accomplished using the Mannich conditions to prepare **22**, which then underwent a three-step sequence to achieve a one-carbon homologation to afford carboxylic acid **23**. Formation of the corresponding acid chloride was followed by the addition of diisopropylamine to deliver alpidem (**1**).[11]

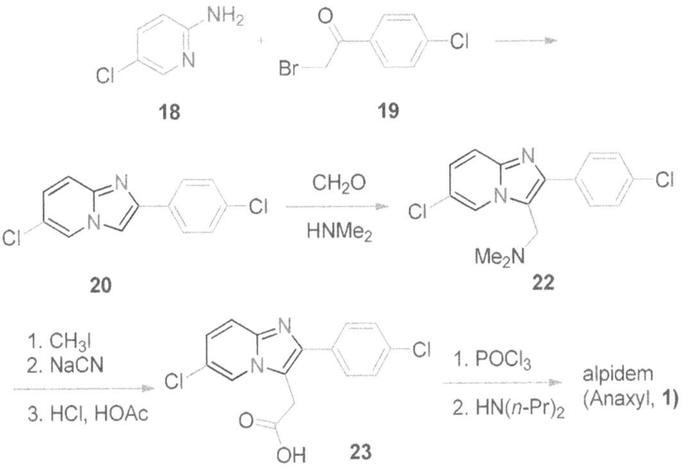

In 2019, Lei and coworkers reported a practical and scalable preparation of zolpidem (**2**) from 2-chloroimidazo[1,2-*a*]pyridine **26**. Therefore, acylation of aminopyridine **24** with maleic acid anhydride was followed by an intramolecular Michael addition to assemble **25**. Methyl ester formation was followed by chlorination to provide the key intermediate, 2-chloroimidazo[1,2-*a*]pyridine **26**. Coupling of **26** with tolylboronic acid was optimally carried out using NiCl$_2$(dppf) as the catalyst to afford adduct **27**. The final step to make zolpidem (**2**) was a straightforward amide formation.[12]

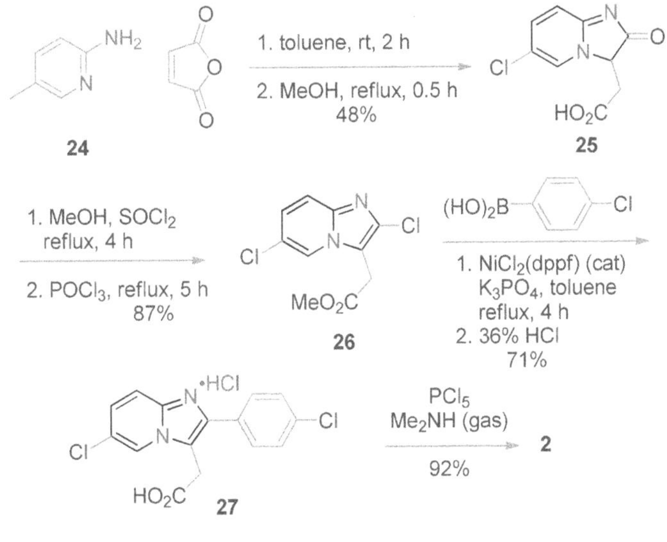

GSK process chemistry reported a kilogram-scale synthesis of their CXCR4 antagonist GSK812397 (**33**). Condensation of 6-bromopyridin-2-amine (**28**) with 1,1,3-trichloropropan-2-one, followed by acidic hydrolysis, led to 5-bromoimidazo[1,2-*a*]pyridine-2-carbaldehyde (**29**). A clever maneuver using lithiated 1-methylpiperazine gave rise to 5-(4-methylpiperazin-1-yl)imidazo[1,2-*a*]pyridine-2-carbaldehyde (**32**) after applying carefully optimized workup conditions. The trick was using intermediates **30** and **31** to serve as a transient protection so that the aldehyde function was conserved without any protection and deprotection. Two additional steps then delivered GSK812397 (**33**).[13]

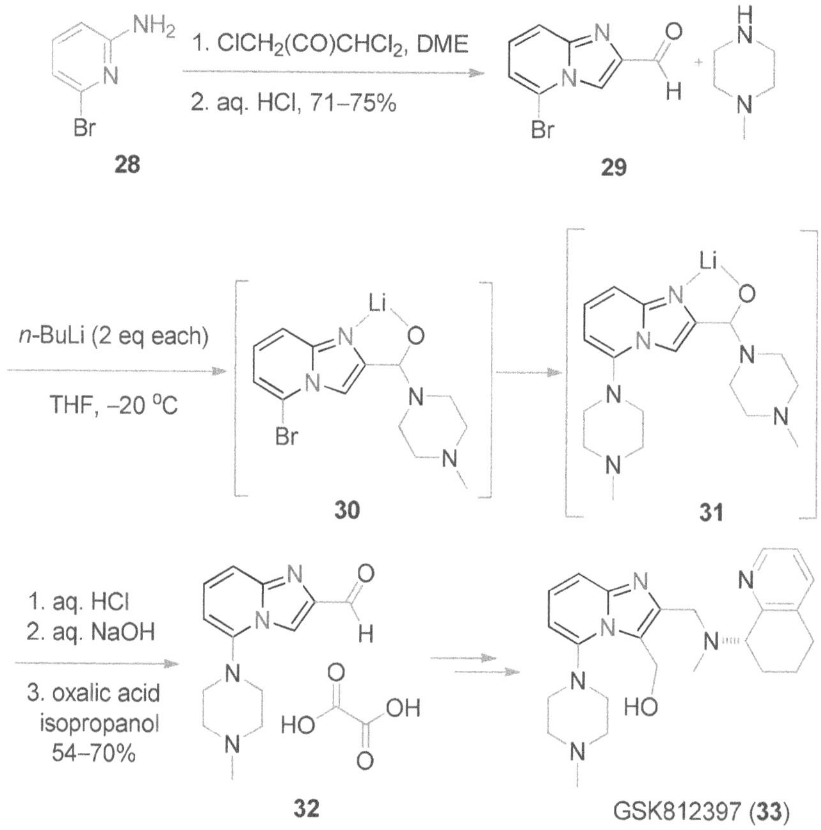

The initial route to prepare pyrazolopyridine **17** used 2-(trifluoromethyl)pyridin-4-ol (**34**) as the starting material. *O*-Methylation was followed by *N*-amination using *O*-mesitylensulfonylhydroxylamine (MSH) as the *N*-amination agent to produce hydrazine **35**. It then underwent a 1,3-dipolar cycloaddition reaction with dimethylacetylene dicarboxylate (DMAD) to give pyrazolopyridine **36** in 49% yield. A three-step sequence from **36** provided ester **37** in 73% yield. An additional four steps of transformations converted **37** to acid **38**, which subsequently was coupled with an amine to deliver amide **17** in excellent yield.[10]

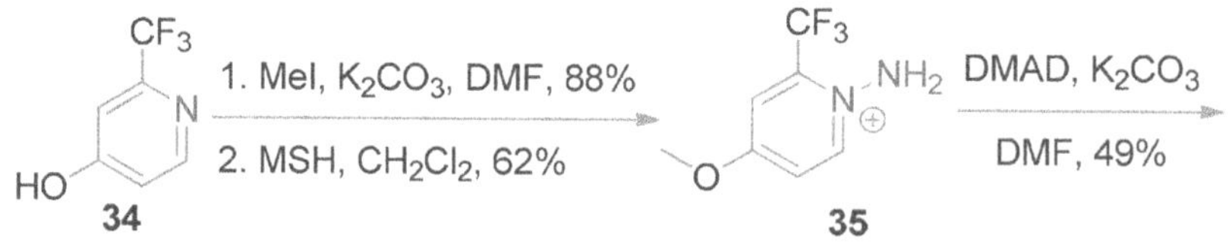

In summary, with four imidazo[1,2-*a*]pyridine-containing drugs and one pyrazolo[1,5-*a*]pyridine-containing drug on the market, they are privileged structures in medicinal chemistry. With two nitrogen atoms with the potential to serve as hydrogen-bond acceptors, imidazopyridines and pyrazolopyridines may boost binding to target proteins and elevate potency. In addition, these structures have found utility in FBDD, covalent inhibitors, reducing metabolic liabilities, and creating novel chemical space and intellectual properties. With many of the advanced intermediates now commercially available, they will find more and more applications in drug discovery.

REFERENCES

1. Yet, L. *Chapter 13. Bicyclic Pyridines Containing Ring-Junction Nitrogen*, In *Privileged Structures in Drug Discovery, Medicinal Chemistry and Synthesis*, Wiley: Hoboken, NJ, 2018, pp 481–535.
2. Durand, A.; Thenot, J. P.; Bianchetti, G.; Morselli, P. L. *Drug Metab. Rev.* **1992**, *24*, 239–266.
3. Mizushige, K.; Ueda, T.; Yukiiri, K.; Suzuki, H. *Cardiovasc. Drug Rev.* **2002**, *20*, 163–174.
4. Leftheris, K.; Satoh, Y.; Schafer, P. H.; Man, H.-W. *Med. Chem. Rev.* **2015**, *50*, 171–184.
5. He, X; Da Ros, S; Nelson, J; Zhu, X; Jiang, T; Okram, B; Jiang, S; Michellys, P-Y; Iskandar, M; Espinola, S.; et al. *ACS Med. Chem. Lett.* **2017**, *8*, 1048–1053.
6. Linton, A.; Kang, P.; Ornelas, M.; Kephart, S.; Hu, Q.; Pairish, M.; Jiang, Y.; Guo, C. *J. Med. Chem.* **2011**, *54*, 7705–7712.
7. Ward, R. A.; Anderton, M.J.; Ashton, S.; Bethel, P. A.; Box, M.; Butterworth, S.; Colclough, N.; Chorley, C. G.; Chuaqui, C.; Cross, D. A. E.; et al. *J. Med. Chem.* **2013**, *56*, 7025–7048.
8. Murray, C. W.; Berdini, V.; Buck, I. M.; Carr, M. E.; Cleasby, A.; Coyle, J. E.; Curry, J. E.; Day, J. E. H.; Day, P. J.; Hearn, K.; et al. *ACS Med. Chem. Lett.* **2015**, *6*, 798–803.
9. Shaw, D. E.; Baig, F.; Bruce, I.; Chamoin, S.; Collingwood, S. P.; Cross, S.; Dayal, S.; Druckes, P.; Furet, P.; Furminger, V.; et al. *J. Med. Chem.* **2016**, *59*, 7901–7914.
10. Miller, J. F.; Chong, P. Y.; Shotwell, J. B.; Catalano, J. G.; Tai, V. W.-F.; Fang, J.; Banka, A. L.; Roberts, C. D.; Youngman, M.; Zhang, H.; et al. *J. Med. Chem.* **2014**, *57*, 2107–2120.
11. Reviews on the synthesis of imidazo[1,2-a]pyridines: (a) Bagdi, A. K.; Santra, S.; Monir, K.; Hajra, A. *Chem. Commun.* **2015**, *51*, 1555–1575. (b) Pericherla, K.; Kaswan, P.; Pandey, K.; Kumar, A. *Synthesis* **2015**, *47*, 887–912.
12. Wang, Y.; Zhang, B.; Zheng, Y.; Ma, Q.; Sui, Q.; Lei, X. *Tetrahedron* **2019**, *75*, 1064–1071.
13. Boggs, S.; Elitzin, V. I.; Gudmundsson, K.; Martin, M. T.; Sharp, M. J. *Org. Process Res. Dev.* **2009**, *13*, 781–785.

5 Bicyclo[1.1.1]pentyl (BCP) as an sp^3 Carbon-Rich Bioisostere for *para*-Phenyl and *tert*-Butyl Groups

THERE IS SOMETHING ABOUT BCP

Bicyclo[1.1.1]pentyl (BCP) may serve as an sp^3 carbon-rich bioisostere for both *para*-phenyl[1] and *tert*-butyl[2] groups in the form of either bicyclo[1.1.1]pentane-1,4-diyl or bicyclo[1.1.1] pentane-1-yl.

Bicyclo[1.1.1]pentane-1,4-diyl (BCP) is one of the three common non-classical *para*-phenyl (*p*-Ph) isosteres. The other two are cubane-1,4-diyl (CUB) and bicyclo[2.2.2]octane-1,4-diyl (BCO). The bridgehead lengths decrease in the following order:[1]

$$p\text{-}Ph\ \left(2.79\ \text{Å},\ 100\%\right) > \text{CUB}\ \left(2.72\ \text{Å},\ 96\%\right) > \text{BCO}\ \left(2.60\ \text{Å},\ 94\%\right) > \text{BCP}\ \left(1.85\ \text{Å},\ 65\%\right).$$

Bicyclo[1.1.1]pentane-1,4-diyl (BCP) stands out in comparison to CUB and BCO. It is the shortest in terms of diagonal distances and the smallest in size, in fact, 35% smaller than *p*-Ph. Since it has the least number of carbons (5) compared to those of BCO (6) and CUB (8), BCP is the least lipophilic. This information is helpful in deciding which particular 3-D isostere to use to replace the 2-D *p*-phenyl moiety for a structure-based drug design (SBDD).

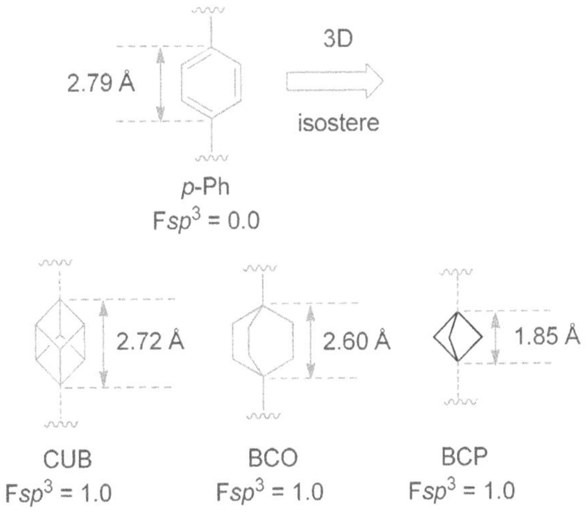

$$F sp^3 = \left(\text{number of sp}^3\text{-hybridized carbon}\right) / \left(\text{total carbon count}\right) \qquad (5.1)^2$$

p-Ph
Fsp^3 = 0.0

BCP
Fsp^3 = 1.0

Due to t-butyl substituent's lipophilicity and thus metabolic instability, many t-butyl bioisosteres have been explored to mitigate its liabilities. They include pentafluorosulfanyl (SF$_5$), bicyclo[1.1.1]pentane-1-yl (BCP), and cyclopropyl-trifluoromethyl (cyclopropyl–CF$_3$).[3] As early as 1993, Barbachyn et al. employed BCP as a t-butyl bioisostere.[4] By installing a BCP group at the N-1 position of fluorofluoroquinolone antibacterial agents, they arrived at U-87947E (2) in an effort to improve upon BMY-40062 (1). The BCP group was expected to exert a unique electronic effect in light of the increased electronegativity of its bridged carbon atom. U-87947E (2) exhibited enhanced activity relative to that of ciprofloxacin (Cipro, 3, the gold standard for the second-generation quinolones) against both Gram-positive aerobic bacteria and anaerobic organisms. Time-kill kinetic studies revealed that U-87947E (2) was exquisitely bactericidal against (greater than 32-fold more active) ciprofloxacin (3)-resistant *Staphylococcus aureus*.[4]

BMY-40062 (1)

U-87947E (2)

ciprofloxacin (Cipro, 3)

Metabotropic glutamate receptors (mGluRs) have been known for over 30 years. mGluR modulators, both agonists and antagonists, hold great promise in treating central nervous system (CNS) disorders such as psychosis and Parkinson's disease. After decades of intense research, regrettably, no drug targeting an mGluR has yet received marketing approval. An mGluR antagonist, (S)-(4-carboxyphenyl)glycine (4), may be considered as an isostere of L-glutamate. Pellicciari and colleagues sought to replace the 2-D p-phenyl ring on 4 with its 3-D brethren as isosteres. Three (S)-BCP-glycine 5 were prepared.[5] When R = CO$_2$H, the BCP–glutamate 5a was tested as a selective mGluR1a antagonist.[5a] Once the carboxylic acid was replaced with a tetrazole, a known carboxylic acid isostere, the resulting BCP–glutamate 5b was a weaker mGluR1a antagonist than 5a.[5b] If the carboxylic acid was replaced by a phosphate, the BCP-glutamate 5c became an mGluR4 and mGluR8 selective ligand.[5c] Interestingly, one-carbon homologation of 5a gave rise to (S)-ω-acidic-BCP-glycine 6 and (R)-ω-acidic-BCP-glycine 6′, respectively. Although both of them were tested inactive for mGluRs, they were found to be N-methyl-D-aspartic acid (NMDA) receptor ligands.[5d]

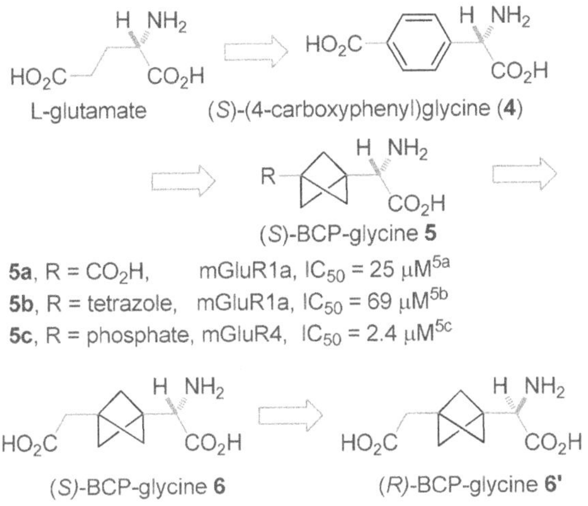

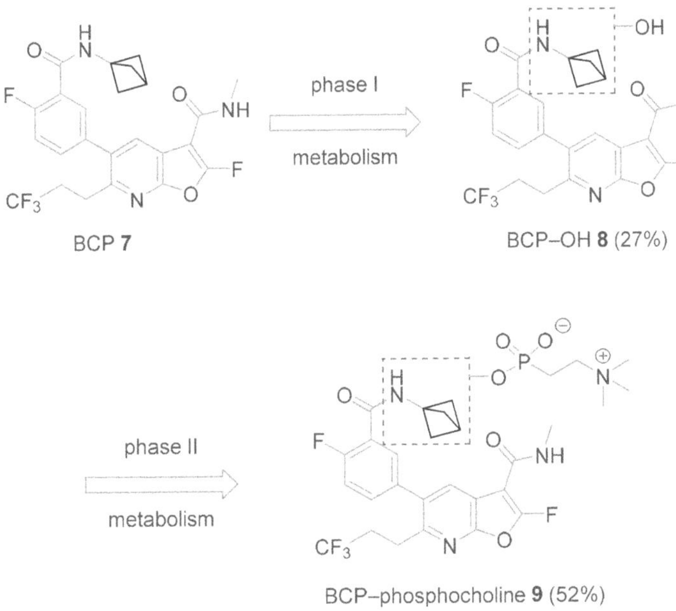

5a, R = CO$_2$H,   mGluR1a, IC$_{50}$ = 25 μM[5a]
5b, R = tetrazole,   mGluR1a, IC$_{50}$ = 69 μM[5b]
5c, R = phosphate, mGluR4,   IC$_{50}$ = 2.4 μM[5c]

Hepatitis C virus non-structural protein 5B (HCV NS5B, an RNA-dependent RNA polymerase, RdRp) inhibitors are successful treatments for HCV infection. In an effort to improve the *t*-butyl group's metabolic stability, a bicyclo[1.1.1]pentane-1-yl (BCP) was introduced into the molecular scaffold as an isostere to afford BCP **7**. Compared to NS5B inhibitors featuring *t*-butyl groups in the same position, the bicyclo[1.1.1]pentane-1-yl analogs such as BCP **7** evaluated in the study retained comparable antiviral activity against genotype (gt) 1a, 1b, and 2a proteins (<10 nM).[6] The absorption, distribution, metabolism, and excretion (ADME) properties were subsequently investigated. Although BCP **7**'s *in vitro* metabolite was BCP–OH **8**, just as expected, its *in vivo* metabolism was unique. The two major metabolites were BCP–OH **8** (27%) and BCP–phosphocholine conjugate **9** (52%). The BCP–phosphocholine conjugate **9** is rare. The other better-known drug to form phosphocholine conjugate with a hydroxyl group is everolimus, a hydroxyethyl prodrug of rapamycin.[6]

BICYCLO[1.1.1]PENTANES IN DRUG DISCOVERY

It has been shown that for some drug candidates, replacing a *p*-phenyl or a *t*-butyl group with BCP could maintain pharmacological efficacy while improving solubility and oral bioavailability.

Darapladib (**10**), a lipoprotein-associated phospholipase A_2 (LpPLA$_2$) inhibitor, was in phase III clinical trials as a treatment of atherosclerosis. But it had suboptimal physicochemical properties including high molecular weight, low solubility, and high property forecast index (PFI) and failed to meet phase III endpoints in 2013 in a trial of 16,000 patients with an acute coronary syndrome (ACS). Substituting one phenyl ring with the fully saturated bicyclo[1.1.1]-pentane-1,4-diyl (BCP) gave rise to BCP-analog **11**. Although it is slightly less potent than the parent darapladib (**10**), BCP-analog **11** is bestowed with superior physicochemical properties with an improved permeability of 705 nm/s from 203 nm/s for darapladib (**10**, AMP, artificial membrane permeability). It has gained a three-fold increase of thermodynamic fasted state simulated intestinal fluid (FaSSIF) solubility and a nine-fold increase in kinetic solubility. As a consequence, analog **11** has a slightly lower PFI value.[7]

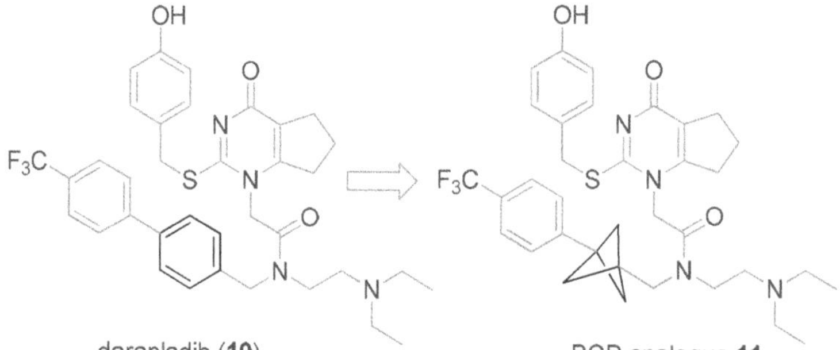

darapladib (**10**)
pIC$_{50}$ = 10.2
FaSSIF (μg/mL) = 399
AMP = 230 nm/s
kinetic solubility = 8 μM
ChromLogD$_{7.4}$ = 6.3
PFI = 10.3

BCP analogue **11**
pIC$_{50}$ = 9.4
FaSSIF (μg/mL) > 1000
AMP = 705 nm/s
kinetic solubility = 74 μM
ChromLogD$_{7.4}$ = 7.0
PFI = 10.0

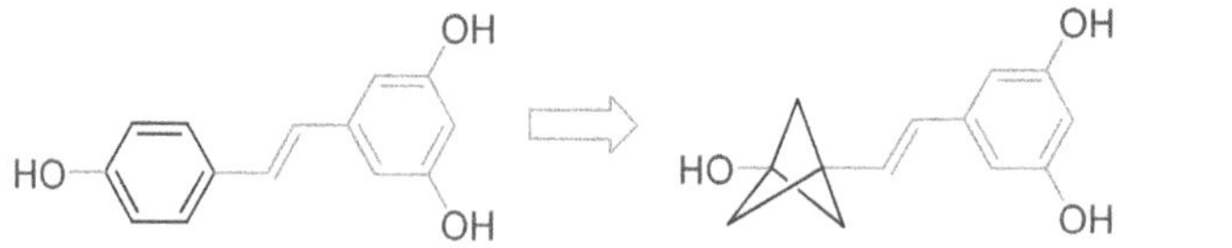

resveratrol (**12**)
Log $D_{7.4}$ = 1.9
solubility at pH7.4 = 19 μg/mL
$t_{1/2}$ = 0.19 h
C_{max} = 273 ng/mL
AUC = 47.5 ng•h/kg
T_{max} (h) = 0.083
Cl/F (L/h/kg) = 409

BCP–resveratrol (**13**)
Log $D_{7.4}$ = 2.9
solubility at pH7.4 = 619 μg/mL
$t_{1/2}$ = 2.6 h
C_{max} = 942 ng/mL
AUC = 587 ng•h/kg
T_{max} (h) = 0.067
Cl/F (L/h/kg) = 21.8

Resveratrol (**12**) has garnered much attention in medical research for the last two decades. But the progress has been hampered since its bioavailability is too low: with three phenol groups, resveratrol (**12**) goes through a rapid first-pass metabolism to its glucuronide and sulfate conjugates. Replacing one of the *p*-phenyl rings with bicyclo[1.1.1]pentane-1,4-diyl (BCP) resulted in BCP–resveratrol (**13**). The alcohol on **13** is nearly neutral in comparison to acidic phenol on **12**. In addition to being more lipophilic, the BCP portion has all sp^3 carbons. BCP–resveratrol (**13**) gained a 32-fold boost of aqueous solubility and a three-fold increase of C_{max}, and it is ten-fold more bioavailable as measured by the AUC values than resveratrol (**12**). Experimental data indicated that the formation of glucuronide and sulfate conjugates in human hepatocytes for BCP–resveratrol (**13**) was significantly reduced since its steric hindrance retards secondary metabolism. More importantly, BCP–resveratrol (**13**) exerted similar biological activities in selected cancer cell lines.[8]

Sulfonamides **14** (BMS-708,163) and **15** are γ-secretase inhibitors with similar potencies, indicating that the main role of the fluorophenyl fragment is that of a *spacer*.[6] Employing bicyclo[1.1.1] pentane-1,4-diyl (BCP) on **15** as a bioisostere for the fluorophenyl moiety on **14**, the Fsp^3 carbon atom count doubled to 0.52 for **15** from 0.25 for **14**. As shown in equation 5.1, Fsp^3 is the fraction of saturated carbons within a molecule, a descriptor of complexity and an alternative to the number of aromatic rings (#Ar).[2] Disruption of aromaticity translated to a higher LipE value of 5.95 for **15** from 4.95 for **14**. More important, this maneuver also translated to practical advantages of improved kinetic and thermodynamic aqueous solubility and increased membrane permeability, probably brought about by a reduction in lipophilicity since the ElogD is reduced by 0.9.[9]

14, IC$_{50}$ (Aβ$_{42}$) = 0.225 nM
ElogD = 4.7
sol pH7.4 = 0.9 μM
PRCK P$_{app}$ (A to B) (10^{-6} cm/s) = 5.52
Fsp^3 = 0.25
LE = 0.40
LipE = 4.95
LELP = 11.75

15, IC$_{50}$ (Aβ$_{42}$) = 0.178 nM
ElogD = 3.8
sol pH7.4 = 29.4 μM
PRCK = 19.3
Fsp^3 = 0.52
LE = 0.43
LipE = 5.95
LELP = 8.83

Imatinib (Gleevec, **17**) was revolutionary in medicine because it was the first kinase inhibitor on the market as the target cancer treatment. It is a Bcr–Abl kinase inhibitor approved to treat a variety of leukemia such as chronic myeloid leukemia (CML) and acute lymphoblastic leukemia (ALL). But with an Fsp^3 value of 0.24, imatinib (**17**) is woefully deficient of sp^3 carbons. As a consequence, it has a high cLog*P* of 4.53, a high melting point of 204°C, and a low aqueous solubility at pH 7.4 of 30.7 μM. Nicolaou et al. prepared several non-classical *p*-phenyl isosteres including cubane, cyclopropane, and cyclobutane to replace the *para*-phenyl ring sandwiched between the amide bond and the piperazine appendage. BCP–imatinib (**18**) was also prepared and investigated for its biopharmaceutical properties. While it had similar biological activities, the Fsp^3 value of BCP–imatinib (**18**) almost doubled that of the prototype imatinib (**17**). The thermodynamic solubility increased by >80-fold from this maneuver, making BCP–imatinib (**18**) more "drug-like".[10]

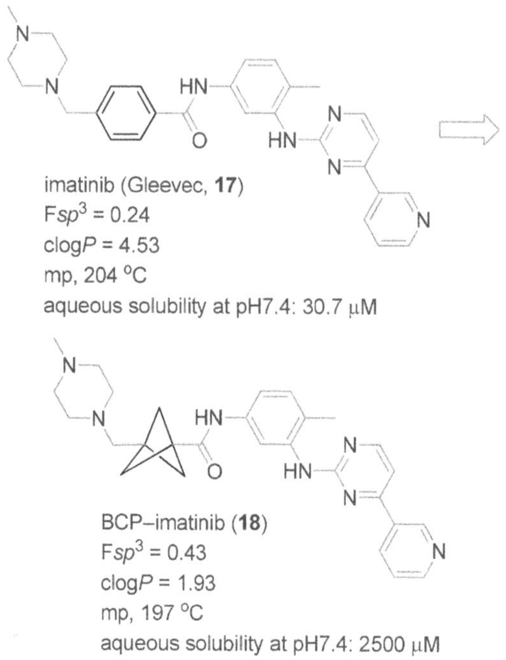
HN
O
HN
N
N
N
N
N
imatinib (Gleevec, **17**)
Fsp^3 = 0.24
clog$P$ = 4.53
mp, 204 °C
aqueous solubility at pH7.4: 30.7 μM
N
N
HN
O
HN
N
N
N
BCP–imatinib (**18**)
Fsp^3 = 0.43
clog$P$ = 1.93
mp, 197 °C
aqueous solubility at pH7.4: 2500 μM

SYNTHESIS OF SOME BICYCLO[1.1.1]PENTANE-CONTAINING DRUGS

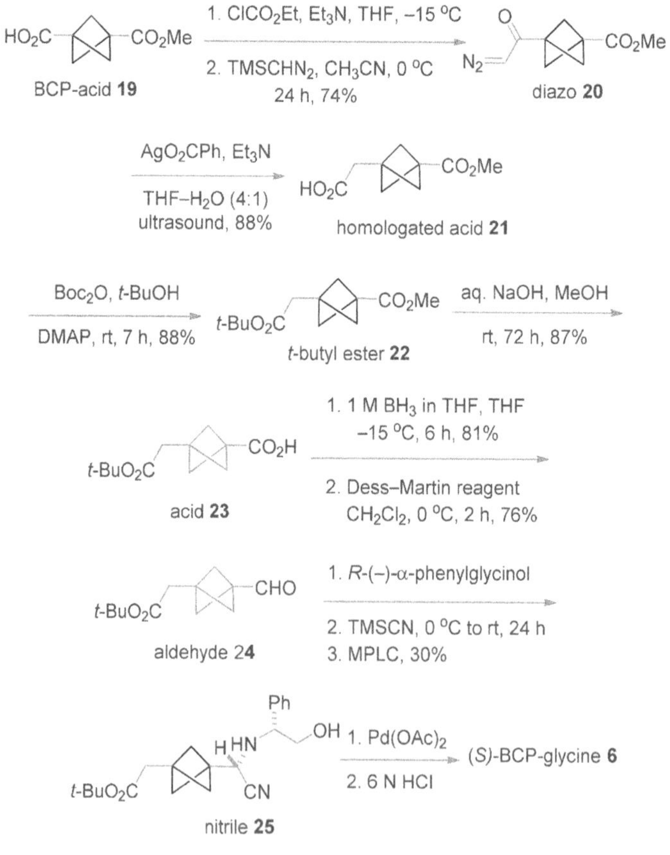
HO_2C
CO_2Me
BCP-acid **19**
1. $ClCO_2Et$, Et_3N, THF, −15 °C
2. $TMSCHN_2$, CH_3CN, 0 °C
24 h, 74%
O
N_2
CO_2Me
diazo **20**
AgO_2CPh, Et_3N
THF–H_2O (4:1)
ultrasound, 88%
HO_2C
CO_2Me
homologated acid **21**
Boc_2O, t-BuOH
DMAP, rt, 7 h, 88%
t-BuO_2C
CO_2Me
t-butyl ester **22**
aq. NaOH, MeOH
rt, 72 h, 87%
t-BuO_2C
CO_2H
acid **23**
1. 1 M BH_3 in THF, THF
−15 °C, 6 h, 81%
2. Dess–Martin reagent
CH_2Cl_2, 0 °C, 2 h, 76%
t-BuO_2C
CHO
aldehyde 2**4**
1. R-(−)-α-phenylglycinol
2. TMSCN, 0 °C to rt, 24 h
3. MPLC, 30%
Ph
OH
H HN
t-BuO_2C
CN
nitrile **25**
1. $Pd(OAc)_2$
2. 6 N HCl
(S)-BCP-glycine **6**

Pellicciari's synthesis of (S)-ω-acidic-BCP-glycine **6** commenced with commercially available ester-BCP-acid **19**. One-carbon homologation of **19** was carried out by a modified Arndt–Eistert reaction via the intermediacy of diazo compound **20**, followed by a Wolff rearrangement to afford homologated acid **21**. Acid **21** was converted to mixed diester **22**, which was readily hydrolyzed to acid **23**. Borane reduction of acid **23** was followed by a Dess–Martin oxidation to produce aldehyde **24**. A Strecker reaction between **24** and R-(–)-α-phenylglycinol gave rise to nitrile **25** after chiral separation. Finally, after cleavage of the chiral auxiliary, hydrolysis delivered (S)-ω-acidic-BCP-glycine **6**.[5d]

Preparation of Pfizer's γ-secretase inhibitor **15** began with the known nitrile-BCP-ester **26**. Simple reduction of **26** gave alcohol **27**, which underwent a Mitsunobu reaction with amide-sulfonamide **28** selectively to afford nitrile **29**. Nitrile **29** was converted to the title oxadiazole **15** via a two-step sequence involving treatment with hydroxylamine to generate an intermediate amide-oxime, which was then transformed to **15** in the presence of triethyl orthoformate and a catalytic amount of boron trifluoride etherate.[9]

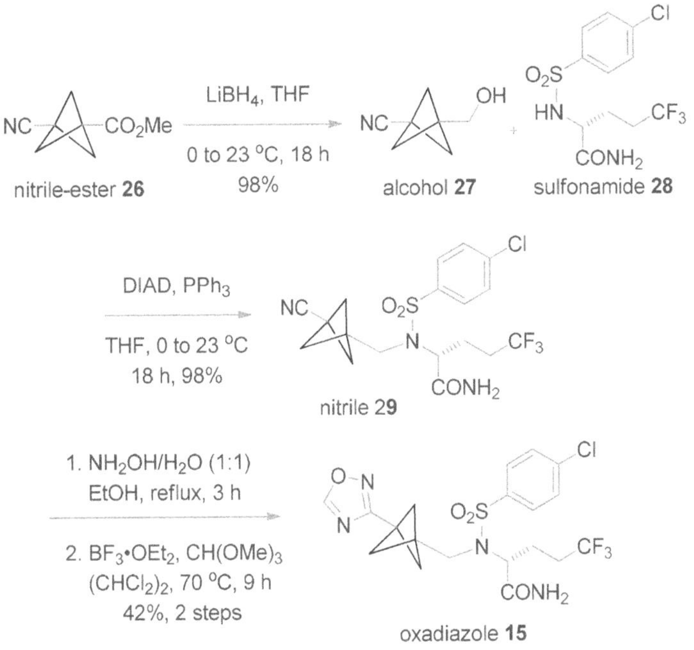

Amino-BCP and di-amino-BCP have been employed as building blocks in the preparation of potential medicines. Aminopyridine **31**, a selective anaplastic lymphoma kinase-2 (ALK-2) inhibitor, was synthesized using amino-BCP **30** as the starting material. It has a potential in the treatment of heterotopic ossification and fibrodysplasia ossificans progressiva.[11] On the other hand, hematopoietic prostaglandin D synthase (H-PGDS) inhibitor **33** was prepared employing diamino-BCP **32** as the starting material.[12]

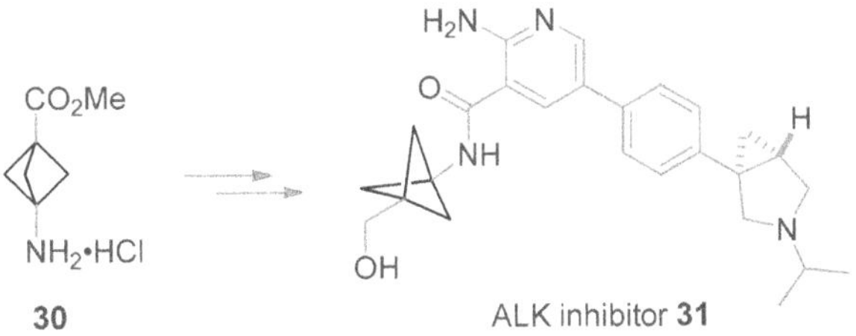

32   H-PGDS inhibitor **33**

To conclude, bicyclo[1.1.1]pentane (BCP) is unique as a non-classical *p*-phenyl isostere as well as a *t*-butyl isostere. In comparison to CUB and BCO, it is the shortest in terms of diagonal distance and the smallest in size. Since it has the least number of carbons vs. BCO and CUB, BCP is the least lipophilic. BCP is an especially useful isostere for the phenyl fragment when it serves largely as a spacer rather than engaging π-stacking interactions with the target protein. While maintaining the biological activities as the *p*-phenyl analog, BCP isosteres provide superior biopharmaceutical properties such as solubility, permeability, and *in vitro* stability. BCP's utility in medicinal chemistry is expected to grow in the future.

REFERENCES

1. Auberson, Y. P.; Brocklehurst, C.; Furegati, M.; Fessard, T. C.; Koch, G.; Decker, A.; La Vecchia, L.; Briard, E. *ChemMedChem* **2017**, *12*, 590–598.
2. (a) Lovering, F.; Bikker, J.; Humblet, C. *J. Med. Chem.* **2009**, *52*, 6752–6756. (b) Lovering, F. *MedChemComm* **2013**, *4*, 515–519.
3. Westphal, M. V.; Wolfstaedter, B. T.; Plancher, J.-M.; Gatfield, J.; Carreira, E. M. *ChemMedChem* **2015**, *10*, 461–469.
4. Barbachyn, M. R.; Hutchinson, D. K.; Toops, D. S.; Reid, R. J.; Zurenko, G. E.; Yagi, B. H.; Schaadt, R. D.; Allison, J. W. *Bioorg. Med. Chem. Lett.* **1993**, *3*, 671–676.
5. (a) Pellicciari, R.; Raimondo, M.; Marinozzi, M.; Natalini, B.; Costantino, G.; Thomsen, C. *J. Med. Chem.* **1996**, *39*, 2874–2976. (b) Costantino, G.; Maltoni, K.; Marinozzi, M.; Camaioni, E.; Prezeau, L.; Pin, J.-P.; Pellicciari, R. *Bioorg. Med. Chem.* **2001**, *9*, 221–227. (c) Filosa, R.; Marinozzi, M.; Costantino, G.; Hermit, M. B.; Thomsen, C.; Pellicciari, R. *Bioorg. Med. Chem.* **2006**, *14*, 3811–3817. (d) Filosa, R.; Fulco, M. C.; Marinozzi, M.; Giacche, N.; Macchiarulo, A.; Peduto, A.; Massa, A.; de Caprariis, P.; Thomsen, C.; Christoffersen, C. T.; et al. *Bioorg. Med. Chem.* **2009**, *17*, 242–250.
6. Zhuo, X.; Cantone, J. L.; Wang, Y.; Leet, J. E.; Drexler, D. M.; Yeung, K.-S.; Huang, X. S.; Eastman, K. J.; Parcella, K. E.; Mosure, K. W.; et al. *Drug Metab. Dispos.* **2016**, *44*, 1332–1340.
7. Measom, N. D.; Down, K. D.; Hirst, D. J.; Jamieson, C.; Manas, E. S.; Patel, V. K.; Somers, D. O. *ACS Med. Chem. Lett.* **2017**, *8*, 43–48.
8. Goh, Y. L.; Cui, Y. T.; Pendharkar, V.; Adsool, V. A. *ACS Med. Chem. Lett.* **2017**, *8*, 516–520.
9. Stepan, A. F.; Subramanyam, C.; Efremov, I. V.; Dutra, J. K.; O'Sullivan, T. J.; DiRico, K. J.; McDonald, W. S.; Won, A.; Dorff, P. H.; Nolan, C. E.; et al. *J. Med. Chem.* **2012**, *55*, 3414–3424.
10. Nicolaou, K. C.; Vourloumis, D.; Totokotsopoulos, S.; Papakyriakou, A.; Karsunky, H.; Fernando, H.; Gavrilyuk, J.; Webb, D.; Stepan, A. F. *ChemMedChem* **2016**, *11*, 31–37.
11. Li, J.; Arista, L.; Babu, S.; Bian, J.; Cui, K.; Dillon, M. P.; Lattmann, R.; Liao, L.; Lizos, D.; Ramos, R.; et al. (Novartis) WO2018014829 (2018).
12. Deaton, D. N.; Guo, Y.; Hancock, A. P.; Schulte, C.; Shearer, B. G.; Smith, E. D.; Stewart, E. L.; Thomson, S. A. (GSK) WO2018069863 (2018).

6 Bicyclo[2.2.2]octane (BCO) as a 3D-Rich Bioisostere for the *para*-Phenyl Group

MERITS OF *SP*³-RICH CARBON BIOISOSTERES—ESCAPE FROM FLATLAND

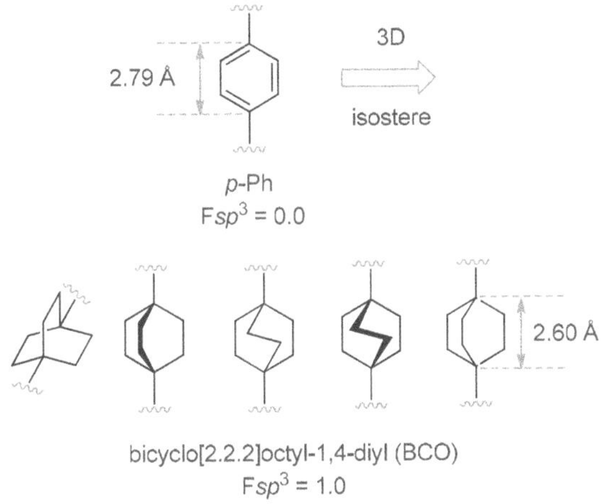

p-Ph
$Fsp^3 = 0.0$

bicyclo[2.2.2]octyl-1,4-diyl (BCO)
$Fsp^3 = 1.0$

The perils of high aromatic ring count are well known with regard to aqueous solubility, lipophilicity, serum albumin binding, CYP450 inhibition, and human ether-a-go-go (hERG) potassium channel activity inhibition.[1] Fully aliphatic bicyclo[2.2.2]octane-1,4-diyl (BCO) is a 3-dimensional bioisostere for the 2-dimensional *para*-phenyl group (*p*-Ph). The fraction of saturated carbon (Fsp^3, defined in equation 1)[2a] for BCO is 1.0 at one extreme of the spectrum, but it is 0 for the aromatic *p*-phenyl ring, at another extreme. From a geometrical point of view, the distance between connecting atoms in the BCO scaffold (2.60 Å) is very similar to the *p*-Ph group (2.82 Å).

$$Fsp^3 = \left(\text{number of } sp^3\text{-hybridized carbon}\right)/\left(\text{total carbon count}\right) \tag{6.1}$$

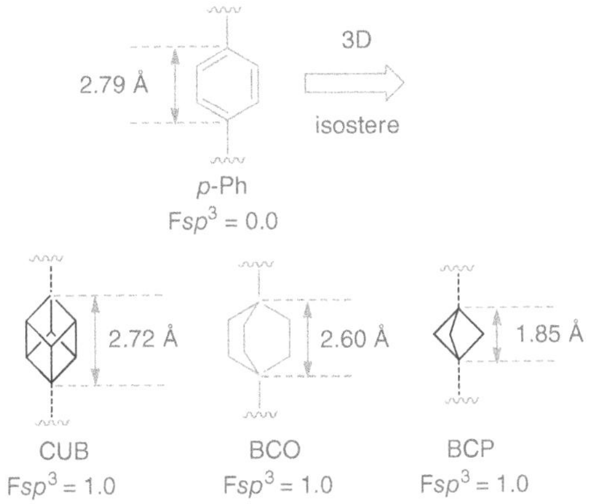

p-Ph
$Fsp^3 = 0.0$

CUB	BCO	BCP
$Fsp^3 = 1.0$	$Fsp^3 = 1.0$	$Fsp^3 = 1.0$

49

In addition to BCO, there are two closely related non-classical *p*-phenyl isosteres: bicyclo[1.1.1] pentane-1,4-diyl (BCP) and cubane-1,4-diyl (CUB). The distances between their bridgeheads are 2.72 and 1.85 Å, respectively. The bridgehead lengths decrease in the following order:

$$p\text{-Ph } (2.79 \text{ Å, } 100\%) > \text{CUB } (2.72 \text{ Å, } 96\%) > \text{BCO } (2.60 \text{ Å, } 94\%) > \text{BCP } (1.85 \text{ Å, } 65\%).$$

This information is useful in deciding which particular 3-D isostere to use to replace the 2-D *p*-phenyl moiety for a structure-based drug design (SBDD).

There is a myriad of advantages for a drug with a higher Fsp^3 value to escape from flatland.

a. A higher degree of saturation for a molecule provides increased opportunity to design out-of-plane substituents and to adjust molecular shape that could increase receptor–ligand complementarity. The molecular complexity might allow the engineering of additional protein–ligand interactions not accessible to a flat aromatic ring and thus improve potency and selectivity to a given target, which should mitigate off-target effects.[1] Compounds with greater specificity and selectivity are expected to show less toxicity due to off-target effects.

b. Compounds containing ionizable amines are more promiscuous than neutral ones. But both amine-containing drugs and neutral drugs tend to have higher promiscuity against drug targets when they have lower Fsp^3 values. In other words, saturation mitigates a drug's promiscuity.[2b]

c. Compounds with higher Fsp^3 values tend to have lower cytochrome P450 (CYP450) inhibitions. Therefore, saturation reduces a drug's tendency to have drug–drug interactions (DDIs).[2b]

d. Disruption of aromaticity reduces the ease with which to form crystalline lattices. Empirically, π stacking enhances the crystallinity of compounds containing aromatic rings. Therefore, disruption of planarity hence aromaticity results in improvement of the melting point and solubility.

The last few decades have seen cLogP getting higher and Fsp^3 getting lower when proper attention was not paid to a drug's physiochemical properties. In order to increase our success rate in drug discovery, it is essential that we get cLogP down and Fsp^3 up. Employing BCO to replace the *para*-phenyl fragment is one step closer to the right direction.[3] All in all, complexity, as measured by both Fsp^3 and the presence of chiral centers, impacts the probability of success in the clinic.

BICYCLO[2.2.2]OCTANES IN DRUG DISCOVERY

It has been shown that for some drug candidates, replacing a phenyl group with BCO could maintain pharmacological efficacy while improving solubility and oral bioavailability.

BMS's daclatasvir (Daklinza, **1**) is the first-in-class hepatitis C virus non-structural protein 5A (HCV NS5A) inhibitor approved by the FDA in 2014 for the treatment of hepatitis C. In an effort to improve its physiochemical properties, constraint cycloalkane–phenyl motifs were employed to replace the flat biphenyl core structure.[4] One of the analogs was phenyl–BCO **2**, which is ten-fold less potent than daclatasvir (**1**) in genotype-1a (gt-1a) replicon, indicating that the biphenyl residue not only serves as a conformationally restrained spacer but also contributes to certain interaction with the NS5A protein. As expected, phenyl–BCO **2** is six-fold more soluble than daclatasvir (**1**) in aqueous solution at pH 7.4, the physiological acidity.[4]

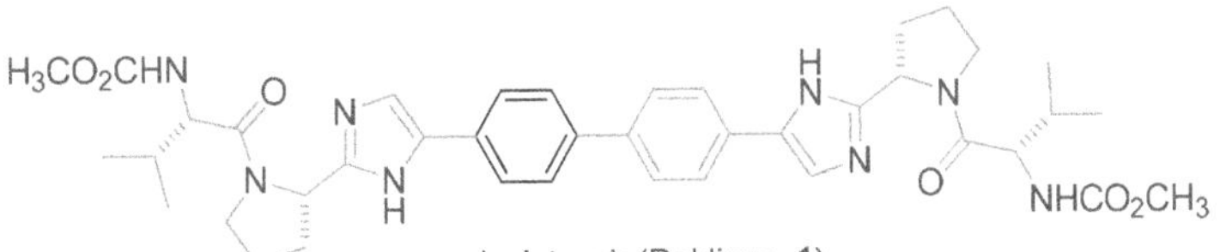

daclatasvir (Daklinza, **1**)

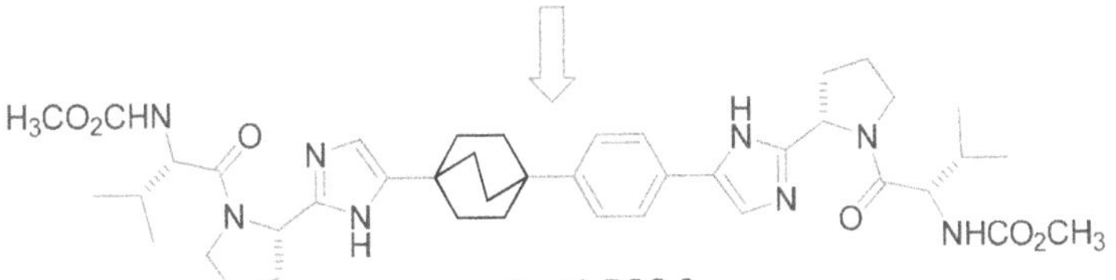

phenyl–BCO **2**

	HCV gt-1a EC$_{50}$ (nM)	HCV gt-1a EC$_{50}$ (nM)	Solubility at pH 7.4 (µg/mL)
Daclatasvir (Daklinza, **1**)	0.14	0.023	6.8
Phenyl–BCO **2**	1.4	0.060	45.0

Pyrimidinooxazinyl *trans*-cyclohexane-acetic acid **3** is a potent, selective, and orally efficacious diacylglycerol acyltransferase-1 (DAT-1) inhibitor. But it undergoes phase II metabolism via conjugation of the acid group to form the acyl glucuronide. Since the reactivity of such metabolites can lead to covalent protein adducts, which may give rise to idiosyncratic toxicity, it is logical to introduce a greater degree of steric crowding around the acid group by replacing the cyclohexane ring with various bi- and tri-cyclic systems. Certain sterically hindered carboxylic acid acyl glucuronides have been shown to be inherently more stable both to hydrolysis and to rearrangement to more reactive isomers, making them less likely to react with proteins *in vivo*. One of the bicyclic derivatives, BCO-acetic acid **4**, was tested as potent as the prototype **3**, yet had superior pharmacokinetic (PK) properties with smaller clearance (Cl$_p$) and volume of distribution (V$_{dss}$), as well as a longer half-life.[5] One added advantage of using BCO to replace the *trans*-cyclohexane is that there is no *cis/trans*-stereochemistry to be of concern.

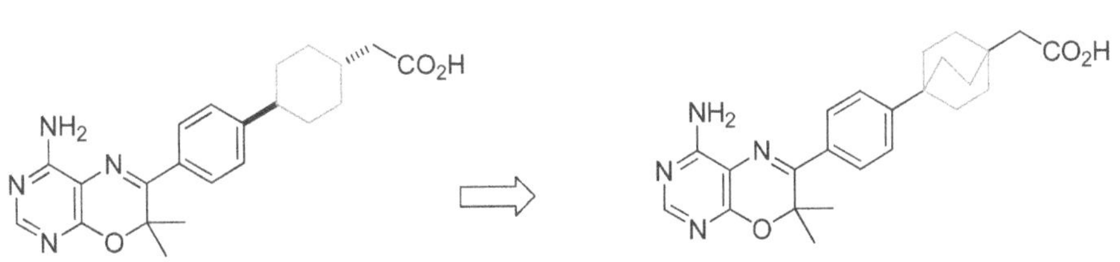

cyclohexane-acetic acid **3**
human DGAT-1 IC$_{50}$ (µM), 0.015
PK parameters in dogs:
Cl$_p$ (mL/min/kg), 12
V$_{dss}$ (L/kg), 2.0
iv half-life (h), 3.9
oral half-life (h), 2.8
bioavailability (%), > 100

BCO-acetic acid **4**
human DGAT-1 IC$_{50}$ (µM), 0.015
PK parameters in dogs:
Cl$_p$ (mL/min/kg), 3.3
V$_{dss}$ (L/kg), 1.0
iv half-life (h), 8.6
oral half-life (h), 9.9
bioavailability (%), > 100

Adenosine is the A on ATP, adenosine triphosphate. As a metabolite of ATP, adenosine exerts a plethora of pharmacologic effects via four G-protein-coupled receptors (GPCRs): A_1, A_{2A}, A_{2B}, and A_3. One of the early adenosine A_1 receptor antagonists was an 8-aryl-substituted xanthine amine congener (XAC, **5**). While being potent, XAC (**5**) has only a moderate aqueous solubility: 90 μM in 0.1 M sodium phosphate at pH 7.2. In an effort to improve its solubility by disrupting aromaticity, hence crystalline lattices, Biogen–Idec employed BCO to replace the phenyl ring on XAC (**5**). One of the BCO analogs, BG9928 (**6**), was tested to be highly potent against human adenosine A_1 receptor and selective: 915-fold vs. hA_{2A} receptor and 90-fold vs. hA_{2B} receptor. With a solubility of 25.4 mg/mL, BG9928 (**6**) showed excellent bioavailability and was tested orally efficacious with an ED_{50} of 0.01 mg/mg in animal diuresis models. BG9928 (**6**) was moved to clinical trials in 2006.[6,7]

XAC (**5**)

BG9928 (**6**)
hA_1, K_i = 7.4 nM
hA_{2A}, K_i = 6410 nM
hA_{2B}, K_i = 90 nM
ED_{50} = 0.01 mg/kg

tricyclic imidazoline **7**
hA_1, K_i = 6 nM
hA_{2A}, K_i = 4400 nM
hA_{2B}, K_i = 580 nM
ED_{50} = 0.01 mg/kg

Inspired by Fujisawa's FK838, a pyrazolopyridine, Biogen–Idec came up with their backup candidate tricyclic imidazoline **7**, which has a biological profile similar to that of the prototype BG9928 (**6**). Different from compounds **1–6**, the BCO fragment of compound **7** is sandwiched by a carbon and an oxygen atom, whereas the BCOs of **1–6** are all sandwiched by two carbon atoms.

One recent drug candidate in clinical trials is AA-115/APG-115 (**10**), which is a potent, selective, non-peptide murine double minute 2 (MDM-2) inhibitor containing a BCO moiety.[2] MDM2 inhibitors blocking the MDM2–p53 interaction [a protein–protein interaction (PPI)] can liberate the tumor suppressor function of p53 and may have potentials as cancer therapies. In the process of improving the chemical stability of a previous clinical development MDM2 inhibitor, Wang's group prepared spirooxindole–phenyl analog **8**. Compound **8** was stable, potent in both enzymatic and cellular (in the SJSA-1 cell line) assays, and capable of achieving partial tumor regression in the SJSA-1 xenograft model. But it had lower C_{max} and AUC than the original MDM2 inhibitor in clinics. In order to improve **8**'s PK properties while retaining its high binding affinity to MDM2 and cellular potency, replacements of the benzoic acid with non-classical benzoic acid mimetics bicyclo[1.1.1] pentyl (BCP) analog **9** and BCO analog **10** were carried out. The BCO analog **10** proved to have a high affinity to MDM and is as potent as **8** in inhibiting cell growth in the SJSA-1 cell line. With an excellent oral pharmacokinetic profile, BCO analog **10** is capable of achieving complete and long-lasting tumor regression *in vivo* and is currently in phase I clinical trials for cancer treatment.[8]

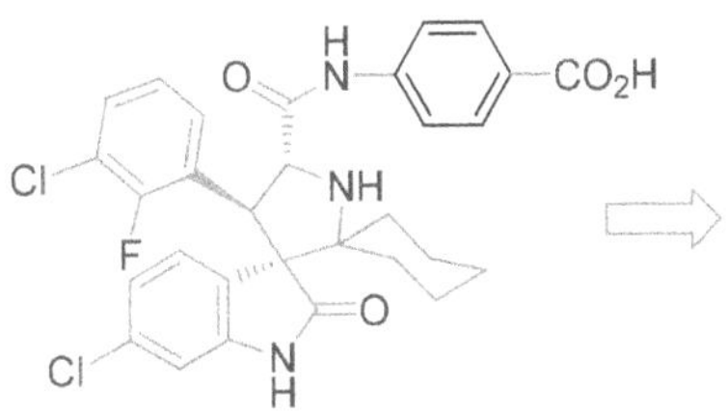

phenyl analogue **8**
MDM2 IC_{50} = 4.4 nM
MDM2 K_i < 1 nM
SJSA-1 IC_{50} = 100 nM

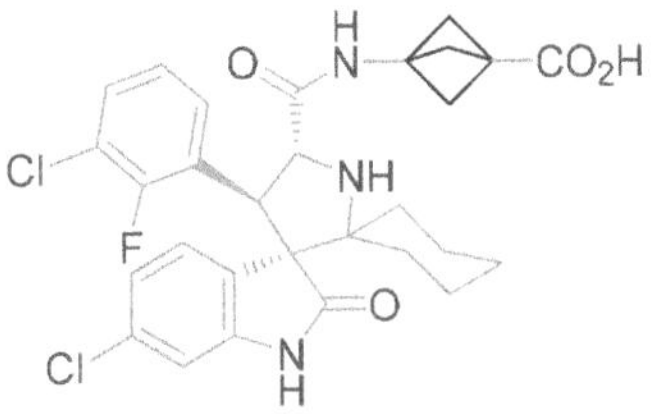

BCP analogue **9**
MDM2 IC_{50} = 6.4 nM
MDM2 K_i < 1 nM
SJSA-1 IC_{50} = 542 nM

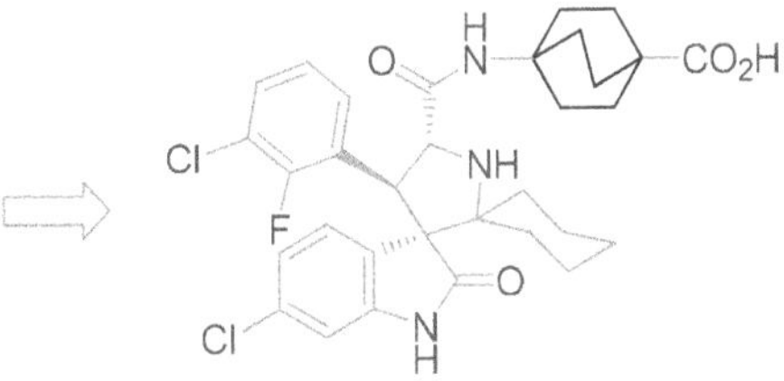

BCO analogue **10**
MDM2 IC_{50} = 3.7 nM
MDM2 K_i < 1 nM
SJSA-1 IC_{50} = 89 nM

SYNTHESIS OF SOME BICYCLO[2.2.2]OCTANE-CONTAINING DRUGS

The synthesis of HCV NS5A inhibitor phenyl–BCO **2** began with dimethyl cyclohexane-1,4-dicarboxylate (**11**) to prepare bromide **12** in four steps, with 47% overall yield. The bromide **12** is now commercially available. An $AlCl_3$-catalyzed Friedel–Crafts alkylation of bromide **12** with benzene led to the formation of phenyl **13**, which underwent an $AlCl_3$-catalyzed Friedel–Crafts acylation with AcCl to produce ketone **14**. Six additional steps converted ketone **14** to keto-ester **15**. After transforming keto ester **15** to bis-imidazole **16**, an additional four steps delivered phenyl–BCO **2**.[4]

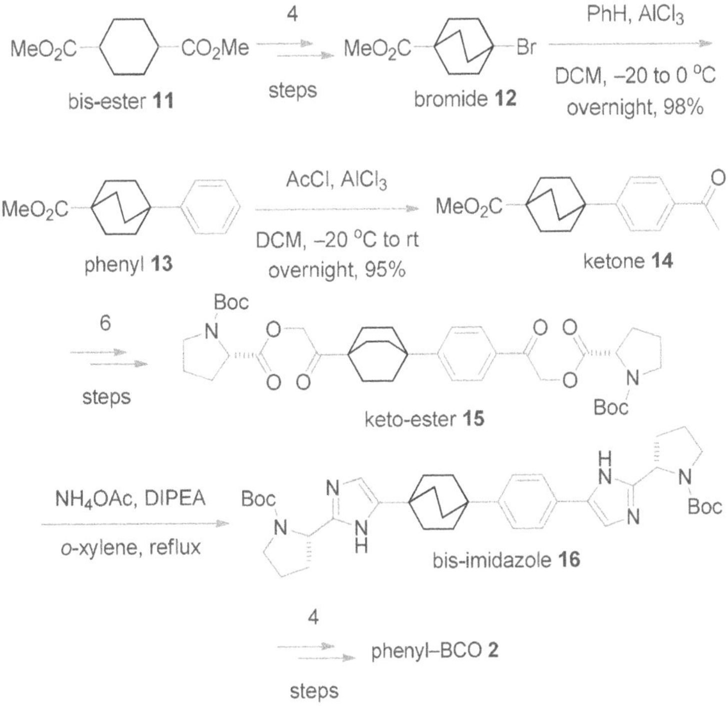

bis-ester **11**
4 steps
bromide **12**
PhH, $AlCl_3$
DCM, −20 to 0 °C
overnight, 98%

phenyl **13**
AcCl, $AlCl_3$
DCM, −20 °C to rt
overnight, 95%
ketone **14**

6 steps
keto-ester **15**

NH_4OAc, DIPEA
o-xylene, reflux
bis-imidazole **16**

4 steps
phenyl–BCO **2**

Preparation of Merck's 11β-hydroxysteroid dehydrogenase type 1 (11β-HSD1) inhibitor BCO–triazole **20** commenced with BCO–ester–acid **17**. Intermediate amide-oxime **18** was produced in three steps from the starting material **17**. Additional three steps transformed **18** to oxadiazole **19** with its ester converted to an amide. Conversion of **19** to BCO–triazole **20** was accomplished in two additional steps.[9]

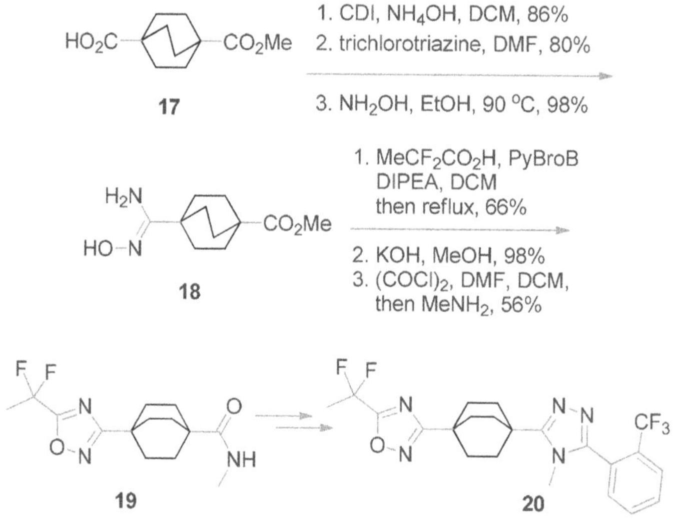

Novartis's synthesis of pyrazolopyridine **27** as an inhibitor of endosomal toll-like receptors (TLRs) for the treatment of autoimmune diseases began with commercially available BCO derivative **21**. After reducing the acid on **21** to the corresponding alcohol **22**, subsequent sulfonylation with 4-(trifluoromethyl)-benzene-1-sulfonyl chloride then afforded tosylate **23**. An S_N2 displacement of tosylate **23** with bicyclic **24** gave rise to adduct **25**. Palladium-catalyzed hydrogenation of **25** removed its Cbz protection to provide secondary amine **26**. A Buchwald–Hartwig coupling between **26** and bromide **27** assembled the core structure, which was readily converted to pyrazolopyridine **27** after acidic removal of the Boc protection.[10]

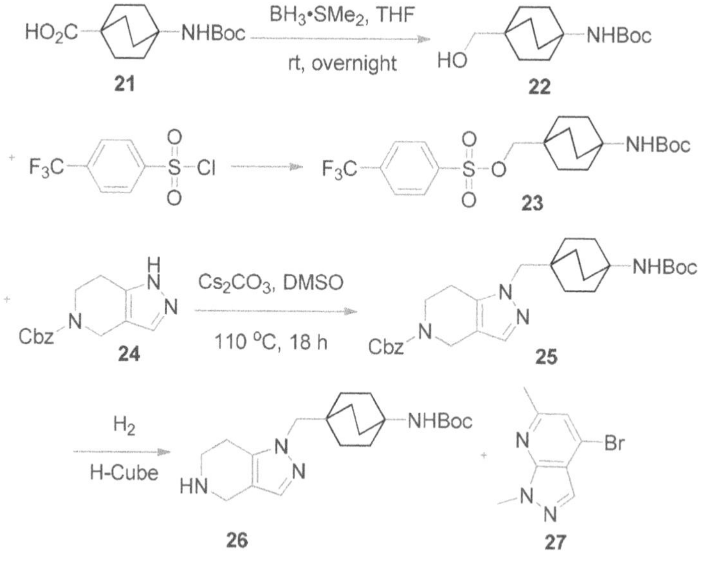

To conclude, along with bicyclo[2.2.2]pentane (BCP) and cubane (CUB), bicyclo[2.2.2]octane (BCO) is one of the three prominent 3-D isosteres for the 2-D phenyl ring. Since BCO has all sp^3 carbons, it could potentially offer several advantages over the flat phenyl group, including possible better binding with the targets, thus higher potency, better selectivity, thus lower promiscuity and lower toxicity, and potentially improved solubility as a consequence of disrupting aromaticity. However, BCO is three-carbon larger and more lipophilic than the bicyclo[2.2.2]pentyl group (BCP). It is sometimes beneficial to consider using BCP in place of BCO as the bioisostere of a *para*-phenyl group.

REFERENCES

1. Ritchie, T. J.; Macdonald, S. J. F. *Drug Disc. Today* **2009**, *14*, 1011–1020.
2. (a) Lovering, F.; Bikker, J.; Humblet, C. *J. Med. Chem.* **2009**, *52*, 6752–6756. (b) Lovering, F. *MedChemComm* **2013**, *4*, 515–519.
3. Auberson, Y. P.; Brocklehurst, C.; Furegati, M.; Fessard, T. C.; Koch, G.; Decker, A.; La Vecchia, L.; Briard, E. *ChemMedChem* **2017**, *12*, 590–598.
4. Zhong, M.; Peng, E.; Huang, N.; Huang, Q.; Huq, A.; Lau, M.; Colonno, R.; Li, L. *Bioorg. Med. Chem. Lett.* **2014**, *24*, 5731–5737.
5. Birch, A. M.; Birtles, S.; Buckett, L. K.; Kemmitt, P. D.; Smith, G. J.; Smith, T. J. D.; Turnbull, A. V.; Wang, S. J. Y. *J. Med. Chem.* **2009**, *52*, 1558–1568.
6. Kiesman, W. F.; Zhao, J.; Conlon, P. R.; Dowling, J. E.; Petter, R. C.; Lutterodt, F.; Jin, X.; Smits, G.; Fure, M.; Jayaraj, A.; et al. *J. Med. Chem.* **2006**, *49*, 7119–7131.
7. Vu, C. B.; Kiesman, W. F.; Conlon, P. R.; Lin, K.-C.; Tam, M.; Petter, R. C.; Smits, G.; Lutterodt, F.; Jin, X.; Chen, L.; et al. *J. Med. Chem.* **2006**, *49*, 7132–7139.
8. Aguilar, A.; Lu, J.; Liu, L.; Du, D.; Bernard, D.; McEachern, D.; Przybranowski, S.; Li, X.; Luo, R.; Wen, B.; et al. *J. Med. Chem.* **2017**, *60*, 2819–2839.
9. Bauman, D. R.; Whitehead, A.; Contino, L. C.; Cui, J.; Garcia-Calvo, M.; Gu, X.; Kevin, N.; Ma, X.; Pai, L.-Y.; Shah, K.; et al. *Bioorg. Med. Chem. Lett.* **2013**, *23*, 3650–3653.
10. Alper, P.; Deane, J.; Jiang, S.; Jiang, T.; Knoepfel, T.; Michellys, P.-Y.; Mutnick, D.; Pei, W.; Syka, P.; Zhang, G.; et al. (Novartis) WO2018047081 (2018).

7 Bicyclo[3.1.0]hexanes

Bicyclo[3.1.0]hexanes are conformationally constrained bioisosteres of cyclohexanes. Although *trans*-bicyclo[3.1.0]hexane does exist, it is so rare that it has never been explored as a drug fragment. Therefore, we focus our attention solely on *cis*-bicyclo[3.1.0]hexanes.

Cyclohexane prefers to adopt the chair conformation, which is thermodynamically most stable. *cis*-Bicyclo[3.1.0]hexane, as a rigidified cyclohexane analog, adopts a *puckered* shape, which closely resembles a boat conformation of cyclohexane. Thus, *cis*-bicyclo[3.1.0]hexane is like a perpetual boat, though a somewhat distorted boat.

cyclohexane

cis-bicyclo[3.1.0]hexane

trans-bicyclo[3.1.0]hexane

Absinthe is a brilliantly green spirit made by adding green anise, grande wormwood, and Florence fennel to liquor. It contains several monoterpenoid thujanes as represented by (–)-α-thujone and (+)-β-thujone, both of which contain the *cis*-bicyclo[3.1.0]hexane core structure. However, thujones are not responsible for the green color.

(–)-α-thujone   (+)-β-thujone

BICYCLO[3.1.0]HEXANE-CONTAINING DRUGS

Only one bicyclo[3.1.0]hexane-containing drug is currently on the market: Schering AG's birth control drug drospirenone (**1**, brand name Yaz when combined with ethinyl estradiol). Structurally related to 17-α-spirolactone, it is a progestin with antimineralocorticoid and antiandrogenic activities.[1]

Several bicyclo[3.1.0]hexane-containing drugs have gone to clinical trials. One recent entry is Arena's cannabinoid receptor type 2 (CB$_2$) agonist APD371 (**2**) for treating chronic pain. Its bicyclo[3.1.0]hexane fragment is fused to a pyrazole ring.[2] The cannabinoid receptor is a G-protein coupled receptor (GPCR) with two isoforms, CB$_1$ and CB$_2$. The structure of APD371 (**2**) bears a certain resemblance to Sanofi's rimonabant (Accomplia), an inverse agonist of CB$_1$, which was approved by the EMA in 2006 as an anorectic antiobesity drug but was withdrawn worldwide in 2008 due to serious psychiatric side effects.

It seems that at some point Arena developed an infatuation toward the bicyclo[3.1.0]hexane building block. Similar to their CB$_2$ agonist APD371 (**2**), their potent GPR109a agonist MK-1903 (**3**), co-developed with Merck, also has its bicyclo[3.1.0]hexane fragment fused to a pyrazole ring. The drug has been in clinical trials for lowering free fatty acids in humans.[3]

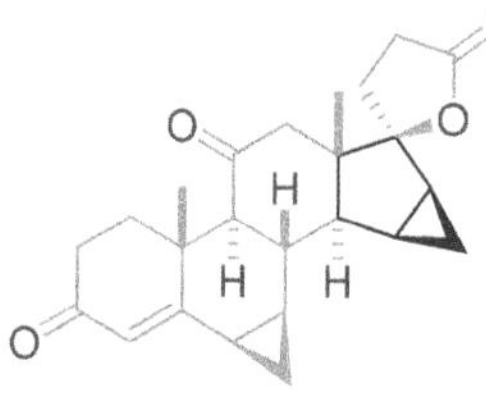

drospirenone
(**1**, Yaz for combination
with ethinyl estradiol)

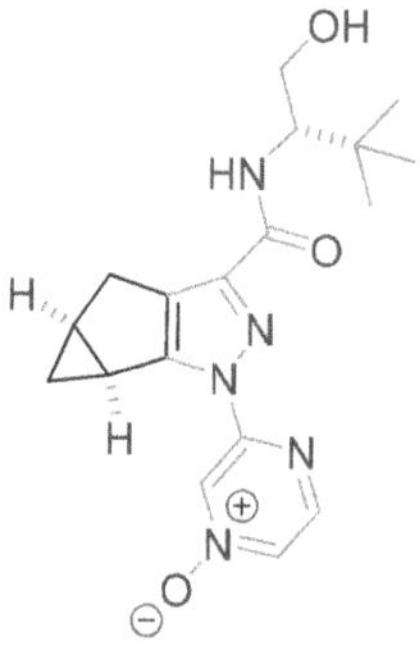

APD371 (**2**)

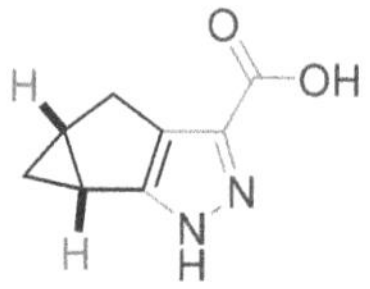

MK-1903 (**3**)

Metabotropic glutamate receptor (mGluR) modulators, both agonists and antagonists, hold great promise in treating central nervous system (CNS) disorders such as psychosis and Parkinson's disease (PD). Lilly's eglumegad (LY354740, **4**) is a potent, selective, and orally active mGluR2/3 agonist that went to clinical trials for treating anxiety and drug addiction.[4] With the bicyclo[3.1.0] hexane core structure, eglumegad (**4**) may be considered as a rigidified analog of L-glutamate. After eglumegad (**4**) was found to have low bioavailability (3%–5%) in humans, its alanine prodrug LY544344 (**5**) was prepared and found to have a 13-fold boost of oral bioavailability in humans in comparison to its prototype eglumegad (**4**).[5] The corresponding dipeptide prodrug was also synthesized and found to have an eight-fold increase of bioavailability in humans relative to eglumegad (**4**).[6] Elevation of bioavailability of the two prodrugs may be attributed to active transports of human amino acid transporters or human peptide transporters.

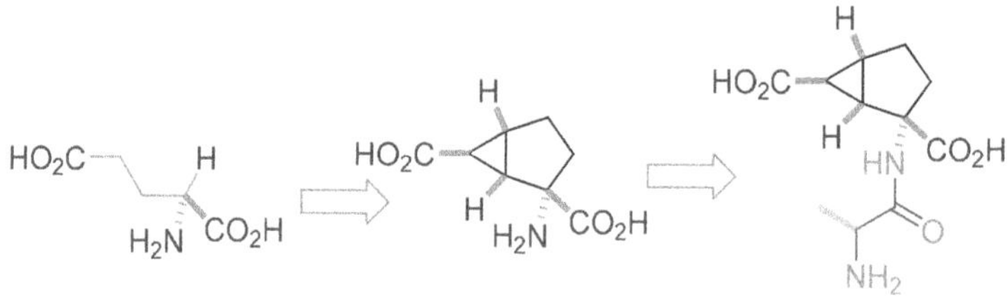

L-glutamate

eglumegad (LY354740, **4**)

LY544344 (**5**)

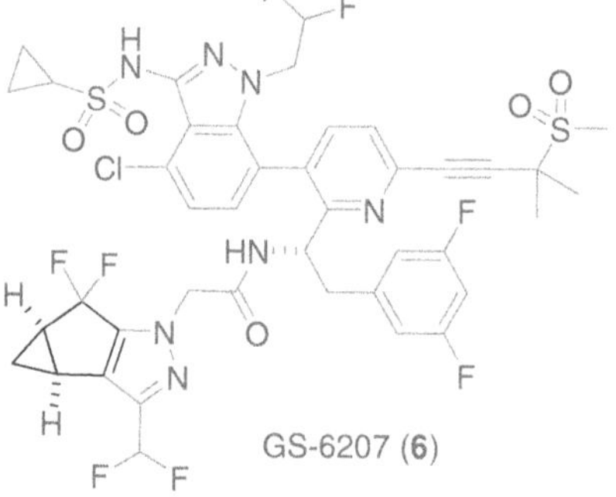

GS-6207 (**6**)

In terms of size, eglumegad (**4**) is a dwarf while Gilead's GS-6207 (GS-CA1, **6**) is a giant. It is a highly potent, selective, and long-acting first-in-class small-molecule HIV-1 capsid inhibitor.[7] With a molecular weight of 958, GS-6207 (**6**) is outside Lipinski's rule of five (Ro5) space, also known as beyond the rule of five (bRo5). It is not orally bioavailable but is injectable.

BICYCLO[3.1.0]HEXANES IN DRUG DISCOVERY

Bicyclo[3.1.0]hexanes as Bioisosteres of Cyclohexanes

Bicyclo[3.1.0]hexanes are frequently employed as isosteres of cyclohexanes.

As early as 1988, an attempt was made to replace phencyclidine (PCP, **7**)'s cyclohexane fragment with conformationally restrained bicyclo[3.1.0]-hexane analogs. The resulting bicyclo[3.1.0]hexane **8** and its diastereomer **8′** were equipotent in binding affinity for the PCP receptor but were only one-seventh as potent as PCP (**7**).[8]

Phencyclidine (7)   Bicyclo[3.1.0]hexane 8   8′

Cyclohexane 9   Bicyclo[3.1.0]hexane 10

There are five neuropeptide Y (NPY) receptors (Y1–Y5). A selective NPY1 antagonist, cyclohexanyl-piperazine **9**, was discovered as a potential therapeutic intervention of obesity. It was noted that the comparatively flexible cyclohexyl ring may contribute to reduced potency and a more constrained analog may offer tighter binding to the receptor. Indeed, the corresponding bicyclo[3.1.0]hexanylpiperazine **10** was proven to be potent ($IC_{50} = 62$ nM) and displayed excellent oral bioavailability in rat ($\%F_{po} = 80$) as well as good brain penetration (B/P ratio = 0.61).[9]

GSK discovered a series of imidazo[1,2-*a*]pyridine hepatitis C virus non-structural protein 4B (HCV NS4B) inhibitors as potential treatments of HCV infection. Cyclohexanol **11** was tested with a high binding affinity (low nanomolar) for both HCV genotype 1b and 1a replicons. However, escalating oral doses of **11** in rats failed to achieve the higher plasma drug exposure required for preliminary safety studies. Furthermore, *in vitro* resistance passaging experiments employing wild-type HCV replicons revealed that single-point mutation within the NS4B protein rendered the virus partially resistant to **11**.[10]

Significant improvements were realized with isosteric modifications in the amide (tail) portion of the series. Bridging the terminal cyclohexyl substituent to form a bicyclo[3.1.0]hexane ring afforded bicyclo[3.1.0]hexanol **12** with $IC_{50} < 1$ nM in the replicon assays. It possessed an improved antiviral profile with no increase in molecular weight and only a modest elevation in lipophilicity. It was also demonstrated that bicyclo[3.1.0]hexanol **12** can inhibit viral replication *in vivo*. This successful proof-of-concept (PoC) study suggests that drugs targeting NS4B may represent a viable treatment option for curing HCV infection.[10]

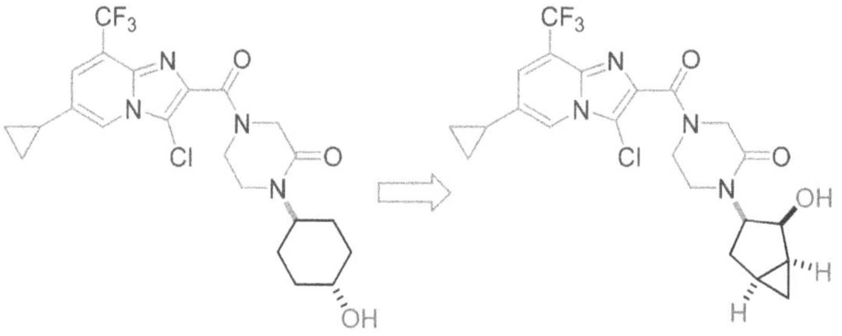

cyclohexanol **11**
NS4B binding affinity
GT1b, IC_{50} = 53 nM
GT1a, IC_{50} = 1.0 nM

bicyclo[3.1.0]hexanol **12**
NS4B binding affinity
GT1b, IC_{50} = 8 nM
GT1a, IC_{50} = 0.6 nM

Gilead's oseltamivir (Tamiflu) is the ethyl ester prodrug of GS-4071 (**13**), a viral neuraminidase (sialidases) inhibitor to treat influenza A. The mechanism of action (MoA) for both oseltamivir and GS-4071 (**13**) is functioning as transition state mimetics. The choice of using cyclohexene was wise since it provided better oral bioavailability. Scaffold hopping to the cyclopentane core structure was also successful, leading to the discovery of peramivir (Rapivab by BioCryst).

Another scaffold hopping from GS-4071 (**13**) led to a novel class of derivatives based on the bicyclo[3.1.0]hexane core structure, proposed as mimics of sialic acid in a distorted boat conformation that is on the catalytic pathway of neuraminidases. As represented by bicyclo[3.1.0]hexane **14**, they demonstrated micromolar inhibition against both group-1 (H5N1) and group-2 (H9N2) influenza neuraminidase subtypes, indicating a good affinity for the α- and β-sialic acid mimics and 150-cavity-targeted derivatives. These results provide a validation of a bicyclo[3.1.0]hexane scaffold as a mimic of a distorted sialic acid bound in the neuraminidase active site during catalysis.[11]

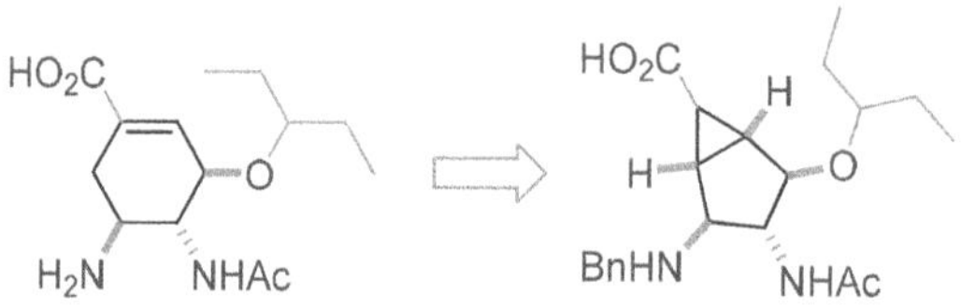

GS-4071 (**13**)

bicyclo[3.1.0]hexane **14**

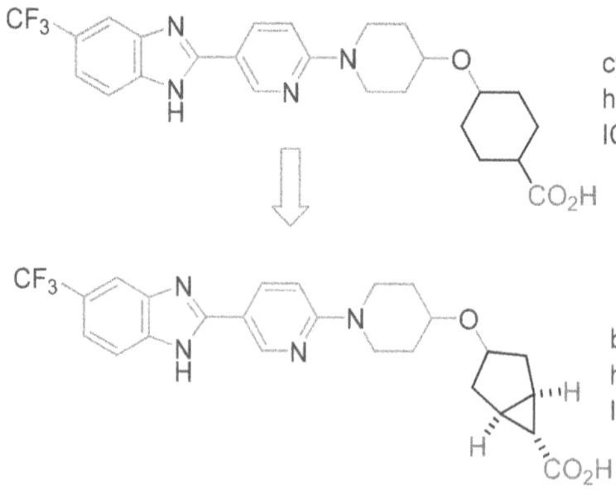

cyclohexane **15**
hDGAT1
IC_{50} = 3.9 nM

bicyclo[3.1.0]hexane **16**
hDGAT1
IC_{50} = 2.0 nM

Diglyceride acyltransferase (DGAT) catalyzes the formation of triglycerides from diacylglycerol and acyl-CoA. DGAT1 inhibitors have the potential for the treatment of obesity. Merck's benzimidazole-based DGAT1 inhibitor **15** containing a cyclohexane carboxylic acid moiety demonstrated excellent potency inhibiting the human DGAT1 enzyme. It was also selective against the A2A receptor and ACAT1. However, *in vivo*, cyclohexane **15** suffered from isomerization at the α-position of the carboxylic acid group, generating active metabolites, which exhibit DGAT1 inhibition comparable to the corresponding parent compound. *The likelihood of generating an active metabolite in humans therefore hampered the advancement of compound* **15**. Replacing the cyclohexane moiety with its isostere bicyclo[3.1.0]hexane led to compound **16**, which maintained *in vitro* and *in vivo* inhibition against the human DGAT1 enzyme. In contrast to the prototype cyclohexane **15**, bicyclo[3.1.0]hexane **16** did not undergo isomerization during the *in vitro* hepatocyte incubation study or the *in vivo* mouse study.[12]

UTILITY OF BICYCLO[3.1.0]HEXANES IN MEDICINAL CHEMISTRY

For decades, Jacobson and coworkers at NIH have focused on replacing the ribose core structure of nucleosides and nucleotides with the bicyclo[3.1.0]-hexane core structure in their drug discovery efforts. By applying structure-based functional group manipulations, rigidified adenosine derivatives (also known as *methanocarba nucleosides*) can be repurposed to satisfy pharmacophoric requirements of various GPCRs (adenosine, P2, and $5HT_{2B}$ serotonin receptors), ion channels, enzymes (kinases and polymerases), and transporters (dopamine transporter), initially detected as off-target activities.[13]

ribose **17**
K_i (hA$_3$AR) = 2,300 pM
Selectivity vs. hA$_1$AR
~ 20

MRS7334 (**18**)
K_i (hA$_3$AR) = 280 pM
Selectivity vs. hA$_1$AR
~ 40,000

The most recent example of methanocarba nucleosides is MRS7334 (**18**) as a highly potent and selective A$_3$ adenosine receptor (A$_3$AR) agonist. In comparison to its corresponding ribose **17**, the bicyclo[3.1.0]hexane-containing **18** has a K$_i$ value of 280 picomolar in the hA$_3$AR assay, an eightfold boost of the binding affinity. MRS7334 (**18**) is also significantly more selective against hA$_1$AR (K$_i$ = 33 nM), a 40,000-fold selectivity. Furthermore, methanocarba nucleoside **18** also displayed a favorable pharmacokinetics profile, as well as an off-target activity profile against 240 GPCRs and 466 kinases. Despite the added synthetic difficulty, the (N)-methanocarba modification has distinct advantages for A$_3$AR, which has translational potential for chronic disease treatment.[14]

Lilly brought several bicyclo[3.1.0]hexane-containing mGluR2/3 *agonists*, such as eglumegad (**4**) and LY544344 (**5**), to clinical trials. Meanwhile, Lilly has pursued mGluR2/3 *antagonists* with rigor during the last decade. Inspired by Taisho's MGS0039 (**19**),[15] an mGluR2/3 antagonist that showed antidepressant and anxiolytic effects in behavioral models in rats, Lilly discovered LY3020371 (**20**) as a potent, selective, and maximally efficacious mGluR2/3 antagonist. LY3020371 (**20**) also demonstrated *in vivo* activity with an antidepressant-like signature in the mouse forced-swim test (mFST) assay when brain levels of this compound exceeded the cellular $mGlu_2$ IC_{50} value.[16]

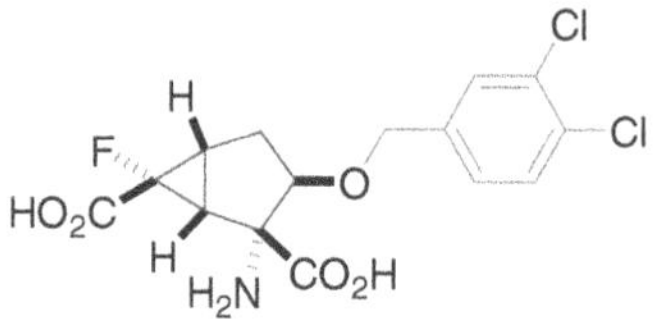

MGS0039 (**19**) 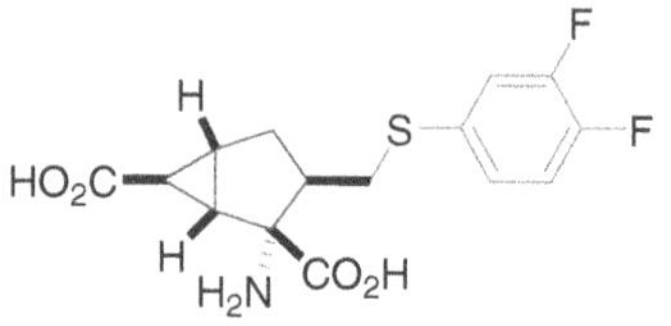   LY3020371 (**20**)

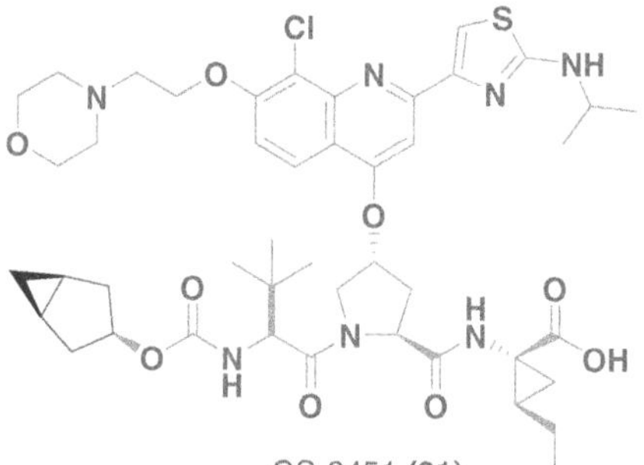

GS-9451 (**21**)

Sheng et al. at Gilead discovered a bicyclo[3.1.0]hexane-containing drug GS-9451 (**21**) as an acid inhibitor of the HCV NS3/4A protease.[17]

Sphingosine-1-phosphate receptor 1 ($S1P_1$) is a GPCR. $S1P_1$ signaling has been associated with the regulation of lymphocyte maturation, migration, and trafficking. $S1P_1$ receptor agonists, on the other hand, showed potential as a treatment of cardiovascular diseases. From high throughput screening (HTS), Actelion found a hit: bicyclo[3.1.0]hexane-fused pyrazole **22**. Through scaffold hopping, thiophene **23** was obtained as an optimized $S1P_1$ receptor agonist with an EC_{50} of 7 nM and was selective against the $S1P_3$ receptor ($EC_{50} = 2{,}880$ nM). It also had favorable pharmacokinetic properties in rats and dogs, distributed well into brain tissue, and efficiently and dose-dependently reduced the blood lymphocyte count in rats. Thiophene **23** affected the heart rate during the wake phase of animals only but showed no effect on the mean arterial blood pressure.[18]

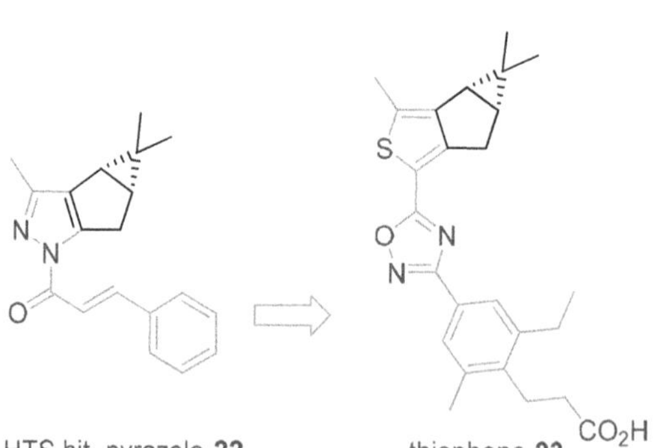

HTS hit, pyrazole **22**   thiophene **23**

dual AKA and CD1 inhibitor **24**   atipamezole (**25**)   F14805 (**26**)

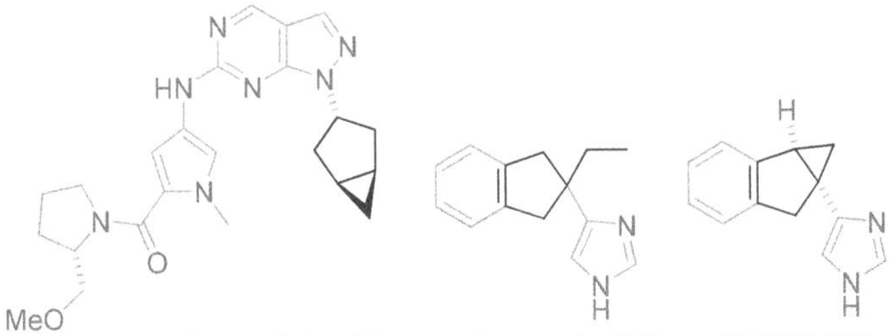

Additional bicyclo[3.1.0]hexane-containing drugs also include Biogen's dual inhibitor **24** of Aurora kinase A (AKA) and cyclin-dependent kinase 1 (CDK1).[19] Rigid analogs, including bicyclo[3.1.0] hexane-containing F14805 (**26**), of the α2-adrenergic blocker atipamezole (**25**), showed that small changes could have big sequences in terms of their pharmacology.[20]

SYNTHESIS OF SOME BICYCLO[3.1.0]HEXANE-CONTAINING DRUGS

Synthesis of AstraZeneca's CB$_2$ agonist APD371 (**2**) began with an intramolecular cyclopropanation of (S)-2-(but-3-en-1-yl)oxirane (**27**) catalyzed by lithium tetramethylpiperidine (LTMP) to prepare bicyclo[3.1.0]hexanol **28**.[21] 2,2,6,6-Tetramethylpiperidine 1-oxyl (TEMPO)-catalyzed bleach oxidation converted alcohol **28** to the corresponding ketone **29**. Acylation of ketone **29** with diethyl oxalate was followed by condensation with 2-hydrazinylpyrazine to assemble tricyclic pyrazinyl-pyrazole ester **30**. Hydrolysis of the ethyl ester using LiOH was followed by selective oxidation employing m-CPBA to form N-oxide acid **31**. Eventually, HATU-mediated amide formation by coupling acid **31** with amino alcohol **32** then delivered APD371 (**2**).[2]

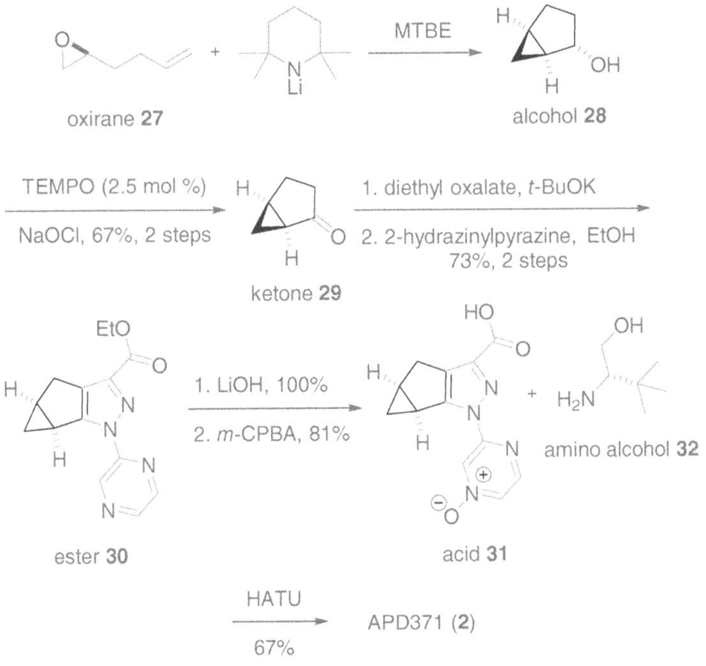

oxirane **27**   alcohol **28**

ketone **29**

ester **30**   acid **31**   amino alcohol **32**

APD371 (**2**)

Lilly's synthesis of their mGluR 2/3 agonist eglumegad (**4**) commenced with a Corey–Chaykovsky carboxycyclopropanation of 2-cyclopentenone with ethyl(dimethylsulfonium)acetate bromide (EDSA) to prepare the bicyclo[3.1.0]-hexane-ketone **33** with excellent diastereoselectivity under optimized conditions. It was then subjected to hydantoin formation under Bucherer–Bergs conditions, yielding **34** as a mixture of two inseparable diastereomers. Exhaustive hydrolysis of hydantoin **34** was followed by esterification to afford bis-ester **35**, which was again hydrolyzed to deliver pure eglumegad (**4**) after ion-exchange chromatography.[4]

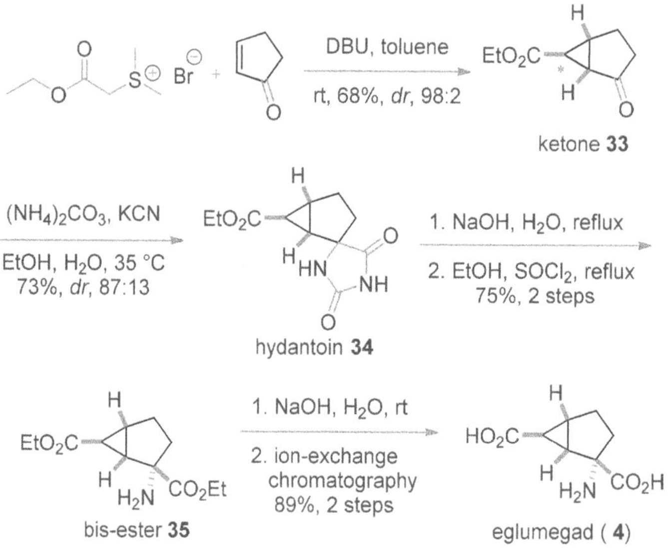

ketone **33**

hydantoin **34**

bis-ester **35**

eglumegad (**4**)

Hu and Luo at BMS began the synthesis of their selective NPY1 receptor antagonist bicyclo[3.1.0]hexanylpiperazine **10** with condensation of 4-methyl-cyclohexanone with N-benzylpiperazine to produce enamine **36**. Chlorination of enamine **36** with NCS at −50°C to room temperature afforded chloroenamine **37**. After filtration of the by-product succinamide, the crude chloroenamine **37** in ether was exposed to the freshly prepared Grignard reagent to assemble the key bicyclo[3.1.0]hexane **38** in a total 82% yield for the last three steps. Palladium-catalyzed debenzylation was followed by an S_NAr replacement reaction with an α-bromopyridine to deliver bicyclo[3.1.0]hexanylpiperazine **10**.[9]

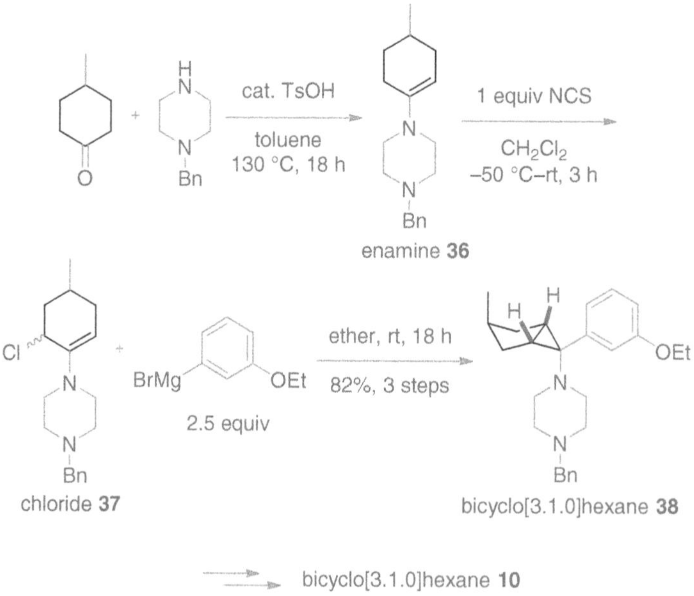

enamine **36**

chloride **37**

bicyclo[3.1.0]hexane **38**

bicyclo[3.1.0]hexane **10**

To make Merck's benzimidazole-based DGAT1 inhibitor **16**, reductive etherification of benzyl 4-oxopiperidine-1-carboxylate (**39**) with cyclopent-3-en-1-ol **40** assembled ether **41**.[22] Rh(II)-catalyzed cyclopropanation of the double bond on alkene **41** furnished the key bicyclo[3.1.0]hexane **42** as a mixture of four diastereomers. After chiral supercritical fluid chromatography (SFC) separation, additional transformations led to the synthesis of bicyclo[3.1.0]hexane **16**.[12]

To conclude, bicyclo[3.1.0]hexanes are conformationally restrained isosteres of cyclohexanes, but with no increase in molecular weight and only a modest elevation in lipophilicity. They may confer tighter binding to the target protein, more resistance to metabolism, and often provide better selectivity, resulting in less off-target effects.

REFERENCES

1. Santhamma, B.; Acosta, K.; Chavez-Riveros, A.; Nickisch, K. *Steroids* **2015**, *102*, 60–64.
2. Han, S.; Thoresen, L.; Jung, J.-K.; Zhu, X.; Thatte, J.; Solomon, M.; Gaidarov, I.; Unett, D. J.; Yoon, W. H.; Barden, J.; et al. *ACS Med. Chem. Lett.* **2017**, *8*, 1309–1313.
3. Boatman, P. D.; Lauring, B.; Schrader, T. O.; Kasem, M.; Johnson, B. R.; Skinner, P.; Jung, J.-K.; Xu, J.; Cherrier, M. C.; Webb, P. J.; et al. *J. Med. Chem.* **2012**, *55*, 3644–3666.
4. Monn, J. A.; Valli, M. J.; Massey, S. M.; Wright, R. A.; Salhoff, C. R.; Johnson, B. G.; Howe, T.; Alt, C. A.; Rhodes, G. A.; Robey, R. L.; et al. *J. Med. Chem.* **1997**, *40*, 528–537.
5. Coffey, D. S.; Hawk, M. K.; Pedersen, S. W.; Vaid, R. K. *Tetrahedron Lett.* **2005**, *46*, 7299–7302.
6. Bueno, A. B.; Collado, I.; de Dios, A.; Domínguez, C.; Martín, J. A.; Martín, L. M.; Martínez-Grau, M. A.; Montero, C.; Pedregal, C.; Catlow J.; et al. *J. Med. Chem.* **2005**, *48*, 5305–5320.
7. Yant, S. R.; Mulato, A.; Hansen, D.; Tse, W. C.; Niedziela-Majka, A.; Zhang, J. R.; Stepan, G. J.; Jin, D.; Wong, M. H.; Perreira, J. M.; et al. *Nat. Med.* **2019**, *25*, 1377–1384.
8. (a) de Costa, B. R.; George, C.; Burke, T. R. Jr.; Rafferty, M. F.; Contreras, P. C.; Mick, S. J.; Jacobson, A. E.; Rice, K. C. *J. Med. Chem.* **1988**, *31*, 1571–1575. (b) de Costa, B R; Mattson, M. V.; George, C.; Linders, J. T. *J. Med. Chem.* **1992**, *35*, 4704–4712.
9. (a) Hu, S.; Huang, Y.; Deshpande, M.; Luo, G.; Bruce, M. A.; Chen, L.; Mattson, G.; Iben, L. G.; Zhang, J.; Russell, J. W.; et al. *ACS Med. Chem. Lett.* **2012**, *3*, 222–226. (b) Luo, G.; Chen, L.; Hu, S.; Huang, Y.; Mattson, G.; Iben, L. G.; Russell, J. W.; Clarke, W. J.; Hogan, J. B.; Antal-Zimanyi, I.; et al. *Bioorg. Med. Chem. Lett.* **2013**, *23*, 3814–3817.
10. Miller, J. F.; Chong, P. Y.; Shotwell, J. B.; Catalano, J. G.; Tai, V. W.-F.; Fang, J.; Banka, A. L.; Roberts, C. D.; Youngman, M.; Zhang, H.; et al. *J. Med. Chem.* **2014**, *57*, 2107–2120.
11. Colombo, C.; Pinto, B. M.; Bernardi, A.; Bennet, A. J. *Org. Biomol. Chem.* **2016**, *14*, 6539–6553.
12. He, S.; Lai, Z.; Hong, Q.; Shang, J.; Reibarkh, M.; Kuethe, J. T.; Liu, J.; Guiadeen, D.; Krikorian, A. D.; Cernak, T. A.; et al. *Bioorg. Med. Chem. Lett.* **2019**, *29*, 1182–1186.
13. Jacobson, K. A.; Tosh, D. K.; Toti, K. S.; Ciancetta, A. *Drug Discov. Today* **2017**, *22*, 1782–1791.
14. Tosh, D. K.; Salmaso, V.; Rao, H.; Campbell, R.; Bitant, A.; Gao, Z.-G.; Auchampach, J. A.; Jacobson, K. A. *ACS Med. Chem. Lett.* **2020**, *11*, 1935–1941.
15. (a) Nakazato, A.; Sakagami, K.; Yasuhara, A.; Ohta, H.; Yoshikawa, R.; Itoh, M.; Nakamura, M.; Chaki, S. *J. Med. Chem.* **2004**, *47*, 4570–4587. (b) Yoshimizu, T.; Shimazaki, T.; Ito, A.; Chaki, S. *Psychopharmacol.* **2006**, *186*, 587–593.

16. Chappell, M. D.; Li, R.; Smith, S. C.; Dressman, B. A.; Tromiczak, E. G.; Tripp, A. E.; Blanco, M.-J.; Vetman, T.; Quimby, S. J.; Matt, J.; et al. *J. Med. Chem.* **2016**, *59*, 10974–10993.
17. Sheng, C. X.; Appleby, T.; Butler, T.; Cai, R.; Chen, X.; Cho, A.; Clarke, M. O.; Cottell, J.; Delaney, W. E.; Doerffler, E.; et al. *Bioorg. Med. Chem. Lett.* **2012**, *22*, 2629–2634.
18. Bolli, M. H.; Velker, J.; Muller, C.; Mathys, B.; Birker, M.; Bravo, R.; Bur, D.; de Kanter, R.; Hess, P.; Kohl, C.; et al. *J. Med. Chem.* **2014**, *57*, 78–97.
19. Le Brazidec, J.-Y.; Pasis, A.; Tam, B.; Boykin, C.; Wang, D.; Marcotte, D. J.; Claassen, G.; Chong, J.-H.; Chao, J.; Fan, J.; et al. *Bioorg. Med. Chem. Lett.* **2012**, *22*, 4033–4037.
20. Vacher, B.; Funes, P.; Chopin, P.; Cussac, D.; Heusler, P.; Tourette, A.; Marien, M. *J. Med. Chem.* **2010**, *53*, 6986–6995.
21. Alorati, A. D.; Bio, M. M.; Brands, K. M. J.; Cleator, E.; Davies, A. J.; Wilson, R. D.; Wise, C. S. *Org. Process Res. Dev.* **2007**, *11*, 637–641.
22. Kuethe, J. T.; Janey, J. M.; Truppo, M.; Arredondo, J.; Li, T.; Yong, K.; He, S. *Tetrahedron* **2014**, *70*, 4563–4570.

8 Bridge-Fused Rings as *m*-Phenyl Bioisosteres

Today, it is widely known that a compound's high aromatic ring count is correlated to its low aqueous solubility, high lipophilicity, high serum albumin binding, high cytochrome protein (CYP)-450 inhibition, and high human ether-a-go-go (hERG) potassium channel activity inhibition.[1] Being three-dimensional, fully aliphatic bioisosteres for the two-dimensional phenyl group may offer superior physiochemical properties. Their fraction of saturated carbon (Fsp^3, defined in equation 8.1)[2] is 1.0 whereas the Fsp^3 for the aromatic phenyl ring is 0.

$$Fsp^3 = \left(\text{number of } sp^3\text{-hybridized carbon}\right) \big/ \left(\text{total carbon count}\right) \tag{8.1}$$

For *para*-substituted phenyl ring (*p*-Ph), three popular non-classical 3-D-rich isosteres are cubane-1,4-diyl (CUB), bicyclo[2.2.2]octane-1,4-diyl (BCO), and bicyclo[1.1.1]pentane-1,4-diyl (BCP). Their bridgehead lengths decrease in the following order:

$$\textit{p-}\text{Ph} \left(2.79 \text{ Å, } 100\%\right) > \text{CUB} \left(2.72 \text{ Å, } 96\%\right) > \text{BCO} \left(2.60 \text{ Å, } 94\%\right) > \text{BCP} \left(1.85 \text{ Å, } 65\%\right).$$

While CUB, BCO, and BCP are reviewed elsewhere, here at the end of this chapter, 2-oxabicyclo[2.2.2]octane (A) as another 3-D isostere of the *p*-phenyl group will be briefly summarized. The additional oxygen atom on A lowers the fragment's lipophilicity (Figure 8.1).

Less known are 3-D-rich bioisosteres for the *meta*-substituted phenyl ring (*m*-Ph) and the *ortho*-substituted phenyl ring (*o*-Ph) (Figure 8.2).

In the literature, at least four bridge-fused rings exist as *m*-phenyl non-classical isosteres: bicyclo[2.2.1]heptane (**B**), bicyclo[2.1.1]hexane (**C**), bicyclo[3.1.1]heptane (**D**), and 2-oxabicyclo[2.1.1]hexane (**E**).[3] Again, the additional oxygen atom on E lowers the lipophilicity of the fragment.

Whereas all bridge-fused aliphatic isosteres **A–E** are potentially subject to the CYP450 metabolism, they have several advantages as drug fragments with higher Fsp^3 values:

FIGURE 8.1   3-D-rich isosteres for the *p*-phenyl fragment.

FIGURE 8.2   3-D-rich isosteres for the *m*-phenyl fragment.

a. A higher degree of saturation for a molecule may increase receptor–ligand complementarity, which should mitigate off-target effects.[1] Saturation (compounds with higher Fsp^3 values) also mitigates a drug's promiscuity even for compounds containing ionizable amines that are more promiscuous than neutral ones.[4]

b. Compounds with higher Fsp^3 values tend to have lower CYP450 inhibitions, thus reducing drug–drug interactions (DDIs) tendency.[4]

c. Disruption of planarity hence aromaticity results in lower melting point and higher solubility.

d. Oxygen-containing isosteres have an added advantage of lower lipophilicity.

Unfortunately, the last few decades saw ClogP getting higher and Fsp^3 getting lower when proper attention was not paid to drugs' physiochemical properties. In order to increase our success rate in drug discovery, it is essential to get cLogP down and Fsp^3 up. Employing the 3-D-rich isosteres to replace the phenyl fragment is one step closer to the right direction.[5] After all, complexity, as measured by both Fsp^3 and the presence of chiral centers, impacts the probability of success in the clinic.

3-D-RICH BIOISOSTERES FOR THE PHENYL RING IN DRUG DISCOVERY

Bicyclo[2.2.1]heptane (**B**) has an interesting geometry. It is most frequently drawn as shown below (left and middle), giving an illusion that the structure is similar to the *para*-phenyl geometry. In fact, it is incorrect because the ring strain renders its geometry closer to the structure that is at right, which is similar to the *meta*-phenyl geometry.

Fragment **B** made an appearance as a substituent of BMS's 11β-HSD-1 inhibitors. 11β-Hydroxysteroid dehydrogenase-type 1 (11β-HSD-1), an enzyme expressed at high levels in the liver and adipose tissue, catalyzes the conversion of inert cortisone to the active glucocorticoid hormone cortisol. Therefore, 11β-HSD-1 inhibitors are actively pursued as pharmacological agents to treat various metabolic diseases. BMS has advanced several 11β-HSD-1 inhibitors with the 1,2,4-triazolopyridine (TZP) core structures to clinical trials for treating patients afflicted with type 2 diabetes, obesity, and the metabolic syndrome. One of them is BMS-823778 (**1**)[6] and the other is BMS-770767 (**2**).[7] Both of them are now in phase II clinical trials. The latter contains fragment B as bicyclo[2.2.1] heptanol, which may be viewed as a 3-D rich isostere of an *m*-phenyl ring.

BMS-823778 (**1**), Phase II   BMS-770767 (**2**), Phase II

bicyclo[2.2.1]heptanyl iodide **4**

WAY-100635 (**3**)

bicyclo[2.2.1]heptanyl methyl fluoride **5**

WAY-100635 (**3**) is a potent and selective 5HT$_{1A}$ receptor antagonist with potential as a drug therapy or marker for studying the pathophysiology of neuropsychiatric disorders. In order to investigate changes of the 5HT$_{1A}$ receptor after binding to WAY-100635 (**3**) using single-photon emission computerized tomography (SPECT), **3**'s bulkier analog bicyclo[2.2.1]heptanyl iodide **4** and the corresponding bridge-fused rings such as admantanyl, cubanyl, and bicycle[2,2,2]octanyl (BCO) analogs were prepared and evaluated. Although compound **4** and the other three bulky analogs showed a low propensity for amide hydrolysis, their brain uptake and the specificity for those radioligands were significantly lower than the parent molecule **3**. Therefore, those designed tracers are not suitable for SPECT imaging.[8]

Further efforts to make radiolabeled 5HT$_{1A}$ receptor ligands led to the synthesis of bicyclo[2.2.1] heptanylmethyl fluoride **5** and its corresponding bridge-fused rings such as admantanyl, cubanyl, and bicycle[2,2,2]octanyl (BCO) analogs. Among the four analogs, compound **5** was reasonably selective and the cubanyl analog showed a suitable metabolic stability. This endeavor provided a promising starting point for the synthesis of the corresponding ^{18}F-labeled positron emission tomography (PET) analogs.[9]

Bicyclo[2.1.1]hexane (**C**) as an *m*-phenyl isostere began to appear only very recently. One of them was among Incyte's phosphoinositide-3-kinase (PI3K) inhibitors. Deregulation of the well-known PI3K pathway has been implicated in numerous pathologies such as cancer, diabetes, thrombosis, rheumatoid arthritis, and asthma. Two PI3K inhibitors have been approved by the FDA. One is Gilead's idelalisib (Zydelig, **6**), which is a PI3Kδ selective inhibitor, and the other is Bayer's copanlisib (Aliqopa, **7**), which is a pan-PI3K inhibitor.[9] In a 2017 patent, Incyte's bicyclo[2.1.1] hexanyl nitrile **8** was claimed to be a selective PI3Kγ inhibitor with an IC$_{50}$ value less than 100 nM.[10] Ironically, bicyclo[3.1.1]heptane (**D**) has made a rare appearance, also in Incyte's 2017 patent on their PI3K inhibitors in the form of bicyclo[3.1.1]heptanyl nitrile **9**.[9] The same goes to 2-oxabicyclo[2.1.1] hexane (**E**), which was represented by 2-oxabicyclo[2.1.1]-hexanyl nitrile **10** on the patent.[11] For direct comparison between compounds **8** and **10**, the latter is likely to have better physiochemical properties because the oxygen atom helps to reduce the molecule's lipophilicity.

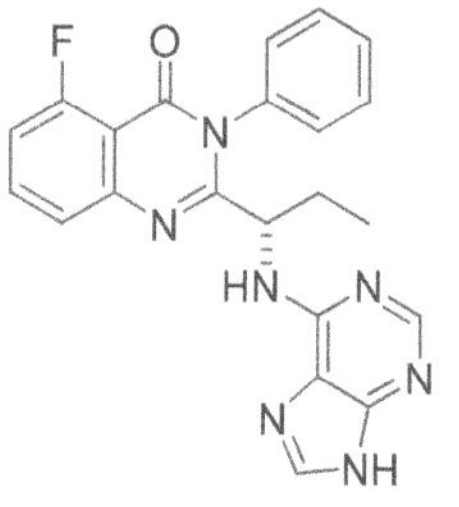

idelalisib (Zydelig, **6**)
Gilead, 2014
PI3Kδ inhibitor

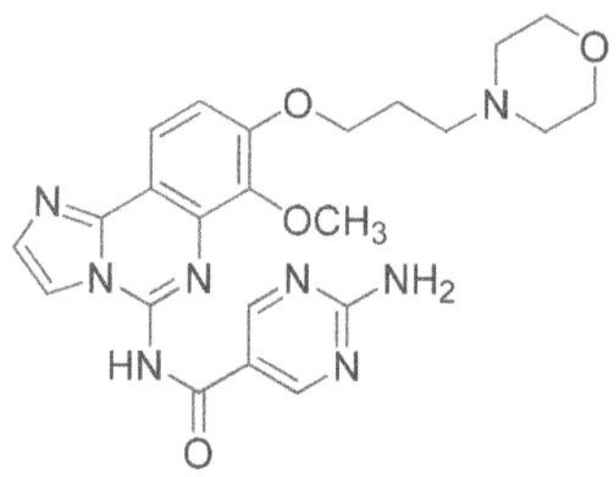

copanlisib (Aliqopa, **7**)
Bayer, 2017
pan-PI3K inhibitor

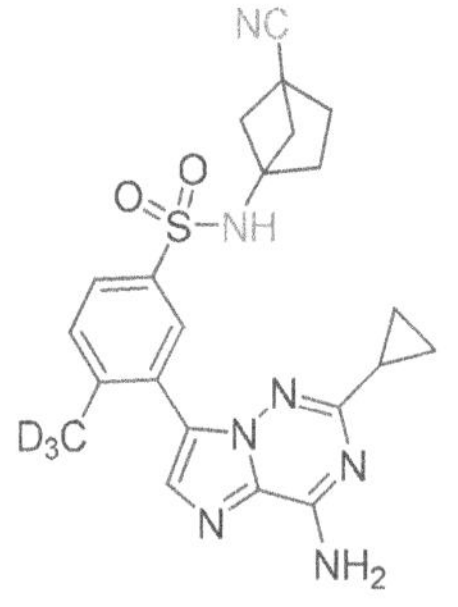

bicyclo[2.1.1]hexanyl nitrile **8**   bicyclo[3.1.1]heptanyl nitrile **9** 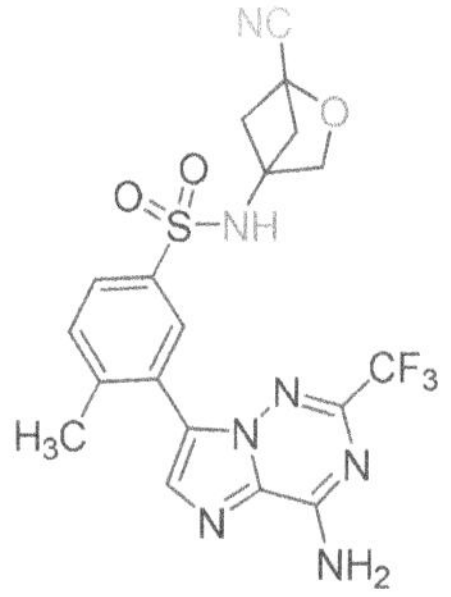   2-oxabicyclo[2.1.1]hexanyl
PI3Kγ IC$_{50}$ < 100 nM   nitrile **10**

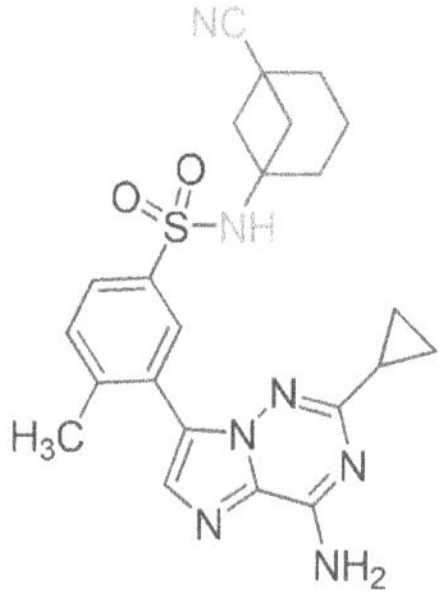

Now, back to *p*-phenyl isostere 2-oxabicyclo[2.2.2]octane (**A**), which is more popular than fragment **B–E** in drug discovery. It is an analog of 2-oxabicyclo[2.1.1]hexane (**E**) with an additional carbon atom and a different substitution trajectory. Fragment **E** was employed as a replacement to piperidine in their programs to discover novel bacterial topoisomerase inhibitors.[12]

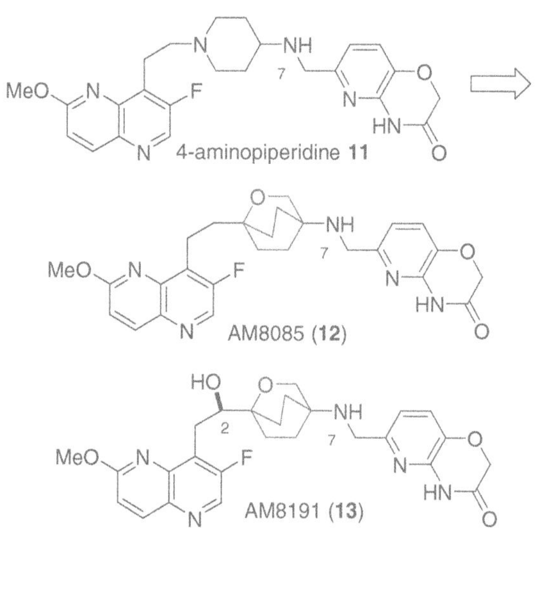

4-aminopiperidine **11**

AM8085 (**12**)

AM8191 (**13**)

The best-known type II topoisomerase inhibitor is probably Bayer's ciprofloxacin (Cipro) among all antibacterials. Much advance has been made to discover new topoisomerase inhibitors, which led to the discovery of 4-aminopiperidine **11**. The hallmark of this linker is the strategic placement of a basic nitrogen atom at position-7 that shows a salt-bridge interaction with Asp_{83} in the X-ray crystal structure. Employing the 2-oxabicyclo[2.1.1]hexane (**E**) linker provided AM8085 (**12**) with reduced basicity and attenuated human ether-a-go-go (hERG) potassium channel activity. Further addition of a hydroxyl group at C-2 gave rise to AM8191 (**13**), which had reduced hERG activity by 30-fold and improved solubility by over 100-fold with a minimum loss of antibacterial potency and spectrum.[12]

SYNTHESIS OF SOME DRUGS CONTAINING 3-D-RICH BIOISOSTERES FOR THE *M*-PHENYL RING

BMS's process procedure to prepare TZP 11HSD-1 inhibitor BMS-770767 (**2**) employed known bicyclo[2.2.0]hexane-1,4-diyldimethanol (**14**)[13] as its starting material. Acid-promoted rearrangement of **14** led to the desired fragment bicyclo[2.2.1]heptane (**B**) in the form of diol **15**, which was readily oxidized to the corresponding acid **16** under environmentally friendly conditions. The union between acid **16** and hydrazine **17** was mediated by 2-chloro-1,3-dimethyl-4,5-dihydro-1*H*-imidazol-3-ium chloride as the coupling agent to provide adduct **18**. Simply treating hydrazide **18** with benzoic acid delivered BMS-770767 (**2**) in good yield.[14]

Synthesis of radiolabeled ligand bicyclo[2.2.1]heptanyl iodide **4** began with the assembly of fragment **B** as intermediate iodide **24**. Thus, $RuCl_3$-mediated oxidation of norbornene (**19**) afforded di-acid **20**, which was subsequently transformed to di-ester **21**. Alkylation of **21** with 1-bromo-2-chloroethane gave rise to bridge-fused di-ester **22**. After mono-saponification to give **23**, its acid functionality was converted to iodide **24** via a hypervalent iodine iodinative decarboxylation. After converting the acid to the corresponding acid chloride, it was coupled with amine **25** to deliver iodide **4**.[8]

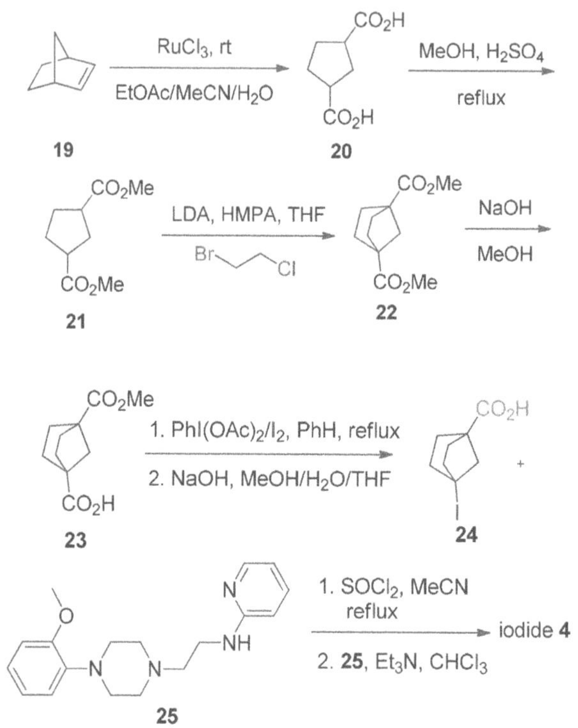

Incyte's preparation of 2-oxabicyclo[2.1.1]hexanyl nitrile **10** as a PI3K inhibitor commenced with the production of fragment E (2-oxabicyclo[2.1.1]hexane) in the form of amino-nitrile **30**. Therefore, after the protection of amino-alcohol **26** to afford **27**, its alcohol functionality was oxidized employing the Dess–Martin reagent to produce aldehyde **28**. After converting aldehyde **28** to oxime **29**, it was transformed to the key intermediate amino-nitrile **30** upon treatment with methanesulfonyl chloride. Simple exposure of amino-nitrile **30** to pre-fabricated sulfonyl chloride **31** then delivered nitrile **10**.[11]

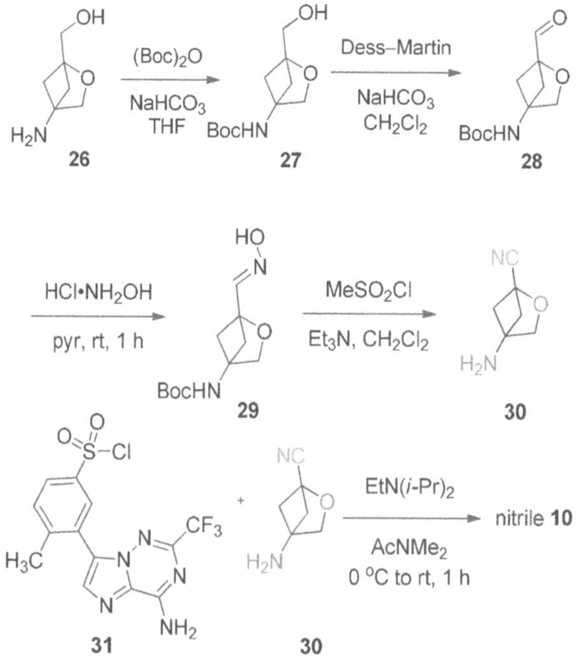

Synthesis of AM8191 (**13**) as a bacterial topoisomerase inhibitor involved a lengthy sequence to prepare fragment **A** (14 steps in total) in the form of aldehyde **34**. As shown below, it took seven steps to assemble intermediate **32** as a bis-tosylate. Exposure of **32** to NaH effected the formation of bridge-fused **33**, which was converted to aldehyde **34** in an additional six steps. After deprotonation of the methyl group on aza-quinoline **35**, the resulting carbanion was quenched with aldehyde **34** to afford two enantiomeric alcohols. After SFC separation, one of the enantiomers was deprotected and coupled with aldehyde **36** via reductive amination to deliver AM8191 (**13**).[12]

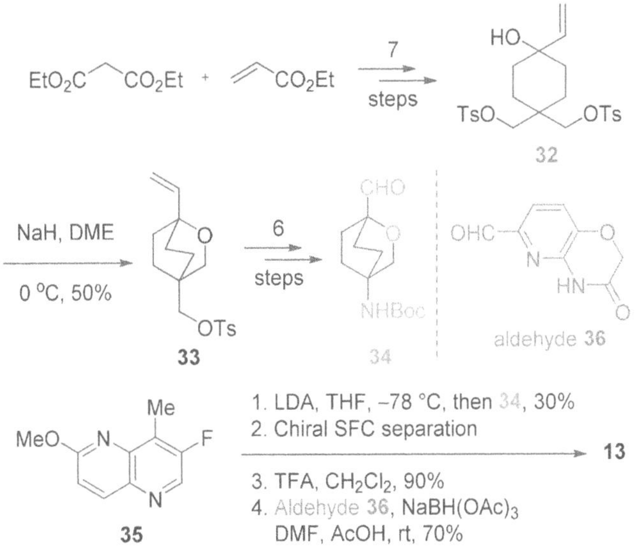

In summary, while saturated *p*-phenyl isosteres are more and more popular in medicinal chemistry, *m*-phenyl and *o*-phenyl fragments do not have many 3-D-rich isosteres. As this review has shown, saturated *m*-phenyl isosteres such as bicyclo[2.2.1]heptane (B), bicyclo[2.1.1]hexane (C), bicyclo[3.1.1]heptane (D), and 2-oxabicyclo[2.1.1]hexane (E) are gaining popularity. As most 3-D-rich isosteres, they (a) have a higher degree of saturation for a molecule, which may increase receptor–ligand complementarity, which should mitigate off-target effects; (b) tend to have lower CYP450 inhibitions, thus reducing DDIs tendency; and (c) may have a lower melting point and higher solubility, and (d) oxygen-containing isosteres have an added advantage of lower lipophilicity.

With many of those bridge-fused intermediates now commercially available, their utility in drug discovery is destined to bear fruits in the future.

REFERENCES

1. Ritchie, T. J.; Macdonald, S. J. F. *Drug Discov. Today* **2009**, *14*, 1011–1020.
2. Lovering, F.; Bikker, J.; Humblet, C. *J. Med. Chem.* **2009**, *52*, 6752–6756.
3. Mykhailiuk, P. K. *Org. Biomol. Chem.* **2019**, *17*, 2839–2849.
4. Lovering, F. *MedChemComm* **2013**, *4*, 515–519.
5. Auberson, Y. P.; Brocklehurst, C.; Furegati, M.; Fessard, T. C.; Koch, G.; Decker, A.; La Vecchia, L.; Briard, E. *ChemMedChem* **2017**, *12*, 590–598.
6. Li, J.; Kennedy, L. J.; Walker, S. J.; Wang, H.; Li, J. J.; Hong, Z.; O'Connor, S. P.; Ye, X.-y.; Chen, S.; Wu, S.; et al. *ACS Med. Chem. Lett.* **2018**, *9*, 1170–1174.
7. Maxwell, B. D.; Bonacorsi, S. J., Jr. *J. Label. Comp. Radiopharm.* **2016**, *58*, 657–664.
8. Al Hussainy, R.; Verbeek, J.; van der Born, D.; Braker, A. H.; Leysen, J. E.; Knol, R. J.; Booij, J.; Herscheid, J. D. M. *J. Med. Chem.* **2011**, *54*, 3480–3491.
9. Al Hussainy, R.; Verbeek, J.; van der Born, D.; Booij, J.; Herscheid, J. D. M. *Eur. J. Med. Chem.* **2011**, *46*, 5728–5735.

10. Perry, M. W. D.; Abdulai, R.; Mogemark, M.; Petersen, J.; Thomas, M. J.; Valastro, B.; Eriksson, A. W. *J. Med. Chem.* **2019**, *62*, 4783–4814.

11. Buesking, A. W.; Sparks, R. B.; Combs, A. P.; Douty, B.; Falahatpisheh, N.; Shao, L.; Shepard, S.; Yue, E. W. (Incyte) WO2017223414 (2017).

12. Singh, S. B.; Kaelin, D. E.; Wu, J.; Miesel, L.; Tan, C. M.; Meinke, P. T.; Olsen, D.; Lagrutta, A.; Bradley, P.; Lu, J.; et al. *ACS Med. Chem. Lett.* **2014**, *5*, 609–614.

13. Lantos, I.; Ginsburg, D. *Tetrahedron* **1972**, *28*, 2507–2519.

14. Qian, X.; Zhu, K.; Deerberg, J.; Yang, W. W.; Yamamoto, K.; Hickey, M. R. (BMS) US9580420 (2016).

9 Cubanes, Are We There Yet?

Cubane was considered too esoteric to be on a drug even though Eaton postulated in 1992 that cubane may serve as a benzene bioisostere.[1] Cubane is stable and has limited toxicity. Moreover, each of its eight carbon atoms may be potentially functionalized to provide a drug with multiple contact points to the target protein, which could improve a drug's specific binding. Moreover, cubane's three-dimensional geometry may also help improve a drug's solubility.[2]

In terms of size, cubane is the closest to benzene in comparison to all isosteric bridged bi-cycloalkyls. The length of the C–C bond of cubane is 1.362 Å, very similar to that of benzene's 1.397 Å. For a *para*-substituted phenyl ring (*p*-Ph), its three popular non-classical 3D bridge-fused isosteres are cubane-1,4-diyl (CUB), bicyclo[2.2.2]octane-1,4-diyl (BCO), and bicyclo[1.1.1]pentane-1,4-diyl (BCP). Their bridgehead lengths decrease in the following order:

$$p\text{-Ph}\ (2.79\ \text{Å},\ 100\%) > \text{CUB}\ (2.72\ \text{Å},\ 96\%) > \text{BCO}\ (2.60\ \text{Å},\ 94\%) > \text{BCP}\ (1.85\ \text{Å},\ 65\%).$$

In 2019, a bolus of five reviews burst to the scene in literature, touting the merits of cubanes in medicinal chemistry and drug discovery.[3] The momentum is gaining.

CUBANE-CONTAINING DRUGS

There is no cubane-containing drug on the market. To the best of our knowledge, no cubane-containing drug has entered phase I clinical trials either thus far.

On the other hand, this also represents an opportunity for innovation. Cubane's stability and low toxicity make it a viable building block for medicines. When three-dimensional interactions are needed, but not the aryl–protein interactions, the cubanyl derivative may show superior affinities to the target proteins in comparison to phenyl and its isosteric bridged bi-cycloalkyl derivatives.

CUBANES IN DRUG DISCOVERY

The cubane motif is making more and more appearances in drug discovery, especially during the last five years.

WAY-100635 (**1**) is a potent and selective $5HT_{1A}$ receptor antagonist with potential as a drug therapy or marker for studying the pathophysiology of neuropsychiatric disorders. In order to investigate changes of the $5HT_{1A}$ receptor after binding to WAY-100635 (**1**) using single-photon emission computerized tomography (SPECT), **1**'s bulkier analog cubanyl iodide **2** and the corresponding bridge-fused rings such as admantanyl, bicyclo[2.2.1]heptanyl (BCH), and bicyclo[2,2,2]octanyl (BCO) analogs were prepared and evaluated. Cubanyl iodide **2** was selective for the $5HT_{1A}$ receptor over other relevant neurotransmitter receptors (such as adrenoceptors, sigma, and dopamine receptors) and serotonin transporters. Its carbon–iodide bond was stable *in vivo*, and, like other three bulky analogs, showed a low propensity for amide hydrolysis. Regrettably, their brain uptake and the specificity for those radio-ligands were significantly lower than the parent molecule **1**. Therefore, those designed tracers are not suitable for SPECT imaging.[4a,b]

Further efforts to make radiolabeled $5HT_{1A}$ receptor ligands led to the synthesis of cubanyl-methyl fluoride **3** and its corresponding bridge-fused rings such as admantanyl, BCH, and BCO analogs. Among the four analogs, the cubanyl analog **3** was reasonably selective and showed suitable metabolic stability. This endeavor provided a promising starting point for the synthesis of the corresponding ^{18}F-labeled positron emission tomography (PET) analogs.[4c]

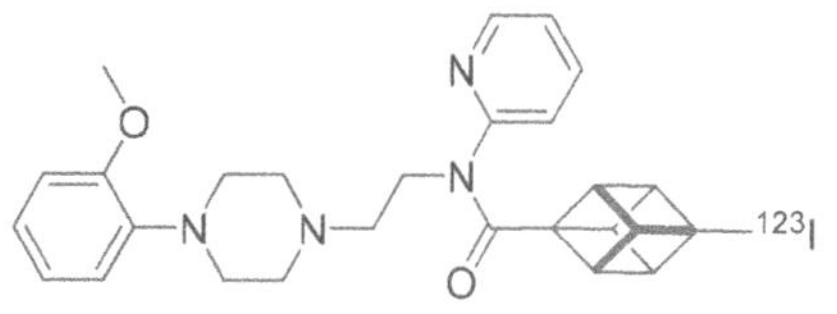

cubanyl iodide **2**

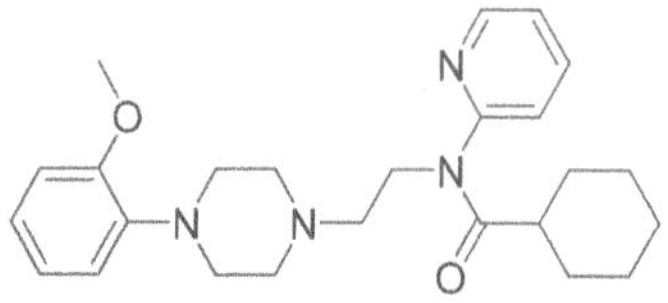

WAY-100635 (**1**)

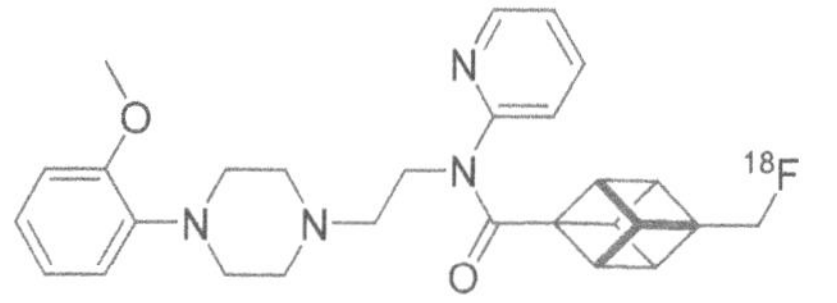

cubanyl methyl fluoride **3**

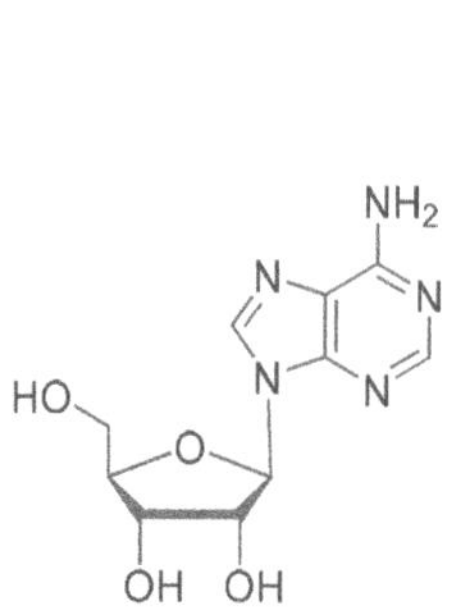

adenosine (**4**)

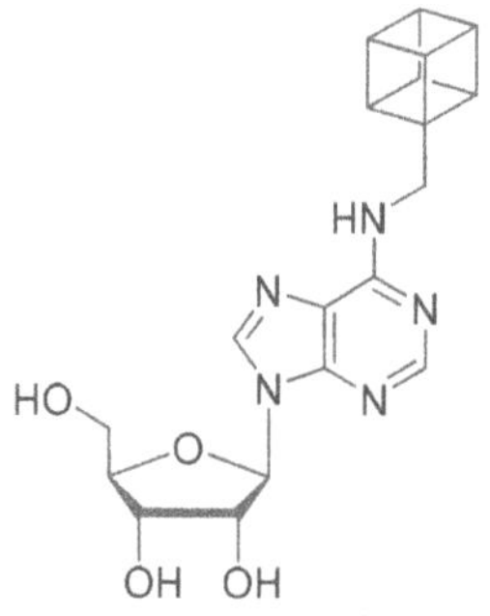

N^6-(cubanylmethyl)adenosine (**5**)

Adenosine (**4**) is an endogenous hormone and adenosine receptor (AR), a G protein-coupled receptor (GPCR), has four subtypes: A_1R, $A_{2A}R$, $A_{2B}R$, and A_3R. Many selective A_1R agonists incorporate various cycloalkyl or heterocyclic substituents at the N^6-position. N^6-(Cubanylmethyl) adenosine (**5**) was found to be a full A_1R agonist ($EC_{50} = 1.1 \, nM$) and activates this receptor selectively over other adenosine receptor subtypes: $A_{2A}R$ (EC_{50}=839 nM), $A_{2B}R$ (EC_{50}=595 nM), and A_3R (EC_{50}=4492 nM). Compound **5** was further evaluated in a simulated ischemia model in cultured cardiomyoblasts, where it was found to impart protective effects under hypoxic conditions that resulted in a significant reduction in cell death.[5]

N-Methyl-D-aspartate (NMDA) receptors are hetero-oligomeric ligand-gated cation channels distributed in the central nervous system (CNS). Glutamate-binding GluN2B subtype-selective NMDA antagonists represent promising therapeutic targets for the symptomatic treatment of multiple CNS pathologies. Kassiou's group discovered that N-substituted 4-(trifluoromethoxy)-benzamidine **6** was a potent binder to the GluN2B-containing NMDA receptors (K_i=2.09 nM). But cubanylamidine **7** lost much affinity (K_i=150 nM), suggesting that steric bulk is not tolerated in this portion of the molecule.[6] There is also an alternative explanation. The benzyl group on **6** not only serves as a space-filler, but also exerts π-stacking with the NMDA protein to afford stronger aryl–protein interactions. The Kassiou group also prepared cubanylamide **8**, which is a novel CNS-active purinergic P2X$_7$ receptor antagonist with antidepressant activity.[7] Since cubanylamide **8** was not as active as the corresponding admantanyl and tris-homocubanyl analogs, further efforts led to cyanoguanidine **9** with significantly boosted biological activities.[8]

6, K_i = 2.09 nM for GluN2B-containing NMDA receptors

7, K_i = 150 nM for GluN2B-containing NMDA receptors

8, hP2X$_7$R pIC_{50} = 6.36 (IC_{50} = 436 nM) LogD$_{7.4}$ = 3.42 (cLogD = 1.46)

9, hP2X$_7$R IC_{50} = 2.9 nM cLogD = 2.53

Back in 1992, Eaton postulated that cubane may serve as a benzene bioisostere. To validate his hypothesis, a group of Australian scientists prepared five cubanyl analogs of five existing biologically active compounds. For example, Merck's vorinostat (SAHA, Zolinza, **10**) is a histone deacetylase (HDAC) inhibitor approved in 2006 for the treatment of cutaneous T-cell lymphoma (CTCL). The corresponding cubanyl analog SUBACUBE (**11**) was determined to be an HDAC inhibitor as well and inhibited the MM96L and MCF7 tumor cell lines with similar IC_{50} values (0.01 for **10** and 0.07 mg/mL for **11**, respectively). *In vitro*, both compounds **10** and **11** were efficient at killing the MyLa2059 cell line. *In vivo*, both compounds showed similar efficacy in tumor growth reduction in a T-cell lymphoma xenograft mouse model. Meanwhile, cubocaine (**13**) was prepared as an analog of anesthetic benzocaine (**12**), which works as a nonselective sodium ion channel blocker. Both benzocaine (**12**) and cubocaine (**13**) demonstrated similar efficacy in an adult male Sprague–Dawley rat pain model. Furthermore, the cubanyl analog of neurotropic compound leteprinim had improved both biological activity and solubility over the parent compound leteprinim itself.[9]

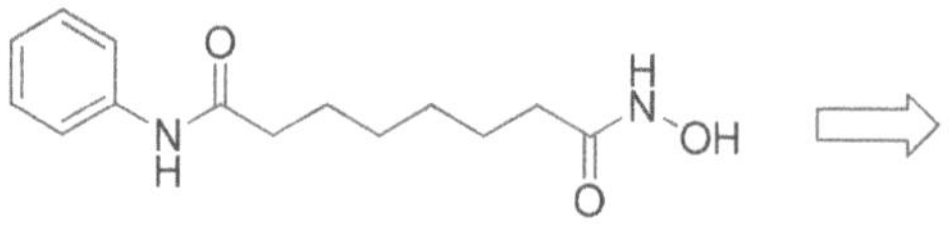

vorinostat (SAHA, Zolinza, **10**)
HDAC inhibitor, IC_{50} = 0.01 μg/mL

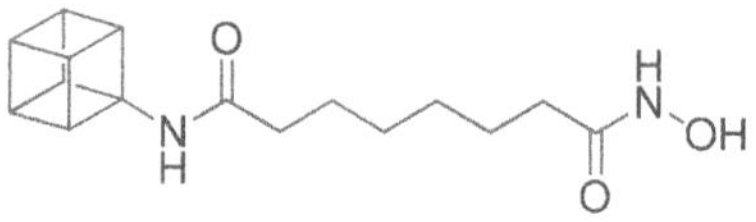

SUBACUBE (**11**)
HDAC inhibitor, IC_{50} = 0.07 μg/mL

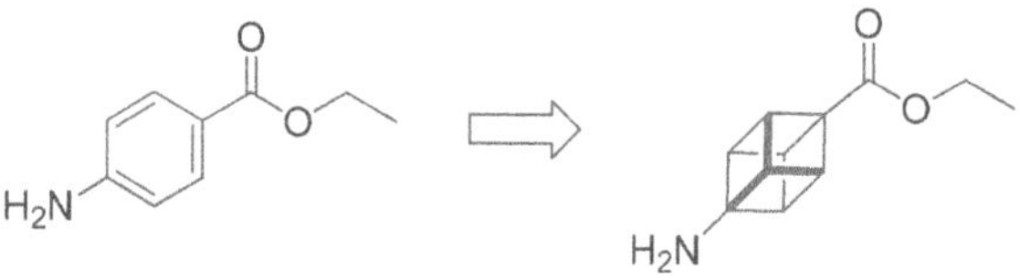

benzocaine (**12**)   cubocaine (**13**)

Imatinib (Gleevec, **14**) was revolutionary in medicine because it was the first kinase inhibitor on the market as the target cancer treatment for cancer. It is a Bcr–Abl kinase inhibitor approved to treat a variety of leukemia such as chronic myeloid leukemia (CML) and acute lymphoblastic leukemia (ALL). But with an Fsp^3 value of 0.24, imatinib (**14**) is woefully deficient of sp^3 carbons. As a consequence, it has a high clogP of 4.53, a high melting point of 204°C, and a low aqueous solubility at pH 7.4 of 30.7 μM. Nicolaou et al. prepared several non-classical *p*-phenyl isosteres including cubane, cyclopropane, BCO, BCP, and cyclobutane to replace the *para*-phenyl ring sandwiched between the amide bond and the piperazine appendage. While CUB–imatinib (**15**) had similar biological activities to the prototype imatinib (**14**), its Fsp^3 value doubled that of **14**. The thermodynamic solubility increased by 12-fold from this maneuver, making CUB–imatinib (**15**) more "drug-like". In fact, CUB–imatinib (**15**) exhibited the most prominent biological activity among the analogs made, reaching 94% cell killing in KU-812 cells with an estimated EC_{50} value of 1.4 μM, and 77% maximum cytotoxicity in MEG-01 cells with an estimated EC_{50} value of 1.8 μM.[10a]

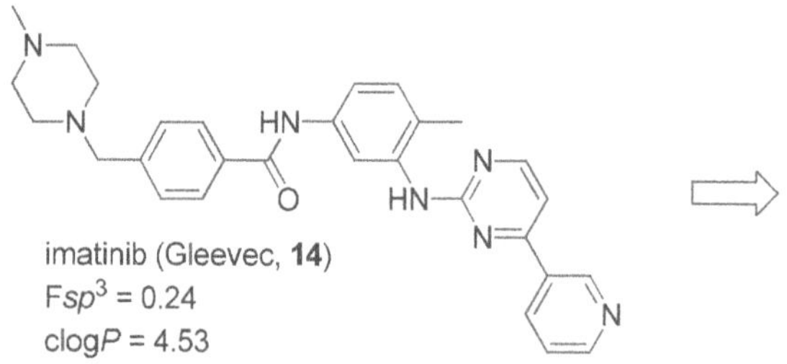

imatinib (Gleevec, **14**)
Fsp^3 = 0.24
clogP = 4.53
aqueous solubility at pH7.4: 30.7 μM

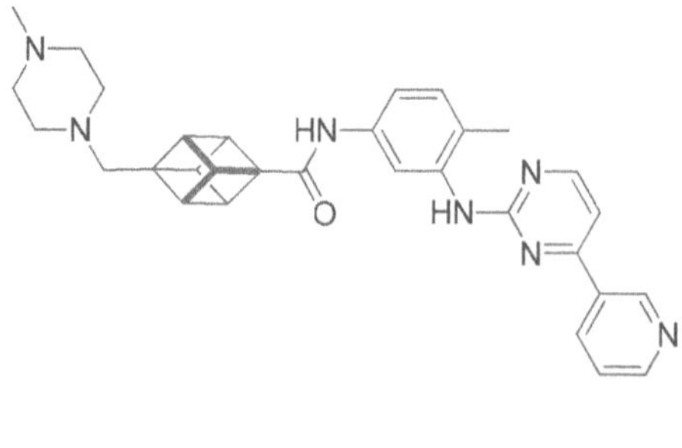

CUB–imatinib (**15**)
Fsp^3 = 0.48
clogP = 1.46
383 μM

Natural tubulysins, such as tubulysin H (**16**), are cytotoxic peptides. Their mechanism of action (MOA), similar to that of vinca alkaloids, involves depolymerization of microtubules thus leading to collapse of the cytoskeleton. Nicolaou's group and Stemcentrx prepared analogs with cubane and other 3D bicycloalkyl linkers. While most of the cubanyl analogs were not tolerated with regard to their biological activities, one particular compound Tb11 (**17**) showed cytotoxic activities for both MES SA ($IC_{50}=0.84$ nM) and HEK 293T ($IC_{50}=0.26$ nM) cell lines.[10b]

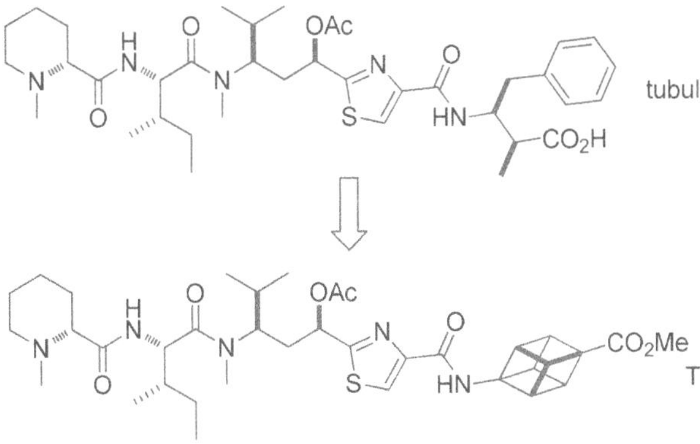

tubulysin H (**16**)

Tb11 (**17**)

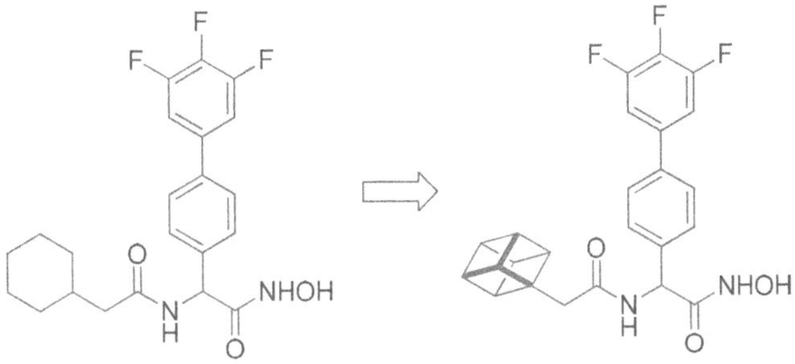

cyclohexyl analogue **18**
Pf-M1, K_i = 812 nM
Pv-M1, K_i = 13.2 nM
Pf-M17, K_i = 288 nM
Pv-M17, K_i = 148 nM

cubanyl analogue **19**
Pf-M1, K_i = 88.9 nM
Pv-M1, K_i = 1.73 nM
Pf-M17, K_i = 318 nM
Pv-M17, K_i = 77.9 nM

Approximately 90% of malaria deaths are caused by *Plasmodium falciparum* (*Pf*), one of the five *Plasmodium* genera. On the other hand, *Plasmodium vivax* (*Pv*) is the predominant parasite outside Africa. An Australian team discovered a series of hydroxamic acids, including **18** and **19**, as cross-species inhibitors of *Plasmodium* M1 and M17 aminopeptidases. The hydroxamic acid group serves as a "warhead" to chelate the catalytic Zn^{2+} ion buried deep within the catalytic domain. For inhibitor **18**, the cyclohexylmethyl amide group occupies the S1′ pockets of the Pf-M1/Pv-M1. Among about 20 permutations investigated to optimize the amide group in the S1′ pocket, the cubanylacetamide **19** (and the norborylacetamide) proved to be the most potent M1 inhibitor of all compounds studied (K_i Pf-M1 = 88.9 nM, Pv-M1 = 1.73 nM).[11] This is a classic case where space-filling hydrophobic interactions are needed so that the bulkiest 3D cubane has a distinct advantage.

SYNTHESIS OF SOME CUBANE-CONTAINING BUILDING BLOCKS

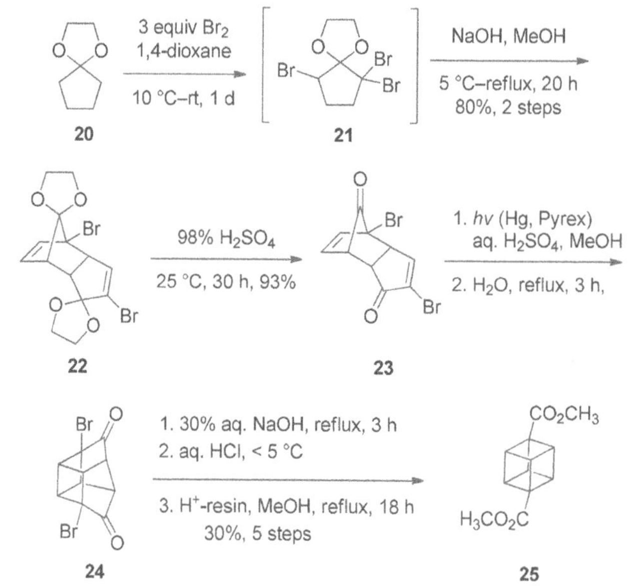

A pilot-scale production of dimethyl 1,4-cubanedicarboxylate (**25**) was published in 2013. As shown below, tribromination of cyclopentanone ethylene ketal (**20**) produced tribromide **21**. The addition of excess methanolic NaOH to the crude bromination mixture not only neutralized the dissolved congenital HBr present but also facilitated double debromination of tribromide **21** to generate the highly reactive 2-bromocyclopentadienone ethylene ketal, which underwent a highly stereoselective Diels–Alder dimerization to assemble bis-ketal **22**. After exhaustive deketalization of **22** to give dione **23**, the key [$2\pi+2\pi$] ene–enone photo-cyclization of dione **23** prepared the caged dione **24** as a mixture of hydrates in a 2.7 kg scale in a 100 L glass reactor. Subsequent double Favorskii ring contraction was accomplished by exposing crude **24** to boiling aqueous NaOH solution to make 1,4-cubanedicarboxylic acid. Eventually, acidification of the reaction mixture was followed by esterification to deliver dimethyl 1,4-cubanedicarboxylate (**25**) in 30% yield for the last five steps.[12]

A series of functionalized cubanes were prepared as building blocks in medicinal chemistry. In one instance, the starting material **26** was first protected as its *t*-butyl ester and the bis-ester was treated with the enolate generated from acetone and LDA to offer dione **27**. Condensation of dione **27** with hydrazine hydrate under acidic conditions was followed by acid cleavage of the *t*-butyl ester to provide cubanyl-pyrazole **28**. Meanwhile, the Weinreb amide from cubanyl acid **26** was treated with the methyl Grignard to produce methyl ketone **29**. Silyl enol ether formation was followed by treatment with NBS to afford α-bromoketone **30**. Condensation of **30** with thioacetamide in methanol under microwave conditions was followed by hydrolysis to deliver cubanyl thiazole **31** in excellent yield.[13]

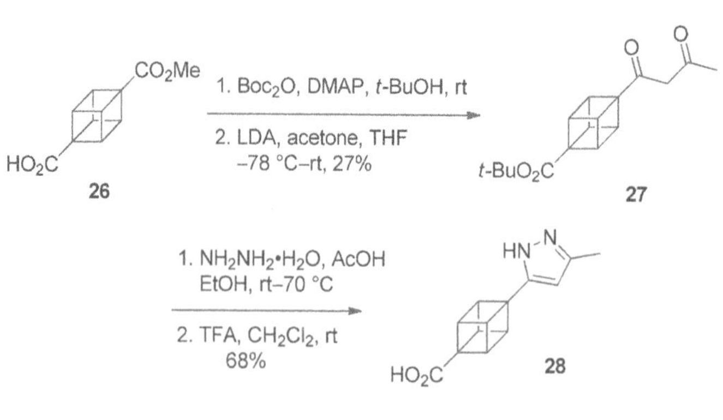

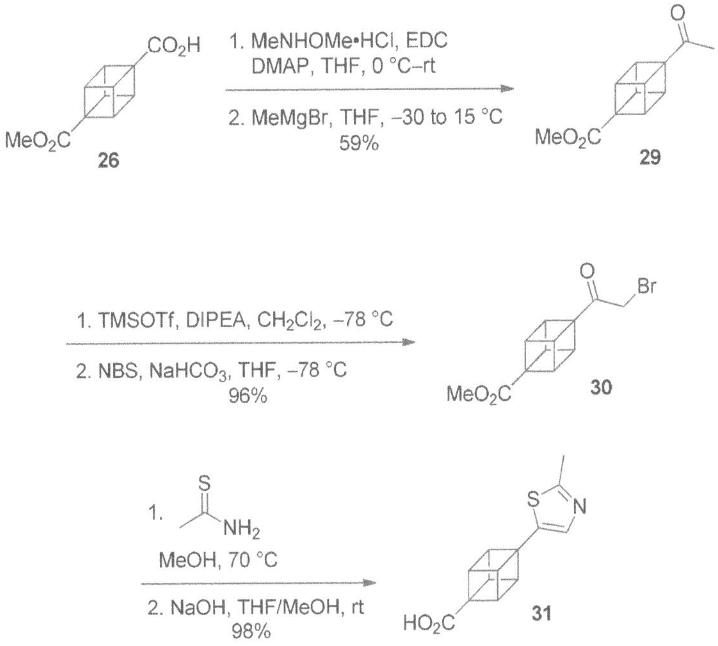

Cubane-containing amino acids have been prepared despite previous difficulties. The Horner–Wadsworth–Emmons coupling between cubane aldehyde **32** and phosphonate **33** gave rise to olefin **34**. Extensively optimized conditions employing H-Cube reduced the double bond on **34** with a minimal decomposition of the starting material to give alkane **35**. Removal of the Boc protection and hydrolysis of the methyl ester delivered the racemic amino acid **36**.[14]

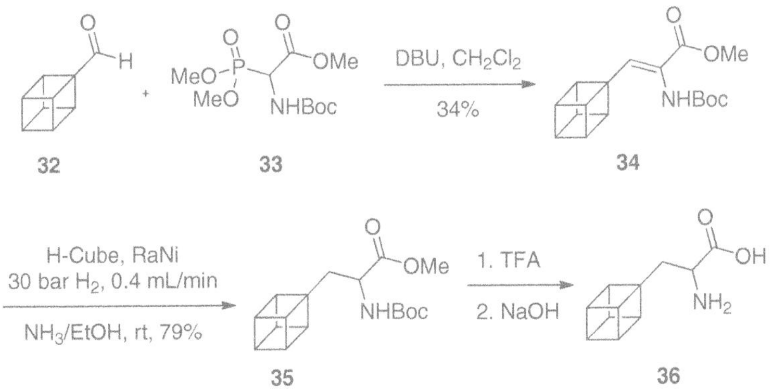

Baran's group described a general amino acid synthesis enabled by innate radical coupling in 2018. Take cubanyl amino acid **40** as an example, cubanyl acid **26** was coupled with *N*-hydroxytetrachlorophthalamide (**37**) to assemble redox-active ester **38** with the aid of *N,N′*-diisopropylcarbodiimide (DIC). Radical coupling between activated substrate **38** and chiral glyoxylate-derived sulfinimine **39** was promoted by $Ni(OAc)_2 \cdot 4H_2O$ and zinc to deliver cubanyl amino acid **40**.[15]

In the realms of cubane-containing drugs, Australian scientists have made important contributions. Historically, Australian Sir Howard Florey won the 1945 Nobel Prize for his role in the development of penicillin. Two Aussies, J. Robin Warren and Barry Marshall, were bestowed the 2005 Nobel Prize for their discovery of *Helicobacter pylori* as a cause of ulcer. Australian scientists also gave the world the first neuraminidase inhibitor zanamivir (Relenza) in 1999 as an inhalation treatment of flu.

In summary, cubane is the closest to benzene in comparison to all isosteric bridged bi-cycloalkyls in terms of size. Therefore, cubane may serve as a bioisostere of benzene as Eaton proposed in 1992 and this conjecture has been confirmed by a number of cubane-containing compounds that showed similar potency to their corresponding phenyl analogs. Cubane would be a good isostere of benzene when the benzene ring acts purely as a conformationally rigid spacer to hold two substituents together with a well-defined distance from each other. Cubane's three-dimensional geometry can provide a drug with multiple contact points to the target protein, which could improve a drug's specific binding. Moreover, the cubane motif has been shown to help improve the drug's solubility as well.

REFERENCES

1. Eaton, P. E. *Angew. Chem. Int. Ed.* **1992**, *31*, 1421–1462.
2. Auberson, Y. P.; Brocklehurst, C.; Furegati, M.; Fessard, T. C.; Koch, G.; Decker, A.; La Vecchia, L.; Briard, E. *ChemMedChem* **2017**, *12*, 590–598.
3. (a) Reekie, T. A.; Williams, C. M.; Rendina, L. M.; Kassiou, M. *J. Med. Chem.* **2019**, *62*, 1078–1095. (b) Houston, S. D.; Fahrenhorst-Jones, T.; Xing, H.; Chalmers, B. A.; Sykes, M. L.; Stok, J. E.; Farfan Soto, C.; Burns, J. M.; Bernhardt, P. V.; De Voss, J. J.; et al. *Org. Biomol. Chem.* **2019**, *17*, 6790–6798. (c) Mykhailiuk, P. K. *Org. Biomol. Chem.* **2019**, *17*, 2839–2849. (d) Locke, G. M.; Bernhard, S. S. R.; Senge, M. O. *Chem. Eur. J.* **2019**, *25*, 4590–4647. (e) Flanagan, K. J.; Bernhard, S. S. R.; Plunkett, S.; Senge, M. O. *Chem. Eur. J.* **2019**, *25*, 6941–6954.
4. (a) Al Hussainy, R.; Verbeek, J.; van der Born, D.; Braker, A. H.; Leysen, J. E.; Knol, R. J.; Booij, J.; Herscheid, J. D. M. *J. Med. Chem.* **2011**, *54*, 3480–3491. (b) Al Hussainy, R.; Verbeek, J.; van der Born, D.; Booij, J.; Herscheid, J. D. M. *Eur. J. Med. Chem.* **2011**, *46*, 5728–5735. (c) Al Hussainy, R.; Verbeek, J.; van der Born, D.; Nolthoff, C.; Knol, R. J.; Booij, J.; Herscheid, J. D. M. *Nucl. Med. Biol.* **2012**, *39*, 1068–1076.
5. Gosling, J. I.; Baker, S. P.; Haynes, J. M.; Kassiou, M.; Pouton, C. W.; Warfe, L.; White, P. J.; Scammells, P. J. *ChemMedChem* **2012**, *7*, 1191–1201.

6. Beinat, C.; Banister, S. D.; Hoban, J.; Tsanaktsidis, J.; Metaxas, A.; Windhorst, A. D.; Kassiou, M. *Bioorg. Med. Chem. Lett.* **2014**, *24*, 828–830.

7. Wilkinson, S. M.; Gunosewoyo, H.; Barron, M. L.; Boucher, A.; McDonnell, M.; Turner, P.; Morrison, D. E.; Bennett, M. R.; McGregor, I. S.; Rendina, L. M.; et al. *ACS Chem. Neurosci.* **2014**, *5*, 335–339.

8. Callis, T. B.; Reekie, T. A.; O'Brien-Brown, J.; Wong, E. C. N.; Werry, E. L.; Elias, N.; Jorgensen, W. T.; Tsanaktsidis, J.; Rendina, L. M.; Kassiou, M. *Tetrahedron* **2018**, *74*, 1207–1219.

9. Chalmers, B. A.; Xing, H.; Houston, S.; Clark, C.; Ghassabian, S.; Kuo, A.; Cao, B.; Reitsma, A.; Murray, C.-E. P.; Stok, J. E.; et al. *Angew. Chem. Int. Ed.* **2018**, *57*, 3580–3585.

10. (a) Nicolaou, K. C.; Vourloumis, D.; Totokotsopoulos, S.; Papakyriakou, A.; Karsunky, H.; Fernando, H.; Gavrilyuk, J.; Webb, D.; Stepan, A. F. *ChemMedChem* **2016**, *11*, 31–37. (b) Nicolaou, K. C.; Yin, J.; Mandal, D.; Erande, R. D.; Klahn, P.; Jin, M.; Aujay, M.; Sandoval, J.; Gavrilyuk, J.; Vourloumis, D. *J. Am. Chem. Soc.* **2016**, *138*, 1698–1708. (c) Nicolaou, K. C.; Erande, R. D.; Yin, J.; Vourloumis, D.; Aujay, M.; Sandoval, J.; Munneke, S.; Gavrilyuk, J. *J. Am. Chem. Soc.* **2018**, *140*, 3690–3711.

11. Vinh, N. B.; Drinkwater, N.; Malcolm, T. R.; Kassiou, M.; Lucantoni, L.; Grin, P. M.; Butler, G. S.; Duffy, S.; Overall, C. M.; Avery, V. M.; et al. *J. Med. Chem.* **2019**, *62*, 622–640.

12. Falkiner, M. J.; Littler, S. W.; McRae, K. J.; Savage, G. P.; Tsanaktsidis, J. *Org. Process Res. Dev.* **2013**, *17*, 1503–1509.

13. Wlochal, J.; Davies, R. D. M.; Burton, J. *Org. Lett.* **2014**, *16*, 4094–4097.

14. Wlochal, J.; Davies, R. D. M.; Burton, J. *Synlett* **2016**, *27*, 919–923.

15. Ni, S.; Garrido-Castro, A. F.; Merchant, R. R.; de Gruyter, J. N.; Schmitt, D. C.; Mousseau, J. J.; Gallego, G. M.; Yang, S.; Collins, M. R.; Qiao, J. X.; et al. *Angew. Chem. Int. Ed.* **2018**, *57*, 14560–14565.

10 Cyclobutanes

Unlike larger and conformationally more flexible cycloalkanes, cyclobutane and cyclopropane have rigid conformations. Due to its ring strain, cyclobutane adopts a rigid puckered (~30°) conformation (like a kite). This unique architecture bestowed certain cyclobutane-containing drugs with unique properties. When applied appropriately, cyclobutyl scaffolds may offer advantages on potency, selectivity, and the pharmacokinetic (PK) profile.

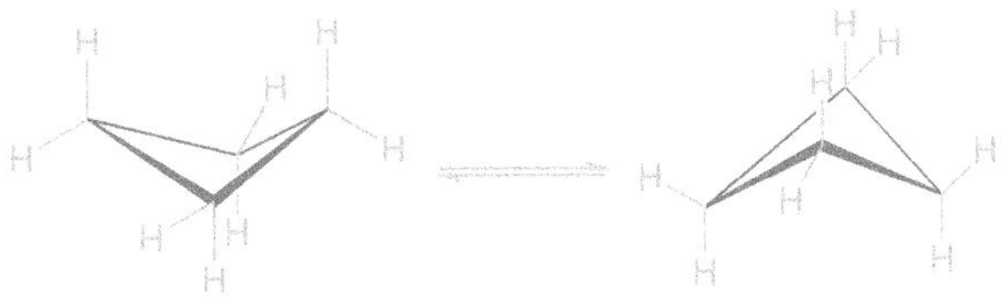

CYCLOBUTANE-CONTAINING DRUGS

At least four cyclobutane-containing drugs are currently on the market. Chemotherapy carboplatin (Paraplatin, **1**) for treating ovarian cancer was prepared to lower the strong nephrotoxicity associated with cisplatin. By replacing cisplatin's two chlorine atoms with cyclobutane-1,1-dicarboxylic acid, carboplatin (**1**) has much lower nephrotoxicity than cisplatin. On the other hand, Schering–Plough/Merck's hepatitis C virus (HCV) NS3/4A protease inhibitor boceprevir (Victrelis, **2**) also contains a cyclobutane group in its P_1 region. It is 3- and 19-fold more potent than the corresponding cyclopropyl and cyclopentyl analogs, respectively.[1]

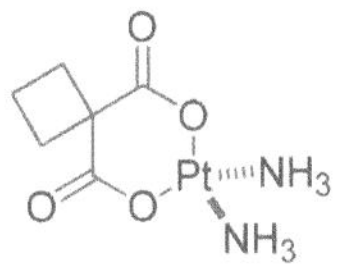

carboplatin (Paraplatin, **1**)
BMS, 1989
DNA intercalator
ovarian cancer

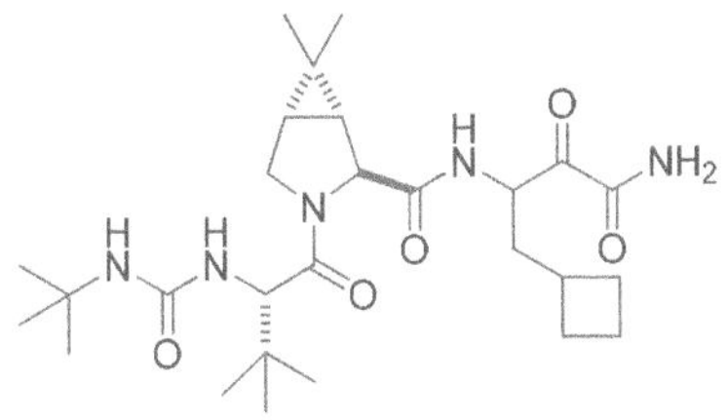

boceprevir (Victrelis, **2**)
Schering-Plough, 2011
HCV NS3/4A protease inhibitor

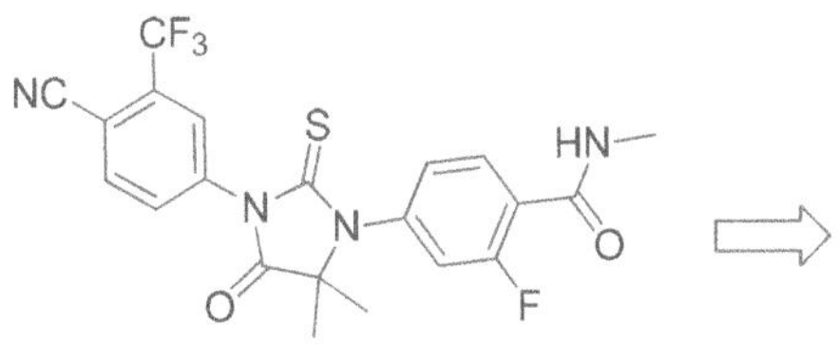

enzalutamide (Xtandi, **3**)
Medivation/Astellas, 2012
androgen receptor antagonist

apalutamide (Erleada, **4**)
Janssen, 2018
androgen receptor antagonist

Androgen receptor (AR) antagonist apalutamide (Erleada, **4**) for treating castration-resistant prostate cancer (CRPC) has a spirocyclic cyclobutane scaffold. It is in the same series as enzalutamide (Xtandi, **3**) discovered by Jung's group at UCLA in the 2000s. The cyclobutyl (**4**) and cyclopentyl derivatives have activities comparable to the dimethyl analog although the corresponding six-, seven-, and eight-membered rings are slightly less active.[2] On a separate note, thiohydantoins are known structural alerts to cause toxicities. However, in this particular case for the particular indication (CRPC), both enzalutamide (**3**) and apalutamide (**4**) have sufficiently large enough therapeutic windows to pass FDA's stringent requirements on safety and efficacy. Being dogmatic about structural alerts would have missed these life-saving medicines!

Agios' ivosidenib (Tibsovo, **16**, *vide supra*), a first-in-class IDH*1* inhibitor, has been recently approved by the FDA in July 2018 for the treatment of IDH1-mutant cancers.[8]

Unlike the conformationally flexible tetrahydrofuran ring found in natural nucleosides, the conformationally more rigid four-membered cyclobutane ring on BMS's lobucavir (**5**) favors a single puckered conformation. It is active against human immunodeficiency virus-1 (HIV-1), hepatitis B virus (HBV), and herpesviruses, suggesting that scaffold conformational flexibility is not essential for antiviral activity and that a rigid scaffold can be compatible with the inhibition of viral replication.[3] Although the phase III clinical trials for lobucavir (**5**) were discontinued in 1999 due to safety concerns, it paved the road for the discovery of BMS's entecavir (Baraclude) as an effective treatment for hepatitis B.

lobucavir (**5**)
RNA-directed DNA
polymerase inhibitors

CYCLOBUTANES IN DRUG DISCOVERY

The cyclobutane motif has been employed to improve a drug's potency, selectivity, and PK profile.

Triazole **6** as a tankyrase (TNKS) inhibitor had a poor PK profile in rats. A structure-guided hybridization approach gave a new series combining triazole **6** and benzimidazolone **7**. Here the *trans*-cyclobutyl linker displayed superior affinity compared to a cyclohexyl and phenyl linker. The resulting hybrid cyclobutane **8** showed favorable activity, selectivity, and *in vitro* ADME profile. Moreover, it was shown to be efficacious in xenograft models.[4]

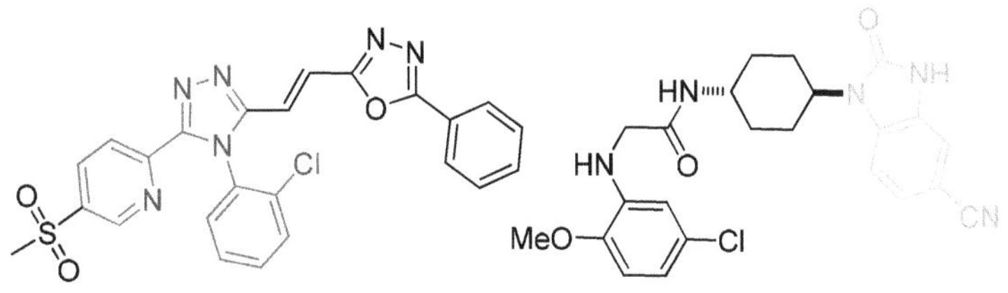

triazole **6**
poor PK in rats

benzimidazolone **7**
TNKS, IC$_{50}$, 25 nM

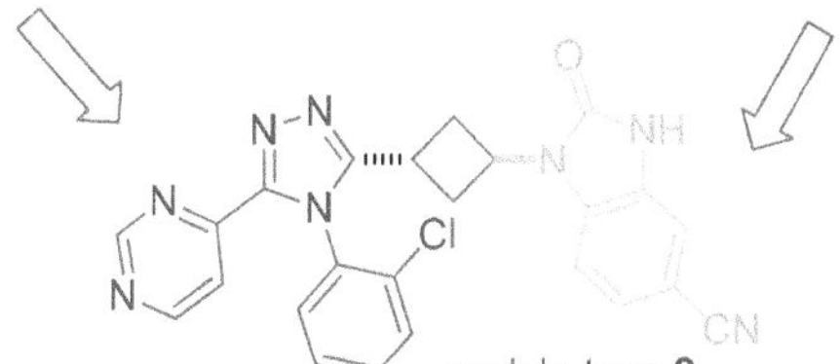

cyclobutane **8**
TNKS, IC$_{50}$, 6.3 nM
good ADME and PK profile

Tetrahydronaphthyridine **9** is a novel inverse agonist for the retinoic acid-related receptor γt (RORγt). Changing the flexible *n*-butanoic acid on **9** to a rigid *cis*-cyclobutane acetic acid on **10** led to an improvement of *in vitro* potency through the reduction of the entropy loss of the carboxylic acid group, which occurs through interaction with amino acid residues in the binding site. Indeed, the resultant TAK-828F (**10**) demonstrated potent RORγt inverse agonistic activity, excellent selectivity against other ROR isoforms and nuclear receptors, and a good PK profile. After testing efficacious in animal models, TAK-828F (**10**) is now in clinical trials for the treatment of Th17-driven autoimmune disease.[5]

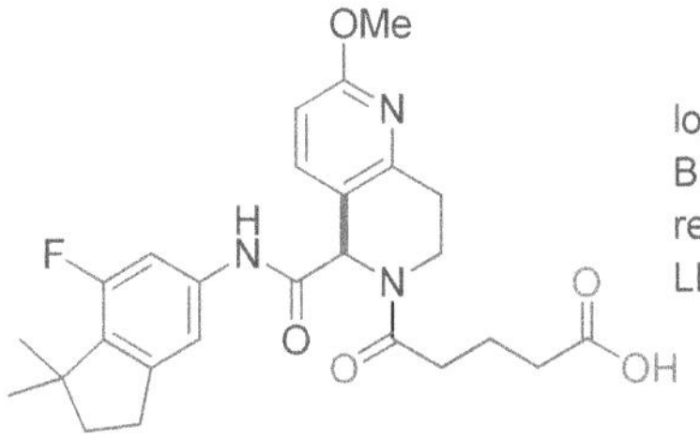

9
logD, 3.39
Binding IC$_{50}$, 3.2 nM
reporter gene, IC$_{50}$, 45 nM
LLE, 3.96

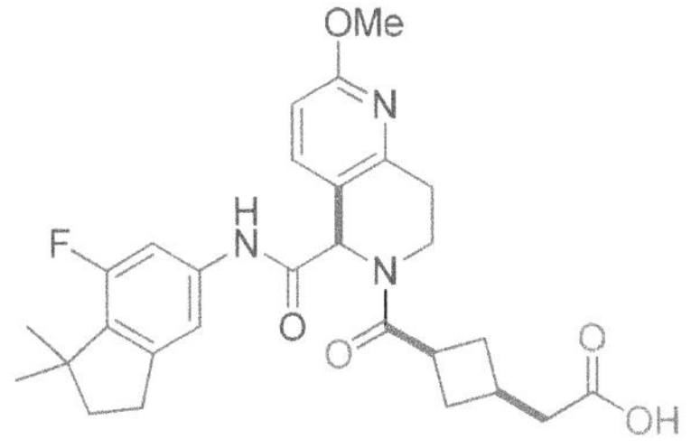

TAK-828F (**10**)
logD, 3.53
Binding IC$_{50}$, 1.9 nM
reporter gene, IC$_{50}$ 6.1 nM
LLE, 4.68

The rigidity offered by the cyclobutane may be useful in improving drug selectivity.

Janus kinases (JAKs) are intracellular tyrosine kinases that mediate the signaling of numerous cytokines and growth factors involved in the regulation of immunity, inflammation, and hematopoiesis. There are four members of the Janus kinase family: JAK1, JAK2, JAK3, and TYK2. Marketed as a treatment of rheumatoid arthritis (RA) since 2012, Pfizer's tofacitinib (Xeljanz, **11**) is a JAK1/JAK3 inhibitor with moderate activity on JAK2. In pursuit of a selective JAK1 inhibitor, Pfizer prepared the corresponding sulfonamides bearing a *cis*-1,3-cyclobutane diamine linker that conferred both excellent potency and excellent selectivity within the JAK family. In particular, PF-04965842 (**12**) has a 28-fold selectivity for JAK1/JAK2. After demonstrating efficacy in a rat adjuvant-induced arthritis (rAIA) model, PF-04965842 (**12**) was nominated as a clinical candidate for the treatment of JAK1-mediated autoimmune diseases.[6]

tofacitinib (Xeljanz, **11**)
Pfizer, 2012
JAK Inhibitor for RA

PF-04965842 (**12**)
JAK1, IC_{50} = 29 nM
JAK2, IC_{50} = 803 nM

Indazole **13** with the methylsulfonamide appendage was found to be a highly selective β_3-adrenergic receptor (β_3-AR) agonist, but it was metabolically unstable because of high clearance. Exchanging the methylsulfonamide to the corresponding cyclobutylsulfonamide and an additional isopropyl/methyl switch at the right portion of the molecules provided indazole **14**. It was not only highly potent and selective as a β_3-AR agonist, but also had desirable metabolic stability and was orally available. Cyclobutylsulfonamide **14** showed dose-dependent β_3-AR-mediated responses in marmoset urinary bladder smooth muscle. It may serve as a candidate drug for the treatment of overactive bladder without off-target-based cardiovascular side effects.[7]

13
β_3-AR EC_{50} = 13 nM
poor pharmacokinetics
(C_{max} and AUC)

14
β_3-AR EC_{50} = 18 nM
improved metabolic
stability
orally available
no cardiovascular side
effects

Point mutations in isocitrate dehydrogenase (IDH) 1 and 2 are found in multiple tumors, including glioma, cholangiocarcinoma, chondrosarcoma, and acute myeloid leukemia (AML). FDA's 2017 approval of Agios/Celgene's mIDH2 inhibitor enasidenib (Idhifa) for treating relapsed/refractory AML fueled much enthusiasm for this novel cancer target. Agios's IDH*1* inhibitor AGI-5198 (**15**) inhibited both biochemical and cellular production of oncometabolite D-2-hydroxyglutarate (2-HG) and was efficacious *in vivo* in a xenograft model. But its poor pharmaceutical properties precluded its use in clinical studies. The major culprits included metabolic instability of the cyclohexane and the imidazole moieties. One key strategy to decrease metabolic clearance was replacing the cyclo-hexyl amine with difluorocyclobutyl amine, which brought the metabolic stability into the medium clearance range. Additional optimizations led to ivosidenib (Tibsovo, **16**), which is potent, selective, and, more importantly, metabolically stable. It is a first-in-class IDH*1* inhibitor now approved by the FDA in July 2018 for the treatment of IDH1-mutant cancers.[8] Agios' enasidenib and ivosidenib represent a novel class of cancer therapy based on cellular differentiation.

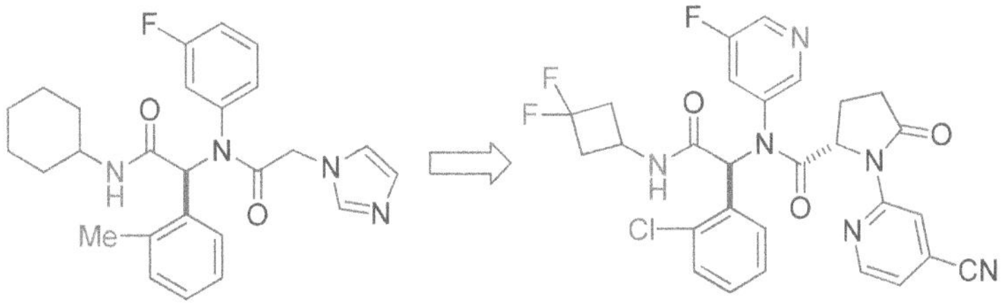

AGI-5198 (**15**)
enzyme IC_{50}, 70 nM
cellular IC_{50}, 497 nM
E_h 0.93

ivosidenib (Tibsovo, **16**)
12 nM
8 nM
0.15

SYNTHESIS OF SOME CYCLOBUTANE-CONTAINING DRUGS

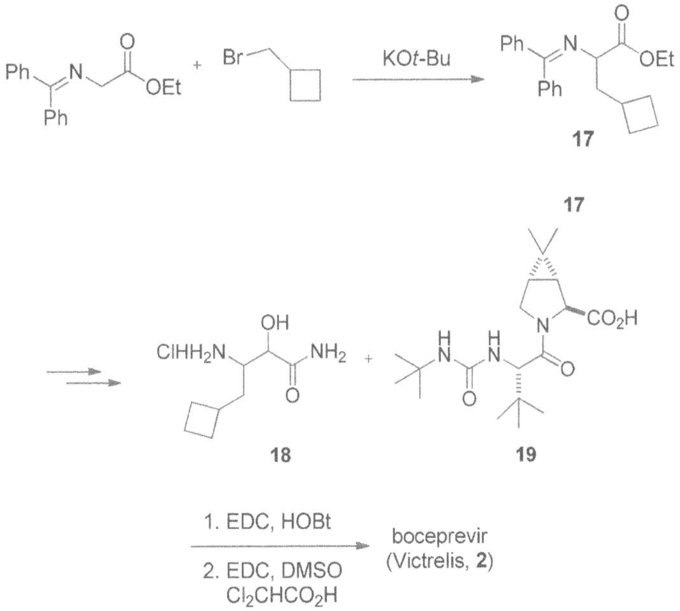

The cyclobutylmethyl fragment on Schering–Plough/Merck's boceprevir (Victrelis, **2**) was incorporated from cyclobutylmethyl bromide. Therefore, alkylation of glycine ethyl ester with cyclobutyl-methyl bromide was aided by KO*t*-Bu to produce adduct **17**. Eight additional steps converted **17** to the desired P_1 intermediate **18**. Amide formation from amine **18** and the P_2–P_3 intermediate as acid **19** was followed by the Moffatt oxidation of the alcohol to ketone to deliver boceprevir (**2**).[1]

In a synthesis of apalutamide (Erleada, **4**), cyclobutanone was employed as the building block to install the spirocyclic cyclobutane motif. A Stecker reaction between cyclobutanone and aniline **20** in the presence of TMS-CN in acetic acid provided cyclobutyl nitrile **21**. Cyclization of **21** with aniline **22** and thiophosgene followed by acidification afforded apalutamide (**4**).[9]

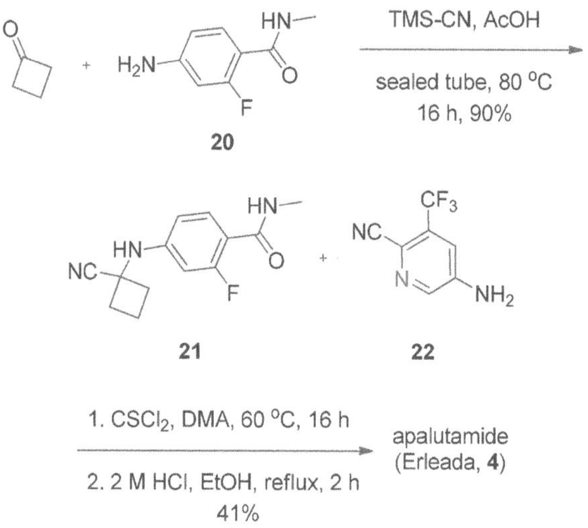

Finally, BMS's synthesis of lobucavir (**5**) relied on a key S_N2 displacement of cyclobutyl tosylate **23**. Thus, the coupling between **23** and 2-amino-6-(benzyloxy)purine (**24**) assembled adduct **25**. Global removal of the three protective groups revealed the desired lobucavir (**5**).[10]

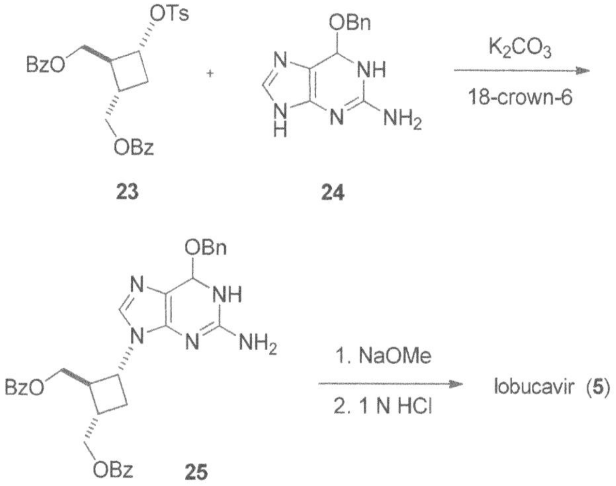

In conclusion, the conformational rigidity has bestowed the cyclobutane derivatives with unique properties. When applied appropriately, cyclobutyl scaffolds may offer advantages on potency, selectivity, and PK profile.

REFERENCES

1. Venkatraman, S.; Njoroge, F. G. Chapter 17, Intervention of Hepatitis C Replication through NS3-4A, The Protease Inhibitor Boceprevir. In *Antiviral Drugs: From Basic Discovery through Clinical Trials*, Kazmierski, W. M., ed., Wiley: Hoboken, NJ, **2011**, pp 239–255.
2. Jung, M. E.; Ouk, S.; Yoo, D.; Sawyers, C. L.; Chen, C.; Tran, C.; Wongvipat, J. *J. Med. Chem.* **2010**, *53*, 2779–2796.
3. Wilber, R.; Kreter, B.; Bifano, M.; Danetz, S.; Lehman-McKeeman, L.; Tenney, D. J.; Meanwell, N.; Zahler, R.; Brett-Smith, H. Chapter 28, Discovery and Development of Entecavir. In *Antiviral Drugs: From Basic Discovery through Clinical T rials*, Kazmierski, W. M., ed., Wiley: Hoboken, NJ, **2011**, pp 401–416.

4. Anumala, U. R.; Waaler, J.; Nkizinkiko, Y.; Ignatev, A.; Lazarow, K.; Lindemann, P.; Olsen, P. A.; Murthy, S.; Obaji, E.; Majouga, A. G.; et al. *J. Med. Chem.* **2017**, *60*, 10013–10025.

5. Kono, M.; Ochida, A.; Oda, T.; Imada, T.; Banno, Y.; Taya, N.; Masada, S.; Kawamoto, T.; Yonemori, K.; Nara, Y.; et al. *J. Med. Chem.* **2018**, *61*, 2973–2988.

6. Vazquez, M. L.; Kaila, N.; Strohbach, J. W.; Trzupek, J. D.; Brown, M. F.; Flanagan, M. E.; Mitton-Fry, M. J.; Johnson, T. A.; TenBrink, R. E.; Arnold, E. P.; et al. *J. Med. Chem.* **2018**, *61*, 1130–1152.

7. Wada, Y.; Nakano, S.; Morimoto, A.; Kasahara, K.-i.; Hayashi, T.; Takada, Y.; Suzuki, H.; Niwa-Sakai, M.; Ohashi, S.; Mori, M.; et al. *J. Med. Chem.* **2017**, *60*, 3252–3265.

8. Popovici-Muller, J.; Lemieux, R. M.; Artin, E.; Saunders, J.O.; Salituro, F. G.; Travins, J.; Cianchetta, G.; Cai, Z.; Zhou, D.; Cui, D.; et al. *ACS Med. Chem. Lett.* **2018**, *9*, 300–305.

9. Pang, X.; Wang, Y.; Chen, Y. *Bioorg. Med. Chem. Lett.* **2017**, *27*, 2803–2806.

10. Bisacchi, G. S.; Braitman, A.; Cianci, C. W.; Clark, J. M.; Field, A. K.; Hagen, M. E.; Hockstein, D. R.; Malley, M. F.; Mitt, T.; et al. *J. Med. Chem.* **1991**, *34*, 1415–1421.

11 Cyclohexanes

As a bioisostere of the *t*-butyl moiety, the cyclohexyl fragment occupies more space, which could be beneficial when binding to a deeper lipophilic pocket on a target protein. On the other hand, as a bioisostere of the flat phenyl group, the cyclohexyl substituent has the advantage of being three-dimensional, which potentially offers more contact points with the target protein.

CYCLOHEXANE-CONTAINING DRUGS

Many drugs isolated from Nature contain the cyclohexyl group either as the core structure such as in dihydroartemisinin (**1**) or on the side-chain as in the case of sirolimus (rapamycin, Rapamune, **2**). Additional cyclohexane-containing drugs from Nature also include steroids, cocaine, FK506, lovastatin (Mevacor), simvastatin (Zocor), morphine and analogs, reserpine, streptomycin, and taxol and its analogs.

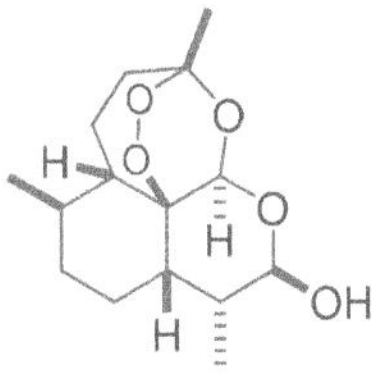

dihydroartemisinin (**1**) 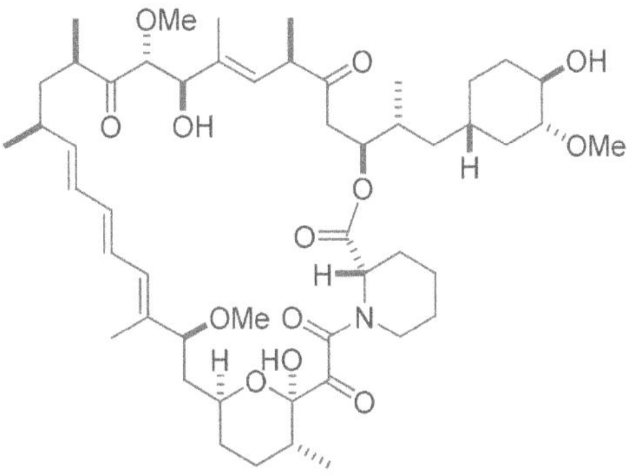   sirolimus (Rapamune, **2**)

Here we focus our attention on synthetic cyclohexane-containing drugs, which encompass nearly all therapeutic areas. Glimepiride (Amaryl, **3**) is a sulfonylurea antidiabetic. Nateglinide (Starlix, **4**), although not a sulfonylurea *per se*, is an antagonist of the sulfonylurea receptor. Both are used to treat type II diabetes mellitus. In terms of CNS drugs, Wyeth's venlafaxine (Effexor, **5**) and its metabolite desvenlafaxine (Pristiq, **6**) are selective serotonin and norepinephrine reuptake inhibitors (SSNRIs), whereas Pfizer's sertraline (Zoloft, **7**) is a selective serotonin reuptake inhibitor (SSRI). All three drugs **5–7** are antidepressants. The most recent entry to antidepressants is Janssen's esketamine (Spravato, **8**, as nasal spray), which modulates the glutamate/GABA neurotransmitter systems, for treatment-resistant depression (TRD). Parke–Davis' gabapentin (Neurontin, **9**) is a relatively older anticonvulsant (its mechanism of action is through inhibiting the α2-δ subunit of calcium channel), Xenoport and GSK co-developed its prodrug, gabapentin enacarbil (Horizant, **10**), which gained the FDA approval for marketing in 2011. Also in 2011, the FDA approved Dainippon's lurasidone (Latuda, **11**), which exhibits significant antagonist effects at the D_2, 5-HT_{2A}, and 7-HT_7 receptors, for the treatment of schizophrenia.

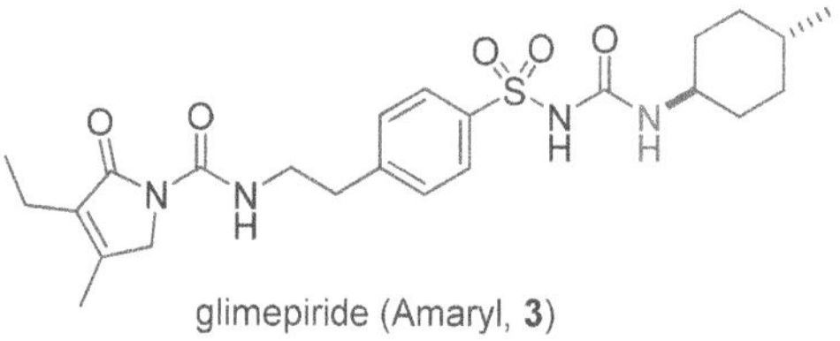

glimepiride (Amaryl, **3**)

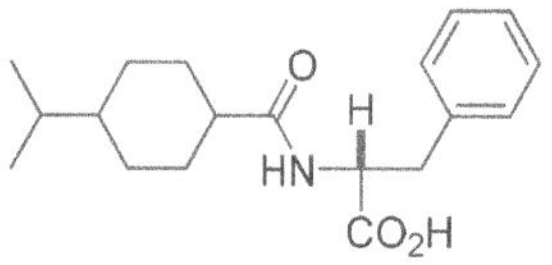

nateglinide (Starlix, **4**)
Novartis, 2000

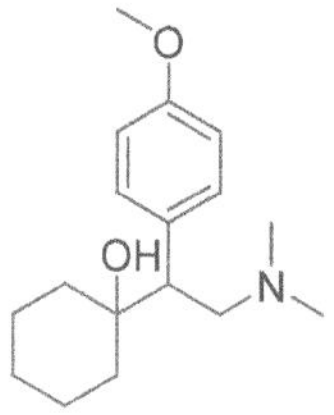

venlafaxine (Effexor, **5**)
Wyeth, 1993

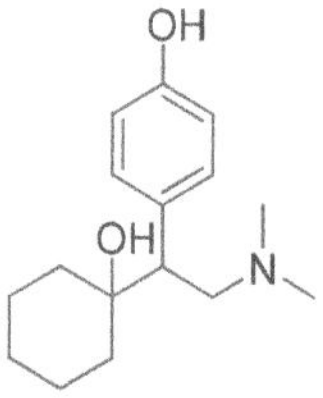

desvenlafaxine (Pristiq, **6**)
Wyeth, 2007

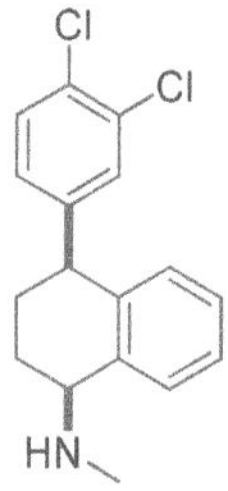

sertraline (Zoloft, **7**)
Pfizer, 1997

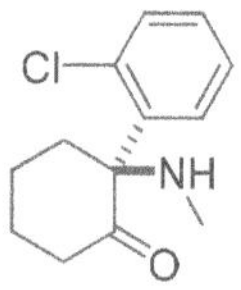

esketamine (Spravato, **8**)
Janssen, 2019

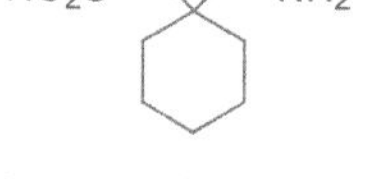

gabapentin (Neurontin, **9**)
Parke-Davis, 1993

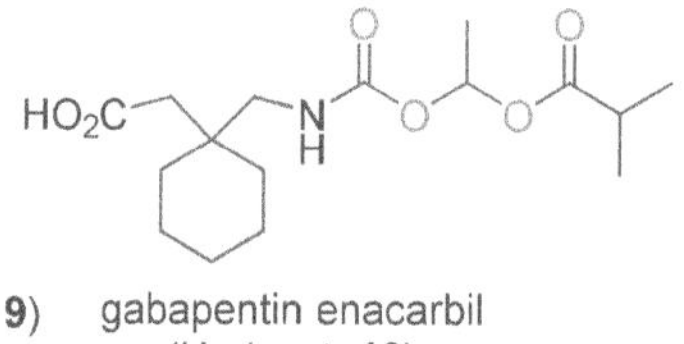

gabapentin enacarbil
(Horizant, **10**)
Xenoport/GSK, 2011

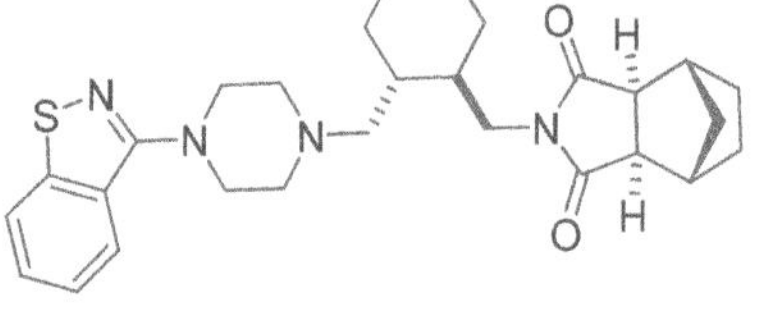

lurasidone (Latuda, **11**)
Dainippon, 2011
D_2, 5-HT_{2A}, 7-HT_7 antagonist

Several cyclohexane-containing cardiovascular drugs exist on the market. Squibb's ACE inhibitor fosinopril (Fozitec, **12**) is one and AstraZeneca's AT_1 receptor antagonist candesartan (Atacand, **13**) is another. For the latter drug, the cyclohexyl group is part of the pro-drug, which is hydrolyzed by esterases *in vivo*. Sankyo's factor Xa inhibitor edoxaban (Savaysa, **14**) has a tri-substituted cyclohexyl moiety as its core structure.

In 1979, Janssen's H_1 receptor antagonist levocabastine (Livostin, **15**), an antihistamine eyedrop, garnered regulatory approval for treating eye allergies. As far as cyclohexane-containing anticancer drugs are concerned, while cisplatin is plagued by renal toxicity, its analog oxaliplatin

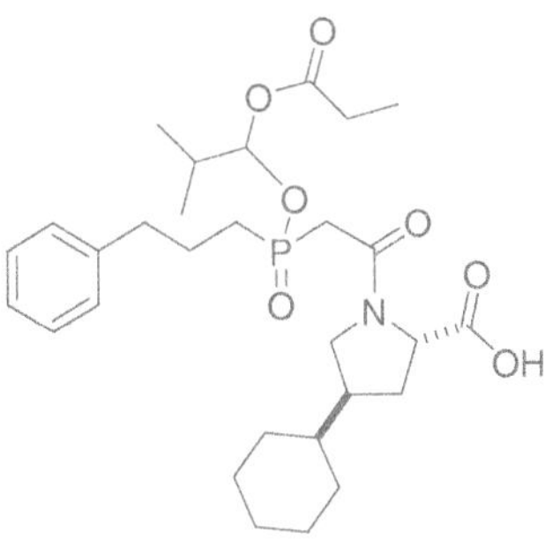

fosinopril (Fozitec, **12**)
Squibb, 1991
ACE inhibitor

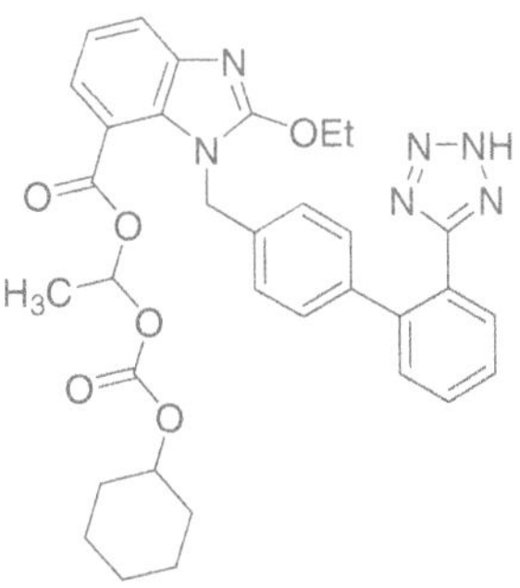

candesartan (Atacand, **13**)
AstraZeneca/Takeda, 1997
AT$_1$ antagonist

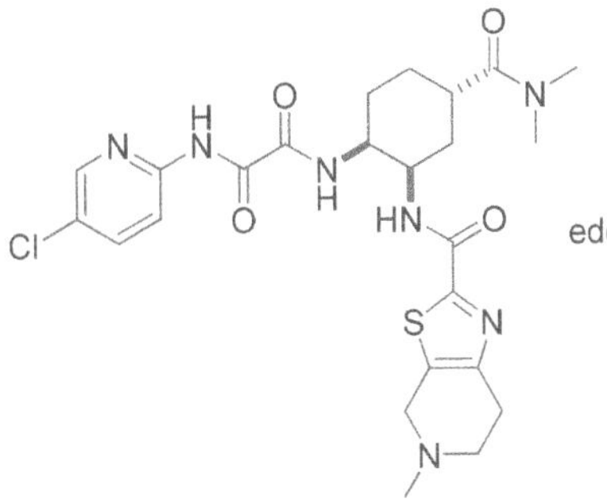

edoxaban (Savaysa, **14**)
Sankyo, 2011
FXa inhibitor

(Eloxatin, **16**) is devoid of nephrotoxicity. Evidently, replacing the two ammonia ligands with a *trans*-diaminocyclohexane is instrumental to the reduction of the drugs' kidney toxicity. In the field of antiviral drugs, Pfizer's CCR5 receptor antagonist maraviroc (Selzentry, **17**) has been on the market to treat HIV infection since 2007.

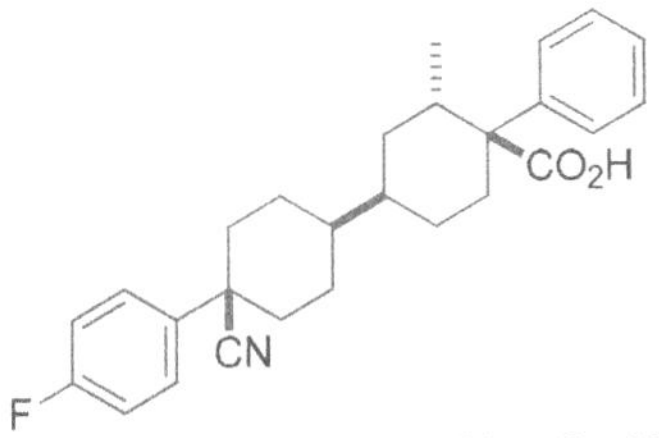

levocabastine (Livostin, **15**)
Janssen, 1979
H1 recptor antagonist

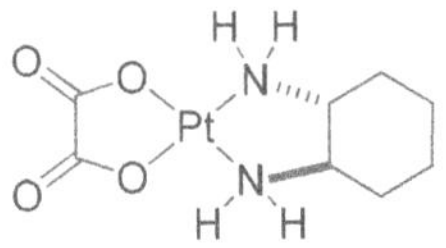

oxaliplatin (Eloxatin, **16**)
Sanofi, 2002
devoid of nephrotoxicity

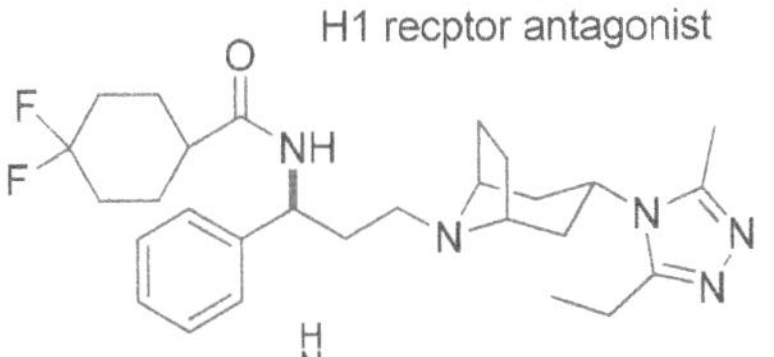

maraviroc (Selzentry, **17**)
Pfizer, 2007
CCR5 receptor antagonist

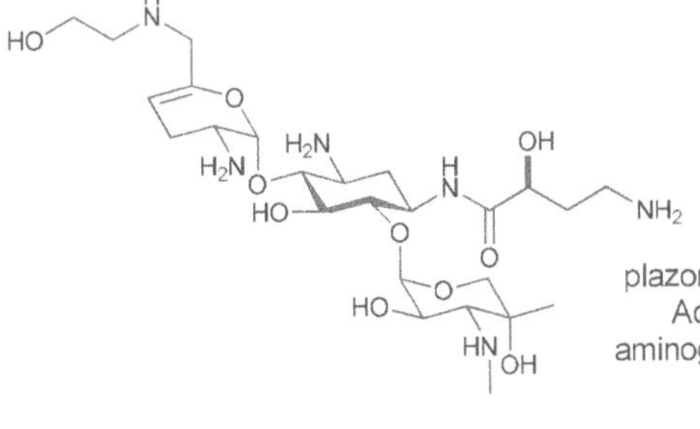

plazomicin (Zemdri, **18**)
Achaogen, 2018
aminoglycoside antibiotic

In 2018, the FDA approved Achaogen's aminoglycoside antibiotic plazomicin (Zemdri, **18**). Regrettably, the drug is a commercial flop and the company went bankrupt recently. Success in science does not always translate to financial success, unfortunately.

CYCLOHEXANES IN DRUG DISCOVERY

The bioisosterism between the cyclohexyl and the *t*-butyl group is amply demonstrated during the structure–activity relationship (SAR) investigations for the two marketed hepatitis C virus (HCV) non-structural protein (NS)3/4A inhibitors telaprevir (Incivek, **20**) and boceprevir (Victrelis, **21**). Both are serine protease reversible covalent inhibitors. Aided by a structure-based drug design (SBDD), Vertex arrived at hexapeptide **19** with a K_i value of 200 nM. Extensive SAR efforts led to the truncation of the P_1' amide. More relevantly, the P_4 position on **19** was an isopropyl fragment (not a *t*-butyl group *per se*, but a close analog). Evidently, the corresponding S_4 pocket on the NS3/4A serine protease protein was deeper. As a consequence, employing a cyclohexyl substituent enhanced the hydrophobic binding, which eventually led to the discovery of a potent and bioavailable covalent telaprevir (Incivek, **20**) on top of P_2 and P_3 optimization.[1] That was a significant achievement of medicinal chemistry considering "Trying to land an inhibitor in the HCV protease target binding site was like trying to land a plane on a piece of pizza—it's flat and greasy and there's nothing to hang onto".[2]

19, K_i = 220 nM
P_1', P_2, P_3, and P_4 optimization

telaprevir (Incivek, **20**)
K_i = 40 nM; K_i* = 7 nM
Vertex, 2011

While veteran drug hunters know intimately the nomenclature of binding pockets of proteases, it might be useful to show the definition here for our novice colleagues. As shown in the scheme below, the active catalytic site serves as the reference point: the catalytic zinc in the scheme for angiotensin-converting enzyme (ACE). But the NS3/4A serine protease protein's catalytic serine residue (Ser139) is the reference point, which attacks the ketoamide "warhead" and causes cleavage of the substrate. Binding pockets on the right of the catalytic site are known as prime pockets (S_1', S_2', S_3', and so on) and binding pockets on the left of the catalytic site are known as non-prime pockets (S_1, S_2, S_3, and so on). Correspondingly, substituents on the endogenous ligands that occupy the prime pockets are known as prime substituents

(P_1', P_2', P_3', and so on) and substituents that occupy the non-prime pockets are known as non-prime substituents (P_1, P_2, P_3, and so on).

$H_3N^{\oplus}$
P_2
P_1
P_1'
P_2'
$CO_2^{\ominus}$
peptide chain
S_2
S_1
Zn^{++}
S_1'
S_2'
enzyme

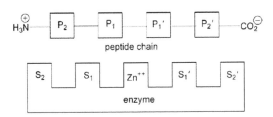

Definition of binding pockets and endogenous ligand for protease (ACE)

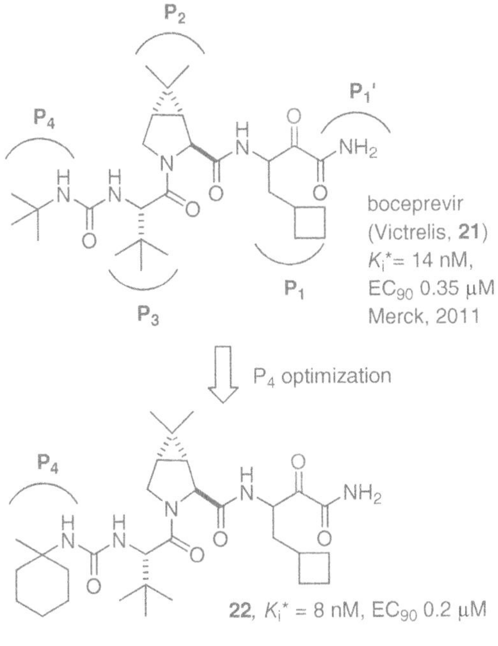

P_2
P_4
P_1'
P_3
P_1
boceprevir
(Victrelis, 21)
K_i* = 14 nM,
EC_{90} 0.35 μM
Merck, 2011
P_4 optimization
P_4
22, K_i* = 8 nM, EC_{90} 0.2 μM

To improve upon boceprevir (Victrelis, **21**), Schering–Plough carried out additional SAR to develop more potent inhibitors with an improved PK profile, particularly in monkeys, to target a once-daily compound. Changing the P_4 group from *t*-butylurea to cyclohexylmethylurea gave rise to compound **22** with higher potency in both an enzyme assay (K_i*) and a cell-based assay (EC_{90}).[3] These efforts eventually led to the marketing of narlaprevir (Arlansa, **61**), a *t*-butylsulfonyl analog of **22**.

Since aripiprazole (Abilify, **23**) as a D_2 partial agonist is probably the most successful antipsychotic, many "me-too" drugs have stemmed from it. Rigidifying aripiprazole (**23**)'s floppy linear linker into a cyclohexyl ring led to the discovery of cariprazine (Vraylar, **24**), which is a D_3 (rD_3 K_i=0.71 nM) selective partial agonist and is 13-fold more selective against the D_2 receptor (rD_2 K_i=9.3 nM) in a rat dopamine receptor assay.[4]

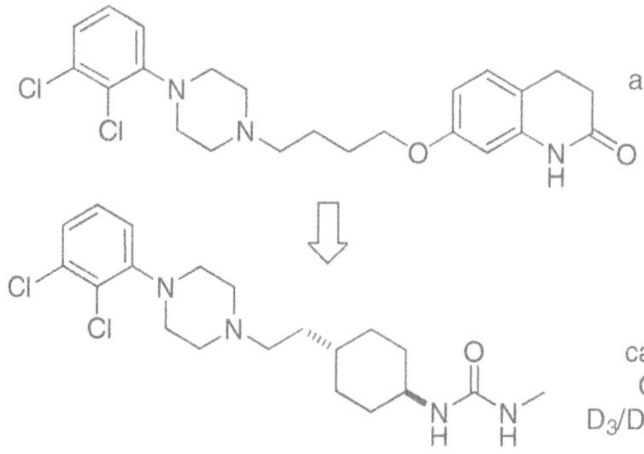

aripiprazole (Abilify, **23**)
Otsuka/BMS, 2002
D$_2$ partial agonist

cariprazine (Vraylar, **24**)
Gedeon Richter, 2015
D$_3$/D$_2$ selective partial agonist

Truncation of the structure of morphine (**25**), a μ opioid receptor agonist, resulted in the discovery of a new pain medicine tramadol (Ultram, **26**) as an opioid analog.[5] It is evident that the cyclohexyl core structure is part of the crucial pharmacophore.

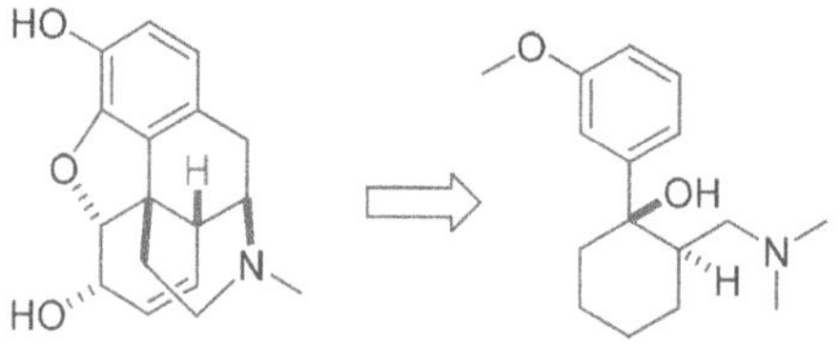

morphine (**25**)
μ opioid receptor agonist

tramadol (Ultram, **26**)
opioid analogue

The cyclohexene ring has been employed as a bioisostere of a furanose ring since its conformational behavior is similar to that of a saturated five-membered ring and it may offer more metabolic stability.[6] The genesis of oseltamivir (Tamiflu, **28**) is a case in point. Gilead wisely chose the cyclohexene ring to replace the tetrahydropyranyl core structure on zanamivir (Relenza, **27**). Zanamivir (**27**) is so polar that it does not cross the cell membrane and thus has to be given via inhalation. The cyclohexene ring was expected to be chemically and enzymatically stable, to be suited for chemical modifications, and, more importantly, to be a suitable bioisostere of the proposed oxonium cation in the transition state of sialic acid cleavage by neuraminidase.[7] Indeed, oseltamivir (Tamiflu, **28**) is orally bioavailable with a bioavailability of 75% and a half-life of 6–10 hours.

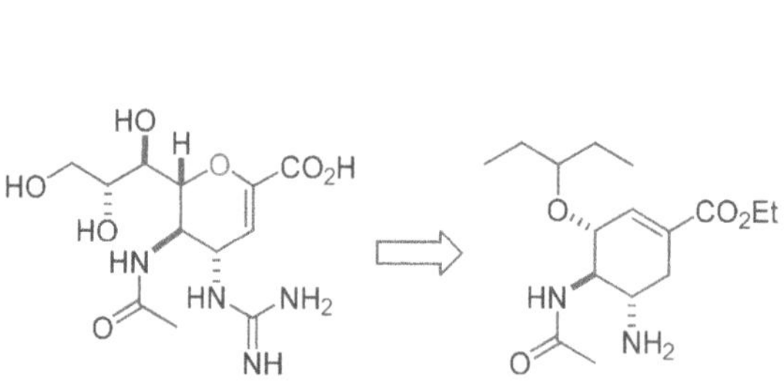

zanamivir (Relenza, **27**)

oseltamivir (Tamiflu, **28**)

In the same vein, *en route* to the discovery of its ground-breaking B-cell lymphoma-2 (BCL-2) inhibitor venetoclax (Venclexta, **31**) via fragment-based drug discovery (FBDD), Abbvie arrived at compound **29** with a rigid biphenyl structure. Switching one of the phenyl rings to cyclohexene provided an opportunity to make ring modifications that were fundamentally different from those that were made to the aromatic ring in its place. Simply bulking up the ring by adding alkyl groups produced the dimethylcyclohexene **30**, which had still higher plasma levels, and also appeared to improve tissue/plasma distributions in various pharmacokinetic models.[8] Addition of the dimethyl group was a great idea since it eliminates the metabolic liability associated with simple cyclohexenes, which are prone to be oxidized to the corresponding aromatic phenyl analog.

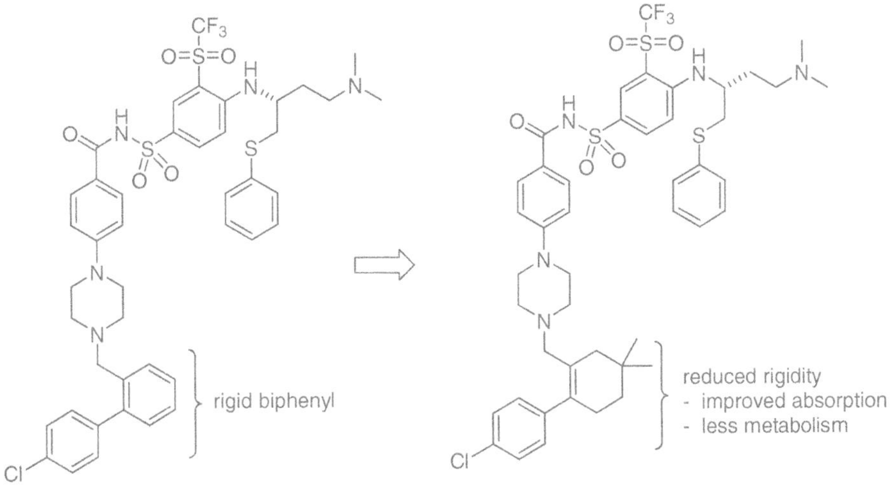

rigid biphenyl

29, $AUC_{0-8 h}$ = 0.87 µg/mL
F%, 8.7
EC_{50} = 39 nM,
H146 SCLC cells, 10% srum
AUC/EC_{50} = 19

reduced rigidity
- improved absorption
- less metabolism

30, $AUC_{0-8 h}$ = 2.9 µg/mL
F%, 14
EC_{50} = 28 nM,
H146 SCLC cells, 10% srum
AUC/EC_{50} = 104

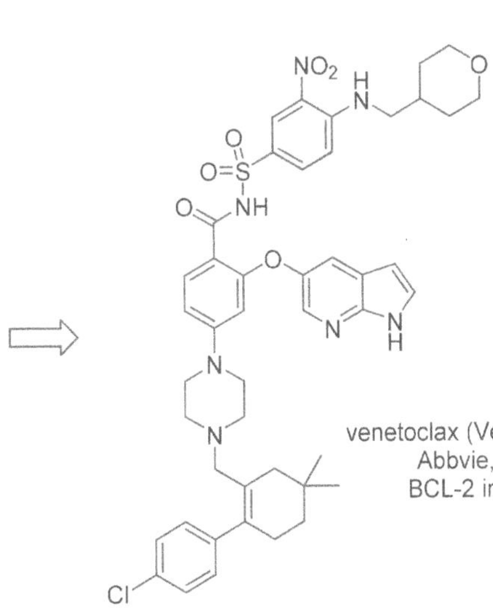

venetoclax (Venclexta, **31**)
Abbvie, 2016
BCL-2 inhibitor

The cyclohexyl fragment helps to break the crystal lattices and boosts aqueous solubility in imidazopyridine antimalarial drugs. Compound **32** as an NF54 inhibitor was rather potent (IC_{50}, 18 nM) but suffered poor aqueous solubility. Replacing the 1-fluoro-2-(trifluoromethyl) phenyl group with a 1,1-difluorocyclohexyl substituent gave rise to compound **33**, which was 3.8-fold less potent than **32** but enjoyed greatly improved aqueous solubility at both pH 2 and 6.5, respectively.[9]

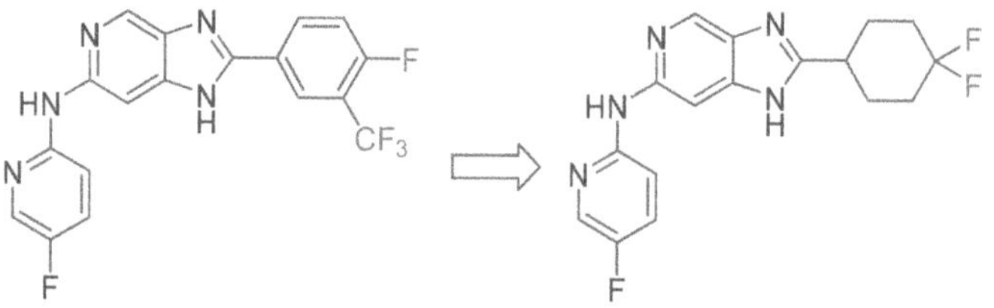

32
aq. solubility at pH2, 13 μM
aq. solubility at pH6.5, < 5 μM

33
aq. solubility at pH2, 200 μM
aq. solubility at pH6.5, 170 μM

Being greasy, the cyclohexyl motif is prone to CYP450 oxidation to the corresponding cyclohexanol. For instance, one of the major metabolites of sulfonylurea drug acetohexamide (Dymelor, **34**, for the treatment of type II diabetes mellitus) is *para*-hydroxylhexyl derivative **35**. Another key metabolite of acetohexamide (**34**) is the secondary alcohol from reduction of the acetyl group by carbonyl reductase.[10] In a similar manner, phencyclidine (PCP, **36**) is metabolized to the corresponding *para*-hydroxyhexyl derivative **37**.[11]

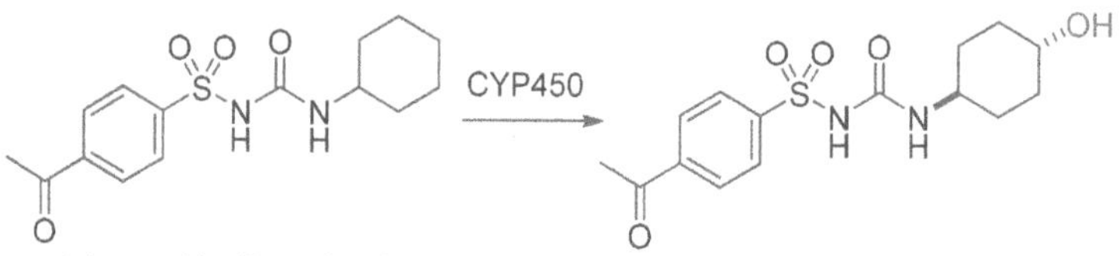

acetohexamide (Dymelor, **34**)   cyclohexanol metabolite **35**

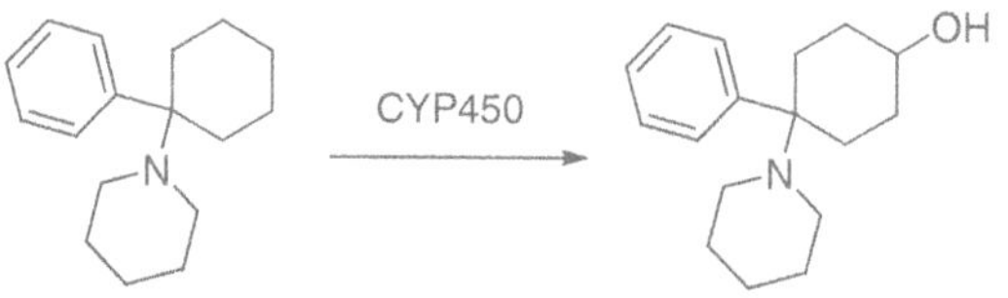

phencyclidine (PCP, **36**)   cyclohexanol metabolite **37**

Proteolysis targeting chimera (PROTAC) as a drug discovery approach has gained much momentum lately since Arvinas' androgen receptor protein degrader ARV-110 advanced to phase I clinical trials in 2019. A similar technique, specific and non-genetic IAP-dependent protein eraser (SNIPER), employs small molecule ligands for E3 ubiquitin ligases cIAP1 (cellular inhibitor of apoptosis protein), which contains a cyclohexyl substituent. For instance, SNIPER(ABL)-062 (**38**) showed binding affinities against ABL1, cIAP1/2, and XIAP and induced potent Bcr-Abl protein degradation.[12] Asciminib is the first allosteric kinase inhibitor in clinical trials.

allosteric Bcr-Abl ligand
asciminib (ABL001)
Novartis
Phase III, 2017

IAP antagonist

SNIPER(ABL)-062 (38)

Schering–Plough employed bicyclo[4.1.0]heptanes as phenyl isosteres for their melanin-concentrating hormone (MCH) receptor antagonists, which have potential as a treatment of obesity. Compound **39** was a potent MCH-R1 antagonist, which exhibited oral efficacy in chronic (28 days) rodent models, reducing cumulative food intake and body weight gain relative to vehicle controls. Unfortunately, the biphenyl amine moiety (in dark gray and light gray was a very potent mutagenic agent as indicated by its strong positive result in an Ames test. Although the biphenylamine itself was not formed *in vivo*, it was deemed unsuitable for development because of the potential risk of exposure to such a highly mutagenic precursor. Replacing the middle phenyl ring with pyrazine, pyrimidine, and saturated derivatives such as piperidines, their MCH-R1 activities were drastically reduced. While the cyclohexenyl replacement was extremely active ($K_i = 3$ nM), the cyclohexenyl fragment has a dual liability of intrinsic metabolic instability associated with the styrene and its potential for generating a biphenylamine via aromatization. Further exploration to discover more stable analogs led to cyclopropanation of the double bond to form a bicyclo[4.1.0]alkyl group and achieved this goal. The bicyclo[4.1.0]heptanyl analog **40** had a comparable binding affinity and similar efficacy in obese animal models and it was devoid of the mutagenicity issue associated with biphenylamine derivatives.[13]

39
MCH-R1 K_i = 2 nM
diet-induced obese (DIO)
at 6 h = 17% @30 mg/kg

40
12 nM
DIO =
10%

SYNTHESIS OF SOME CYCLOHEXANE-CONTAINING DRUGS

Merck's vernakalant (Kynapid, **44**) is an atrial potassium channel blocker. In one of the synthetic routes leading to vernakalant (**44**), racemic cyclohexyl epoxide (**41**) was opened with protected prolinol **42** as the nucleophile in hot water. The resulting mixture of diastereomers was separated by classical resolution of the corresponding tartrate salt to afford *cis*-isomer **43**.

Subsequent ether formation from **43** was followed by de-benzylation to deliver the desired active pharmaceutical ingredient (API) **44**.[14]

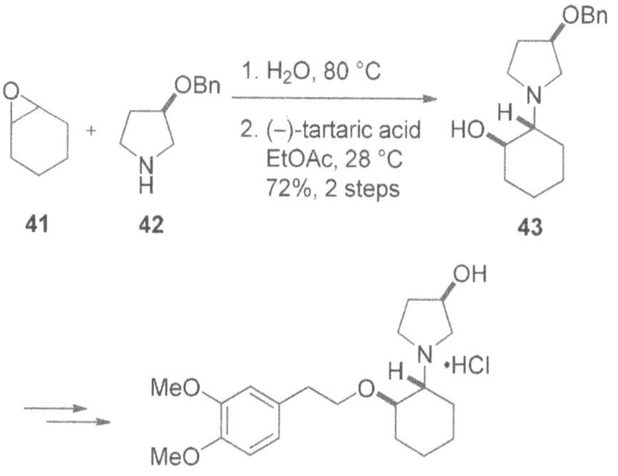

vernakalant (Kynapid, **44**)

Dainippon's preparation of lurasidone (Latuda, **11**) commenced with mesylation of commercially available diol **45** to give bis-mesylate **46**. Condensation of bis-electrophile **46** with aryl-piperizine **47** offered dialkylation product as ammonium salt **48**. Since **48** is such a reactive intermediate, its S_N2 reaction with succinimide **49** readily took place to deliver lurasidone (**11**), which was conveniently converted to lurasidone hydrochloride as the API.[15]

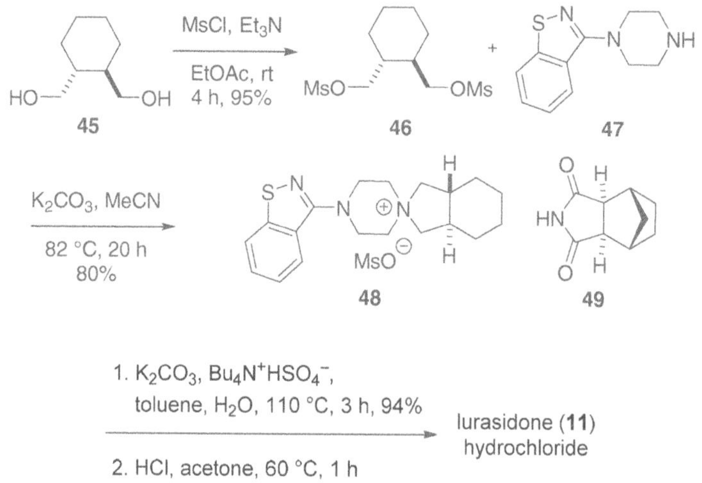

Production of Sankyo's FXa inhibitor edoxaban (Savaysa, **14**) began with cyclohexyl epoxide **50** as the starting material. Regio-specific S_N2 reaction with sodium azide gave the corresponding hydroxyazide intermediate, which was converted to alcohol **51** via palladium-catalyzed hydrogenation in the presence of Boc$_2$O. Mesylation of **51** and another S_N2 reaction with sodium azide produced azide **52**, which underwent another palladium-catalyzed hydrogenation and protection sequence to offer, this time, Cbz carbamate-protected amine **53**. Saponification of the ester group on **53** produced the acid, which was coupled with dimethylamine-HCl salt to form amide **54**. Removal of the Cbz protection and reaction of the exposed primary amine with oxalic acid gave rise to oxalate salt **55**, which was eventually transformed to the API edoxaban (Savaysa, **14**) after five additional steps.[16]

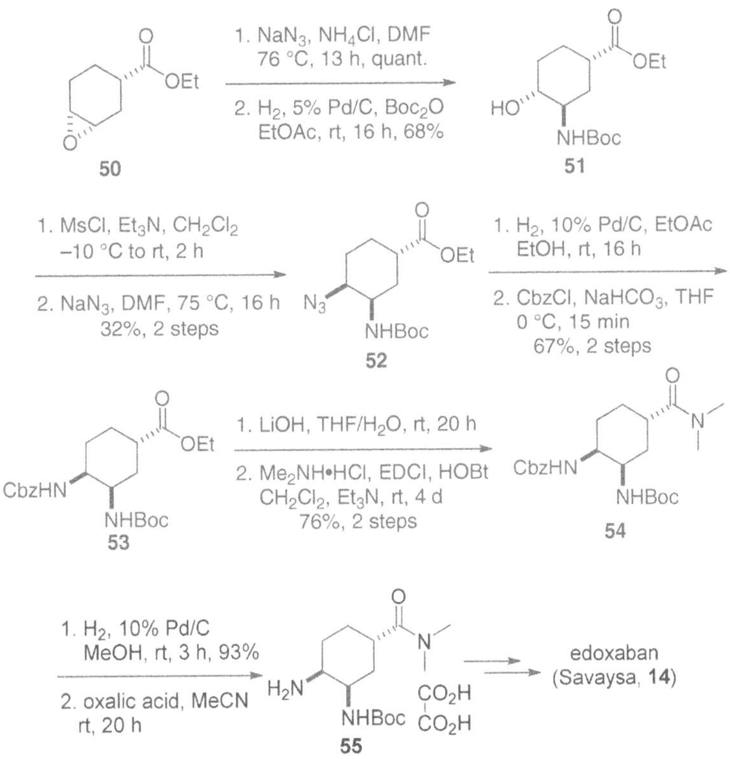

Methyl cyclohexylcarboxylate (**56**) was employed as the starting material for the synthesis of HCV NS3 serine protease inhibitor narlaprevir (Arlansa, **61**). Silyl enol ether **57** was generated *in situ* by treating **54** with freshly prepared LDA followed by quenching with TMSCl. It was immediately treated chloride **58** under Lewis acid catalysis to assemble adduct **59**. Subsequently, the ester on **59** was hydrolyzed to the corresponding acid and the sulfide was oxidized by Oxone to the corresponding sulfone **60**. Coupling of the key cyclohexyl intermediate **60** with three amino acid fragments delivered narlaprevir (**61**) in another additional six steps.[17]

narlaprevir (Arlansa, **61**)
HCV NS3 serine protease inhibitor

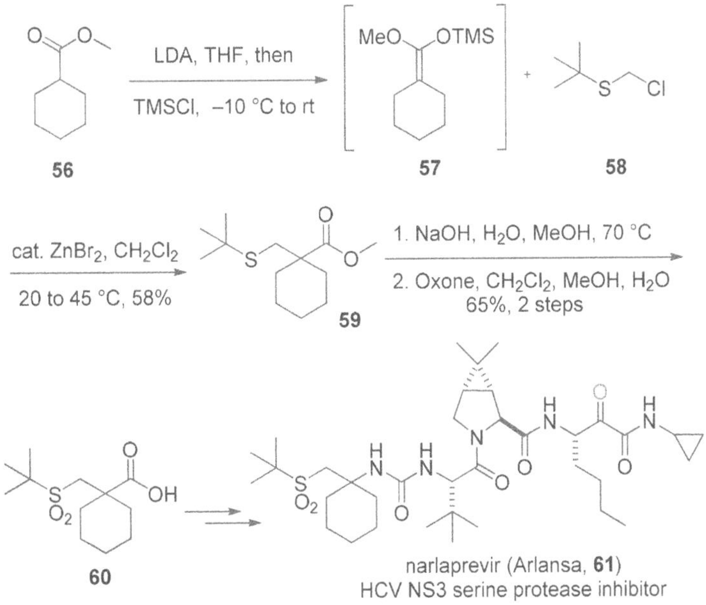

Schering–Plough's synthesis of the bicyclo[4.1.0]heptanyl analog **40** involved a modified Simmons–Smith reaction. Thus, cyclopropanation of cyclohexene **62** was achieved by treating **62** with CH_2I_2 and Et_2Zn in the presence of TFA to give bicyclo[4.1.0]heptanyl **63**. After removal of the ketal protection, the resulting ketone **64** underwent a reductive amination with amine **65** to afford adduct **66**. To avoid direct ketone reduction by $NaBH_4$, the imine intermediate was pre-formed with the aid of $Ti(Oi\text{-}Pr)_4$ before adding $NaBH_4$. Coupling between **66** and isocyanate **67** then delivered the final product **40** after chiral separation.[13]

To conclude, the cyclohexyl fragment is a popular building block in both natural and synthetic drugs, serving either as the core structure or as part of a peripheral side chain. The cyclohexyl group may function as a bioisostere of the *t*-butyl group for a deeper hydrophilic pocket on a target protein. As a rigid version of the floppy alkyl chain, the cyclohexyl replacement reduces entropy and may offer better affinity. As a bioisostere of the flat phenyl group, the cyclohexyl substituent has the advantage of being three-dimensional, which potentially offers more contact points with the target protein. This concept has been proven in the discovery of venetoclax (Venclexta). In addition, the cyclohexenyl motif is a metabolically more stable bioisostere of furanose and this concept has been demonstrated by the success of oseltamivir (Tamiflu).

REFERENCES

1. Lin, C.; Kwong, A. D.; Perni, R. B. *Infect. Disord. Drug Targets* **2006**, *6*, 3–16.
2. Kwong, A. D.; Kauffman, R. S.; Hurter, P.; Mueller, P. *Nat. Biotechnol.* **2011**, *29*, 993–1003.
3. Bennett, F.; Huang, Y.; Hendrata, S.; Lovey, R.; Bogen, S. L.; Pan, W.; Guo, Z.; Prongay, A.; Chen, K. X.; Arasappan, A.; et al. *Bioorg. Med. Chem. Lett.* **2010**, *20*, 2617–2621.
4. Ágai-Csongor, E.; Domány, G.; Nógrádi, K.; Galambos, J.; Vágó, I.; Keserü, G. M.; Greiner, I.; Laszlovszky, I.; Gere, A.; Schmidt, E.; et al. *Bioorg. Med. Chem. Lett.* **2012**, *22*, 3437–3440.
5. Bravo, L.; Berrocoso, E.; Mico, J. A. *Exp. Opin. Drug Discov.* **2017**, *12*, 1281–1291.

6. Herdewijn, P.; De Clercq, E. D. *Bioorg. Med. Chem. Lett.* **2001**, *11*, 1591–1597.

7. Kim, C. U.; Lew, W.; Williams, M. A.; Liu, H.; Zhang, L.; Swaminathan, S.; Bischofberger, N.; Chen, M. S.; Mendel, D. B.; Tai, C. Y.; et al. *J. Am. Chem. Soc.* **1997**, *119*, 681–690.

8. Park, C.-M.; Bruncko, M.; Adickes, J.; Bauch, J.; Ding, H.; Kunzer, A.; Marsh, K. C.; Nimmer, P.; Shoemaker, A. R.; Song, X.; et al. *J. Med. Chem.* **2008**, *51*, 6902–6915.

9. Le Manach, C.; Paquet, T.; Wicht, K.; Nchinda, A. T.; Brunschwig, C.; Njoroge, M.; Gibhard, L.; Taylor, D.; Lawrence, N.; Wittlin, S.; et al. *J. Med. Chem.* **2018**, *61*, 9371–9385.

10. Kishimoto, M.; Kawamori, R.; Kamada, T.; Inaba, T. *Drug Metab. Dispos.* **1994**, *22*, 367–370.

11. Carroll, F. I.; Brine, G. A.; Boldt, K. G.; Cone, E. J.; Yousefnejad, D.; Vaupel, D. B.; Buchwald, W. F. *J. Med. Chem.* **1981**, *24*, 1047–1051.

12. Shimokawa, K.; Shibata, N.; Sameshima, T.; Miyamoto, N.; Ujikawa, O.; Nara, H.; Ohoka, N.; Hattori, T.; Cho, N.; Naito, M. *ACS Med. Chem. Lett.* **2017**, *8*, 1042–1047.

13. (a) Xu, Ru.; Li, S.; Paruchova, J.; McBriar, M. D.; Guzik, H.; Palani, A.; Clader, J. W.; Cox, K.; Greenlee, W. J.; Hawes, B. E.; et al. *Bioorg. Med. Chem.* **2006**, *14*, 3285–3299. (b) Su, J.; McKittrick, B. A.; Tang, H.; Burnett, D. A.; Clader, J. W.; Greenlee, W. J.; Hawes, B. E.; O'Neill, K.; Spar, B.; Weig, B.; et al. *Bioorg. Med. Chem.* **2007**, *15*, 5369–5385.

14. Chuo, D. T. H.; Jung, G; Plouvier, B.; Yee, J. G. K. WO2006138673A2 (2006).

15. Ae, N.; Fijiwasa, Y. U.S. Patent US20110263847 A1 (2011).

16. (a) Kawanami, K. WO2010104106A1 (2010). (b) Kawanami, K. WO20122017932A1 (2012).

17. Arasappan, A.; Bennett, F.; Bogen, S. L.; Venkatraman, S.; Blackman, M.; Chen, K. X.; Hendrata, S.; Huang, Y.; Huelgas, R. M.; Nair, L.; et al. *ACS Med. Chem. Lett.* **2010**, *1*, 64–69.

12 Cyclopentanes

To minimize torsional strain, cyclopentane puckers adopt an "envelope" conformation. Cyclopentanes on drugs may serve as either the core scaffold or an appendage to occupy a hydrophobic pocket of the target such as an enzyme or a receptor.

CYCLOPENTANE-CONTAINING DRUGS

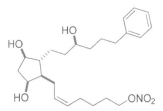

latanoprostene (Vyzulta, **1**)
Bausch and Lomb, 2017
a nitric oxide (NO)-donating
prostaglandin $F_{2\alpha}$ analogue

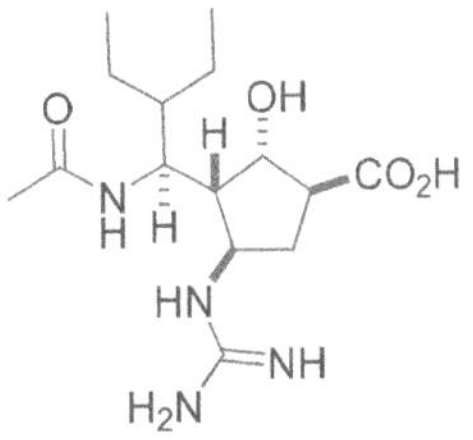

peramivir (Rapivab, **2**)
BioCryst, 2014
neuraminidase inhibitor

All steroid-based drugs have a fused cyclopentane ring as an integral part of the steroidal architecture. The majority of prostaglandins have a cyclopentane core structure as the consequence of oxidation of arachidonic acid by cyclooxygenases (COXs). The latest example of prostaglandin analogs is Bausch and Lomb's latanoprostene (Vyzulta, **1**) for treating open-angle glaucoma or ocular hypertension. It is a nitric oxide (NO)-donating prostaglandin $F_{2\alpha}$ analog. BioCryst's peramivir (Rapivab, **2**) is a neuraminidase inhibitor to treat influenza. Abbvie/Enanta's glecaprevir (**3**, with pibrentasvir, Mavyret), with a di-substituted cyclopentane motif, is a hepatitis C virus nonstructural protein (HCV NS)-3/4A protease inhibitor. Similarly, simeprevir (Olysio, **4**), Janssen/Medivir's HCV NS3/4A protease inhibitor, has a tri-substituted cyclopentane scaffold.

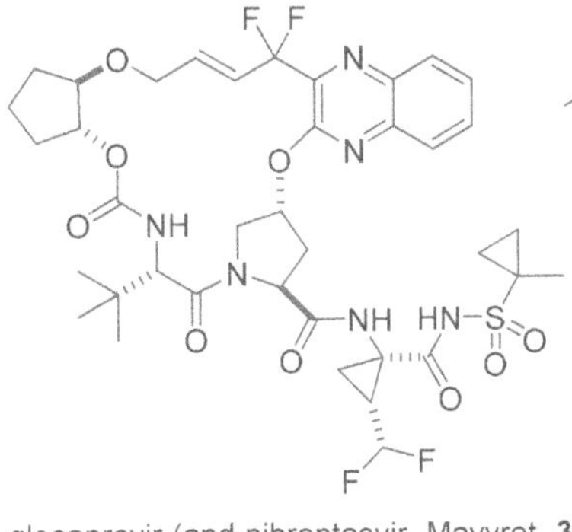

glecaprevir (and pibrentasvir, Mavyret, **3**)
Abbvie/Enanta, 2017
HCV NS 3/4A protease inhibitor

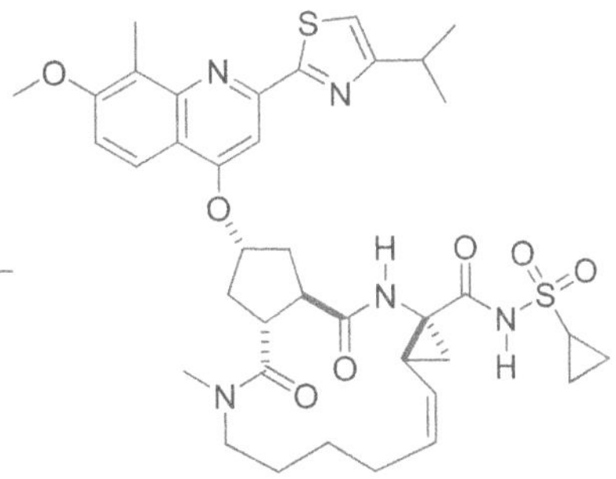

simeprevir (Olysio, **4**)
Janssen/Medivir, 2013
HCV NS3/4A protease inhibitor

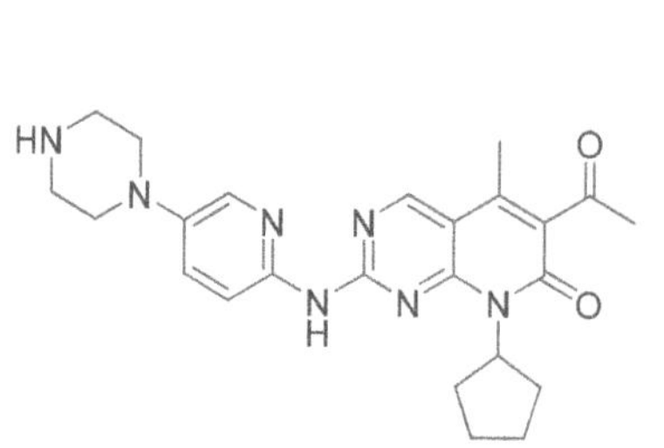

palbociclib (Ibrance, **5**)
Pfizer, 2015
CDK4/6 inhibitor

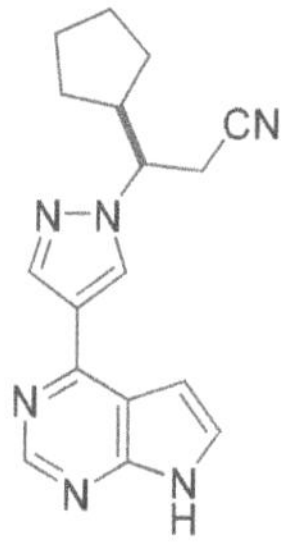

ruxolitinib (Jakafi, **6**)
Incyte, 2011
JAK1/2 inhibior

Pfizer's palbociclib (Ibrance, **5**) is a cyclin-dependent kinase (CDK) 4/6 inhibitor approved by the FDA in 2015 for treating ER-positive and HER2-negative breast cancer. On the other hand, Incyte's ruxolitinib (Jakafi, **6**), a Janus kinase (JAK) 1/2 inhibitor, was approved in 2011 for the treatment of bone marrow cancer. AstraZeneca's ticagrelor (Brilinta, **7**), a synthetically challenging P2Y$_{12}$ platelet inhibitor as an anticoagulant, has a tetra-substituted cyclopentane as its sidechain. Cyclopentanes occasionally show up on drugs in the form of spirocyclic bicycles. In addition to BMS's buspirone (Buspar, not shown), Sanofi/BMS's irbesartan (Avapro, **8**), an angiotensin II receptor blocker (ARB) for treating hypertension, also has a spirocyclic cyclopentane.

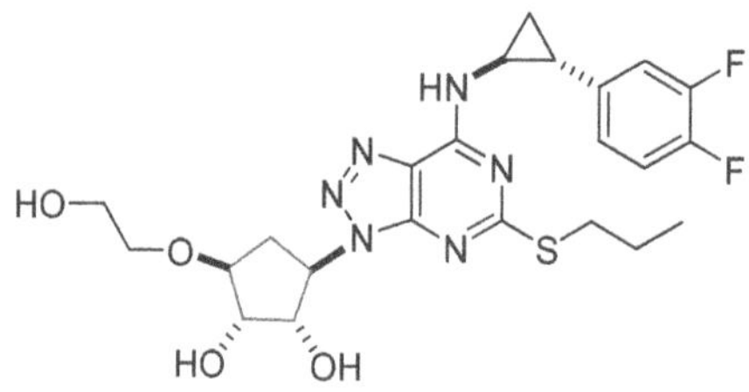

ticagrelor (Brilinta, **7**)
AstraZeneca, 2010
P2Y$_{12}$ platelet inhibitor

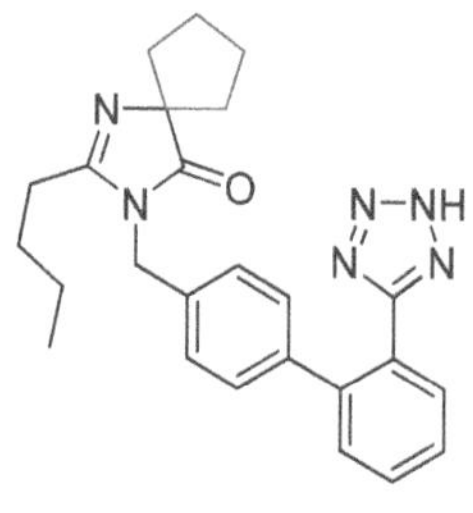

irbesartan (Avapro, **8**)
Sanofi/BMS, 2001
angiotensin II receptor antagonist

CYCLOPENTANES IN DRUG DISCOVERY

In comparison to the ubiquitous cyclohexanes, there are fewer examples of cyclopentanes in medicines. However, nothing succeeds like success. When it works, cyclopentane may be the optimal fragment on a drug.

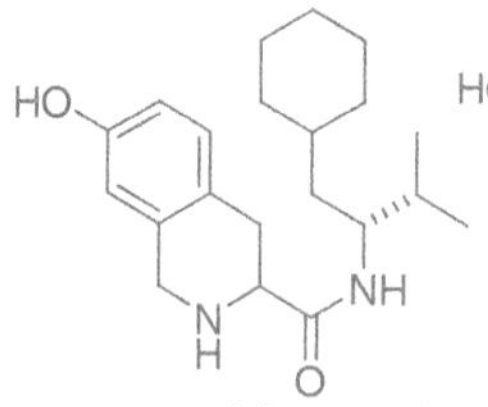

cyclohexane **9**
$Ke(\kappa) = 0.14$ nM
$\mu/\kappa = 1,730$
$\delta/\kappa = 24,570$

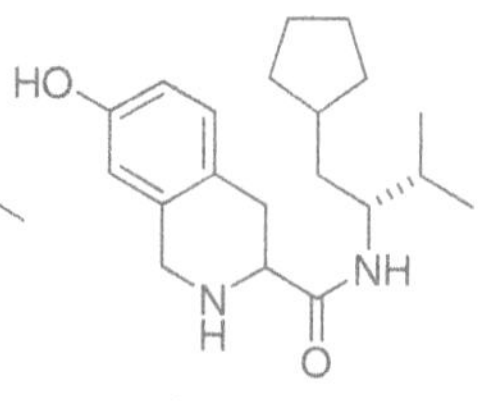

cyclopentane **10**
$Ke(\kappa) = 0.058$ nM
$\mu/\kappa = 5,900$
$\delta/\kappa = 27,000$

In drug discovery, sizes matter. In pursuit of kappa (κ) opioid receptor antagonists as potential pharmacotherapies for treating depression, anxiety, and substance abuse, cyclopentane proved to be the right size. The lead compound cyclohexane **9** was a potent κ opioid receptor antagonist ($K_e = 0.14$ nM) and selective against mu (μ) and delta (δ) subtypes. Changing the six-membered ring on **9** to cyclopentane **10**, cyclobutane **11**, and cycloheptane **12** revealed that cyclopentane **10** had the optimal profile. Not only was cyclopentane **10** the most potent ($K_e = 0.048$ nM), it was more selective against mu (μ) and delta (δ) subtypes as well.[1]

cyclobutane **11**
$Ke(\kappa) = 2.61$ nM
$\mu/\kappa = 102$
$\delta/\kappa = 1,245$

cycloheptane **12**
$Ke(\kappa) = 0.20$ nM
$\mu/\kappa = 292$
$\delta/\kappa = 3,760$

adenosine (**13**)
$t_{1/2} < 2$ min (iv)
%F = 0%

aristeromycin (**14**)
$t_{1/2} = 23$ min
CL = 7.9 ml/min
$V_{ss} = 260$ mL

In the field of nucleosides, *carbocyclic nucleosides (carbanucleosides)* are important. The substitution of the endocyclic oxygen atom by a CH_2 moiety increases not only the chemical stability of the *N*-glycosidic bond but also makes these derivatives metabolically resistant to the action of several enzymes such as pyrimidine and purine nucleoside phosphorylases.[2] For instance, aristeromycin (**14**) is a direct carbon analog of adenosine (**13**). The *N*-glycosidic bond between ribose and adenine (A) on adenosine (**13**) is subjected to metabolism and cleavage by both phosphorylases and hydrolases. In contrast, the carbocyclic isostere aristeromycin (**14**) is unaffected by metabolism and cleavage by those two classes of enzymes but still maintains similar biological activities.[3]

Building upon aristeromycin (**14**)'s improved pharmacokinetic profile, two cyclopentane-containing carbanucleosides gained FDA's approvals. Abacavir (Ziagen, **15**) as an HIV reverse-transcriptase inhibitor, initially discovered by Vince and later developed by GSK, has been on the market since 1998 to treat AIDS.[4] BMS's entecavir (Baraclude, **16**) for the treatment of hepatitis B, has a remarkably improved bioavailability in comparison to the furanose counterparts.[5]

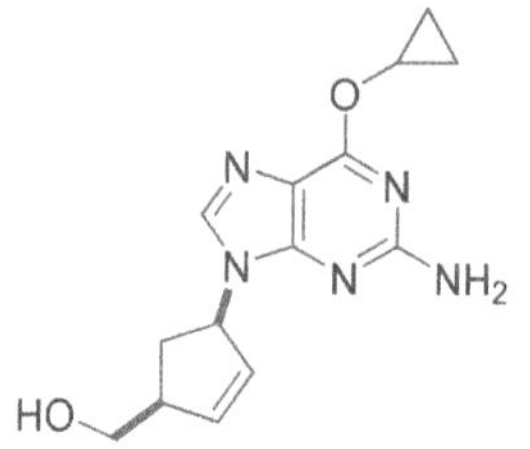

abacavir (Ziagen, **15**)
GSK, 1998
carbocyclic nucleoside
$t_{1/2}$ = 1.5 h
%F = 83%
PPB = 50%

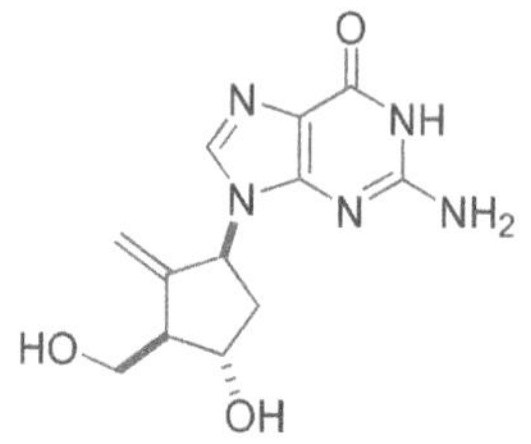

entecavir (Baraclude, **16**)
BMS, 2005
carbocyclic nucleoside
$t_{1/2}$ = 4–9 h
%F = 70%
PPB = 13%

Influenza neuraminidase inhibitor zanamivir (Relenza, **17**) is not bioavailable and has to be given as an oral inhalation. Switching its pyranose scaffold to cyclohexene, among other optimizations, gave rise to oseltamivir (Tamiflu, **18**), which is orally bioavailable. As an ester prodrug, it has a bioavailability of 75% for the corresponding carboxylic acid. X-Ray crystal structures of complexes of the neuraminidase enzyme and its inhibitors indicated that potent inhibition of the enzyme is determined by the relative positions of the interacting inhibitor substituents (carboxylate, glycerol, acetamido, and hydroxyl) rather than by the absolute position of the central ring. To that end, BioCryst designed peramivir (**2**) with a cyclopentane core scaffold. It has several common functionalities as both zanamivir (**17**) and oseltamivir (**18**) including carboxylate, glycerol, acetamido, and hydroxyl groups.[6]

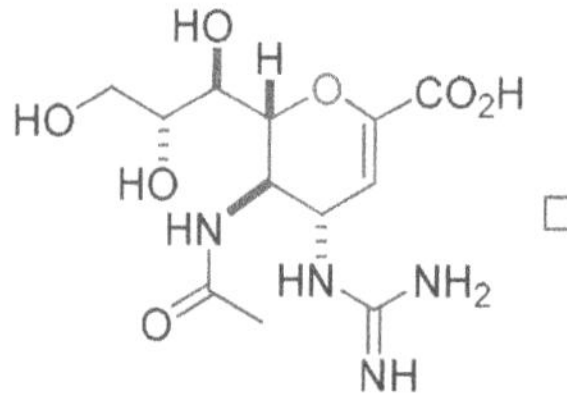

zanamivir (Relenza, **17**)
$t_{1/2}$ = 2.6–5.1 h
%F = 2%

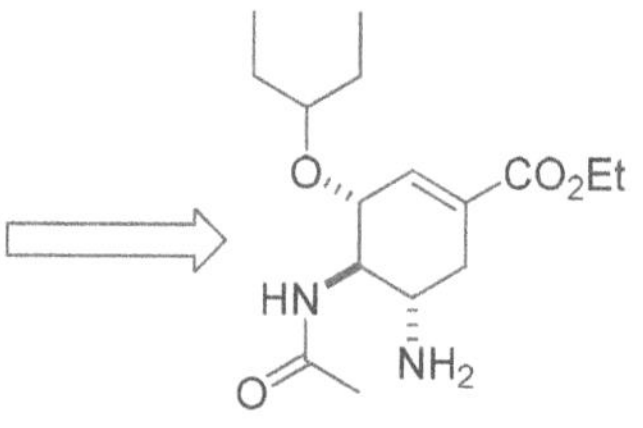

oseltamivir (Tamiflu, **18**)
$t_{1/2}$ = 6–10 h (iv)
%F = 75%

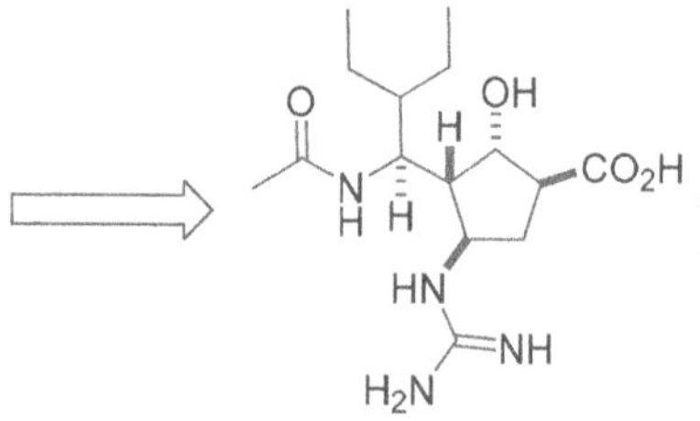

peramivir (Rapivab, **2**)
$t_{1/2}$ = 8–21 h (iv)
%F, low

SYNTHESIS OF SOME CYCLOPENTANE-CONTAINING DRUGS

Pfizer's process route for preparing palbociclib (**5**) commenced with an S_NAr displacement of 5-bromo-2,4-dichloropyrimidine (**19**) with cyclopentylamine. The resultant adduct **20** was then converted to palbociclib (**5**) in additional three steps.[7]

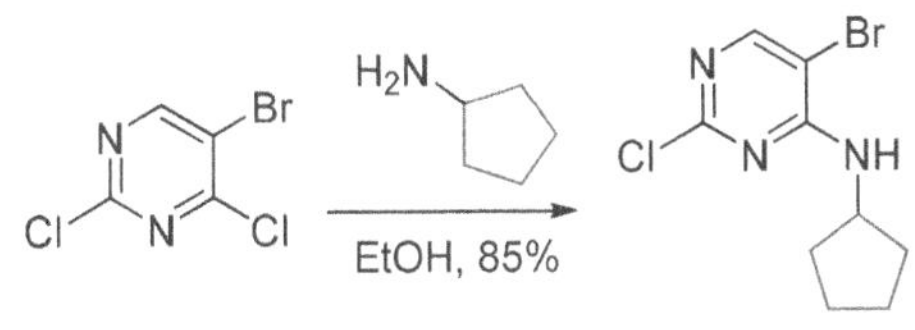

pyrimidine **19**   adduct **20**

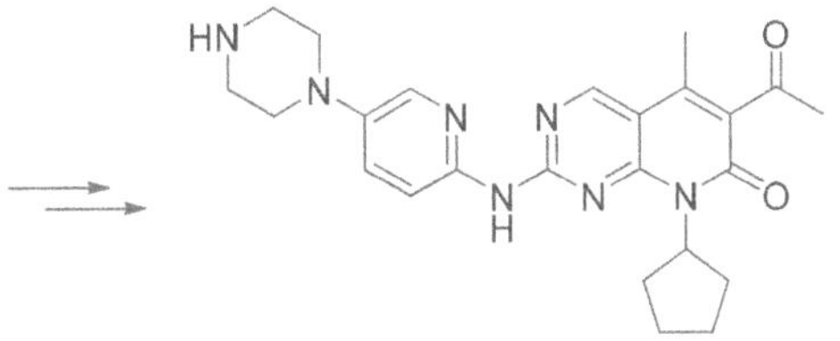

palbociclib (Ibrance, **5**)

Synthesis of tetra-substituted cyclopentane with four contiguous chiral centers on BMS's ente-cavir (**16**) is not trivial. An advanced intermediate cyclopentyl epoxide **21** was prepared in five steps from sodium cyclopentadienide. An S_N2 displacement of **21** with purine-amine **22** assembled adduct **23**, which was transformed to entecavir (**16**) in three additional steps.[5]

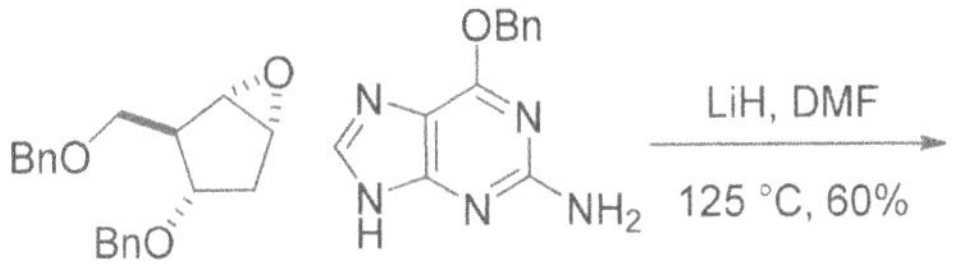

21   **22**

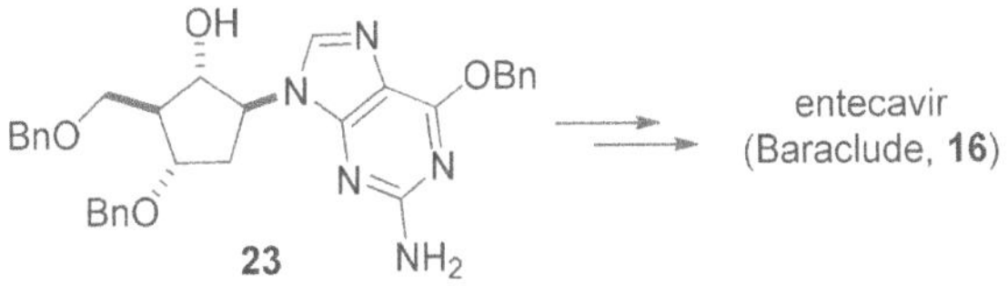

entecavir
(Baraclude, **16**)

23   NH$_2$

HCV NS3/4A protease inhibitor glecaprevir (**3**) has a di-substituted cyclopentane moiety. Its synthesis started with racemic cyclopentane-1,2-diyl diacetate [(±)-**24**], simply prepared from the corresponding diol. Chiral resolution using amino lipase provided (1*R*,2*R*)-2-hydroxycyclopentyl acetate (**25**). The cyclopentane fragment **26** was obtained from allylation of **25** and hydrolysis of the acetate. Coupling between alcohol **26** and L-*tert*-butyl-leucine led to carbamate **27**, which was converted to glecaprevir (**3**) after five additional steps.[8]

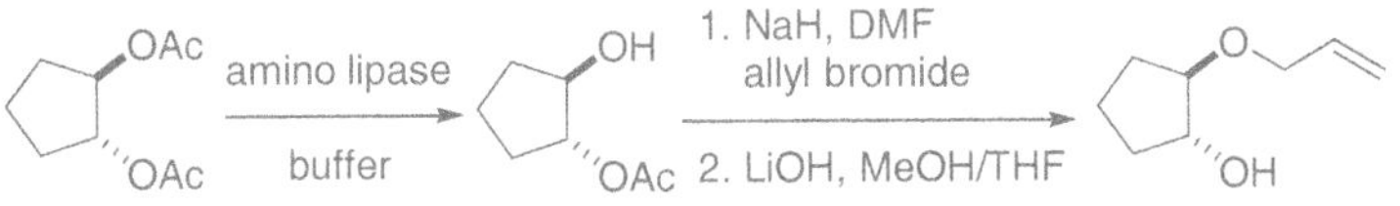

(±)-**24**   **25**   **26**

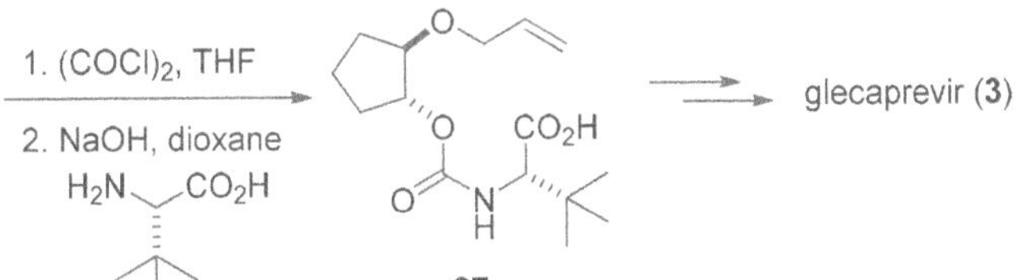

glecaprevir (**3**)

27

BMS's short process route to irbesartan (**8**) began with the reaction of 1-amino-cyclopentane-carboxylic acid ester **28** with ethyl pentanimidate (**29**) in the presence of acetic acid in refluxing xylene to assemble dihydroimidazolone **30**. An S_N2 alkylation of **30** with phenylbenzyl bromide **31** in the presence of sodium hydride in DMF gave **32**. Finally, the synthesis of irbesartan (**8**) was completed by the tetrazole formation from the reaction of the nitrile group of **32** with tributyltin azide in refluxing xylene.[9]

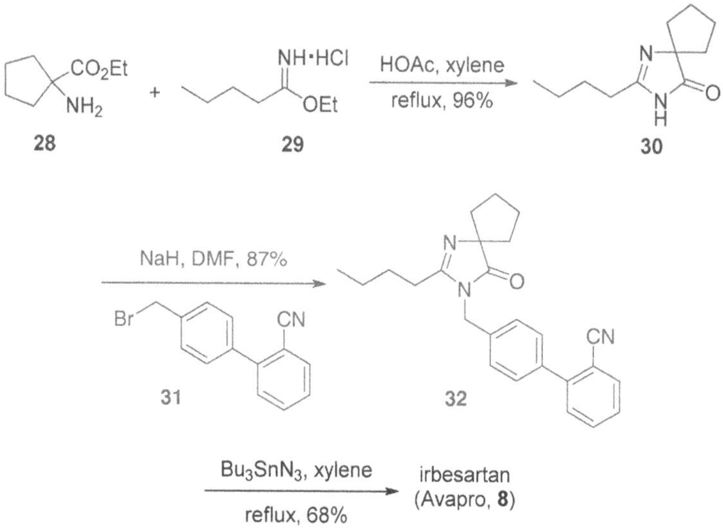

In summary, cyclopentanes may serve as either the core scaffold or an appendage on drugs to occupy a target's hydrophobic pocket. One of its more important utility is serving as a bioisostere of furanose to prepare carbanucleosides to improve drugs' pharmacokinetic profiles.

REFERENCES

1. Kormos, C. M.; Ondachi, P. W.; Runyon, S. P.; Thomas, J. B.; Mascarella, S. W.; Decker, A. M.; Navarro, H. A.; Fennell, T. R.; Snyder, R. W.; Carroll, F. I. *J. Med. Chem.* **2018**, *61*, 7546–7559.
2. Boutureira, O.; Matheu, M. I.; Diaz, Y.; Castillón, S. *Chem. Soc. Rev.* **2013**, *42*, 5056–5072.
3. Schaddelee, M. P; Dejongh, J.; Collins, S. D.; de Boer, A. G; Ijzerman, A. P.; Danhof, M. *Eur. J. Pharmacol.* **2004**, *504*, 7–15.
4. Vince, R. *Chemtracts* **2008**, *21*, 127–134.
5. Ruediger, E.; Martel, A.; Meanwell, N.; Solomon, C.; Turmel, B. *Tetrahedron Lett.* **2004**, *45*, 739–742.
6. Chand, P.; Kotian, P. L.; Dehghani, A.; El-Kattan, Y.; Lin, T.-H.; Hutchison, T. L.; Babu, Y. S.; Bantia, S.; Elliott, A. J.; Montgomery, J. A. *J. Med. Chem.* **2001**, *44*, 4379–4392.
7. Duan, S.; Place, D.; Perfect, H. H.; Ide, N. D.; Maloney, M.; Sutherland, K.; Price Wiglesworth, K. E.; Wang, K.; Olivier, M.; Kong, F.; et al. *Org. Process Res. Dev.* **2016**, *20*, 1191–1202.
8. Or, Y. S.; Ma, J.; Wang, G.; Wang, B. (Enanta), WO2012040167 (2012).
9. Miranda, E. I.; Vlaar, Co.; Zhu, J. (BMS) US7211676 (2007).

13 Cyclopropane Derivatives as Metabolically More Robust Bioisosteres for Linear Alkyl Substituents

Cyclopropane is unique. On the one hand, all three of its carbons are aliphatic with sp^3 hybridization. On the other hand, the C–H bond closely resembles an aromatic C–H bond because the external orbitals are more similar to sp^2 orbitals due to the ring strain. Some people call it a "fat double bond". It is widely known that cyclopropane is more resistant to CYP450 metabolic oxidation than its linear brethren, n-propyl, n-ethyl, or methyl group.

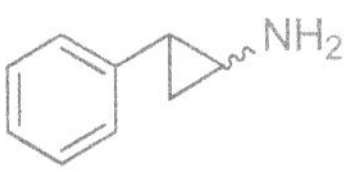

tranylcypromine (Parnate, **1**)
SK&F, 1961
monoamine oxidase inhibitor

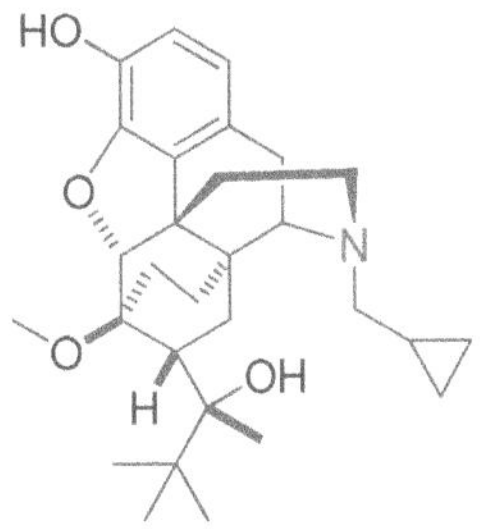

buprenorphine (Subutex, **2**)
Reckitt & Colman, 1981
opioid

Dozens of FDA-approved drugs contain a cyclopropyl moiety, spanning across all therapeutic areas. Conspicuous examples include tranylcypromine (Parnate, **1**), buprenorphine (Subutex, **2**), efavirenz (Sustiva, **3**), prasugrel (Effient, **4**), cabozantinib (Cometriq, **5**), trametinib (Mekinist, **6**), lenvatinib (Lenvima, **7**), and olaparib (Lynparza, **8**).

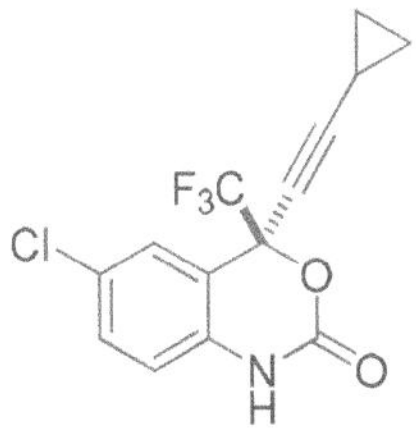

efavirenz (Sustiva, **3**)
BMS/Merck, 1998
reverse transcriptase inhibitor

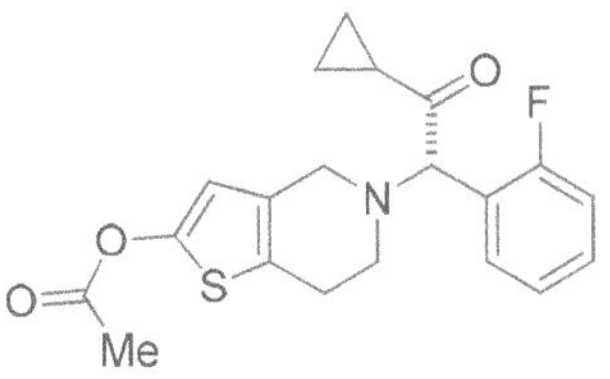

prasugrel (Effient, **4**)
Daiichi Sankyo/Lilly, 2009
P2Y12 inhibitor

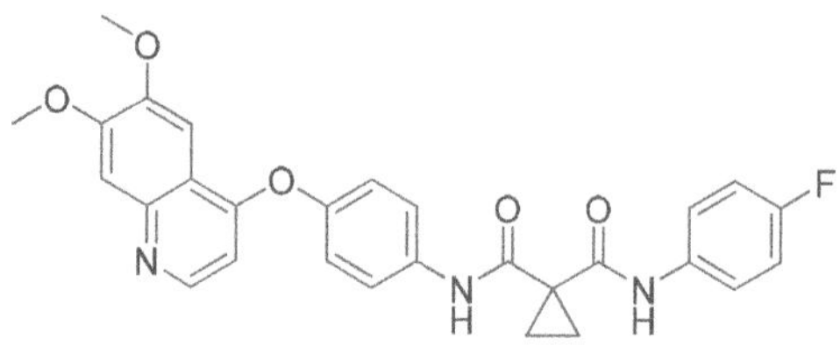

cabozantinib (Cometriq, **5**)
Exelixis, 2012
VEGFR2 and C-Met inhibitor

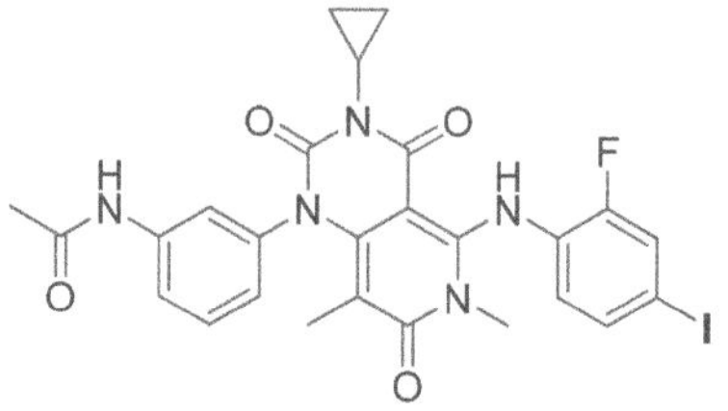

trametinib (Mekinist, **6**)
GSK, 2013
MEK1/2 inhibitor

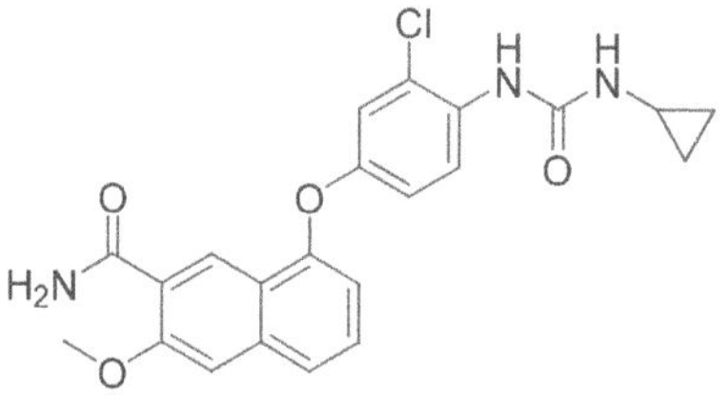

lenvatinib (Lenvima, **7**)
Eisai, 2015
VEGFR and FGFR inhibitor

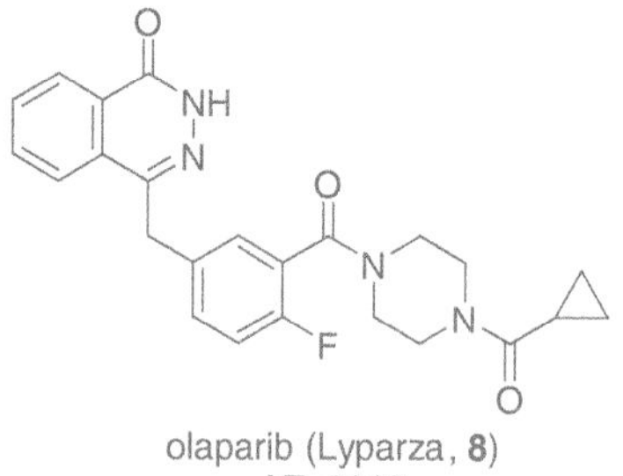

olaparib (Lyparza, **8**)
AZ, 2018
PARP1 inhibitor

The cyclopropylmethyl group was employed as a metabolically more stable isostere of a methyl group in the context of beta-blockers (β_1 adrenoceptor antagonists). Metoprolol (**9**) has a relatively short duration of action (half-life, 3 hours) and an elevated "first pass" hepatic deactivation that is responsible for its low bioavailability (F%, 38%). CYP450 2D6 is known to be the major isoform to carry out the metabolism of the methyl group. Cyclopropylmethyl as an isosteric replacement for the methyl group on metoprolol (**9**) gave rise to betaxolol (Kerlon, **10**). It was found to exhibit an appropriate preclinical pharmacological and human pharmacokinetics with elevated oral bioavailability (F%, 89%) and prolonged half-life (14–22 hours) for the treatment of chronic cardiovascular diseases such as hypertension and angina.[1]

metoprolol (**9**, log *P* 2.0)
$t_{1/2}$ 3 h, *F* 38%

betaxolol (Kerlon, **10**, log *P* 2.8)
$t_{1/2}$ 14–22 h, *F* 89%

BI-RH-414 (**11**)
nevirapine (Viramune, **12**)

	BI-RH-414 (**11**)	nevirapine (Viramune, **12**)
HIV-1 RT enzymatic assay,	IC_{50} = 35 nM	IC_{50} = 84 nM
cellular culture assay,	IC_{50} = 30 nM	IC_{50} = 40 nM
solubility,	0.17 mg/mL	0.10 mg/mL
bioavailability	+	++
toxicology	not significant	not significant

In choosing a drug candidate, a cyclopropyl derivative may be advantageous in terms of pharmacokinetics over its linear alkyl counterparts. Boehringer Ingelheim's tricyclic diazepinone **11** is a potent non-nucleoside reverse transcriptase inhibitor (NNRTI). Another potential drug candidate was the cyclopropyl analog **12**. Not only was ethyl derivative **11** more potent than **12** in both enzymatic and cellular assays, but also it was more soluble. Nonetheless, the cyclopropyl analog **12** was nominated as the drug candidate because it was more bioavailable due to the fact that cyclopropane was more resistant to metabolism while the ethyl group was prone to undergo dealkylation.[2] Nevirapine (Viramune, **12**) was the first NNRTI approved by the FDA in 1996 for the treatment of AIDS patients infected by the HIV. NNRTIs represented a giant step forward, thanks to their improved efficacy and PK profiles over initial nucleoside reverse transcriptase inhibitors (NRTIs) that were invariably associated with high toxicity and low bioavailability.

The cyclopropane carboxylate motif facilitated to overcome paclitaxel (Taxol, **13**)'s drug-resistant issue. Because paclitaxel (**13**) is a Pgp substrate, the Ojima group carried out extensive structure–activity relationship (SAR) investigations to address the problem. A new taxoid SB-T-1214 (**14**) is not a Pgp substrate and is 25-fold more potent than paclitaxel (**13**). The exceptional activity may be ascribed to an effective inhibition of Pgp binding replacing acetate C-10 with the cyclopropane carboxylate. The magic of success stems from the astute observation that modifications at C-10 are tolerated for the activity against normal cancer cell lines, but the activity against a drug-resistant human breast cancer cell line expressing multidrug resistance (MDR) phenotype MCF7-R is highly dependent on the structure of the C-10 modifier. The other variation on **13** is the replacement of the original C-3′ phenyl group with the isobutenyl substituent.[3]

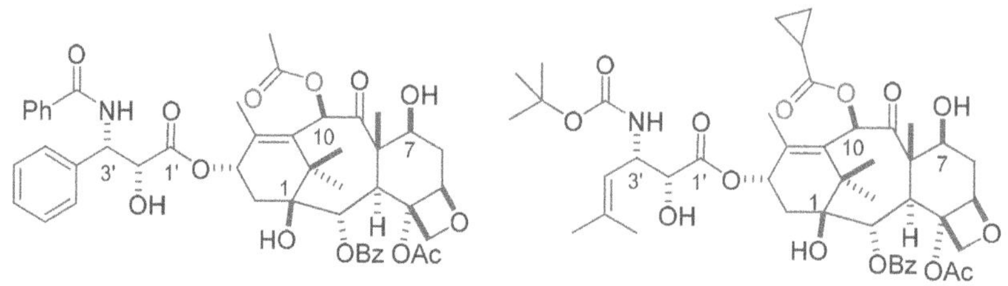

paclitaxel (Taxol, **13**)
potent
Pgp substrate

SB-T-1214 (Taxol, **14**)
25X more potent
Not a Pgp substrate

Cyclopropane may serve as a bioisostere of a benzene ring to lower the lipophilicity and elevate the bioavailability. Biphenyl **15** was potent as a factor Xa (FXa) inhibitor with a K_i value of 0.3 nM. But the biphenyl motif contributed to the molecule's lipophilicity (log P, 5.96). When one of the phenyl rings was replaced with a cyclopropyl ring, the resulting compound **16** had markedly improved potency (K_i value: 0.021 nM) compared to the biphenyl **15**, a general phenomenon across several paired analogs. Moreover, cyclopropyl **16** also exhibited lower lipophilicity (log P, 4.99). X-ray co-crystal of **16** with FXa confirmed a perpendicular conformation, increased potency appeared to be a function of optimized hydrophobic interactions with S_4 and slightly reduced strain in the bound geometry.[4]

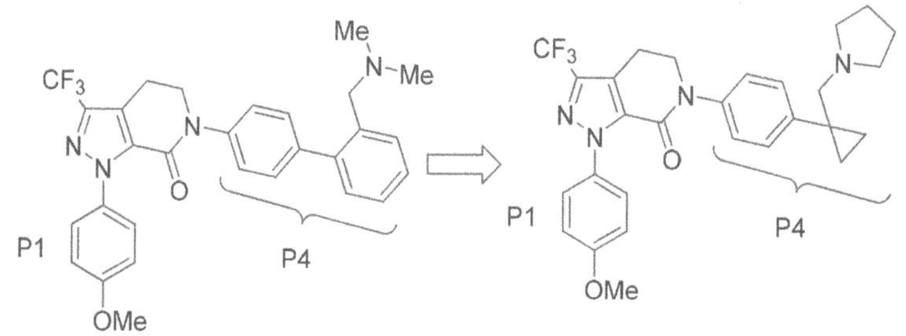

biphenyl **15**
FXa: K_i = 0.3 nM
log P: 5.94
cLog P: 5.095

cyclopropyl **16**
FXa: K_i = 0.021 nM
log P: 4.99
cLog P: 4.08

In summary, cyclopropane derivatives play an important role in medicinal chemistry. Their most popular utility is to serve as metabolically more stable isosteres of linear aliphatic substituents.

REFERENCES

1. Manoury, P. M.; Binet, J. L.; Rousseau, J.; Lefevre-Borg, F. M.; Cavero, I. G. *J. Med. Chem.* **1987**, *30*, 1003–1011.
2. Grozinger, K.; Proudfoot, J.; Hargrave, K. *Discovery and Development of Nevirapine*, In Chorghade, M. S. ed.; *Drug Discovery and Development, Vol. 1: Drug Discovery*, Wiley: Weinheim, 2006, pp 353–363.
3. Ojima, I.; Slater, J. C.; Michaud, E.; Kuduk, S. D.; Bounaud, P.-Y.; Vrignaud, P.; Bissery, M.-C.; Veith, J. M.; Pera, P.; Bernacki, R. J. *J. Med. Chem.* **1996**, *39*, 3889–3896.
4. Qiao, J. X.; Cheney, D. L.; Alexander, R. S.; Smallwood, A. M.; King, S. R.; He, K.; Rendina, A. R.; Luettgen, J. M.; Knabb, R. M.; Wexler, R. R.; et al. *Bioorg. Med. Chem. Lett.* **2008**, *18*, 4118–4123.

14 Deazapurines

Four deazapurines exist: 1-deazapurine, 3-deazapurine, 7-deazapurine, and 9-deazapurine.

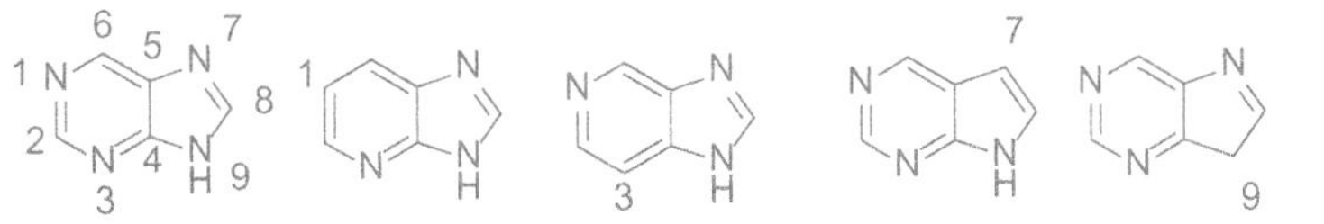

purine   1-deazapurine   3-deazapurine   7-deazapurine   9-deazapurine

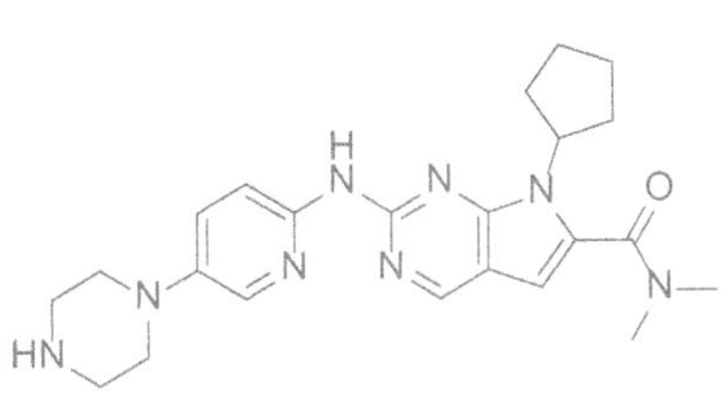

ribociclib (Kisqali, 1)
Novartis, 2017
CDK4/6 inhibitor

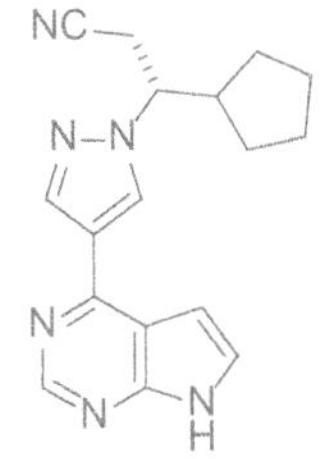

ruxolitinib (Jakafi, 2)
Incyte, 2011
bone marrow cancer

Deazapurines are *bona fide* bioisosteres of purines and exist in several marketed drugs. Novartis's ribociclib (Kisqali, **1**), approved by the FDA in 2017 for treating breast cancer, is a cyclin-dependent kinase (CDK) 4/6 inhibitor. Incyte's ruxolitinib (Jakafi, **2**), a Janus kinase (JAK) 1/2 inhibitor, was approved in 2011 for the treatment of bone marrow cancer. Lilly's pemetrexed (Alimta, **3**), a folate analog metabolic inhibitor initially discovered by Ed Taylor, was approved in 2004 for treating pleural mesothelioma. BioCryst's forodesine (Mundesine, **4**), a transition-state analog inhibitor of purine nucleoside phosphorylase, won the FDA approval in 2017 for treating leukemia. Notably, the four representative deazapurine-containing drugs are all 7-deazapurines.

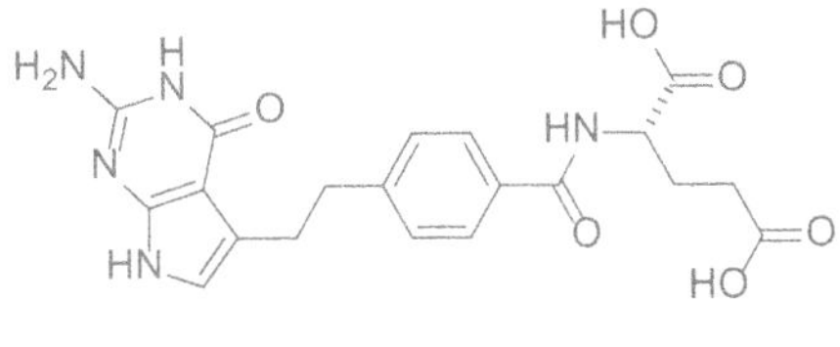

pemetrexed (Alimta, 3)
Lilly, 2004
folate analogue metabolic inhibitor
for pleural mesothelioma

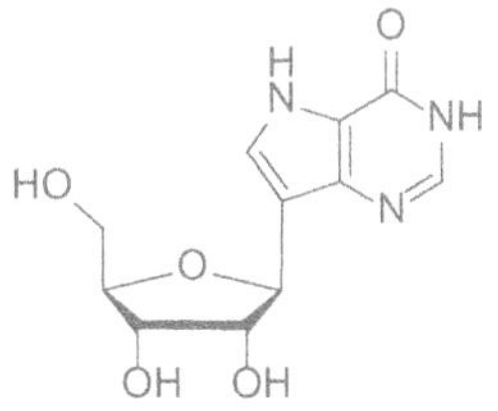

fododesine (Mundesine, 4)
BioCryst Pharmaceuticals, 2017
transition-state analog inhibitor
of purine nucleoside phosphorylase

One of the Incyte's syntheses of ruxolitinib (Jakafi, **2**) employed 6-chloro-7-deazapurine (**8**) as a starting material. As shown below, (*R*)-4-bromopyrazole **5** underwent a Suzuki–Miyaura coupling with bis(pinacolato)diboron (**6**) to produce pyrazole-boronate **7** *in situ*. Subsequently, pyrazole-boronate **7**, in turn, underwent another Suzuki coupling with 6-chloro-7-deazapurine (**8**) to deliver ruxolitinib (Jakafi, **2**) in 64% yield for two steps.[1]

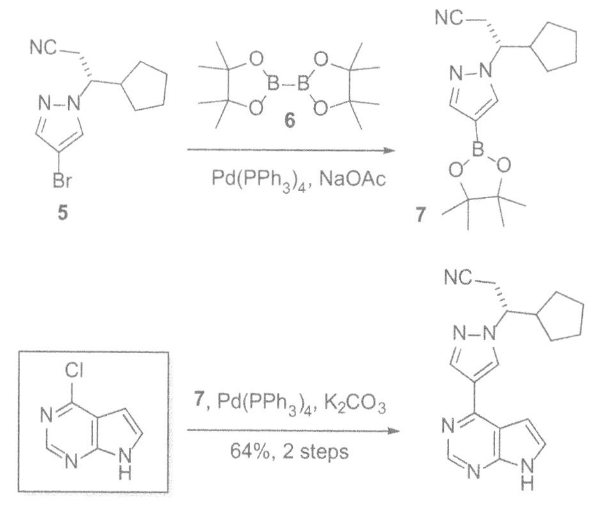

8   ruxolitinib (Jakafi, **2**)

Many 7-deazapurines have found utility in drug discovery as purine's isosteres.

7-Deazapurine (pyrrolo[2,3-*d*]pyrimidine) nucleosides are important analogs of biogenic purine nucleosides with diverse biological activities. Replacement of the N_7 atom with a carbon atom makes the five-membered ring more electron-rich and brings a possibility of attaching additional substituents at the C_7 position. This often leads to derivatives with increased base-pairing in DNA or RNA or better binding to enzymes.[2] Not surprisingly, many 7-deazapurine derivatives found utility as antiviral and anticancer drugs.

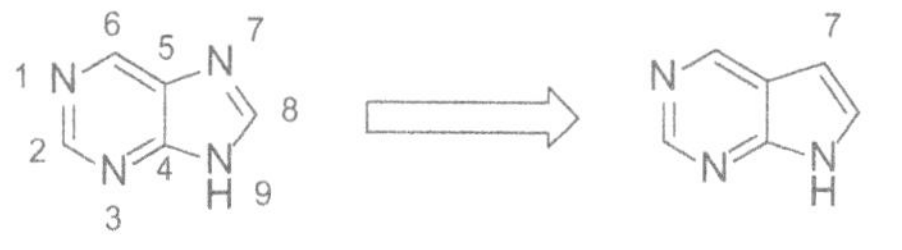

purine   7-deazapurine

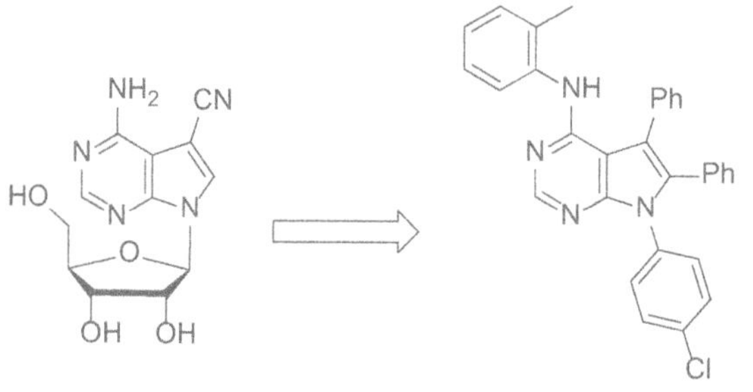

toyocamycin (**9**)   7-deazapurine **10**

7-Deazapurine **10** was derived using naturally occurring toyocamycin (**9**) as a starting point in an effort to search for non-nucleoside hepatitis C virus (HCV) NS5B polymerase inhibitors. The non-toxic doses of 7-deazapurine **10** on Huh 7.5 cell line were determined and its antiviral activity against the HCVcc genotype was examined. The percent of reduction for the non-toxic dose of **10** was 90%.[3]

A hit-to-lead (H2L) exercise led to 7-deazapurines as potent, selective, and orally bioavailable TNNI3K inhibitors.

From GSK's screening effort, purine **11** was identified as a good hit as a type I inhibitor of troponin I-interacting kinase (TNNI3K). But it was only moderately potent and had an extremely poor aqueous solubility. An X-ray structure of hit **11** bound to the ATP-binding site of TNNI3K confirmed its type I binding mode and was used to rationalize the structure–activity relationship (SAR). Identification of the 7-deazapurine heterocycle as a superior template (vs. purine) and its elaboration by the introduction of C_4-benzenesulfonamide and 7-deazapurine substituents produced compounds with substantial improvements in potency and general kinase selectivity. 7-Deazapurine **12** has increased aqueous solubility (50 µM) and offers an improved rat pharmacokinetics profile (Cl=68 mL/min/kg, F=34%, >10-fold increase in poDNAUC, DNAUC, dose-normalized area under the curve). It has properties suitable for use in *in vitro* and *in vivo* experiments aimed at elucidating the role of TNNI3K in cardiac biology and serves as leads for developing novel heart failure medicines.[4]

hit **11**

TNNI3K IC_{50} = 500 nM
calcd pKa BH^+ = 3.2
insoluble in IV dosing vehicle
poDNAUC = 0.0 h-kg/L
> 10X selectivity against
96% (195/203) of kinome
B-Raf V600E IC_{50} = 32 nM

lead **12**

TNNI3K IC_{50} = 80 nM
calcd pKa BH^+ = 5.2

Cl = 68 mL/min/kg, F = 34%
poDNAUC = 0.08 h-kg/L
> 10X selectivity against
97% (180/185) of kinome
B-Raf V600E IC_{50} = 50 nM

A class of *9-deazapurines* have been synthesized and investigated for their inhibitory power against three ATP-binding cassette (ABC) transport proteins such as P-gp, MRP1, and BCRP.

P-gp (permeability glycoprotein) is the most prevalent drug efflux transporter. In addition, multidrug resistance-associated protein 1 (MRP1) and breast cancer resistance protein (BCRP) also contribute to multidrug resistance. The development of potent and selective inhibitors of one of the three transport proteins was the focus of medicinal chemistry in the last four decades, but triple inhibitors are rare. Systemic SAR effort led to such a 9-deazapurine (7H-pyrrolo[3,2-d]pyrimidine) triple inhibitor **13**, which was active in a very low micromolar concentration range against all three transporters and restored sensitivity toward related cells. More important, it was a non-competitive inhibitor of calcein AM (P-gp), daunorubicin (MRP1), and pheophorbide A (BCRP) transport.[5]

Pgp (calcein AM) IC_{50}, 1.46 µM

MRP1 (daunorbicin) IC_{50}, 0.50 µM

BCRP (pheophorbide A) IC_{50}, 1.69 µM

9-deazapurine triple inhibitor **13**

Whereas 7-deazapurines occupy the most prominent position as purine isosteres, *1-deazapurines* (3*H*-imidazo[4,5-*b*]pyridines) and *3-deazapurines* (1*H*-imidazo[4,5-*c*]pyridines) also find applications in medicinal chemistry. For instance, 1-deazapurine **14** was found to be a potent compound against carbonic anhydrase-II, α-glucosidase, and β-glucuronidase.[6] Meanwhile, a novel halogenated 3-deazapurine derivative of ascorbic acid **15** showed both antitumor and antiviral activities.[7]

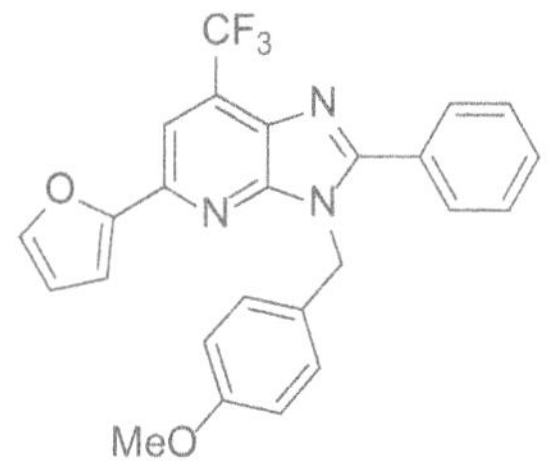

1-deazapurine **14** 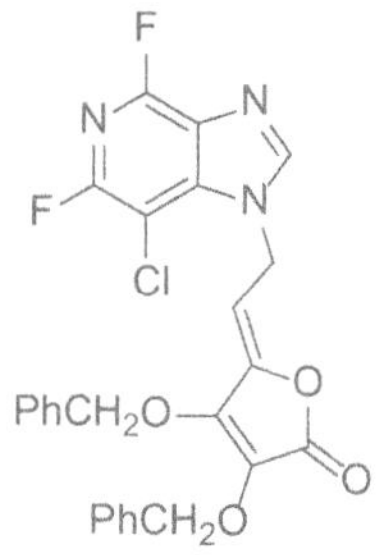   3-deazapurine **15**

In conclusion, the deazapurines have been extensively employed as purine's bioisosteres. Although 7-deazapurines are most frequently encountered including four drugs **1–4** on the market, 1-deazapurines, 3-deazapurines, and 9-deazapurines all find utility in discovering new drugs.

REFERENCES

1. Lin, Q.; Meloni, D.; Pan, Y.; Xia, M.; Rodgers, J.; Shepard, S.; Li, M.; Galya, L.; Metcalf, B.; Yue, T.-Y.; et al. *Org. Lett.* **2009**, *11*, 1999–2002.
2. Perlíková, P.; Hocek, M. *Med. Res. Rev.* **2017**, *37*, 1429–1460.
3. Mohamed, M. S.; Sayed, A. I.; Khedr, M. A.; Soror, S. H. *Bioorg. Med. Chem. Lett.* **2016**, *24*, 2146–2157.
4. Lawhorn, B. G.; Philp, J.; Zhao, Y.; Louer, C.; Hammond, M.; Cheung, M.; Fries, H.; Graves, A. P.; Shewchuk, L.; Wang, L.; et al. *J. Med. Chem.* **2015**, *58*, 7431–7448.
5. Stefan, K.; Schmitt, S. M.; Wiese, M. *J. Med. Chem.* **2017**, *60*, 8758–8780.
6. Ali, I; Khan, A; Hussain, A.; Umar Farooq, U.; Ismail, M.; Hyder, V.; Ahmad, V. U.; Iaroshenko, V. O.; Hussain, H.; Langer, P. *Med. Chem. Res.* **2016**, *25*, 2599–2606.
7. Stipković Babić, M.; Makuc, D.; Plavec, J.; Martinović, T.; Kraljević Pavelić, S.; Pavelić, K.; Snoeck, R.; Andrei, G.; Schols, D.; Wittine, K.; et al. *Eur. J. Med. Chem.* 2015, *102*, 288–302.

15 Furopyridines

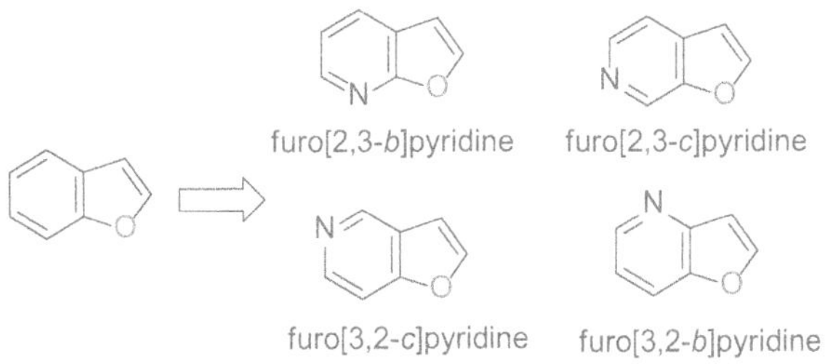

furo[2,3-b]pyridine   furo[2,3-c]pyridine

benzofuran ⟹

furo[3,2-c]pyridine   furo[3,2-b]pyridine

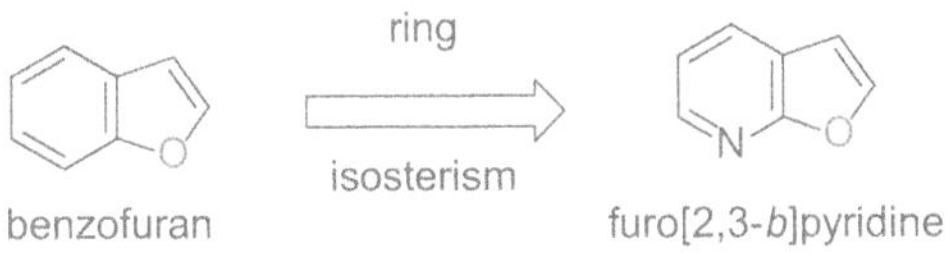

benzofuran   ring isosterism ⟹   furo[2,3-b]pyridine

Structure	cLogP	cLogS	HBA	HBD	TPSA	MW
Benzofuran	2.11	−2.78	1	0	13Å	118
Furopyridine	1.39	−2.82	2	0	26Å	119

As bioisosteres to benzofuran, four possible furopyridines exist: furo[2,3-b]pyridine, furo[2,3-c]pyridine, furo[3,2-c]pyridine, and furo[3,2-b]pyridine. Collectively they are also known as azabenzofurans. In comparison to their progenitor benzofuran, furopyridines possess an additional nitrogen atom, which may function as a hydrogen bond acceptor. As in the case of a phenyl–pyridine switch, when aligned appropriately with the target protein, azabenzofurans may gain additional protein–ligand interactions in comparison to the parent benzofurans. Furthermore, the presence of an additional nitrogen atom also lowers the molecule's lipophilicity and thus impacts its physicochemical properties such as aqueous solubility.[1]

FUROPYRIDINE-CONTAINING DRUGS

Ironically, no furopyridine-containing drugs, *per se*, are on the market. This may reflect past difficulty in synthesizing them either as core structures or as peripheral attachments. Antihypertensive/diuretic agent cicletanine (Tenstaten, **1**) has a tetrahydrofuropyridine core structure.[2] A tetrahydrofuran moiety is more stable than the electron-rich furan toward metabolic oxidation by CYP450 (*vide infra*).

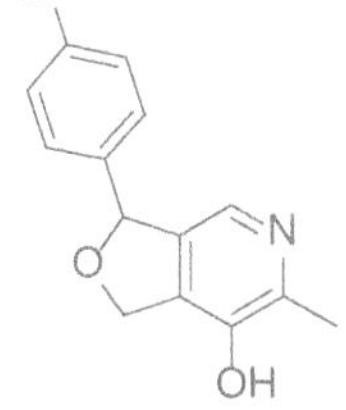

cicletanine
(Tenstaten, **1**)

Merck's indinavir (Crixivan, **2**) was one of the first HIV protease inhibitors approved by the FDA in 1996. When its pyridine fragment was replaced with furo[2,3-*b*]pyridine, the resulting protease inhibitor L-754,394 (**3**) was tested to be an *unusually* highly potent and selective mechanism-based inhibitor (MBI, also known as suicide substrate inhibitor) of cytochrome P450 according to *in vitro* studies on its metabolic activation.[3]

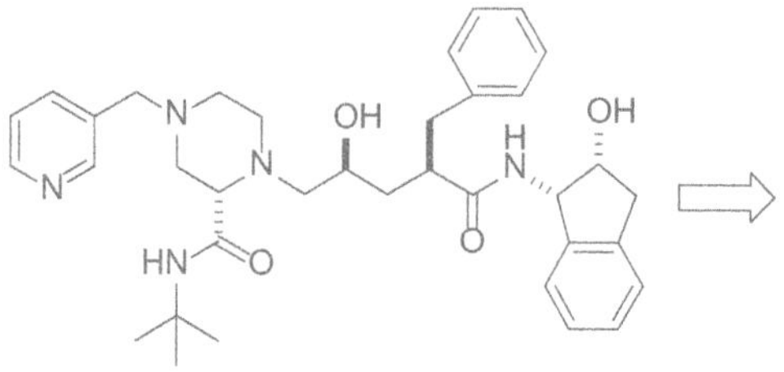

indinavir (Crixivan, **2**)

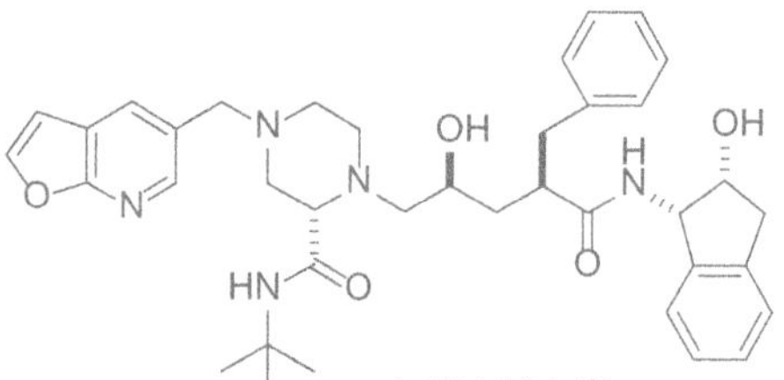

L-754,394 (**3**)

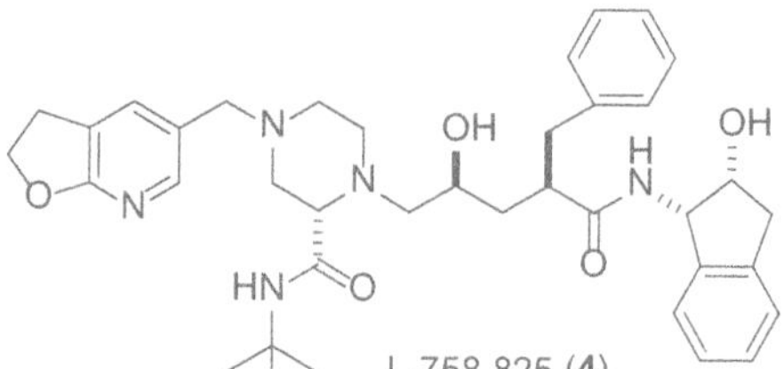

L-758,825 (**4**)

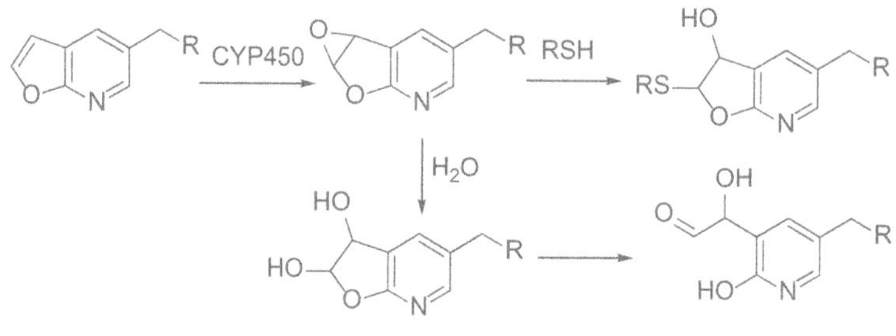

Similar to furan, the furo[2,3-*b*]pyridine motif on L-754,394 (**3**) is readily oxidized by CYP450 3A enzymes to the corresponding epoxide ring, which may be opened by nucleophiles such as water and glutathione (GSH). In hepatic microsomal preparations from rats, dogs, rhesus monkeys, and humans, L-754,394 (**3**) underwent NADPH-dependent metabolic activation to generate electrophilic intermediates, which became covalently bound to cellular proteins, causing the destruction of CYP450 enzymes. In contrast, neither indinavir (Crixivan, **2**), which lacks the furan ring, nor L-758,825 (**4**), which is a dihydrofuran derivative was found to act as suicide substrate inhibitors of liver microsome CYP450. Therefore, the furan ring is responsible for the metabolic activation of L-754,394 (**3**).[3] In the end, although furopyridines are prone to CYP metabolic oxidation, they are probably less reactive than just furan or benzofuran rings because pyridine is an electron-deficient heterocycle.

FUROPYRIDINES IN DRUG DISCOVERY

In one case, 7-aminofuro[2,3-*c*]pyridine **5** was one of OSI's HTS hits of TAK1 inhibitors.[4a] Transforming growth factor β receptor-associated kinase 1 (TAK1 or MAP3K7) is a serine/threonine kinase, which forms a key part of canonical immune and inflammatory signaling pathways. TAK1 inhibitors have the potential to treat cancer and inflammatory diseases.

As a mere HTS hit, compound **5** was already reasonably potent (IC$_{50}$, 1.2 µM). Single crystal X-ray structure revealed that the nitrogen atom on the furo[2,3-*c*]pyridine formed a hydrogen bond with the NH function provided by alanine-106 (Y106) of the TAK target protein, which helped to promote the protein–ligand binding. The geometrical vicinity between the sulfur atom on benzothiophene and the oxygen atom on furo[2,3-*c*]pyridine (2.8 Å, well within the van der Waals contact distance) suggests that polarization of the sulfur atom led to positive interactions between sulfur and oxygen. In essence, polarized sulfur behaved like an NH group (S=NH!) to form an intramolecular hydrogen bond with the oxygen atom.[5]

In addition to being a promiscuous kinase inhibitor, showing >50% inhibition of 42/192 kinases, 7-aminofuro[2,3-*c*]pyridine **5** has serious pharmacokinetic liabilities as well. The metabolic vulnerability of the sulfur atom on the benzothiophene fragment was addressed by switching to the benzothiadiazole, which still maintained the positive interaction between the sulfur and the oxygen atoms. Furthermore, as shown in the figure on top of the next page, having three hydrogen bond donors (HBDs) is detrimental to cell permeability. Capping the piperidine NH with an acetyl group led to TAK1 inhibitor **6**, which was potent (IC$_{50}$=4 nM) and selective against several potentially troublesome kinases such as KDR/VEGFR2 and the cell cycle kinase Aurora B and CHK1.[4a]

Since compound **6** still had high extraction ratios (ERs) in both mouse and human liver microsomes: 0.70 and 0.68, respectively, which presaged significant clearance and metabolism issues *in vivo*. Frustrated by their inability to replace the primary amine group without losing activity, OSI chose to install an electron-withdrawing group chlorine at the C-3 position of the furo[2,3-*c*]pyridine core structure. The maneuver killed two birds with one stone: it helped to solve the metabolism and kinase selectivity problems at once. In the end, they transformed a series of potent but relatively poorly kinase selective 7-aminofuro[2,3-*c*]pyridine inhibitors of TAK1 with poor PK as represented by **5** into more selective inhibitors with excellent oral exposure, as represented by TAK1 inhibitor **7**.[4b]

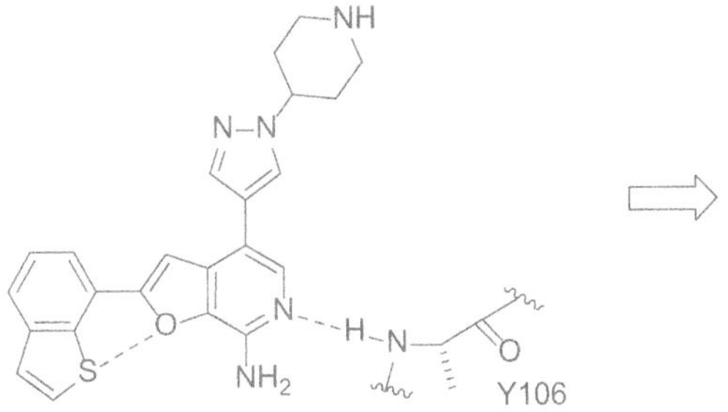

HTS hit **5**, TAK1 IC$_{50}$, 1.2 µM

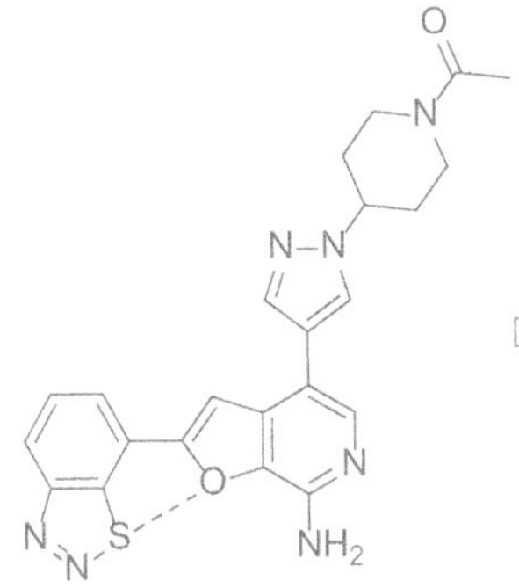

TAK1 inhibitor **6**, IC$_{50}$, 4 nM
ER = 0.70 (MLM), 0.68 (HLM)

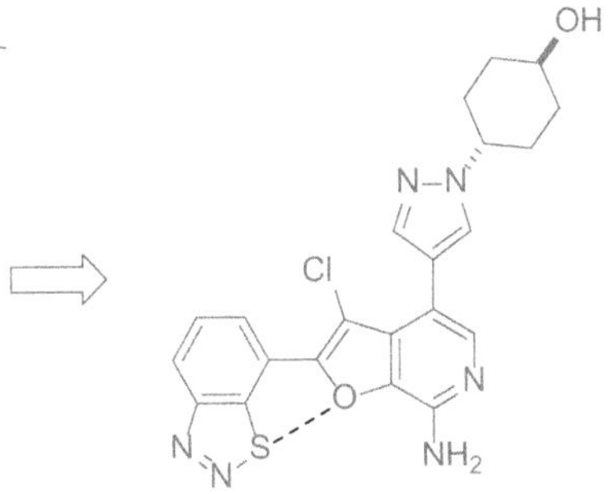

TAK1 inhibitor **7**, IC$_{50}$, 28 nM
ER = 0.39 (MLM), 0.40 (HLM)

OSI also had success with another furopyridine core structure on their kinase inhibitors: they discovered a series of 6-aminofuro[3,2-c]pyridines as potent and orally efficacious inhibitors of cMET and RON kinases.[6]

Pfizer's anaplastic lymphoma kinase (ALK)/cMET/RON inhibitor crizotinib (Xalkori, **8**) was approved by the FDA in 2011. Its pyridine nitrogen on the 2-aminopyridine core acts as a hydrogen bond acceptor for a backbone NH of the hinge region, while the 2-amino group donates a hydrogen bond to the interior hinge carbonyl, thus interacting with the same residues as ATP in a mutually exclusive fashion. OSI opted to employ 6-aminofuro[3,2-c]pyridine as an isostere of 2-aminopyridine and the fused additional furan retained the hinge binding but provided different vectors for substituents to interaction with the target protein. One of the derivatives, OSI-296 (**9**) was tested potent and selective with a good PK profile (>70% bioavailability in rodents). More importantly, it showed significant tumor growth inhibition (TGI) in multiple cMET-driven xenograft models in mice at once-daily doses of 50 mg/kg or less.[6]

crizotinib (Xalkori, **8**)
ALK cell IC$_{50}$, 80 nM

OSI-296 (**9**)
cMET biochem, IC$_{50}$, 26 nM

Furopyridines have found many applications in the kinase field. Furo[2,3-c]pyridine-based indanone oximes were discovered as potent and selective B-Raf inhibitors.[7] Meanwhile, furo[3,2-b] pyridine was revealed to be a privileged scaffold for highly selective kinase inhibitors, namely CDC-like kinase (CLK) inhibitors.[8]

In another case, one particular isomer furo[2,3-c]pyridine helped to provide rapid brain penetration and high oral bioavailability in rat.[9]

PNU-282,987 (**9'**) is a potent and selective α7 neuronal nicotinic acetylcholine receptor (α7 nAChR) agonist with the potential to treat cognitive deficits in schizophrenia. Regrettably, it possesses significant human ether-a-go-go (hERG) potassium channel activity. Efforts to improve its safety led to replacing the p-chlorophenyl group with 6,5-fused analogs to afford benzofuran **10**, among others. Although benzofuran **10** stood out for its potency and stability in rat liver microsomes (RLM), furan is notorious for its tendency for metabolic activation since it is so electronic rich. Therefore, all four furopyridines including furo[2,3-c]pyridine **11** were prepared to mitigate the liability. Among the four furopyridines, only the furo[2,3-c]pyridine **11** was potent enough as an (α7 nAChR) agonist. Both **10** and **11** had reduced hERG activity. Compound **11** was also tested selective with an excellent *in vitro* profile. Moreover, it is characterized by rapid brain penetration and high oral bioavailability in rats and demonstrates *in vivo* efficacy in auditory sensory gating and novel object recognition in an *in vivo* model to assess cognitive performance.[9]

PNU-282,987 (**9'**)
α7-5HT$_3$, EC$_{50}$, 128 nM
α7 K_i 24 nM
hERG %block @ 20 μM, 57%
rat F (%)   74%
brain/plasma ratio 5.0

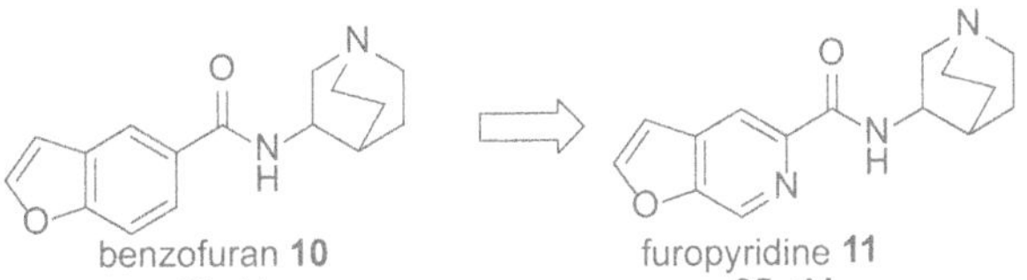

benzofuran **10**
47 nM
1.6 nM
33%
88%
1.24

furopyridine **11**
65 nM
8.8 nM
29%
63%
1.5

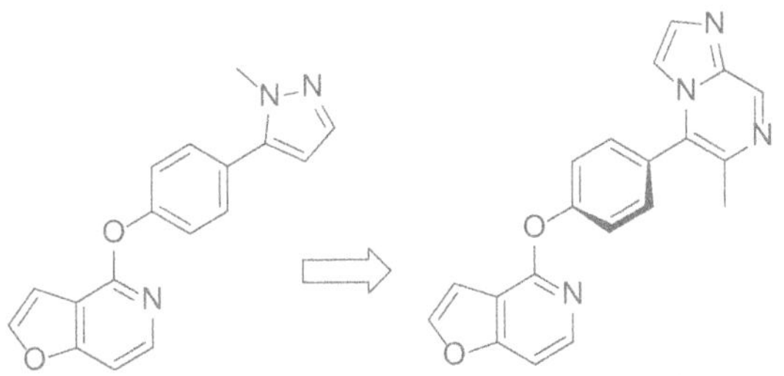

HTS hit **12**
D_1 EC_{50} (cAMP):
2.3 µM, 37% E_{max}
D_1 binding K_i: 2.2 µM

PF-6256142 (**13**)
D_1 EC_{50} (cAMP):
33 nM, 91% E_{max}
D_1 binding K_i: 12 nM

In the past, nearly all known D_1 selective agonists are catecholamines including the only one on the market, fenoldopam (Corlopam). To avoid catechol and phenol-containing D_1 selective agonists, Pfizer carried out an HTS of three million compounds and found only one hit that fitted their requirements. The hit was furo[3,2-*c*]pyridine **12**. Extensive hit-to-lead (H2L) efforts eventually led to the discovery of atropisomer PF-6256142 (**13**), a potent and selective orthosteric agonist of the D_1 receptor that has reduced receptor desensitization relative to dopamine and other catechol-containing agonists. PF-6256142 (**13**) also has an excellent pharmacokinetics profile with an F% value of 85%. It merits clinical study because in chronic diseases, such as schizophrenia and Parkinson's disease, the duration of therapeutic effect is an important component of patient quality of life.[10]

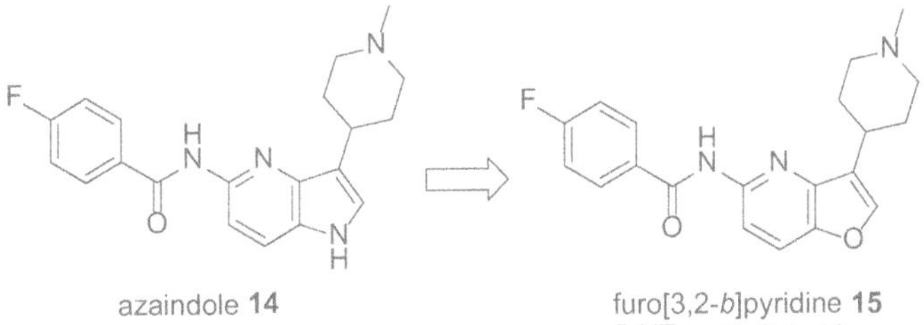

azaindole **14**
$5\text{-}HT_{1F}$, K_i, 7.6 nM
ratio, 1A/1F, 7.3
ratio, 1B/1F, 160
ratio, 1D/1F, 960

furo[3,2-*b*]pyridine **15**
$5\text{-}HT_{1F}$, K_i, 3.1 nM
ratio, 1A/1F, 134
ratio, 1B/1F, > 1000
ratio, 1D/1F, > 1000

On one occasion, furopyridines helped to boost the selectivity for $5\text{-}HT_{1F}$ receptor agonists as represented by azaindole **14**. In comparison to **14**, its furo[3,2-*b*]pyridine bioisosteres such as **15** possessed a similar affinity for the $5\text{-}HT_{1F}$ receptor and had improved selectivity for $5\text{-}HT_{1A}$, $5\text{-}HT_{1B}$, and $5\text{-}HT_{1D}$. Furo[3,2-*b*]pyridine **15** may have potential as a therapeutic for acute treatment of migraine.[11]

SYNTHESIS OF SOME FUROPYRIDINE-CONTAINING DRUGS

OSI's synthesis of their TAK1 inhibitor **7** commenced by using (*E*)-3-(furan-3-yl)acrylic acid (**16**) as the starting material. It was converted to furopyridone **17** via the intermediacy of the corresponding acyl azide, followed by the Curtius rearrangement. After conversion to the 7-chloride **18** by the action of POCl₃, the chloride was displaced with methoxylamine. Reduction using zinc in acetic acid provided amine **19**. Protection of the amine was followed by deprotonation by LDA and quench with hexachloroethane to afford dichloride **20** when more equivalents of LDA and hexachloroethane were used. A simple removal of the Boc protection revealed the amine group, which was subsequently converted to iodide **21** by the action of NIS. TAK1 inhibitor **7** was prepared from iodide **21** in several additional transformations.[4]

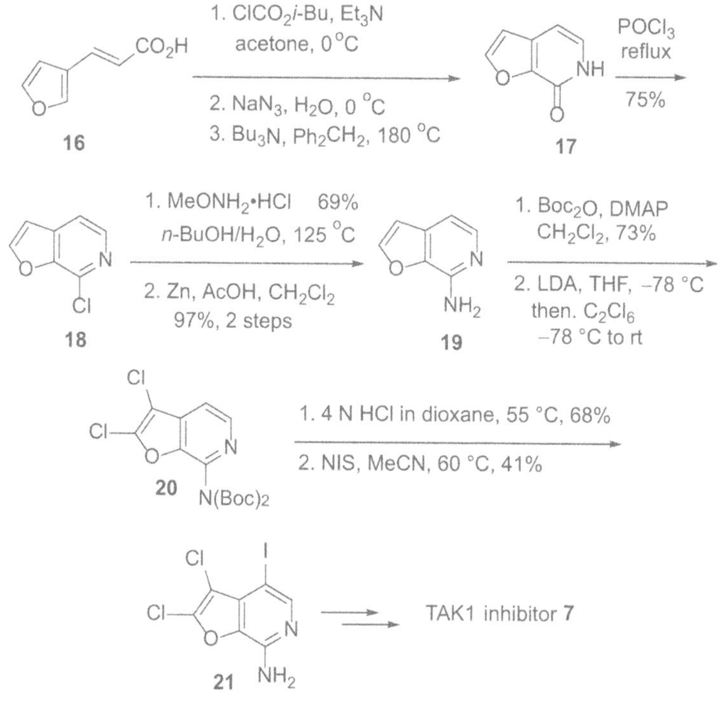

In the preparation of OSI-296 (**9**), furopyridine **22** was used as the starting material. A three-step sequence involving treatment with NBS, followed by POBr₃, and selective reduction of the 4-bromine substituent converted furopyridine **22** to 7-bromofuro[3,2-*c*]pyridine (**23**). Conversion of the 7-Br group to 7-OH and subsequent nitration at C6 provided **24**, unto which the ether bond was forged by Mitsunobu coupling with alcohol **25**. The resulting ether was then transformed to OSI-296 (**9**) in several additional steps.[6]

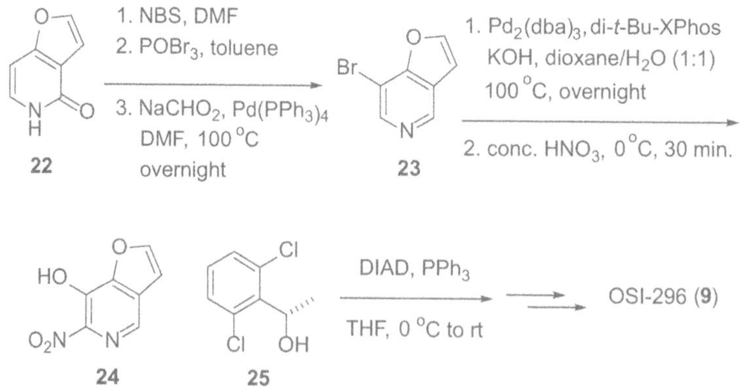

Lilly's synthesis of the 5-HT$_{1F}$ agonist **15** employed 6-chloro-2-iodopyridin-3-ol (**26**) as the starting material. A Mitsunobu coupling with allylic alcohol **27** yielded ether **28**. A Larock indole synthesis afforded furo[3,2-*b*]pyridine **29** in 60% yield. Reduction of the Boc protection produced the desired methyl-piperidine **30**, which was converted to furo[3,2-*b*]pyridine **15** in a few additional steps.[11]

In summary, as in the case of a phenyl–pyridine switch, furopyridines have some advantages over the parent benzofurans. The nitrogen atom may serve as a hydrogen bond acceptor and result in additional protein–ligand interactions. Furthermore, the presence of an additional nitrogen atom also lowers the molecule's lipophilicity and thus impacts its physicochemical properties such as aqueous solubility. Furopyridine building blocks have found a wide utility in drug design and drug synthesis.

REFERENCES

1. Fumagalli, F.; de Melo, S. M. G.; Ribeiro, C. M.; Solcia, M. C.; Pavan, F. R.; da Silva Emery, F. *Bioorg. Med. Chem. Lett.* **2019**, *29*, 974–977.
2. Kalinowski, L.; Szczepanska-Konkel, M.; Jankowski, M.; Angielski, S. *Gen. Pharmacol.* **1999**, *33*, 7–16.
3. Sahali-Sahly, Y.; Balani, S. K.; Lin, J. H.; Baillie, T. A. *Chem. Res. Toxicol.* **1996**, *9*, 1007–1012.
4. (a) Hornberger, K. R.; Berger, D. M.; Crew, A. P.; Dong, H.; Kleinberg, A.; Li, A.-H.; Medeiros, M. R.; Mulvihill, M. J.; Siu, K.; Tarrant, J.; et al. *Bioorg. Med. Chem. Lett.* **2013**, *23*, 4517–4522. (b) Hornberger, K. R.; Chen, X.; Crew, A. P.; Kleinberg, A.; Ma, L.; Mulvihill, M. J.; Wang, J.; Wilde, V. L.; Albertella, M.; Bittner, M.; et al. *Bioorg. Med. Chem. Lett.* **2013**, *23*, 4511–4516.
5. For a superb survey of the role of noncovalent sulfur interactions in drug design, see: Beno, B. R.; Yeung, K.-S.; Bartberger, M. D.; Pennington, L. D.; Meanwell, N. A. *J. Med. Chem.* **2015**, *58*, 4383–4438.
6. Steinig, A. G.; Li, A.-H.; Wang, J.; Chen, X.; Dong, H.; Ferraro, C.; Jin, M.; Kadalbajoo, M.; Kleinberg, A.; Stolz, K..; et al. *Bioorg. Med. Chem. Lett.* **2013**, *23*, 4381–4387.
7. Buckmelter, A. J.; Ren, L.; Laird, E. R.; Rast, B.; Miknis, G.; Wenglowsky, S.; Schlachter, S.; Welch, M.; Tarlton, E.; Grina, J.; et al. *Bioorg. Med. Chem. Lett.* **2011**, *21*, 1248–1252.
8. Němec, V.; Hylsová, M.; Maier, L.; Flegel, J.; Sievers, S.; Ziegler, S.; Schröder, M.; Berger, B.-T.; Chaikuad, A.; Valcikova, B.; et al. *Angew. Chem. Int. Ed.* **2019**, *58*, 1062–1066.
9. Wishka, D. G.; Walker, D. P.; Yates, K. M.; Reitz, S. C.; Jia, S.; Myers, J. K.; Olson, K. L.; Jacobsen, E. J.; Wolfe, M. L.; Groppi, V. E.; et al. *J. Med. Chem.* **2006**, *49*, 4425–4436.

10. Davoren, J. E.; Nason, D.; Coe, J.; Dlugolenski, K.; Helal, C.; Harris, A. R.; LaChapelle, E.; Liang, S.; Liu, Y.; O'Connor, R.; et al. *J. Med. Chem.* **2018**, *61*, 11384–11397.
11. Mathes, B. M.; Hudziak, K. J.; Schaus, J. M.; Xu, Y.-C.; Nelson, D. L.; Wainscott, D. B.; Nutter, S. E.; Gough, W. H.; Branchek, T. A.; Zgombick, J. M.; et al. *Bioorg. Med. Chem. Lett.* **2004**, *14*, 167–170.

16 Indazoles

Indazole has two isomers: 1*H*-indazole and 2*H*-indazole. Both of them may potentially serve as bioisosteres for indole. Like indole, indazole has an NH group to serve as a hydrogen bond donor. Unlike indole, indazole has an additional nitrogen atom to serve as a hydrogen bond acceptor. This may provide improved affinity to the target protein when appropriate. Indazole is a privileged scaffold for many kinase inhibitors, presumably forming pivotal interactions with hinge residues.

indole   1*H*-indazole   2*H*-indazole

Indazole is important in at least three major areas in medicinal chemistry. (a) Indazole is a good bioisostere of phenol. Generally speaking, phenol's heterocyclic bioisosteres, including indazole, tend to be more lipophilic and less vulnerable to phase I and II metabolisms in comparison to phenol. (b) Indazole is a privileged fragment in fragment-based drug discovery (FBDD). (c) Indazole has been proven to be a privileged scaffold in scaffold hopping exercises, especially for protein kinase inhibitors.

INDAZOLE-CONTAINING DRUGS

There are about six indazole-containing drugs on the market. Several aza-indazole-containing drugs have also gained regulatory approval. 1*H*-Indazole bendazac (**1**) is an old nonsteroidal anti-inflammatory drug (NSAID). Bendazac (**1**) and its main metabolite, the 5-hydroxyl derivative, provide antioxidant effects as scavengers of oxygen-derived free radicals. Bendazac lysine salt as an antidenaturant is nowadays used as an eye drop to treat cataracts. Benzyldamine (**2**), a bendazac (**1**) analog, is also an NSAID used for treating cataracts. The third analog, lonidamine, has only been explored at the preclinical stages. Roche's granisetron (Kytril, **3**), a serotonin 5-HT$_3$ receptor antagonist used as an antiemetic to treat nausea and vomiting following chemotherapy, has a 1*H*-indazole core structure. Pfizer's axitinib (Inlyta, **4**), also a 1*H*-indazole, is a dual vascular endothelial growth factor receptor (VEGFR) and platelet-derived growth factor receptor (PDGFR) inhibitor.

GSK's pazopanib (Votrient, **5**), a 2*H*-indazole, is a VEGFR and c-Kit inhibitor. Tesaro's poly(ADP-ribosyl) polymerase (PARP) inhibitor niraparib (Zejula, **6**) is also a 2*H*-indazole.

bendazac (**1**)   benzyldamine (**2**)   granisetron (Kytril, **3**)
Beecham/Roche, 1994
5-HT$_3$ receptor antagonist

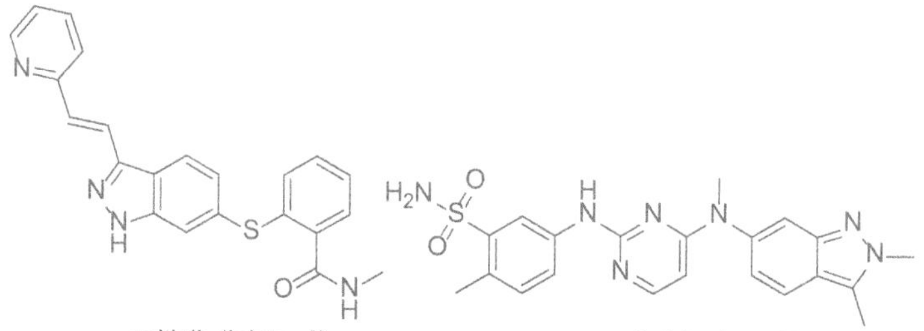

axitinib (Inlyta, **4**)
Pfizer, 2012
VEGF and PDGFR inhibitor

pazopanib (Votrient, **5**)
GSK, 2009
VEGFR and c-Kit inhibitor

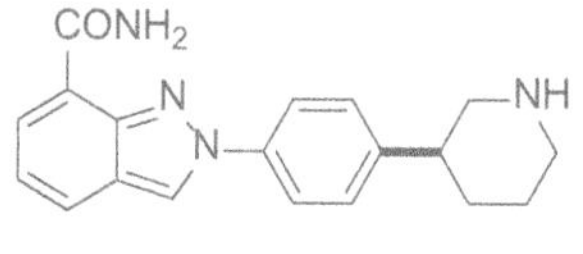

niraparib (Zejula, **6**)
Tesaro, 2017
PARP inhibitor

INDAZOLES IN DRUG DISCOVERY

Indazole has been used in at least three major areas in drug discovery: (a) as the bioisostere of phenol and indole, (b) as a fragment for fragment-based drug discovery (FBDD), and (c) as a scaffold in scaffold hopping.

a. Indazole as a bioisostere of phenol

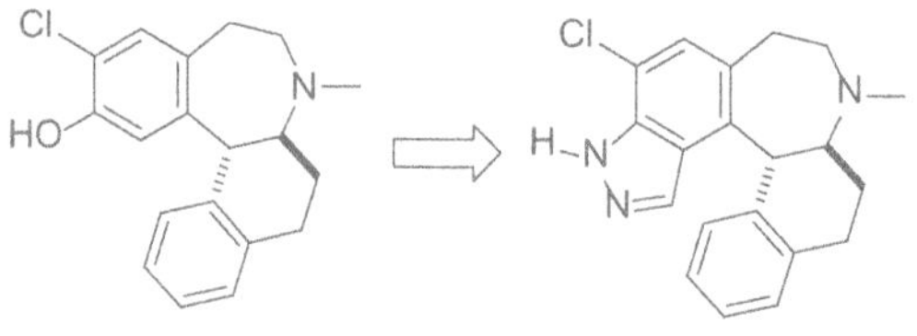

SCH 23390 (**7**)
D_1 K_i = 1.4 nM
D_5 K_i = 2.8 nM
rat PK (10 mg/kg, po)
AUC = 0 h·µg/mL
C_{max} = 0 ng/mL
T_{max} = 0 h

indazole **8**
D_1 K_i = 14 nM
D_5 K_i = 30 nM
rat PK (10 mg/kg, po)
AUC = 0.06 h·µg/mL
C_{max} = 15 ng/mL
T_{max} = 2 h

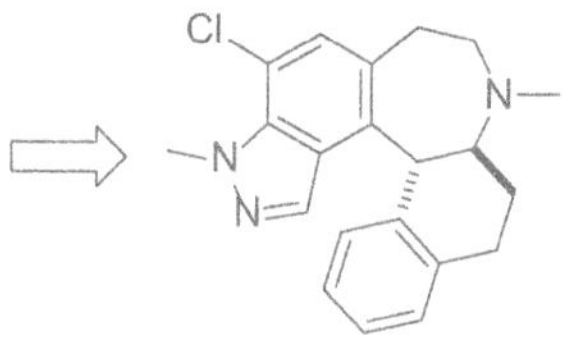

methylated indazole **9**
D_1 K_i = 183 nM

Back in the 1980s, Schering–Plough reported the discovery of the first high-affinity and selective D_1/D_5 antagonist SCH 23390 (**7**). However, benzazepine **7** was inactive in rhesus monkeys with a very short duration of action. Pharmacokinetic (PK) evaluation revealed that extensive O-glucuronidation of the phenol and N-dealkylation of the N–Me group *in vivo* may have contributed to its poor PK profile. Schering–Plough chose to replace the metabolically problematic phenol with indazole and other heterocycles with a hydrogen bond donating NH functionality. *Generally speaking, phenol's heterocyclic bioisosteres tend to be more lipophilic and less vulnerable to phase I and II metabolisms in comparison to phenol.* Indeed, indazole analog **8** began to show an improved PK profile although it was tested about ten-fold less potent than SCH 23390 (**7**). As a testimony to the importance of the hydrogen bond donating NH functionality, methylated indazole **9** had a significantly decreased (6×) affinity for the D_1 receptor.[1]

phenol **10**
Lck IC_{50} = 8.5 nM
IV Cl_p = 65.5 mL/min/kg
IV V_{ss} = 0.3 L/kg
IV $T_{1/2}$ = 0.12 h
IV AUC_{0-t} = 635 ng·h/mL
mouse p.o., F % < 1%

indazole **11**
Lck IC_{50} = 19 nM
IV Cl_p = 22.4 mL/min/kg
IV V_{ss} = 0.7 L/kg
IV $T_{1/2}$ = 0.53 h
IV AUC_{0-t} = 743 ng·h/mL
mouse p.o., F % = 25%

Lymphocyte-specific protein tyrosine kinase (Lck) is a member of the Src family of tyrosine kinases and Lck inhibitors may offer an approach to treat T cell-mediated inflammatory disorders. One of GSK's Lck inhibitors, phenol **10**, was potent with an IC_{50} value of 8.5 nM in an Lck biochemical assay. In the docking model, the oxygen of phenol **10**'s 5-hydroxyl group is 2.3 Å away from the backbone NH of Lck's Asp_{382}, indicating that the oxygen here serves as a hydrogen bond acceptor. Meanwhile, the phenol's hydrogen is 2.7 Å away from the sidechain of Glu_{288}, implying that either Glu_{288} moves closer when the compound binds or a water-mediated interaction is present. Unfortunately, phenol **10** has a poor pharmacokinetic profile, exhibiting high clearance, short half-life, and low oral bioavailability. This is not all that surprising since phenols are prone to phase I and II metabolism.[2]

Indazole as an isostere of phenol again proved to be fruitful here and phenol **10**'s indazole analog **11** was tested potent with an improved protein kinase selectivity profile. Docking modeling indicated that their hydrogen bonding vectors mimic each other very closely, maintaining the direct interaction with Asp_{382} and the direct or water-mediated hydrogen bond to Glu_{288}. More importantly, indazole **11** demonstrated a superior PK profile for both IV and p.o. routes of administration with a 25% oral bioavailability, (F%), whereas the value for phenol **10** was 0%.[2]

Phosphatidylinositol 3-kinases (PI3K) play a central role in broad cellular functions including cell growth, proliferation, differentiation, survival, and intracellular trafficking. Gilead's PI3Kδ selective inhibitor idelalisib (Zydelig) was approved in 2014 and Bayer's pan-PI3K inhibitor copanlisib (Aliqopa) was approved in 2017.

Using Hayakawa's 4-morpholino-2-phenylquinazoline PI3Kα p110α inhibitors as a starting point,[3a] Genentech arrived at phenol **12**. But its bioavailability was low in mice and rats, even though it displayed enhanced metabolic stability in human and mouse microsomes. Again, this is mainly due to the *O*-glucuronidation of the phenol group. The metabolic liability was addressed once again by employing the indazole isostere. In the end, indazole **13** (GDC-0941, pictilisib) was not only potent and selective for PI3Kα over PI3Kβ, δ, and γ subtypes, but also exhibited acceptable oral bioavailability in all species tested. The crystal structure of **13** bound to p110α revealed that the two nitrogen atoms on indazole are in hydrogen bonding distance to the phenol oxygen of Tyr_{867} and the carbonyl group of Asp_{841}. Pictilisib (**13**) was advanced to phase I and II clinical trials but is no longer progressing at the moment.[3b]

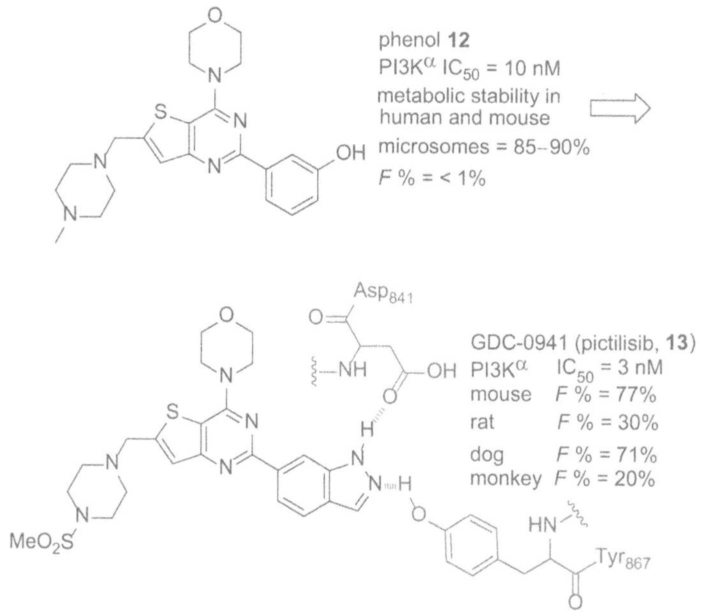

phenol **12**
PI3K$^\alpha$ IC_{50} = 10 nM
metabolic stability in human and mouse microsomes = 85–90%
F % = < 1%

GDC-0941 (pictilisib, **13**)
PI3K$^\alpha$ IC_{50} = 3 nM
mouse F % = 77%
rat F % = 30%
dog F % = 71%
monkey F % = 20%

The Ras-MAP kinase pathway has been implicated in tumor progression for a variety of human cancers. The Raf kinases, which are components of this cascade, are serine/threonine kinases that activate MEK1/2. Mutant B-Raf containing a V600E substitution (where B-Raf protein's 600th amino acid valine is replaced by glutamic acid) causes aberrant constitutive activation of this pathway and has a high occurrence in several human cancers. Three B-Raf kinase inhibitors have been approved by the FDA to treat cancer: vemurafenib (Zelboraf, 2011), dabrafenib (Tafinlar, 2013), and encorafenib (Braftovi, 2018), with vemurafenib (Zelboraf) being the first marketed drug that was discovered employing a fragment-based drug discovery (FBDD) strategy.

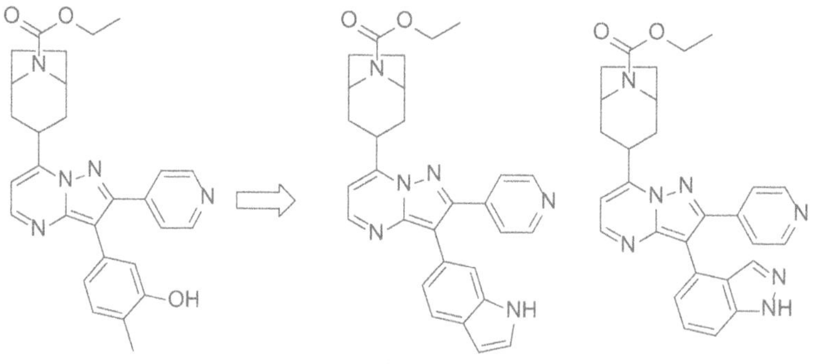

phenol **14**
B-raf, IC_{50} = 0.3 nM
A375, IC_{50} = 570 nM
WM266-4, IC_{50} = 230 nM

indole **15**
B-raf, IC_{50} = 36 nM
A375, IC_{50} = 7.5 μM
WM266-4, IC_{50} > 10 μM

indazole **16**
B-raf, IC_{50} = 2 nM
A375, IC_{50} = 380 nM
WM266-4, IC_{50} = 300 nM

Phenol **14** is a B-raf inhibitor (IC$_{50}$, 0.3 nM). Bioisosterism was deployed to address the potential metabolic soft spot posed by the phenol. Although replacing the phenol functional group with hydrogen-bond *acceptors*, not surprisingly, did not work (low potency), isosteric substitution with hydrogen bond *donors* offered indole **15** (IC$_{50}$, 36 nM) and indazole **16** (IC$_{50}$, 2 nM), respectively, that were active. More important, **16** potently inhibited cell proliferation at sub-micromolar concentrations in the B-Raf V600E human melanoma cell lines A375 and WM266. Subsequent docking suggested that the indazole N and NH form two hydrogen bonds with B-Raf's Glu$_{501}$ and Asp$_{594}$, respectively. This exercise demonstrated that indazole is an effective isostere of phenol.[4]

b. Indazole as a fragment in FBDD

In the field of kinase inhibitors, indazoles and azaindoles, among many heteroarenes, have been fruitful fragments in FBDD.

Screening "lead-like" fragment library with an average molecular weight of 160 provided fragment bromoaminoindazole **17** as a JAK-2 inhibitor with an IC$_{50}$ value of 40.9 μM. The co-crystal structure of indazole **17** and JAK2 revealed that they make two hydrogen-bonded contacts with the hinge residues Glu$_{930}$ and Leu$_{932}$, respectively. Guided by X-ray co-crystal structures, structure–activity relationship (SAR) investigations led to aminoindazole **18** with an IC$_{50}$ value of 78 nM while maintaining a 0.41 ligand efficiency (LE = −RT ln IC$_{50}$/HAC, HAC = heavy atom count) value.[5]

bromoaminoindazole **17**
JAK-2, IC$_{50}$ = 40.9 μM
LE = 0.54

aminoindazole **18**
JAK-2, IC$_{50}$ = 78 nM
LE = 0.41

Also in the JAK arena, fragment hit **19** was obtained from screening a kinase-targeted library of 500 fragments against JAK2 at a single concentration of 100 μM. Similar to that of bromoaminoindazole **17**, the co-crystal structure of indazole **19** and JAK2 revealed that they make two hydrogen-bonded contacts with the hinge residues Glu$_{930}$ and Leu$_{932}$ as well. Growing fragment hit **19** by installing a phenol moiety at the 6-position afforded greatly improved potency. Fine-tuning the substituents on the phenol and sulfonamide moieties gave rise to "lead-like" indazole **20**.[6]

fragment hit **19**
JAK1 pIC$_{50}$ = 4.5 (30 μM)
JAK2 pK$_D$ = 4.9 (13 μM)
JAK1 LE = 0.45

lead **20**
JAK1 pIC$_{50}$ = 8.4
cell pIC$_{50}$ = 7.0
LE = 0.45

fragment **21**
MW 226
Hsp90 IC$_{50}$ = 45 µM
LE = 0.35
BEI = 19.2

fragment **22**
MW 269
Hsp90 IC$_{50}$ = 485 µM
LE = 0.28
BEI = 15.1

Hsp90 inhibitor **23**
Hsp90 binding assay,
IC$_{50}$ = 60 nM
A2780 viability assay,
IC$_{50}$ = 57 nM

It has been observed that subtle changes in the chemistry of a fragment can induce a change in compound orientation in the crystal structure with the target protein. The crystal structure of fragment **21** and heat shock protein-90 (Hsp90) showed that they interact via hydrogen bonds of both nitrogen atoms to Asp$_{93}$ and the conserved water molecule. Interestingly, an identical interaction pattern was found for the phenolic OH group of fragment **22**, but not its indazole NH! A similar phenomenon was observed for hydroxyindazole fragments before.[7] The reversed orientation might be influenced by the substitution pattern of the hydroxyindazole core as the cyclohexyl ring of fragment **22** is bound in the lipophilic pocket of the helical form. In contrast, this lipophilic pocket is not addressed by fragment **21**, but its phenolic substituent forms π-interactions to Met$_{98}$ toward the solvent entrance of the ATP-site. In due course, SAR investigations via fragment elaboration/decoration led to the discovery of a series of hydroxy-indazole-carboxamides as potent Hsp90 inhibitors. Compound **23**, for example, possessed significantly improved affinity and antiproliferative effects in different human cancer cell lines as demonstrated by cell viability assays.[8]

Fragment **24** was obtained from an Aurora kinase A inhibition assay. Docking studies indicated that the N1-H on **24** formed a hydrogen bond with the carbonyl of Glu$_{211}$, whereas the N2 atom is hydrogen bonded to the backbone of Ala$_{213}$. A subsequent *in silico* FBDD approach identified aryloyl analog **25** with the highest ligand efficiency (LE). Eventually, knowledge-based drug design provided inhibitor **26**, which had a 60-fold improvement in potency over **25**. Molecular docking analysis of **26**'s binding information revealed that its indazole core forms hydrogen bonds with the hinge residues Glu$_{211}$ and Ala$_{213}$ as in the case of fragment **24**.[9]

in silico
fragment
screening

inhibitor **24**
Aurora A IC$_{50}$ = 13 µM

inhibitor **25**
Aurora A IC$_{50}$ = 1.66 µM

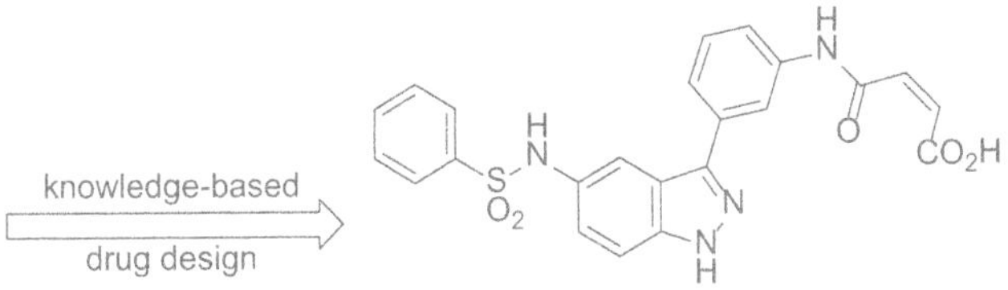

inhibitor **26**
Aurora A IC$_{50}$ = 26 nM

Indazole **27** was generated via an *in silico* high throughput screen (HTS) as an Unc-51-like kinase 1 (ULK1) inhibitor. Docking analysis indicated that the indazole ring system makes two hydrogen bond interactions with the amide backbone of the hinge region, specifically Glu$_{93}$ and Cys$_{95}$. Subsequent structure-guided rational drug design produced ULK1 inhibitor **28**, which had not only increased activity against ULK1, but also showed certain stability in human microsomes with negligible CYP inhibition.[10]

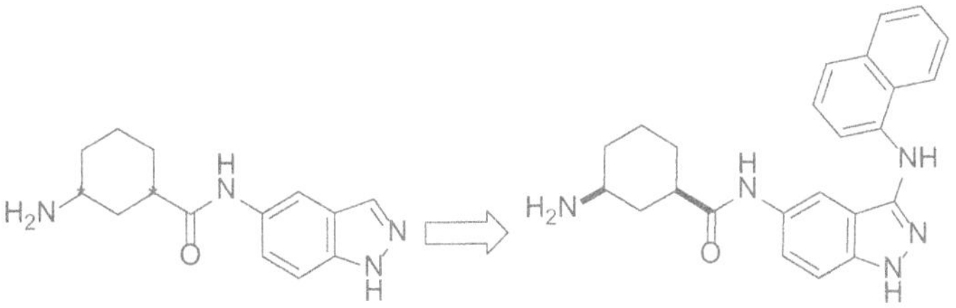

in silico HTS hit **27**
ULK1 IC$_{50}$ = 22.4 μM

ULK1 inhibitor **28**
IC$_{50}$ = 45 nM, t$_{1/2}$ = 224 min (human)

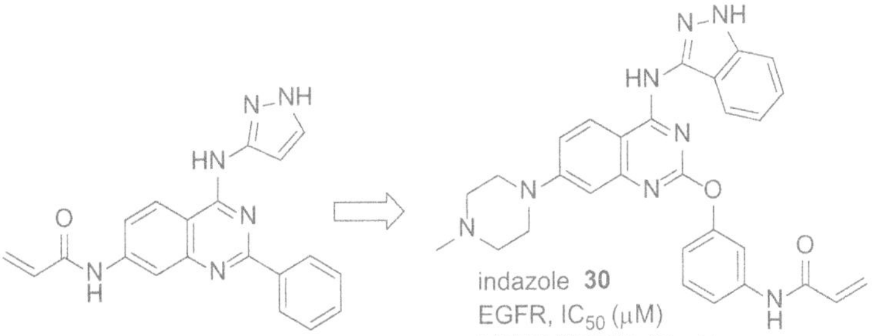

hit compound pyrazole **29**

indazole **30**
EGFR, IC$_{50}$ (μM)
L858R/T790M, 70 nM
L858R,   0.5 μM
wild-type,   1.7 μM

Indazole **30**, as a covalent epidermal growth factor receptor (EGFR) inhibitor, was obtained employing hit compound pyrazole **29** as the starting point. It showed a strikingly increased inhibitory effect on the drug-resistant mutant of EGFR. Structurally, as a hinge-binding element, the indazole scaffold provided suitable chemical and spatial features to accommodate the space between the hinge region and the gatekeeper residues, without suffering from steric clashes with the methionine side chain. Additionally, the indazole moiety played a prominent role in forming hydrogen bonds with the peptide backbone of residues Glu$_{339}$ and Met$_{341}$, respectively. The phenyl part of the indazole furnished a favorable hydrophobic gatekeeper interaction, resulting in improved protein–ligand interactions (PPIs).[11]

c. Indazole in scaffold hopping

At least five nonnucleoside HIV reverse transcriptase inhibitors (NNRTIs) are now on the market to combat HIV/AIDS. But HIV can become resistant to any single antiretroviral drug, so a combination of drugs is required. On the other hand, a single drug with features of two drugs may offer advantages against the clinically relevant mutations of reverse transcriptase, particularly K103N and Y181C. Using molecular hybridization based on crystallographic overlays of efavirenz (Sustiva, **31**) and the second-generation NNRTI capravirine **32**, a series of indazoles, as represented by **33**, were discovered as novel NNRTIs with excellent metabolic stability and mutant resilience.[12]

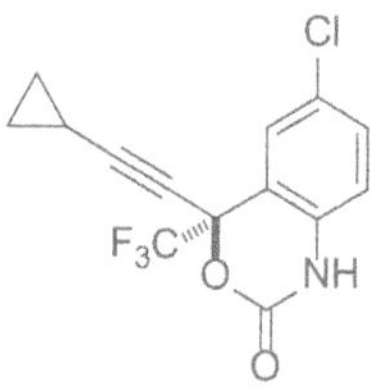

efavirenz (Sustiva, **31**)
BMS/Merck, 1998
NNRTI, wt IC_{50} = 14 nM

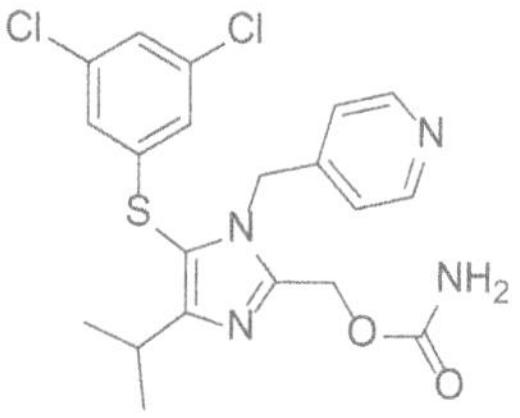

capravirine **32**
NNRTI
wt IC_{50} = 47 nM

molecular hybridation

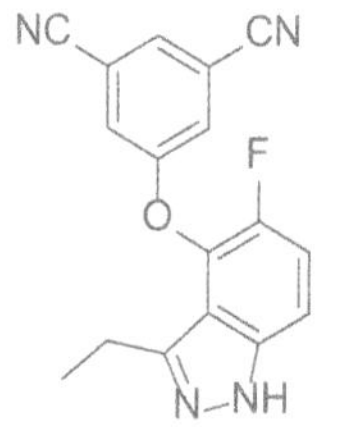

indazole **33**
NNRTI
wt IC_{50} = 25 nM

Bi-phenyl **34** was a reasonably potent bradykinin B1 (BB_1) receptor antagonist, but it suffered from a high permeability glycoprotein (Pgp) directional transporter ratio (it is considered high if the ratio is higher than 2.5). Scaffold hopping gave rise to a series of phenyl-indazoles as novel BB_1 receptor antagonists. Compound **35** in the series showed an acceptable Pgp ratio and rat pharmacokinetic profile.[13]

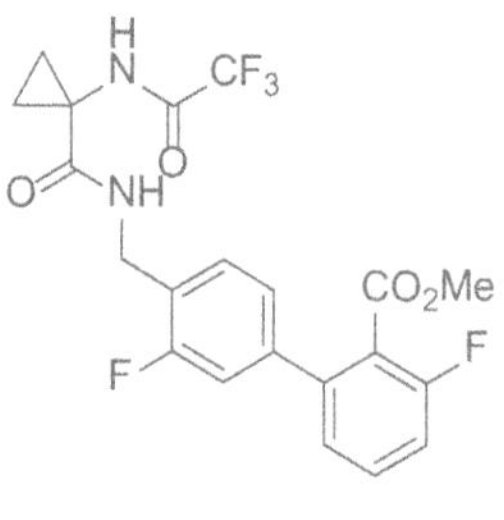

bisphenyl **34**
hBK1 K_i = 1.5 nM
Pgp ratio = 4.1

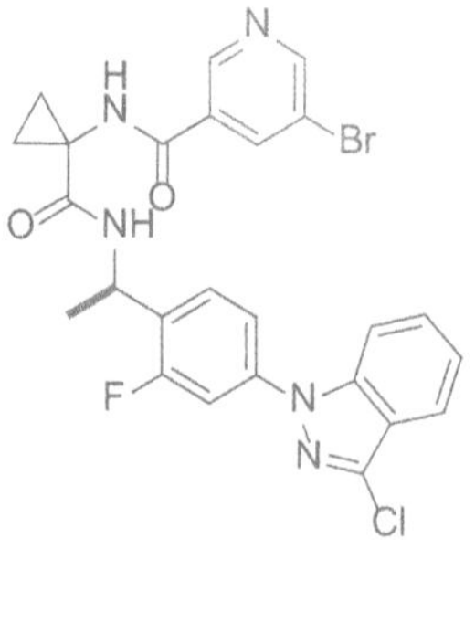

phenyl-indazole **35**
hBK1 K_i = 0.12 nM
Pgp ratio = 2.3
P_{app} = 13
rat PK:
F% = 24
$t_{1/2}$ = 3.1 h
Cl = 2.5 mL/min/kg

Bacterial DNA gyrase is a clinically validated target. Despite the excellent *Sa* GyrB binding potency, gyrase inhibitor pyrazolopyridone **36** does not show the whole cell antibacterial activity. The lack of minimal inhibitory concentration (MIC) could be due to poor cell membrane penetration resulted from polar functional groups, therefore, a low log D. In an effort to increase log D to improve cell penetration, a series of indazoles were prepared. Indazole **37**, in particular, possessed an excellent Gram-positive MIC profile in addition to retaining enzymatic activity.[14]

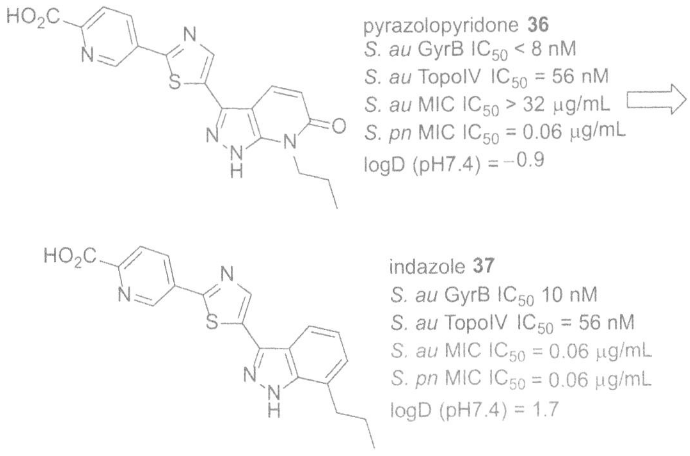

pyrazolopyridone **36**
S. au GyrB IC_{50} < 8 nM
S. au TopoIV IC_{50} = 56 nM
S. au MIC IC_{50} > 32 μg/mL
S. pn MIC IC_{50} = 0.06 μg/mL
logD (pH7.4) = −0.9

indazole **37**
S. au GyrB IC_{50} 10 nM
S. au TopoIV IC_{50} = 56 nM
S. au MIC IC_{50} = 0.06 μg/mL
S. pn MIC IC_{50} = 0.06 μg/mL
logD (pH7.4) = 1.7

Scaffold hopping has been a fruitful strategy to create novel intellectual properties (IP). For instance, pyrazole **38**, from the patent literature, was a G-protein-coupled receptor GPR120 agonist. AstraZeneca scientists speculated that rigidification of the ether side chain by connecting from oxygen to the pyrazole through installation of an indazole ring as in **39** would present a series of novel GPR120 agonists, and it did. Indazole **39** was potent for GPR120 with good selectivity against GPR40. It also had a favorable pharmacokinetic profile and was tested efficacious in an oral glucose tolerance test (OGTT) mouse model.[15]

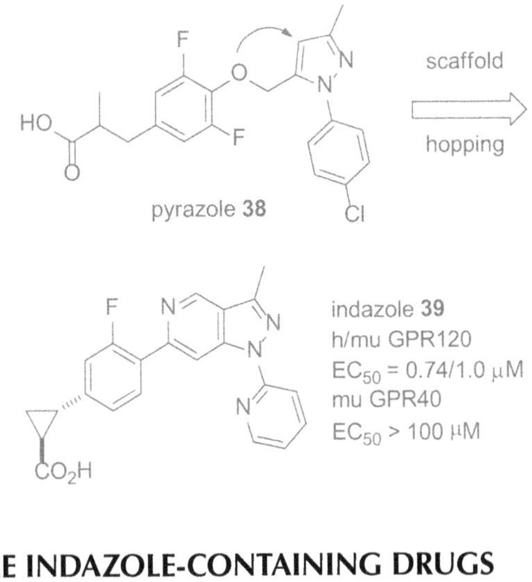

pyrazole **38**

scaffold hopping

indazole **39**
h/mu GPR120
EC_{50} = 0.74/1.0 μM
mu GPR40
EC_{50} > 100 μM

SYNTHESIS OF SOME INDAZOLE-CONTAINING DRUGS

In the synthesis of the NSAID bendazac (**1**), benzylaniline **40** was used as its starting material. Nitrosolation was accomplished using nitrous acid. The resulting *N*-nitroso intermediate **41** was reduced with sodium thiosulfate to the corresponding hydrazine, which simultaneously cyclized to the indazolone **42**. Subsequent alkylation with methyl chloroacetate was followed by hydrolysis to deliver bendazac (**1**).[16]

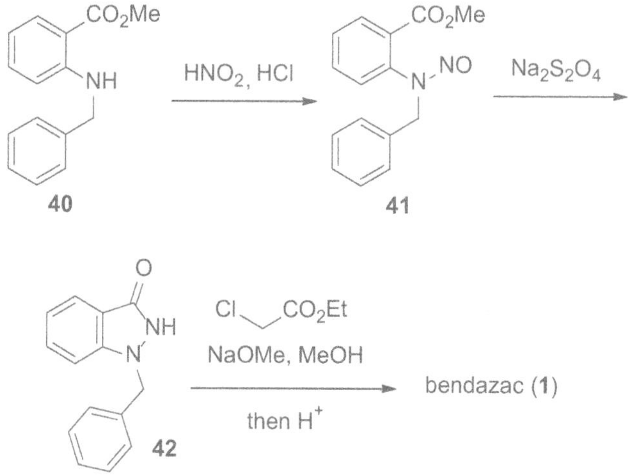

For the synthesis of Roche's granisetron (Kytril, **3**), 5-methoxyindazole (**43**) was protected as its SEM derivative **44**, which was then installed in the ethyl ester on the three-position to give **45**. Switching the SEM group with a methyl group was accomplished by acidic hydrolysis to afford indazole **46** followed by methylation to produce **47**. Basic hydrolysis of **47** provided the 5-methoxy-1-methyl-indazole-3-carboxylic acid (**48**). Coupling **48** with the bicyclic amine **49** produced amide **50** and the methyl ether was converted into the benzyl ether. The end result was the benzyloxyl derivative of granisetron **51**.[17]

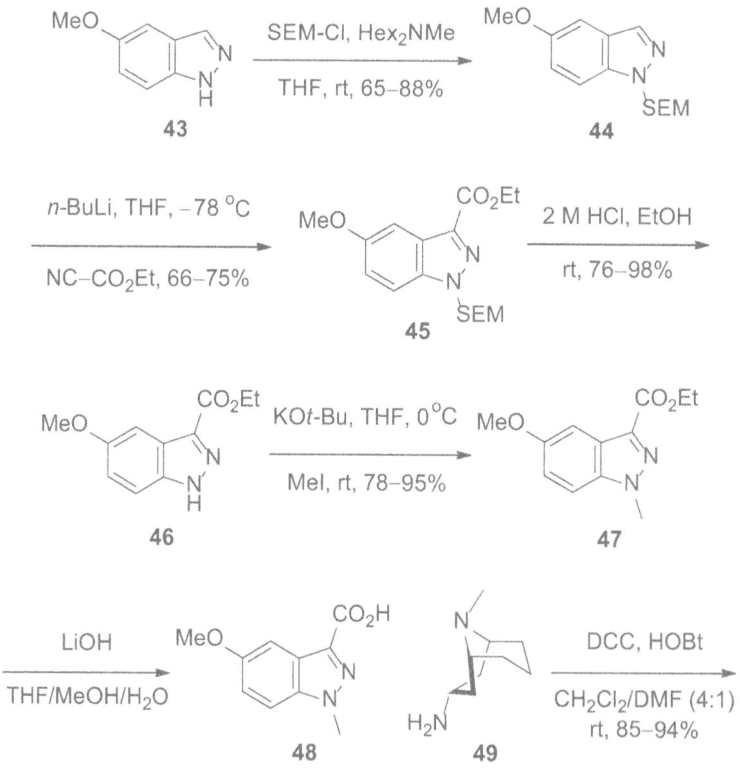

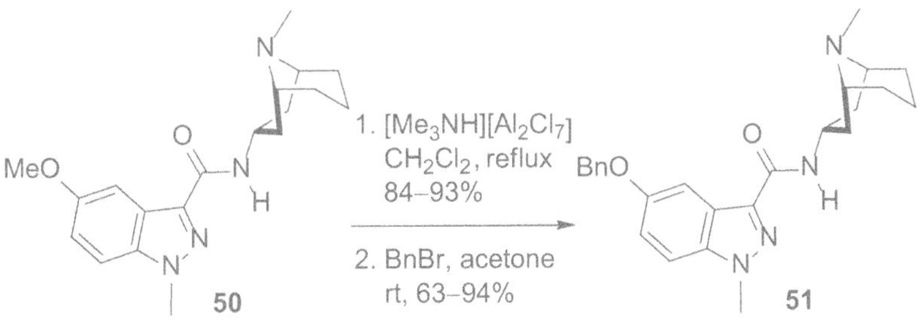

A synthesis of Merck's PARP inhibitor niraparib (Zejula, **6**) began with amidation of acid **52** with *t*-butylamine to make **53**. The C–N coupling between indazole **53** and phenyl bromide **54** assembled adduct **55**. Finally, double deprotection delivered niraparib (**6**).[18]

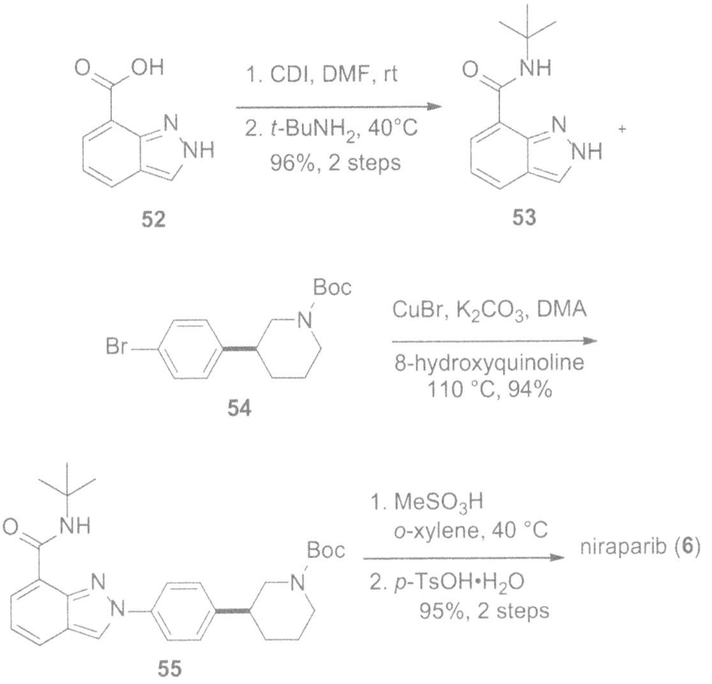

The synthesis of Genetench's GDC-0941 (pictilisib, **13**) provides a good opportunity to appreciate indazole's isomerism during its preparation.

Diazotization of 3-bromo-2-methylaniline (**56**) was followed by basic cyclization to construct 4-chloroindazole (**57**). Carefully optimized conditions provided selective protection of **57** to give the desired isomer **58b** in 98% yield, which underwent a halogen–metal exchange followed by borylation to afford boronic acid **59**. Nickel-catalyzed coupling between **59** and benzotriazolyl-piperazine chloride **60** was proven to be superior to palladium catalysis to assemble adduct **61** since expensive scavenging residual palladium was circumvented. Acidic removal of the THP protection then delivered pictilisib (**13**).[19]

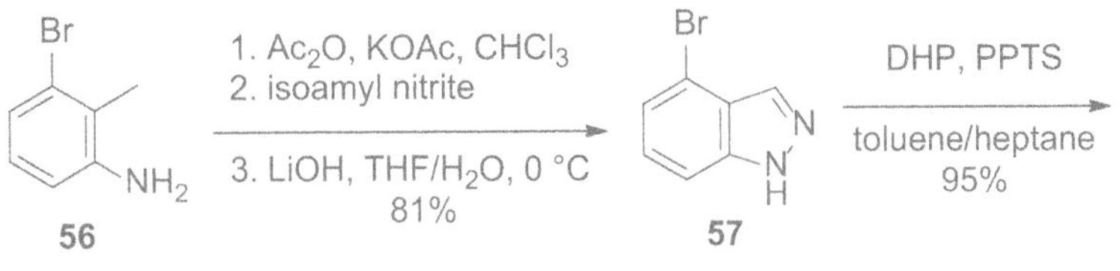

In summary, indazole is important in at least three major areas in medicinal chemistry. (a) Indazole is a good bioisostere of phenol. Generally speaking, phenol's heterocyclic bioisosteres, including indazole, tend to be more lipophilic and less vulnerable to phase I and II metabolisms in comparison to phenol itself. (b) Indazole is a privileged fragment in FBDD. (c) Indazole has been proven to be a privileged scaffold in scaffold hopping exercises, especially for protein kinase inhibitors.

REFERENCES

1. Wu, W.-L.; Burnett, D. A.; Spring, R.; Greenlee, W. J.; Smith, M.; Favreau, L.; Fawzi, A.; Zhang, H.; Lachowicz, J. E. *J. Med. Chem.* **2005**, *48*, 680–693.
2. Bamborough, P.; Angell, R. M.; Bhamra, I.; Brown, D.; Bull, J.; Christopher, J. A.; Cooper, A. W. J.; Fazal, L. H.; Giordano, I.; Hind, L.; et al. *Bioorg. Med. Chem. Lett.* **2007**, *17*, 4363–4368.
3. (a) Hayakawa, M.; Kaizawa, H.; Moritomo, H.; Koizumi, T.; Ohishi, T.; Okada, M.; Ohta, M.; Tsukamoto, S.-I.; Parker, P.; Workman, P.; et al. *Bioorg. Med. Chem. Lett.* **2006**, *14*, 6847–6858. (b) Folkes, A. J.; Ahmadi, K.; Alderton, W. K.; Alix, S.; Baker, S. J.; Box, G.; Chuckowree, I. S.; Clarke, P. A.; Depledge, P.; Eccles, S. A.; et al. *J. Med. Chem.* **2008**, *51*, 5522–5532.
4. Di Grandi, M. J.; Berger, D. M.; Hopper, D. W.; Zhang, C.; Dutia, M.; Dunnick, A. L.; Torres, N.; Levin, J. I.; Diamantidis, G.; Zapf, C. W.; et al. *Bioorg. Med. Chem. Lett.* **2009**, *19*, 6957–6961.
5. Antonysamy, S.; Hirst, G.; Park, F.; Sprengeler, P.; Stappenbeck, F.; Steensma, R.; Wilson, M.; Wong, M. *Bioorg. Med. Chem. Lett.* **2009**, *19*, 279–282.
6. Ritzén, A.; Sørensen, M. D.; Dack, K. N.; Greve, D. R.; Jerre, A.; Carnerup, M. A.; Rytved, K. A.; Bagger-Bahnsen, J. *ACS Med. Chem. Lett.* **2016**, *7*, 641–646.
7. Roughley, S. D.; Hubbard, R. E. *J. Med. Chem.* **2011**, *54*, 3989–4005.
8. Buchstaller, H.-P.; Eggenweiler, H.-M.; Sirrenberg, C.; Graedler, U.; Musil, D.; Hoppe, E.; Zimmermann, A.; Schwartz, H.; Maerz, J.; Bomke, J.; et al. *Bioorg. Med. Chem. Lett.* **2012**, *22*, 4396–4403.
9. Chang, C.-F.; Lin, W.-H.; Ke, Y.-Y.; Lin, Y.-S.; Wang, W.-C.; Chen, C.-H.; Kuo, P.-C.; Hsu, J. T. A.; Uang, B.-J.; Hsieh, H.-P. *Eur. J. Med. Chem.* **2016**, *124*, 186–199.
10. Wood, S. D.; Grant, W.; Adrados, I.; Choi, J. Y.; Alburger, J. M.; Duckett, D. R.; Roush, W. R. *ACS Med. Chem. Lett.* **2017**, *8*, 1258–1263.

11. (a) Engel, J.; Richters, A.; Getlik, M.; Tomassi, S.; Keul, M.; Termathe, M.; Lategahn, J.; Becker, C.; Mayer-Wrangowski, S.; Gruetter, C.; et al. *J. Med. Chem.* **2015**, *58*, 6844–6863. (b) Tomassi, S.; Lategahn, J.; Engel, J.; Keul, M.; Tumbrink, H. L.; Ketzer, Ju.; Muehlenberg, T.; Baumann, M.; Schultz-Fademrecht, C.; Bauer, S.; et al. *J. Med. Chem.* **2017**, *60*, 2361–2372.

12. Jones, L. H.; Allan, G.; Barba, O.; Burt, C.; Corbau, R.; Dupont, T.; Knochel, T.; Irving, S.; Middleton, D. S.; Mowbray, C. E.; et al. *J. Med. Chem.* **2009**, *52*, 1219–1223.

13. Bodmer-Narkevitch, V.; Anthony, N. J.; Cofre, V.; Jolly, S. M.; Murphy, K. L.; Ransom, R. W.; Reiss, D. R.; Tang, C.; Prueksaritanont, T.; Pettibone, D. J.; et al. *Bioorg. Med. Chem. Lett.* **2010**, *20*, 7011–7014.

14. Zhang, J.; Yang, Q.; Romero, J. A. C.; Cross, J.; Wang, B.; Poutsiaka, K. M.; Epie, F.; Bevan, D.; Wu, Y.; Moy, T.; et al. *ACS Med. Chem. Lett.* **2015**, *6*, 1080–1085.

15. McCoull, W.; Bailey, A.; Barton, P.; Birch, A. M.; Brown, A. J. H.; Butler, H. S.; Boyd, S.; Butlin, R. J.; Chappell, B.; Clarkson, P.; et al. *J. Med. Chem.* **2017**, *60*, 3187–3197.

16. Shen, H.; Gou, S.; Shen, J.; Zhu, Y.; Zhang, Y.; Chen, X. *Bioorg. Med. Chem. Lett.* **2010**, *20*, 2115–2118.

17. (a) Bermudez, J.; Fake, C. S.; Joiner, G. F.; Joiner, K. A.; King, F. D.; Miner, W. D.; Sanger, G. J. *J. Med. Chem.* **1990**, *33*, 1924–1929. (b) Vernekar, S. K. V.; Hallaq, H. Y.; Clarkson, G.; Thompson, A. J.; Silvestri, L.; Lummis, S. C. R.; Lockner, M. *J. Med. Chem.* **2010**, *53*, 2324–2328.

18. Hughes, D. L. *Org. Process Res. Dev.* **2017**, *21*, 1227–1244.

19. Tian, Q.; Cheng, Z.; Yajima, H. M.; Savage, S. J.; Green, K. L.; Humphries, T.; Reynolds, M. E.; Babu, S.; Gosselin, F.; Askin, D.; et al. *Org. Process Res. Dev.* **2013**, *17*, 97–107.

17 Indoles

The most prominent feature of the molecule indole is its NH group. Since the lone pair of electrons on the nitrogen atom take part in maintaining indole's aromaticity, the NH is acidic (pK$_a$ ~ 17) and often serves as a hydrogen bond donor to the target protein. In terms of ligand–protein interactions, the indole fragment of a drug may offer π–π stacking or cation–π stacking with the target protein.

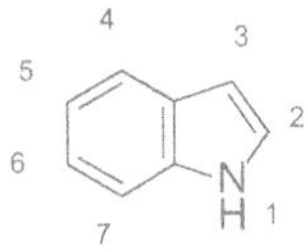

indole 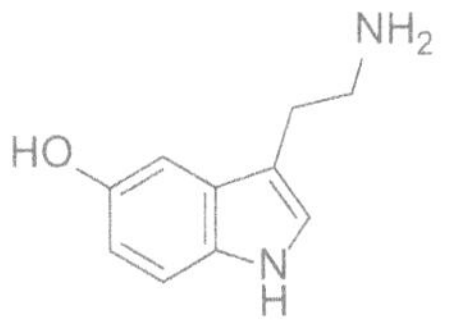   5-hydroxytryptamine (serotonin) 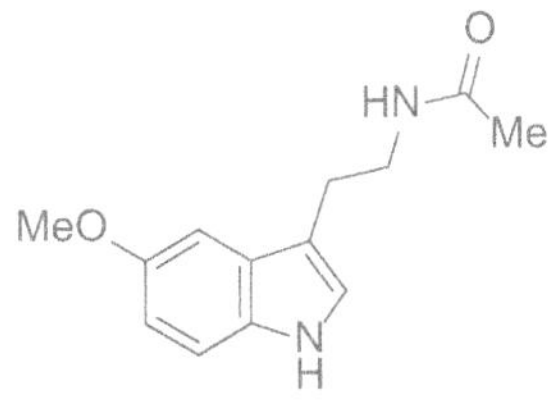   melatonin

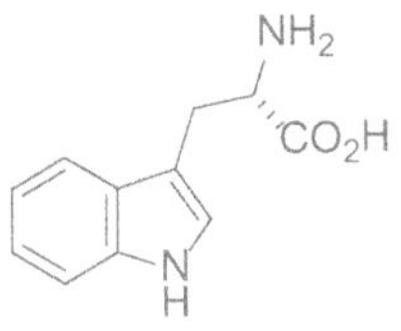

L-tryptophan 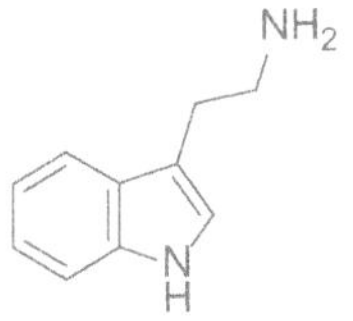   tryptamine

Indole embodies a myriad of natural products and pharmaceutical agents. In the human body, endogenous serotonin, 5-hydroxytryptamine (5-HT), is a monoamine neurotransmitter primarily found in the gastrointestinal tract (GIT) and central nervous system (CNS). It modulates vasoconstriction and many brain activities by binding to the serotonin receptors 5HT1–7. Another indole-containing endogenous ligand, melatonin, regulates circadian rhythms, most noticeably, sleep. In addition, indole-containing tryptamine is closely related to melatonin and the amino acid tryptophan.

INDOLE-CONTAINING DRUGS

In addition to the hundreds of well-known indole plant alkaloids (e.g., yohimbine, reserpine, strychnine, ellipticine, lysergic acid, physostigmine, and so on), the indole ring is present in dozens of the FDA-approved drugs. The central importance of indole derivatives such as serotonin and tryptophan in living organisms has inspired medicinal chemists to design and synthesize thousands of indole-containing pharmaceuticals.

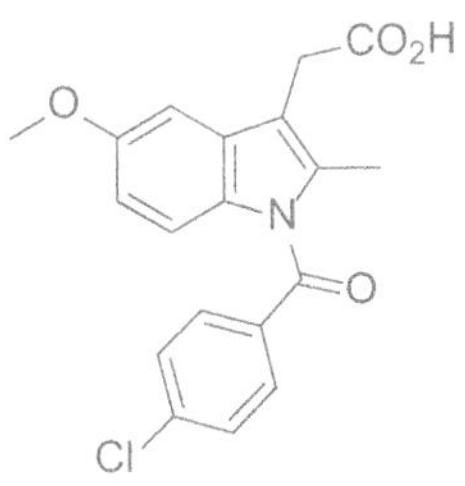

indomethacin (Indocin, **1**)
Merck, 1964
NSAID

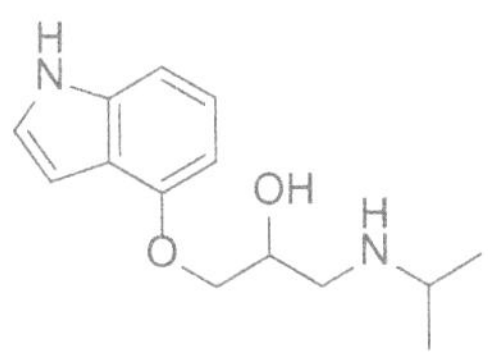

pindolol (Visken, **2**)
Sandoz, 1977
non-selective β-blocker

143

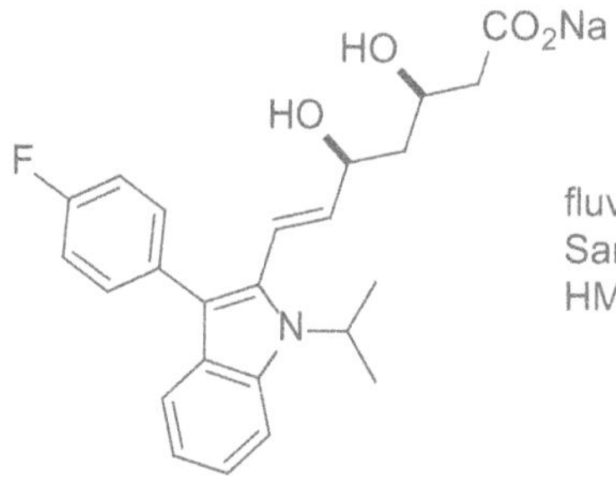

fluvastatin (Lescol, **3**)
Sandoz, 1984
HMG-CoA reductase inhibitor

One of the early drugs containing an indole ring is Merck's indomethacin (Indocin, **1**), a nonsteroidal anti-inflammatory drug (NSAID). Another early indole-containing drug was Sandoz's nonselective β-blocker pindolol (Visken, **2**). Among over 20 β-blockers on the market, "me-too" drug pindolol (**2**) has the same pharmacophore as the rest of them but its indole moiety provided Sandoz with novel intellectual properties (IP). Also by Sandoz, fluvastatin sodium (Lescol, **3**) is an HMG-CoA reductase inhibitor (a statin) for lowering cholesterol.

A class of indole-containing "triptans" are serotonin receptors 5-HT_{1B} and 5-HT_{1D} dual agonists used to treat migraines. The prototype was Glaxo's sumatriptan (Imitrex, **4**). Following sumatriptan (**4**)'s clinical and commercial success, three "me-too" indole-containing triptan antimigraine drugs were put on the market. They are naratriptan (Amerge, **5**), zolmitriptan (Zomig, **6**), and rizatriptan (Maxalt, **7**).

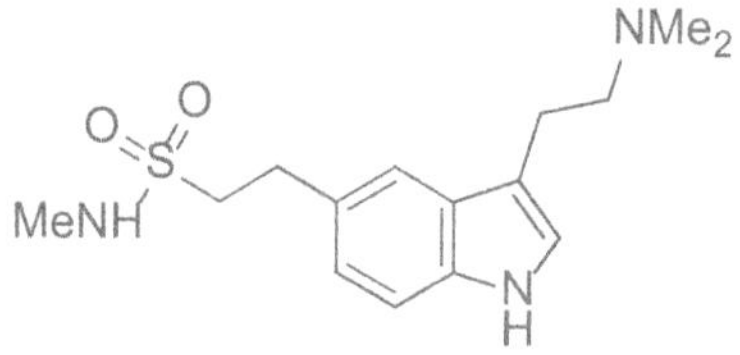

sumatriptan (Imitrex, **4**)
Glaxo, 1991
$5HT_{1D}$ and $5HT_{1B}$ agonist

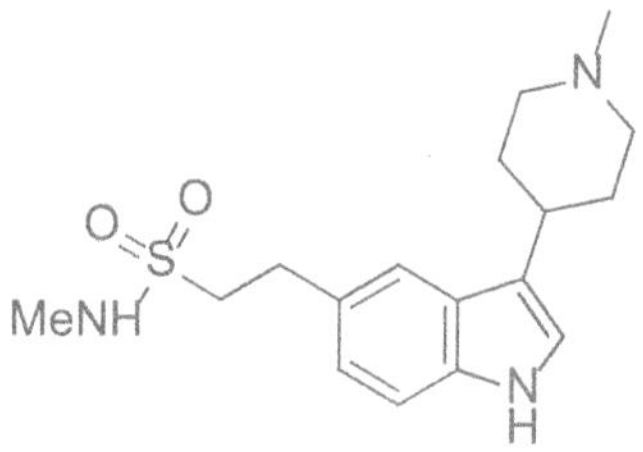

naratriptan (Naramig, **5**)
GSK, 1997
$5HT_{1D}$ and $5HT_{1B}$ agonist

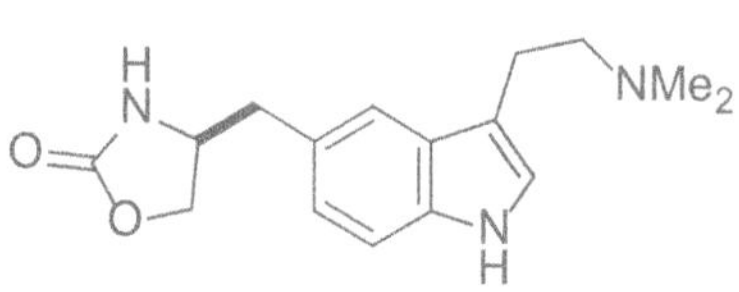

zolmitriptan (Zomig, **6**)
AstraZeneca, 1997
$5HT_{1D}$ and $5HT_{1B}$ agonist

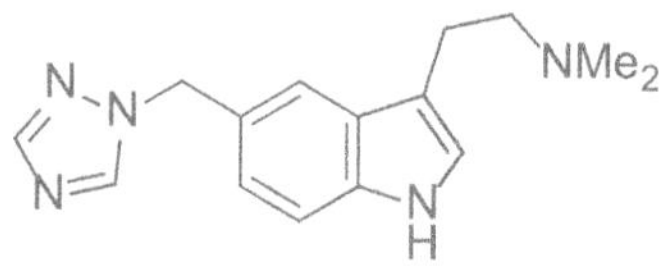

rizatriptan (Maxalt, **7**)
Merck, 1998
$5HT_{1D}$ and $5HT_{1B}$ agonist

Furthermore, Pfizer's zafirlukast (Accolate, **8**) is indicated for the treatment of mild-to-moderate asthma and chronic obstructive pulmonary disease (COPD). It is a selective and competitive antagonist of cysteinyl leukotriene-receptors (LTC_4, LTD_4, and LTE_4, LTRA). Meanwhile, Pfizer's delavirdine (Rescriptor, **9**) is a nonnucleoside reverse transcriptase inhibitor (NNRTI) for the treatment of HIV-positive patients.

Antiemetics ramosetron (Nasea, **10**) and dolasetron (Anzemet, **11**) are potent and highly selective 5-HT_3 receptor antagonists for the treatment of chemotherapy-induced nausea and vomiting.

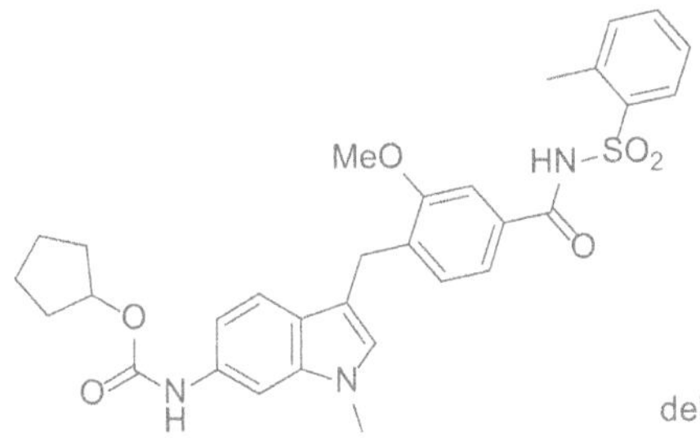

zafirlukast (Accolate, **8**)
Pfizer, 1997
LTRA

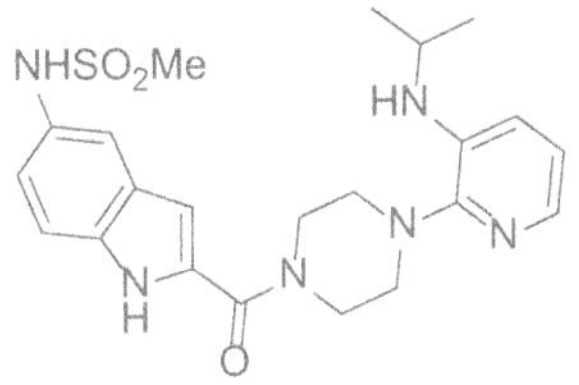

delavirdine (Rescriptor, **9**)
Pfizer, 1997
NNRTI

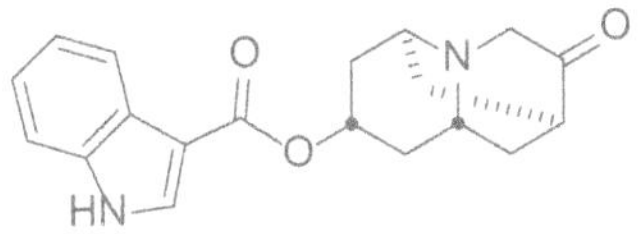

dolasetron (Anzemet, **10**)
Pfizer, 2002
5-HT$_3$ antagonist

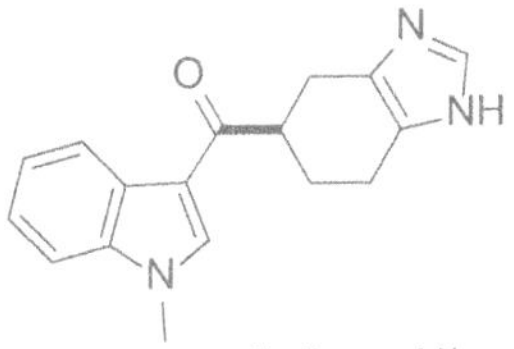

ramosetron (Irribow, **11**)
Astellas, 2008
5-HT$_3$ antagonist

Indole-containing vilazodone (Viibryd, **12**) bears some structural resemblance to antipsychotic drug ziprasidone (Geodon, a D$_2$ receptor antagonist). But in reality, it is a selective serotonin reuptake inhibitor (SSRI) and a 5TH$_{1A}$ agonist in terms of its mechanism of action (MOA). On the other hand, Novartis's panobinostat (Farydak, **13**) is a histone deacetylase (HDAC) inhibitor. While its hydroxamic acid motif is the pharmacophore/warhead bound to the catalytic zinc of HDAC, its indole fragment largely serves as a space-filler.

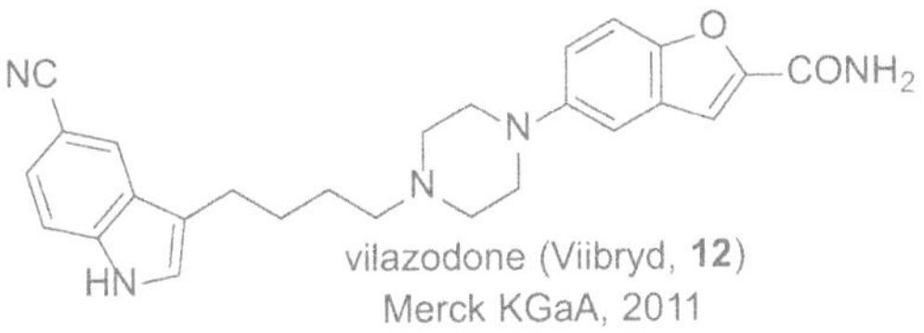

vilazodone (Viibryd, **12**)
Merck KGaA, 2011
SSRI and 5TH$_{1A}$ agonist

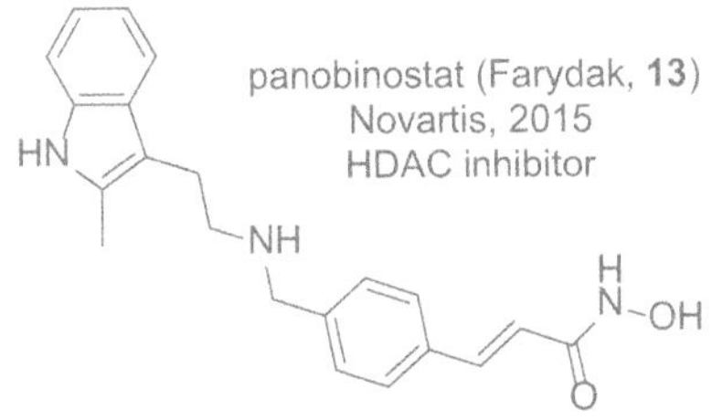

panobinostat (Farydak, **13**)
Novartis, 2015
HDAC inhibitor

Epidermal growth factor receptor (EGFR) inhibitors are among the earliest kinase inhibitors on the market. But resistance invariably developed and covalent inhibitors have been invented to combat the L858R and, more significantly, T790M mutations. AstraZeneca's third-generation EGFR inhibitor osimertinib (Tagrisso, **14**) is a covalent inhibitor expressly designed to overcome the T790M mutation by taking advantage of Cys-797 at EGFR's active site. It has been shown to have the efficacy and safety to treat patients with T790M-positive EGFR-TKI resistant nonsmall cell lung cancer (NSCLC). In addition, Clovis' selective poly(ADP-ribosyl) polymerase-1 (PARP-1) inhibitor rucaparib (Rubraca, **15**) contains a unique tricyclic indole core structure.

osimertinib (Tagrisso, **14**)
AZ, 2015
EGFR inhibitor (3rd generation)
C797S mutation

rucaparib (Rubraca, **15**)
Clovis, 2016
PARP1 inhibitor

Vertex has made great strides in the field of cystic fibrosis (CF) medicines. Its tezacaftor (**16**, Symdeko when combined with ivacaftor) helps move the cystic fibrosis transmembrane conductance regulator (CFTR) protein to the correct position on the cell surface and is designed to treat people with the F508del mutation. Finally, the latest approval of an indole-containing drug by the FDA is BMS's beclabuvir (Ximency, **17**), a hepatitis C virus nonstructural protein 5B (HCV NS5B, an RNA-dependent RNA polymerase, RdRp) inhibitor for the treatment of HCV infection. It has an interesting octacyclic indole core structure.

tezacaftor (**16**, Symdeko when
combined with ivacaftor)
Vertex, 2018
CFTR modulator

beclabuvir (Ximency, **17**)
BMS, 2019
HCV NS5B inhibitor

INDOLES IN DRUG DISCOVERY

Indole derivatives are ubiquitous in drug discovery. Rather than covering them exhaustively, only several representative bioisostere examples and potential safety liabilities of drugs containing a 3-methylindole moiety are reviewed here for brevity.

REPRESENTATIVE BIOISOSTERES

The NH group on indole often functions as a hydrogen bond donor with the drug target protein. The absence of the NH group often weakens the binding affinity exponentially. For instance, indole gloxylylamine **18** is a partial agonist of the benzodiazepine receptor with a reasonably good binding affinity ($K_i = 85$ nM). Isosteric replacement of the indole nucleus with benzothiophene led to compound **19**, which saw a 40-fold decrease of binding affinity toward the benzodiazepine receptor ($K_i = 3.37$ μM).[1]

The perils of anilines have been well known as they are frequently oxidized by CYP450 *in vivo* to the corresponding iminoquinones as reactive metabolites. In contrast, the corresponding indole analogs are more benign.

Tertiary aniline **20** is a selective androgen receptor degrader (SARD) but suffered from poor metabolic stability, thus lacking *in vivo* activity when administered orally. Fusion of the tertiary aniline of **20** into the B-ring led to a series of indole derivatives, as represented by **21** that retained both the AR inhibition and SARD activities with very high potency but also with improved ligand efficiency and *in vitro* stability. Interestingly, even though the enantiomer of indole **21** had a very low binding affinity for AR ($K_i > 10\,\mu M$), it demonstrated comparably high SARD activities, possibly because it works as a proteolysis-targeting chimera (PROTAC).[2]

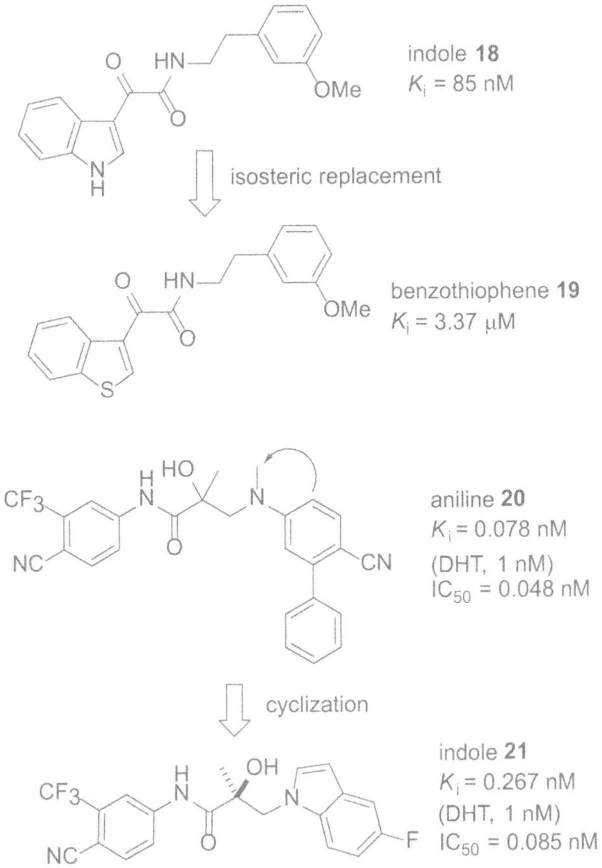

Inhibition of HCV NS5B has been proven to be a fruitful MOA for treating HCV infection. Validation came when the FDA approved Gilead/PharmaTech's sofosbuvir (Sovaldi) in 2013. It quickly became a blockbuster drug and has contributed to the process of curing HCV.

Benzimidazole-5-carboxamide **22** is a potent *allosteric* HCV NS5B inhibitor, binding to the polymerase's thumb pocket 1 finger loop. Although it has high intrinsic potency against polymerase ($IC_{50} = 40\,nM$), its cell-based sub-genomic 1b replicon activity is low ($EC_{50} = 1.86\,\mu M$). Replacement of the benzimidazole core of **22** (ClogP = 4.4) with more lipophilic indole-5-carboxamide analogs led to a series of inhibitors with up to 30-fold improvements in cell-based sub-genomic 1b replicon assay. Optimization of C-2 substitution on the indole core led to the identification of analogs such as indole-5-carboxamide **23** (ClogP = 4.9) with EC_{50} of 60 nM (1b replicon). Furthermore, they showed improved pharmacokinetic properties as well.[3]

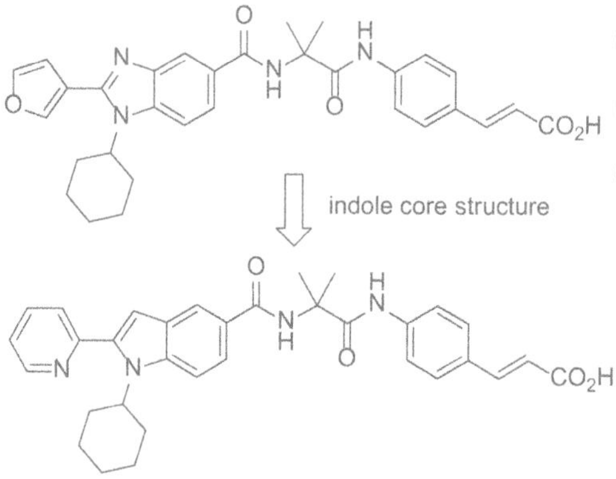

benzimidazole **22**
IC_{50} = 40 nM
EC_{50} = 1.86 μM
(1b replicon)
Clog P = 4.4

indole core structure

indole **23**
IC_{50} = 40 nM
EC_{50} = 60 nM
(1b replicon)
Clog P = 4.9

POSSIBLE LIABILITIES OF DRUGS CONTAINING 3-METHYLINDOLE

The indole-ring system exists in a plethora of endogenous amino acids, neurotransmitters, and drugs. The metabolic 2,3-oxidation of the indole ring by CYP450 takes place from time to time (by indole oxygenase, IDO), but its correlation to *in vivo* toxicity is not often observed.

3-Methylindole, unfortunately, has been associated with a higher risk of adverse outcomes, namely, pneumotoxin in animals. Evidence was found to support the formation of 2,3-epoxy-3-methylindoline by IDO as a reactive intermediate of the pneumotoxin 3-methylindole.[4] 3-Methylindole has been shown to form adducts with glutathione (GSH), proteins, and DNA using *in vitro* preparations.[5]

The CYP450-mediated bioactivation of 3-methylindole may be summarized here: oxidation of the 3-methyl group occurs either directly via deoxygenation or via epoxidation of the 2,3-double bond leading to 2,3-epoxy-3-methylindole, the reactive intermediate that can be trapped by endogenous nucleophiles, such as GSH. The presence of a leaving group on the C3-methyl increases the likelihood of the formation of electrophilic reactive intermediates.

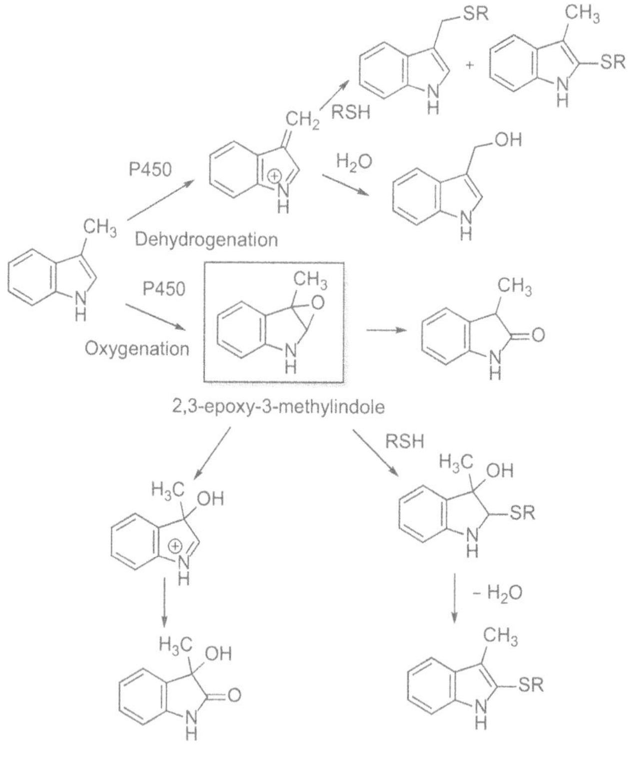

2,3-epoxy-3-methylindole

Zafirlukast (**8**) has been associated with occasional idiosyncratic hepatotoxicity. Structurally, zafirlukast (**8**) is similar to 3-methylindole because it contains an *N*-methylindole moiety that has a 3-alkyl substituent on the indole ring. The results presented here describe the metabolic activation of zafirlukast (**8**) via a similar mechanism to that described for 3-methylindole. NADP(H)-dependent biotransformation of zafirlukast (**8**) by hepatic microsomes from rats and humans afforded a reactive metabolite, which was detected as its GSH adduct **24**.[6] The formation of this reactive metabolite in human liver microsomes (HLM) was shown to be exclusively catalyzed by CYP3A enzymes. Evidence for *in vivo* metabolic activation of zafirlukast (**8**) was obtained when the same GSH adduct **24** was detected in the bile of rats given an *i.v.* or oral dose of the drug.

The observation of *in vitro* metabolic activation of the 3-benzylindole moiety in zafirlukast (**8**) to give the glutathione adduct **24** is an indication that the 3-methyl-indole activation pathway applies to other activated 3-alkyl indoles as well.[7]

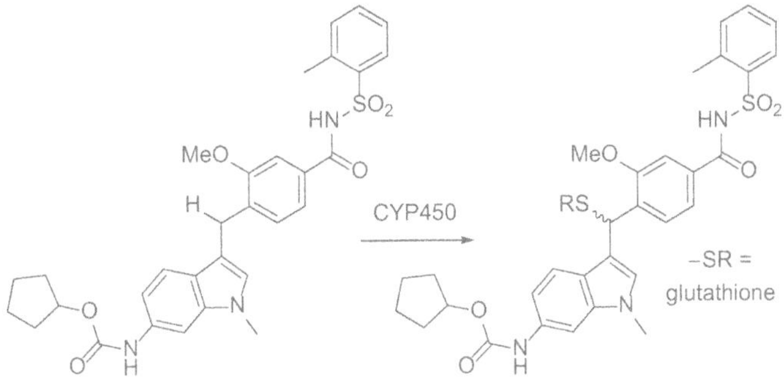

zafirlukast (**8**)   zafirlukast-GSH adduct **24**

SYNTHESIS OF SOME INDOLE-CONTAINING DRUGS

AstraZeneca's synthesis of osimertinib (**14**) commenced with condensation of 2,4-dichloro-pyrimidine (**25**) with N-methylindole to assemble adduct **26** with the help of $FeCl_3$. Further S_NAr coupling of **26** with aniline **27** in the presence of TsOH afforded 2-phenylamino-3-indolylpyrimidine **28**, which was converted to the desired osimertinib (**14**) in three additional steps.[8]

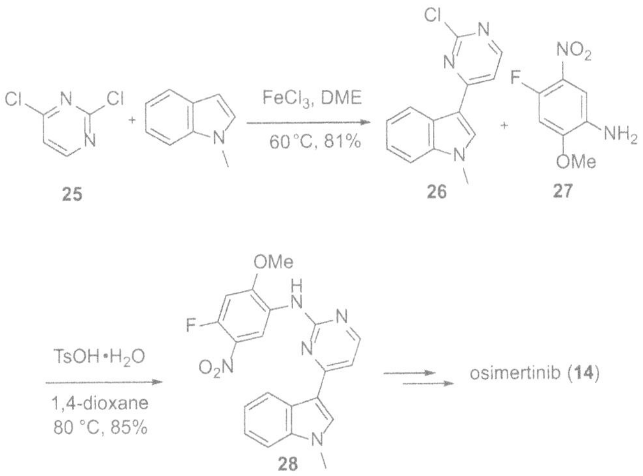

Clovis' synthesis of rucaparib (**15**) involved a reductive condensation of methyl 6-fluoro-1*H*-indole-4-carboxylate (**29**) and freshly released aldehyde **30** (to avoid polymerization) with the aid of triethylsilane and TFA to assemble tryptamine **31**. Exposure of phthalimide **31** to aqueous

methylamine released the primary amine, which underwent a simultaneous intramolecular cyclization to produce the intermediate lactam, which was smoothly brominated at the C-2 position of the
indole to afford bromide **32** in 74% yield for two steps. Subsequently, a Suzuki coupling between
indolyl-2-bromide **32** and the requisite arylboronic acid was followed by reductive amination with
another molecule of methylamine to deliver rucaparib (**15**) in good yields.[9]

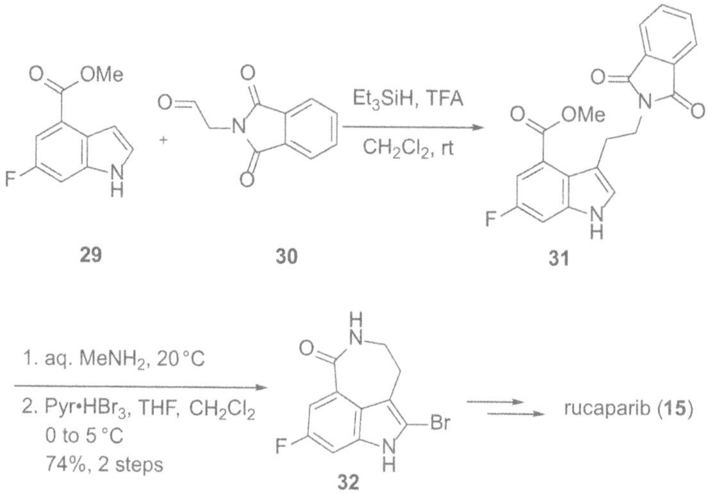

BMS's preparation of their HCV NS5B inhibitor beclabuvir (**17**) began with condensation of
1*H*-indole-6-carboxylic acid with cyclohexanone to prepare indolyl cyclohexene **33** in quantitative yield. Palladium-catalyzed hydrogenation of **33** was followed by esterification to afford
indole **34**. Again, bromination was accomplished with pyridine · HBr_3 complex to provide indolyl-
2-bromide **35**. Eventually, the octacyclic indole was assembled after seven additional steps to
deliver beclabuvir (**17**).[10]

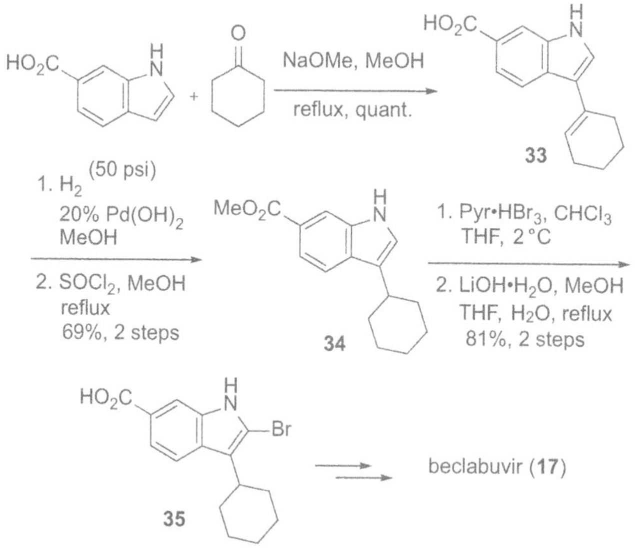

In summary, the key interactions of the indole nucleus in a drug include NH serving as a hydrogen bond donor and π–π stacking or cation–π stacking with the target protein. The indole motif is

a privileged fragment in drug discovery, boasting more than 17 marketed indole-containing drugs, with the majority of them as kinase inhibitors. In particular, amino- and diamino-indole fragments are especially prevalent.

REFERENCES

1. Settimo, F. D.; Lucacchini, A.; Marini, A. M.; Martini, C.; Primofiore, G.; Senatore, G.; Taliani, S. *Eur. J. Med. Chem.* **1996**, *31*, 951–956.
2. Hwang, D.-J.; He, Y.; Ponnusamy, S.; Mohler, M. L.; Thiyagarajan, T.; McEwan, I. J.; Narayanan, R.; Miller, D. D. *J. Med. Chem.* **2019**, *62*, 491–511.
3. (a) Beaulieu, P. L.; Gillard, J.; Jolicoeur, E.; Duan, J.; Garneau, M.; Kukolj, G.; Poupart, M.-A. *Bioorg. Med. Chem. Lett.* **2011**, *21*, 3658–3663. (b) Beaulieu, P. L.; Chabot, C.; Duan, J.; Garneau, M.; Gillard, J.; Jolicoeur, E.; Poirier, M.; Poupart, M.-A.; Stammers, T. A.; Kukolj, G.; et al. *Bioorg. Med. Chem. Lett.* **2011**, *21*, 3664–3670.
4. Skordos, K.; Skiles, G. L.; Laycock, J. D.; Lanza, D. L.; Yost, G. S. *Chem. Res. Toxicol.* **1998**, *11*, 741–749.
5. Regal, K. A.; Laws, G. M.; Yuan, C.; Yost, G. S.; Skiles, G. L. *Chem. Res. Toxicol.* **2001**, *14*, 1014–1024.
6. Kassahun, K.; Skordos, K.; McIntosh, I.; Slaughter, D.; Doss, G. A.; Baillie, T. A.; Yost, G. S. *Chem. Res. Toxicol.* **2005**, *18*, 1427.
7. Blagg, J. *Structural Alerts*, In Abraham, D. J.; Rotella, D. P., eds.; *Burger's Medicinal Chemistry, Drug Discovery and Development*, 7th ed., Wiley: Hoboken, NJ, **2010**, *Vol. 3*, pp 301–334.
8. (a) Finlay, M. R. V.; Anderton, M.; Ashton, S.; Ballard, P.; Bethel, P. A.; Box, M. R.; Bradbury, R. H.; Brown, S. J.; Butterworth, S.; Campbell, A.; et al. *J. Med. Chem.* **2014**, *517*, 8249–8267. (b) Butterworth, S.; Finlay, M. R.; Verschoyle; Ward, R. A.; Kadambar, V. K.; Chandrashekar, R. C.; Murugan, A.; Redfearn, H. M. WO2013014448 (2013).
9. (a) Gillmore, A. T.; Badland, M.; Crook, C. L.; Castro, N. M.; Critcher, D. J.; Fussell, S. J.; Jones, K. J.; Jones, M. C.; Kougoulos, E.; Mathew, J. S.; et al. *Org. Process Res. Dev.* **2012**, *16*, 1897–1904. (b) Li, W.; Li, J.; De Vincentis, D.; Mansour, T. S. *Tetrahedron* **2008**, *64*, 7871–7876.
10. (a) Bender, J. A.; Ding, M.; Gentles, R. G.; Hewawasam, P. WO20070270405 (2007). (b) Gentles, R. G.; Ding, M.; Bender, J. A.; Bergstrom, C. P.; Grant-Young, K.; Hewawasam, P.; Hudyma, T.; Martin, S.; Nickel, A.; Regueiro-Ren, A.; et al. *J. Med. Chem.* **2014**, *57*, 1855–1879.

18 Oxetanes

Oxetane adopts a rigid and slightly puckered (8.7°) conformation. As a bioisostere of dimethyl and carbonyl groups, oxetane is more metabolically stable and lipophilicity neutral.[1] Since oxetane is an electron-withdrawing group, it reduces the basicity of its adjacent nitrogen atom and the subtle modulation of the basicity may lower the drug's overall lipophilicity. The last two decades have seen a flurry of oxetane's utility in medicinal chemistry. Over a dozen oxetane-containing drugs have now progressed to different phases of clinical trials.

OXETANE-CONTAINING DRUGS

Three FDA-approved oxetane-containing drugs are taxol (**1**) and its two semisynthetic brethren: Sanofi's docetaxel (Taxotere, **2**) and cabazitaxel (Jevtana, **3**), all chemotherapies for treating cancer. Their mechanism of action (MOA) is disrupting protein microtubule functions in the cell, which pull apart the chromosomes before cell division (mitosis). Computational studies showed that the oxetane moiety provides: (a) rigidification of the overall structure and (b) a H-bond acceptor for a threonine-OH group in the binding pocket. Furthermore, any permutation of the oxetane ring such as replacing the oxygen atom with sulfur or nitrogen resulted in lower activities.

paclitaxel (Taxol, **1**)
BMS, 1993
Disrupt microtubules

docetaxel (Taxotere, **2**)
Sanofi, 1995
Disrupt microtubule function

cabazitaxel (Jevtana, **3**)
Sanofi, 2010
Disrupt microtubule function

OXETANES IN DRUG DISCOVERY

Oxetanes are frequently utilized to improve the physiochemical properties of drugs.

An oxetane fragment conferred a profound impact on Pfizer's zeste homolog 2 (EZH2) inhibitors, especially their oral bioavailability while still maintaining the activity.[2] The initial lead compound bicyclic lactam **4** was active in both enzymatic (the Y641N mutant form of the enzyme) and cellular-based assays and displayed impressive tumor growth inhibition effects in the Karpas-422 xenograft model. Regrettably, **4** was extensively metabolized (HLM Cl = 169 µL/min/mg in protein), had poor permeability in the MDCK-LE assay, and had low thermodynamic solubility due to its high crystal-linity (mp, 246°C). One of the fundamental tactics to improve a drug's solubility is to break its aroma-ticity by adding more sp^3 carbons and oxetane fits the bill. After extensive optimization using ligand and property-based design strategies (especially lipophilic ligand efficiency, LipE), Pfizer arrived at the oxetane-containing PF-06821497 (**5**) where the aromatic dimethylisoxazole on **4** was replaced by all sp^3 centers. PF-06821497 (**5**) displayed the best combination of EZH2 inhibitory activity, LipE, *in vitro* metabolic stability, and permeability characteristics. More importantly, it had a drastically improved (150-fold) thermodynamic solubility over **4**. In comparison, the corresponding tetrahydro-furan analogs had similar potency as **4** but were less permeable. The corresponding tetrahydropyran analogs were more lipophilic, had lower kinetic solubility, and were relatively more labile in the *in vitro* metabolic stability assessments without any appreciable boost of potency. After thorough pro-filing of PF-06821497 (**5**) in terms of PK/PD as a drug candidate, it has been advanced to clinical trials after it was shown to display robust tumor growth inhibition activity in mouse xenograft models along with strongly associated pharmacodynamic effects such as reduction of H3K27me3 in tumors.[2]

lactam-isoxazole **4**
EZH2 Y641N K_i = 1.2 nM
HLM Cl 169 μL/min/mg protein
LipE = 5.8
Solubility = 2 μg/mL

PF-06821497 (**5**)
EZH2 Y641N K_i < 0.1 nM
HLM Cl 39 μL/min/mg protein
LipE = 8.3
Solubility = 315 μg/mL

Inhibitors of dual leucine zipper kinase (DLK, MAP3K12), prominent in the regulation of neu-ronal degradation, have potential as a treatment of neurodegenerative diseases such as Alzheimer's disease (AD). Using an initial hit from high-throughput screening (HTS) as a starting point, Genentech arrived at a potent DLK inhibitor **6** (K_i=33 nM). Since it was extensively metabolized, a scaffold-hopping campaign produced piperidine-oxetane **7**. Here, an oxetane was successfully used to reduce the basicity of piperidine to limit efflux, important for a brain penetrant, while maintain-ing good metabolic stability.[3] Further efforts to improve **7**'s potency, kinase selectivity, and drug-like properties befitting a brain-penetrant therapeutic delivered azabicyclo[3.1.0]-hexane pyrazole **8** (K_i=3 nM). With favorable *in vitro* safety properties and *in vivo* tolerability and efficacy in animal models, DLK inhibitor **8** has been advanced to clinical trials.[4]

6
DLK K_i = 33 nM (LipE 2.9)
pJNK IC_{50} = 300 nM
Rat CLu = 4500 mL/min/kg

7
DLK K_i = 42 nM (LipE 3.7)
pJNK IC_{50} = 536 nM
Rat CLu = 327 mL/min/kg

8
DLK K_i = 3 nM (LipE 5.8)
Clog P = 2.8
tPSA = 78

Oxetan-3-ol has been evaluated as a bioisostere of carboxylic acid.[5] The acid functionality on the household analgesic ibuprofen (**9**, pKa=4.64) is negatively ionized under physiological conditions that are responsible for an insufficient passive diffusion across biological membranes. In stark contrast, as an analog of ibuprofen (**9**), oxetan-3-ol **10** is mostly neutral (pKa>2) at physiological pH and comparatively more lipophilic and more permeable in comparison to ibuprofen (**9**). Given the relatively low acidity and high permeability, oxetan-3-ol **10** may be useful in the context of central nervous system (CNS) drug design. Moreover, oxetan-3-ol **10** also inhibits 5-lipoxygenase-derived leukotriene B_4 (LTB$_4$) while ibuprofen (**9**) is inactive against this target.[5]

ibuprofen (**9**)
pKa = 4.64
PGE$_2$/D$_2$ IC$_{50}$ = 0.6 µM
LTB$_4$ IC$_{50}$ > 100 µM

oxetan-3-ol **10**
pKa > 12
PGE$_2$/D$_2$ IC$_{50}$ = 34.1 µM
LTB$_4$ IC$_{50}$ = 8.4 µM

As a cathepsin S inhibitor, Lilly's lead tetrahydronaphthalene **11** was not very potent (IC$_{50}$=4 µM). An extensive structure–activity relationship (SAR) campaign was exercised to optimize the lead compound. A key aspect of this modification process was replacing the N-methyl group on **11**'s piperazine with an N-oxetanyl unit to modulate the basicity of the nitrogen atom and to lower the overall lipophilicity. This modification, along with others, led to the development of clinical candidate LY3000328 (**12**) for the treatment of abdominal aortic aneurysm (AAA).[6]

11
hCat S IC$_{50}$ = 4,000 nM
mCat S IC$_{50}$ = 3,930 nM

LY3000328 (**12**)
hCat S IC$_{50}$ = 7.7 nM
mCat S IC$_{50}$ = 1.7 nM

Trifluoromethyl oxetane may serve as a less lipophilic bioisostere of the *tert*-butyl group. AstraZeneca's G-protein coupled receptor 119 (GPR119) agonist **13** as a potential diabetes treatment, although potent, suffered from poor aqueous solubility (24 µM). After extensive experimentations, trifluoromethyl oxetane derivative **14** had a desirable aqueous solubility of 116 µM and was found to be the best combination of potency boost and metabolic stability. While the corresponding ethyl and isopropyl-oxetane derivatives proved unstable in human liver microsomes (HLM), no metabolism of the trifluoromethyl oxetane was detected. Similarly, the ring expanded homolog of **14**, trifluoromethyl tetrahydrofuran, had a similar potency, higher lipophilicity, which resulted in lower solubility, higher clearance in HLM, and an erosion of ligand-lipophilicity efficiency (LLE = pEC_{50} − log D).[7]

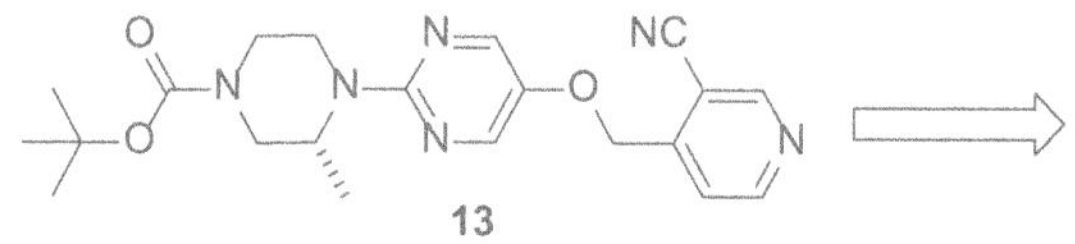

13

aqueous solubility, 24 μM

$Log D_{7.4} = 3.3$

mp = 91–93°C

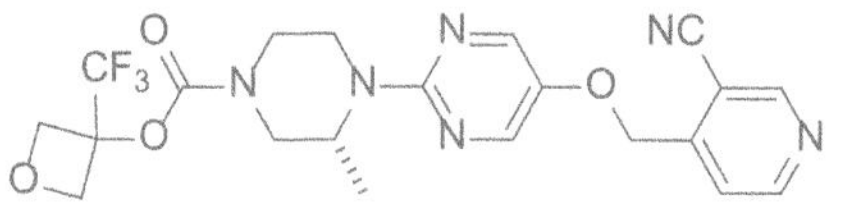

14

aqueous solubility, 110 μM

$Log D_{7.4} = 2.5$

mp = 53–55°C

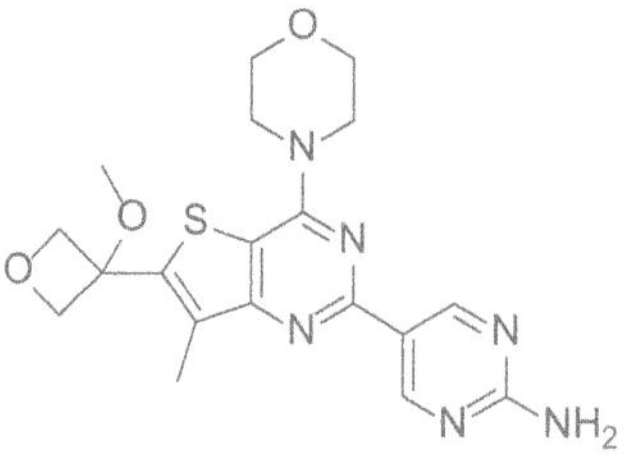

PI3K inhibitor **15**

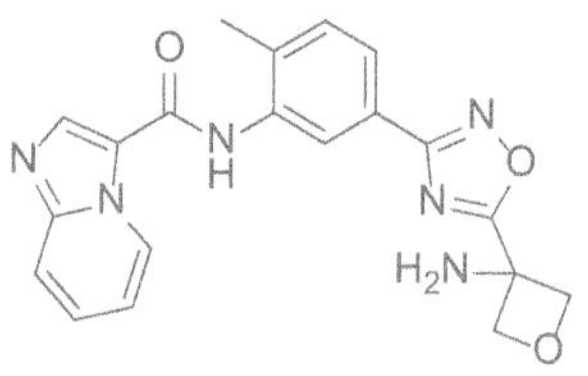

C-kit kinase inhibitor **16**

Not all oxetane moieties on drugs are in the simplest form as on **10** and **12**. More substituted oxetanes have made their way to be parts of drugs. For instance, Genentech's phosphatidylinositol 3-kinase (PI3K) inhibitor **15** contains a 3-methoxy-oxetane, which is helpful to boost brain penetration.[8] A C-kit kinase inhibitor **16** possesses a 3-amino-oxetane motif.[9] Merck's tyrosine kinase MET inhibitor **17** has a 3-fluorooxetane fragment[10] and AstraZeneca's melanin-concentrating hormone receptor 1 (MCHr1) antagonist AZD1979 (**18**) is appended with a spirocyclic azetidine-oxetane substituent, which imparts favorable physiochemical properties.[11]

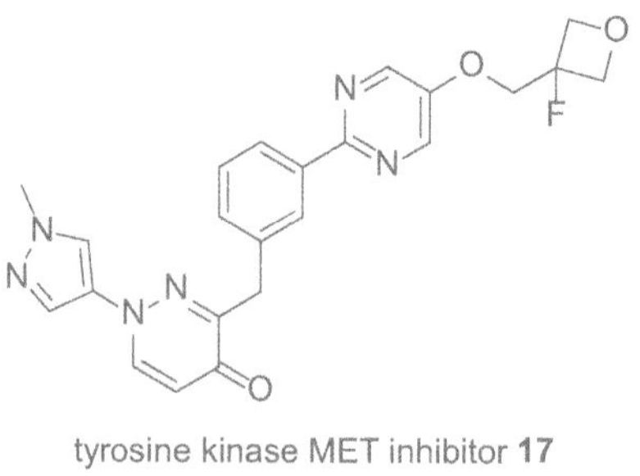

tyrosine kinase MET inhibitor **17**

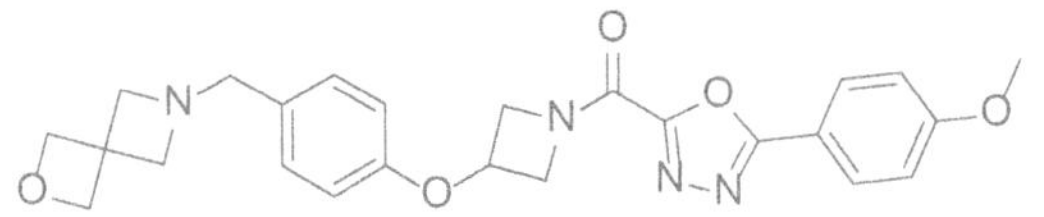

MCHr1 antagonist **18**

SYNTHESIS OF SOME OXETANE-CONTAINING DRUGS

Genentech's synthesis of its PI3K inhibitor **15** installed the 3-methoxy-oxetane group using oxetanone as the starting material. Thus, lithiation of the thiophene moiety on morpholinopyrimidine **19** was followed by the addition of oxetanone to give rise to 3-oxetanol **20**. Methylation of **20** was straight-forward to afford 3-methoxy-oxetane **21** and a subsequent Suzuki coupling with 2-aminopyrimidine-5-boronic acid pinacol ester (**22**) assembled the desired **15**.[8]

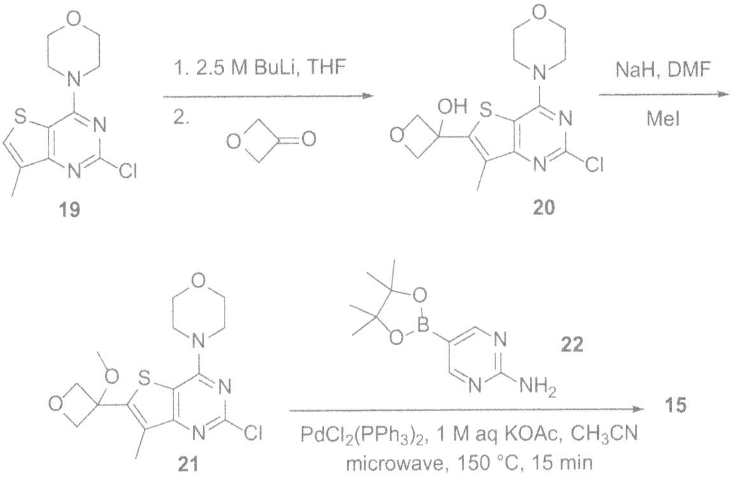

Synthesis of Pfizer's crenolanib (**25**), an inhibitor of FMS-like tyrosine kinase 3 (FLT3) and platelet-derived growth factor receptor-α/β (PDGFRα/β), was also uneventful. A simple S_N2 reaction between the advanced intermediate phenol **23** and 3-(bromomethyl)-3-methyloxetane (**24**) prepared crenolanib (**25**) after acidic removal of the Boc protection.[12]

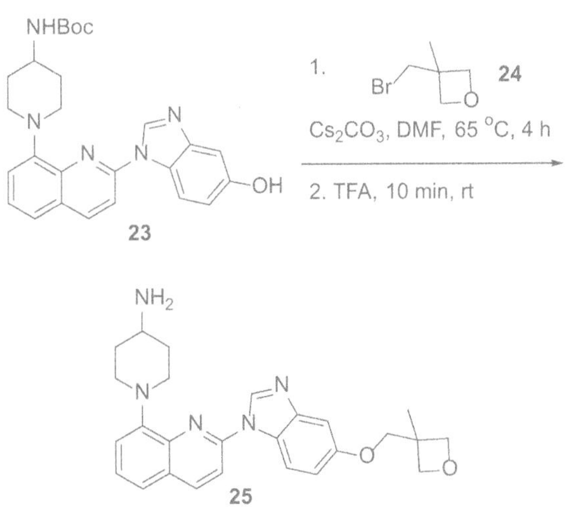

A synthesis of C-kit kinase inhibitor **16** employed 3-aminooxetane-3-carboxylic acid (**26**) as the source of its 3-amino-oxetane motif. After protection of **26** with Boc, the resultant **27** was then condensed with intermediate **28** to produce **16** after acidic removal of the Boc protection.[9]

To conclude, oxetanes have been employed to improve the physiochemical properties of drugs. Currently, over a dozen oxetane-containing drugs have progressed to different phases of clinical trials. Once one of them gains the FDA approval, the enthusiasm toward its utility in drug discovery will grow exponentially.

REFERENCES

1. Bull, J. A.; Croft, R. A.; Davis, O. A.; Doran, R.; Morgan, K. F. *Chem. Rev.* **2016**, *116*, 12150–12233.
2. Kung, P.-P.; Bingham, P.; Brooun, A.; Collins, M.; Deng, Y.-L.; Dinh, D.; Fan, C.; Gajiwala, K. S.; Grantner, R.; Gukasyan, H. J.; et al. *J. Med. Chem.* **2018**, *61*, 650–665.
3. Patel, S.; Harris, S. F.; Gibbons, P.; Deshmukh, G.; Gustafson, A.; Kellar, T.; Lin, H.; Liu, X.; Liu, Y.; Ma, C.; et al. *J. Med. Chem.* **2015**, *58*, 8182–8199.
4. Patel, S.; Meilandt, W. J.; Erickson, R. I.; Chen, J.; Deshmukh, G.; Estrada, A. A.; Fuji, R. N.; Gibbons, P.; Gustafson, A.; Harris, S. F.; et al. *J. Med. Chem.* **2017**, *60*, 8083–8102.
5. Lassalas, P.; Oukoloff, K.; Makani, V.; James, M.; Tran, V.; Yao, Y.; Huang, L.; Vijayendran, K.; Monti, L.; Trojanowski, J. Q.; et al. *ACS Med. Chem. Lett.* **2017**, *8*, 864–868.
6. Jadhav, P. K.; Schiffler, M. A.; Gavardinas, K.; Kim, E. J.; Matthews, D. P.; Staszak, M. A.; Coffey, D. S.; Shaw, B. W.; Cassidy, K. C.; Brier, R. A.; et al. *ACS Med. Chem. Lett.* **2014**, *5*, 1138–1142.
7. Scott, J. S.; Birch, A. M.; Brocklehurst, K. J.; Brown, H. S.; Goldberg, K.; Groombridge, S. D.; Hudson, J. A.; Leach, A. G.; MacFaul, P. A.; McKerrecher, D.; et al. *MedChemComm* **2013**, *4*, 95–100.
8. Heffron, T. P.; Salphati, L.; Alicke, B.; Cheong, J.; Dotson, J.; Edgar, K.; Goldsmith, R.; Gould, S. E.; Lee, L. B.; Lesnick, J. D.; et al. *J. Med. Chem.* **2012**, *55*, 8007–8020.
9. Liu, X.; Li, X.; Loren, J.; Molteni, V.; Nabakka, J.; Nguyen, B.; Petrassi, H. M. J.; Yeh, V. (IRM LLC). WO2013033116 (2013).
10. Young, J.; Czako, B.; Altman, M.; Guerin, D.; Martinez, M.; Rivkin, A.; Wilson, K.; Lipford, K.; White, C.; Surdi, L.; et al. (Merck) WO2011084402 (2011).
11. Johansson, A.; Loefberg, C.; Antonsson, M.; von Unge, S.; Hayes, M. A.; Judkins, R.; Ploj, K.; Benthem, L.; Linden, D.; Brodin, P.; et al. *J. Med. Chem.* **2016**, *59*, 2497–2511.
12. Kath, J. C.; Lyssikatos, J. P.; Wang, H. F. (Pfizer) WO2004020431 (2014).

19 Piperidine, the Enchanted Ring

PIPERIDINE-CONTAINING DRUGS

Thanks to its ubiquitous presence in drugs, piperidine is a truly "enchanted" ring. Like many nitrogen-containing drugs, when charged, piperidine may enhance solubility and offer additional binding to the targets. Meanwhile, when neutral, piperidine-containing drugs may cross the cell membrane more readily.

Paul Janssen bestowed us with two powerful, piperidine-containing drugs. One is haloperidol (Haldol, 1), a typical antipsychotic. The other is fentanyl (Duragesic, 2), which is 100-fold more potent than morphine. Fentanyl (2) has contributed much to today's opioid epidemic.

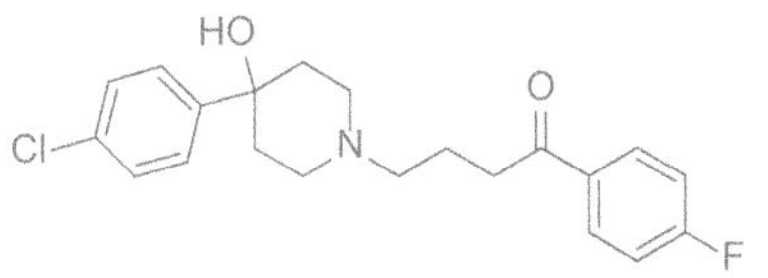

haloperidol (Haldol, 1)
Janssen, 1959
typical anti-psychotic

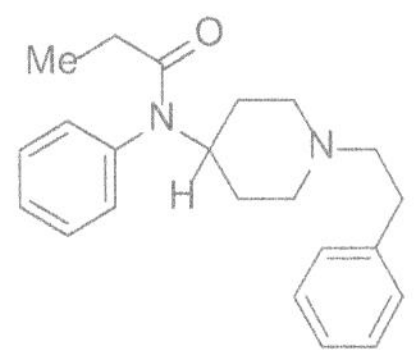

fentanyl (Duragesic, 2)
Janssen, 1959
100X more potent than morphine

Piperidine is one of the "privileged scaffolds", present in drugs encompassing all therapeutic areas. In addition to the alkylated forms such as haloperidol (1) and fentanyl (2), some piperidines exist in the "naked" form, i.e., the NH form. Paroxetine (Paxil, 3) is a selective serotonin reuptake inhibitor (SSRI) for treating depression and niraparib (Zejula, 4) is a poly(ADP-ribosyl) polymerase (PARP) inhibitor for treating ovarian cancer. Furthermore, two kinase inhibitors also have the "naked" form of the piperidine ring. Pfizer's crizotinib (Xalkori, 5) is an anaplastic lymphoma kinase (ALK) inhibitor and Exelixis's cobimetinib (Cotellic, 6) is a mitogen-activated protein kinase-1/2 (MEK1/2) inhibitor. Both crizotinib (5) and cobimetinib (6) are targeted cancer therapies.

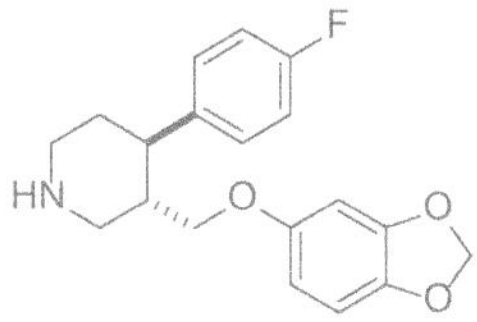

paroxetine (Paxil, 3)
GSK, 1992
SSRI

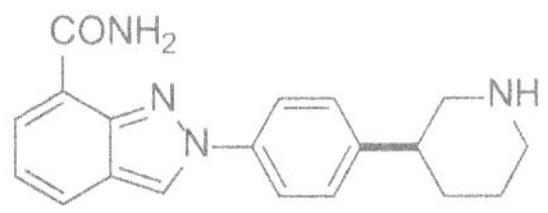

niraparib (Zejula, 4)
Tesaro, 2017
PARP inhibitor

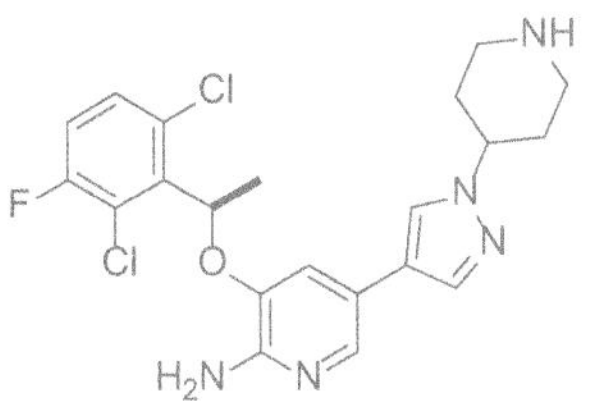

crizotinib (Xalkori, 5)
Pfizer, 2011
ALK/ROS Inhibitor

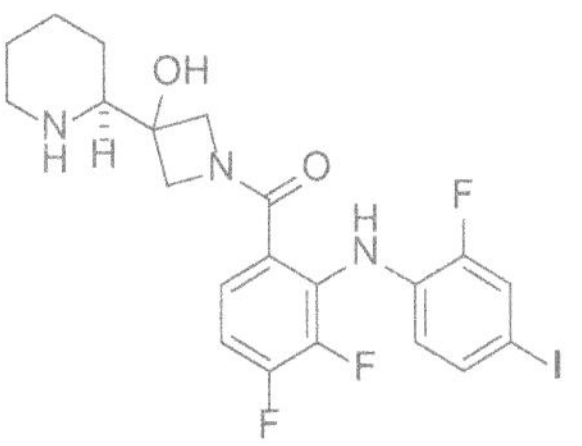

cobimetinib (Cotellic, 6)
Exelixis/Genentech, 2015
MEK1/2 inhibitor

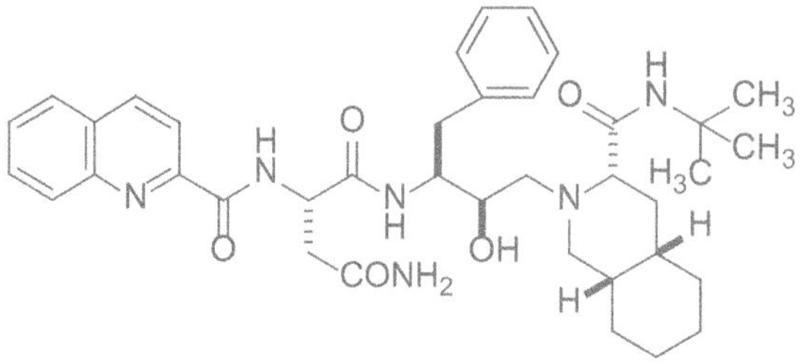

saquinavir (Invirase, **7**)
Merck, 1995
HIV protease inhibitor

Many more complicated, substituted piperidine rings also exist in drugs and potential drugs. Merck's HIV protease inhibitor saquinavir (Invirase, **7**) contains piperidine as part of a bicyclic architecture.

While drugs **1–7** are marketed drugs, many piperidine-containing drugs are in the discovery or development stages. For example, 3-fluoro-1,4-substituted piperidine **8** is a selective T-type calcium channel inhibitor[1] and 3,5-disubstituted piperidine **9** is an orally active renin inhibitor with an improved pharmacokinetic profile over older ones.[2]

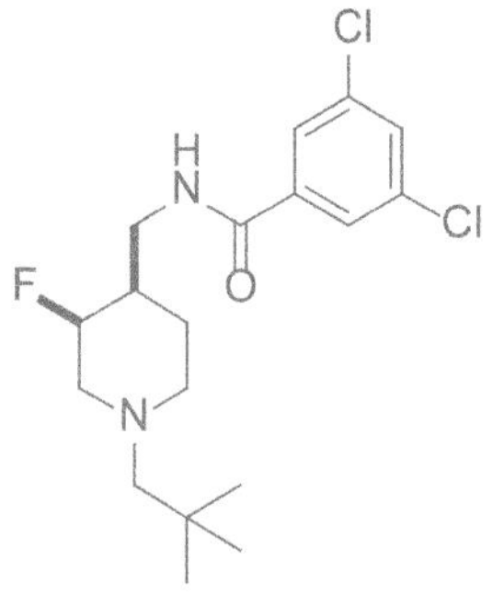

3-fluoro-1,4-substituted piperidine **8**
selective T-type
calcium channel inibitor

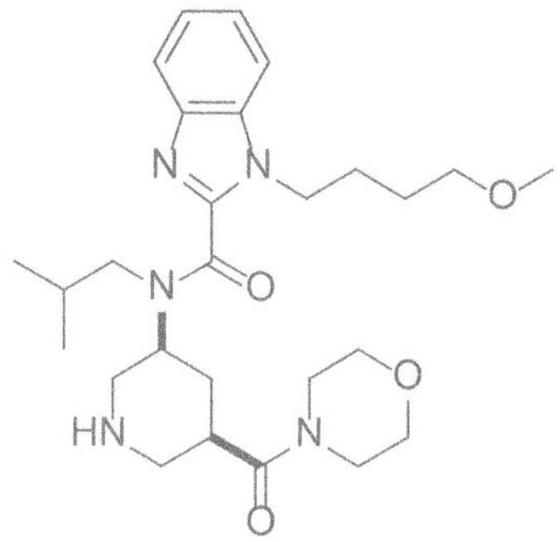

3,5-disubstituted piperidine **9**
renin inhibitor

PIPERIDINE IMPROVES AQUEOUS SOLUBILITY OF DRUGS

In addition to being a part of a drug's pharmacophore, piperidines have been used to improve the drug's aqueous solubility.

With a pKa of 11.22 for piperidine *per se*, *N*-alkylated piperidines have a pKa of approximately 9.5. Installation of piperidine rings has been routinely employed to boost the drug's aqueous solubility. For instance, 4-aminoquinazoline **10** was a potent kinase insert domain receptor (KDR) inhibitor with a poor solubility. A basic piperidine ring was installed on the side chain to replace the triazole, which resulted in **11** with up to a 500-fold improvement of solubility at pH 7.4, the physiological acidity.[3]

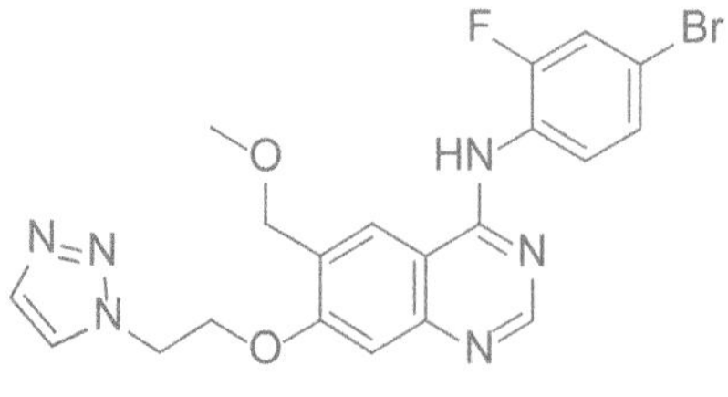

10, pKa = 5.3
sol, 0.7 μM

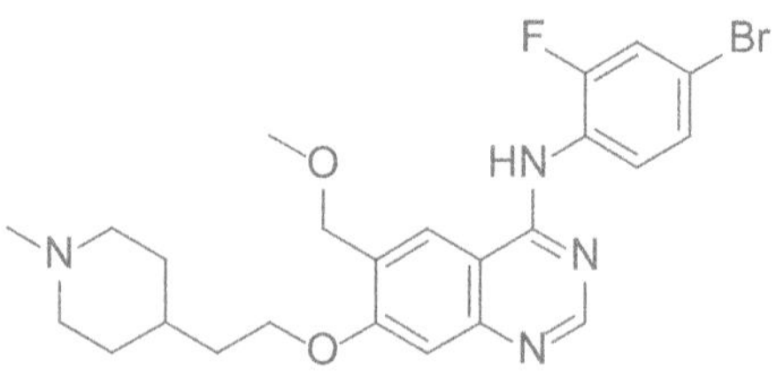

11, pKa = 9.4, 5.3
sol, 330 μM

PIPERIDINE ADDRESSES DRUG-RESISTANCE ISSUES

Pgp (permeability glycoprotein), the most prevalent drug efflux transporter, is often overexpressed in tumor cells and is implicated as a cause of multidrug resistance. Half of the marketed drugs are Pgp substrates. One of the tactics of addressing the Pgp issue is modifying log P to reduce penetration into the lipid bilayer where binding to Pgp occurs.

Tetracyclic compound **12** is chemotherapy plagued with cytotoxic drug resistance as a consequence of being a Pgp substrate.[4] A Mannich reaction of **12** offered the corresponding 3-aminomethyl-piperidine derivative **13**. The maneuver conferred a salient feature to the resulting piperidine compound, namely, the potency for tumor cells otherwise resistant to a variety of anticancer drugs. It is likely that the steric hindrance of bicyclic piperidine **13** minimized the hydrogen bonding-donating potential of the adjacent phenol group.

12, Pgp substrate   13, **NOT** a Pgp substrate

SYNTHESIS OF SOME PIPERIDINE-CONTAINING DRUGS

Pfizer's tofacitinib (Xeljanz, **17**) is the first-in-class Janus kinase (JAK) inhibitor for the treatment of rheumatoid arthritis (RA). One of its syntheses began with trisubstituted piperidine **14**.[5] An S_NAr coupling between **14** and 2,4-dichloro-7H-pyrrolo[2,3-d]pyrimidine (**15**) produced adduct **16**. Debenzylation and concurrent dechlorination of **16** were followed by amidation using ethyl cyanoacetate to deliver tofacitinib (**17**).[6]

14   15

16   tofacitinib (Xeljanz, **17**)
                      Pfizer, 2012
                      JAK Inhibitor for RA

Pharmacyclics' ibrutinib (Imbruvica, **21**) is the first-in-class Bruton's tyrosine kinase (Btk) inhibitor for treating mantle cell lymphoma, chronic lymphocytic leukemia, and Waldenstrom's macroglobulinemia. Inhibition of Btk activity prevents downstream activation of the B-cell receptor (BCR) pathway and subsequently blocks cell growth, proliferation, and survival of malignant B cells. Therefore, Btk inhibitors are good targeted cancer therapies.

One of the syntheses of ibrutinib (**21**) involves a Mitsunobu reaction between 1*H*-pyrazolo[3,4-*d*] pyrimidine **18** and 3-hydroxyl-piperidine **19** to afford adduct **20** after removal of the Boc protection. The reaction between piperidine **20** and acryloyl chloride then assembled ibrutinib (**21**).[7]

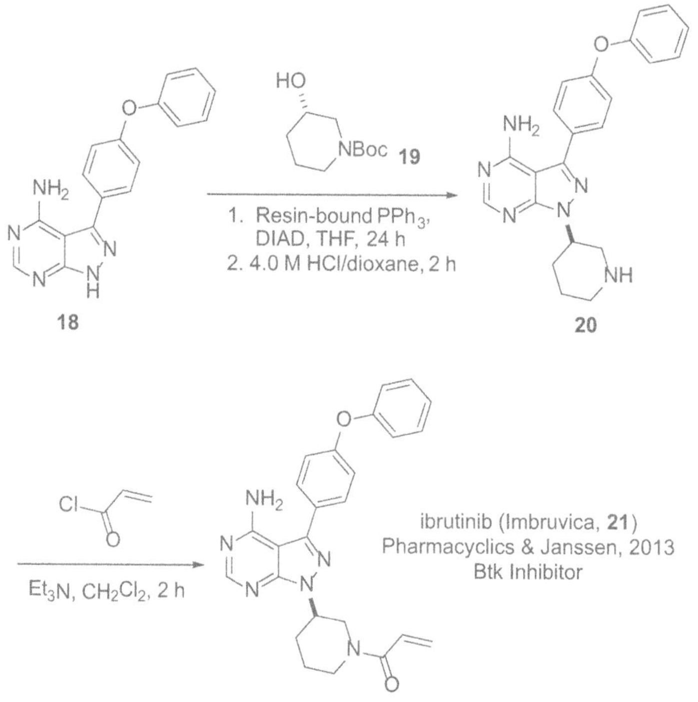

In summary, piperidine is a privileged scaffold. It contributes to pharmacology via tighter binding to the enzymes or the receptors. Its nitrogen atom is responsible for the elevation of the drug's aqueous solubility. With many exquisitely decorated piperidine-containing building blocks now commercially available, they will find more and more utility in medicinal chemistry.

REFERENCES

1. Yang, Z.-Q.; Barrow, J. C.; Shipe, W. D.; Schlegel, K.-A. S.; Shu, Y.; Yang, F. V.; Lindsley, C. W.; Rittle, K. E.; Bock, M. G.; Hartman, G. D.; et al. *J. Med. Chem.* **2008**, *51*, 6571–6477.
2. Tokuhara, H.; Imaeda, Y.; Fukase, Y.; Iwanaga, K.; Taya, N.; Watanabe, K.; Kanagawa, R.; Matsuda, K.; Kajimoto, Y.; Kusumoto, K.; et al. *Bioorg. Med. Chem.* **2018**, *26*, 3261–3286.
3. Hennequin, L. F.; Stokes, E. S.; Thomas, A. P.; Johnstone, C.; Plé, P. A.; Ogilvie, D. J.; Dukes, M.; Wedge, S. R.; Kendrew, J.; Curwen, J. O. *J. Med. Chem.* **2002**, *45*, 1300–1312.
4. Shchekotikhin, A. E.; Shtil, A. A.; Luzikov, Y. N.; Bobrysheva, T. V.; Buyanov, V. N.; Preobrazhenskaya, M. N. *Bioorg. Med. Chem.* **2005**, *13*, 2285–2291.
5. Ripin, D. H. B.; Abele, S.; Cai, W.; Blumenkopf, T.; Casavant, J. M.; Doty, J. L.; Flanagan, M.; Koecher, C.; Laue, K. W.; McCarthy, K.; et al. *Org. Process Res. Dev.* **2003**, *7*, 115–120.
6. Vaidyanathan, R. *Development of a Robust, Environmentally Responsible Process for the Manufacture of Tofacitinib Citrate*, In *Scalable Green Chemistry*; Koenig, S., ed.; Pan Stanford Publishing Pte. Ltd.: Singapore, **2013**, pp 185–205.
7. Pan, Z.; Scheerens, H.; Li, S. Schultz, B. E.; Sprengeler, P. A.; Burrill, L. C.; Mendonca, R. V.; Sweeney, M. D.; Scott, K. C.; Grothaus, P. G.; et al. *ChemMedChem* **2007**, *2*, 58–61.

20 Pyrazines

Structurally, pyrazine is a planar hexagon, similar to benzene, in both bond angles and lengths. Its C–N bonds are shorter and C–N–C bond angles are smaller than their phenyl counterparts.

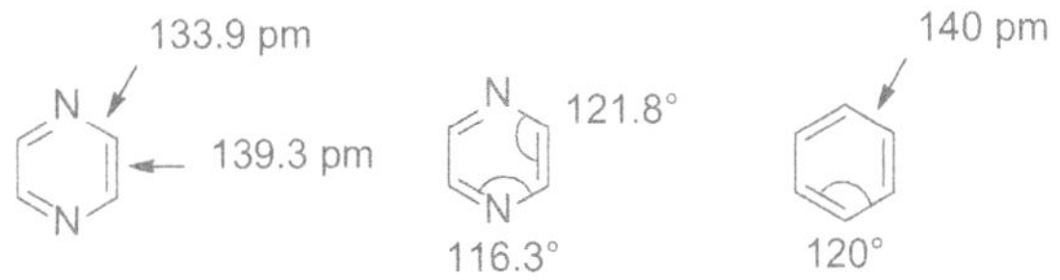

Pyrazine's Bond Lengths and Angles

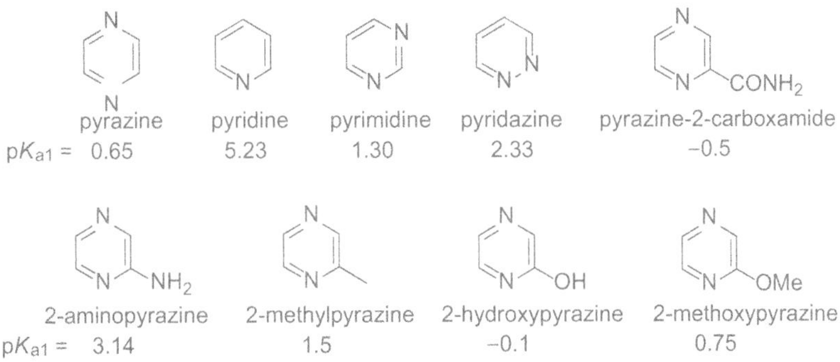

With six π-electrons, pyrazine is an electron-*deficient* (also known as electron-*poor*) aromatic heterocycle because of the increased electronegativity of the nitrogen atoms. Due to the presence of the electronegative nitrogen atoms, the electron density on the ring carbons is less than one. The lone-pair electrons do not take part in the delocalization for aromaticity, so this molecule can act as a mild base. Its $pK_{a1} = 0.65$ and $pK_{a2} = 5.78$. Basic dissociation constants (pK_{a1}) of pyrazine, some pyrazine derivatives, and other diazines are listed above.[1]

PYRAZINE-CONTAINING DRUGS

There are approximately eight pyrazine-containing drugs on the market in the US. For instance, Pfizer's glipizide (Glucotrol, **1**) is an old sulfonylurea antidiabetic, which works by stimulating insulin secretion to metabolize carbohydrates. Its mechanism of action (MOA) is serving as a potassium channel blocker.

Vertex's bortezomib (Velcade, **2**) was the first proteasome inhibitor approved to treat relapsed multiple myeloma (MM) and mantle cell lymphoma. Mechanistically, the boron atom binds, with high affinity and specificity, to the catalytic site of the 26S proteasome, which maintains the immortal type of myeloma.

Rhone–Poulenc Rorer's older sleeping pill, zopiclone (Imovane, **3**), functions as a GABA$_A$ modulator. It is no longer available in the US because of its dubious benefit/risk profile. Having realized that the *S*-enantiomer is significantly more active and less toxic than the *R*-enantiomer, Separacor separated them and arrived at eszopiclone (Lunesta, **4**), the *S*-isomer.

Also in the CNS arena, Pfizer's varenicline (Chantix, **5**) works as a partial agonist of α4β2 nicotinic acetylcholine receptor (nAChR) for helping smoke cessation.

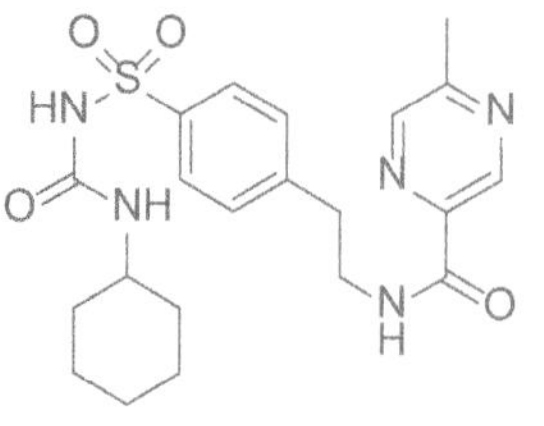

glipizide (Glucotrol, **1**)
Pfizer, 1984
sulfonylurea antidiabetic

(potassium channel blocker)

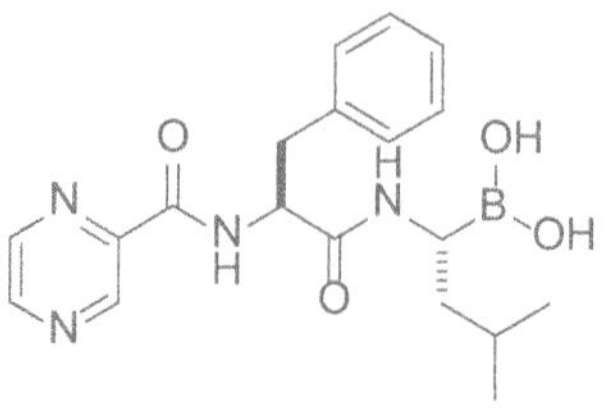

bortezomib (Velcade, **2**)
Vertex, 2003
Malignancy of bone marrow

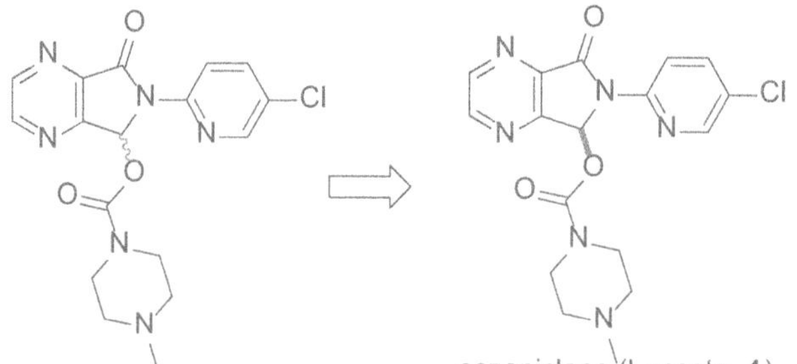

zopiclone (Imovane, **3**)
not available in US

GABA$_A$ modulators for insomnia

eszopiclone (Lunesta, **4**)
Separacor, 2004

GABA$_A$ modulators for insomnia

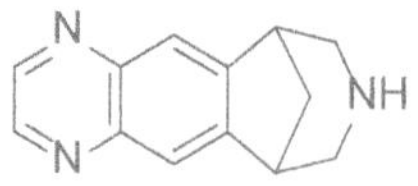

varenicline (Chantix, **5**)
Pfizer, 2006
partial agonist of α4β2
nicotinic acetylcholine receptor (nAChR)

Vertex's telaprevir (Incivek, **6**) is an HCV NS3/4A serine protease inhibitor. It is a reversible covalent inhibitor—the protease's serine$_{139}$ forms an acetal tetrahedral intermediate with the ketone functional group of the terminal keto-amide "warhead."

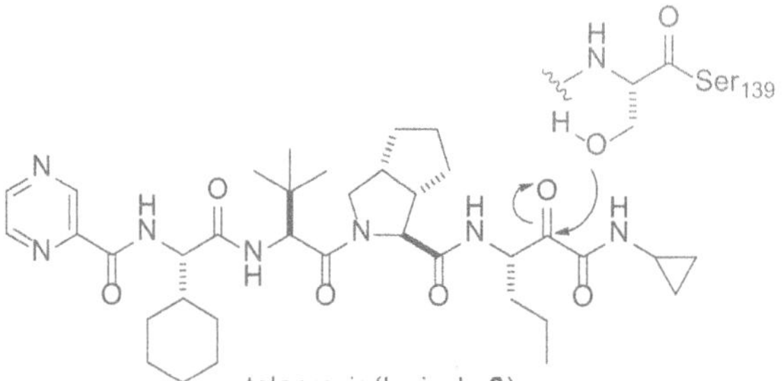

telaprevir (Incivek, **6**)
Vertex, 2011
HCV NS3/4A serine protease inhibitor

Nippon Shinyaku licensed its PGI-2 receptor agonist selexipag (Uptravi, **7**) to Actelion. The drug is an oral treatment of pulmonary arterial hypertension (PAH). Interestingly, the *N*-methylsulfonamide is hydrolyzed to the corresponding carboxylic acid *in vivo* to provide a slow-release pharmacological effect.

The quest for a new treatment of Alzheimer's disease (AD) has met with failures too numerous to count. Eisai/Biogen's elenbecestat (**8**) is a β-secretase-1 (BACE1, also known as beta-site amyloid precursor protein cleaving enzyme) inhibitor that failed phase III clinical trials for lack of efficacy.

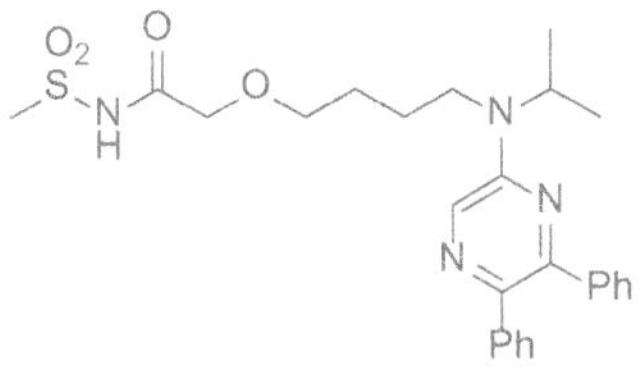

selexipag (Uptravi, **7**)
Nippon Shinyaki/Actelion, 2015
PGI-2 receptor agonist

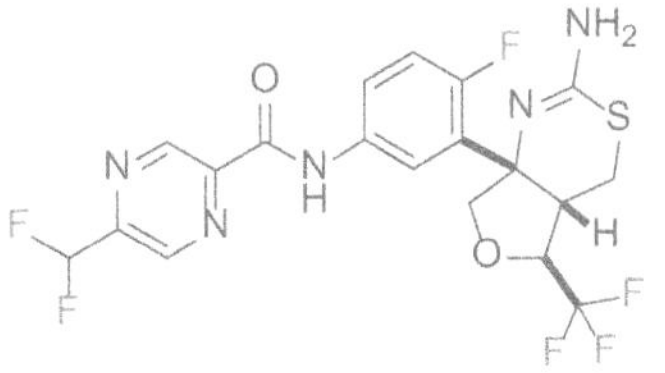

elenbecestat (**8**)
Eisai/Biogen
BACE1 inhibitor

PYRAZINES IN DRUG DISCOVERY

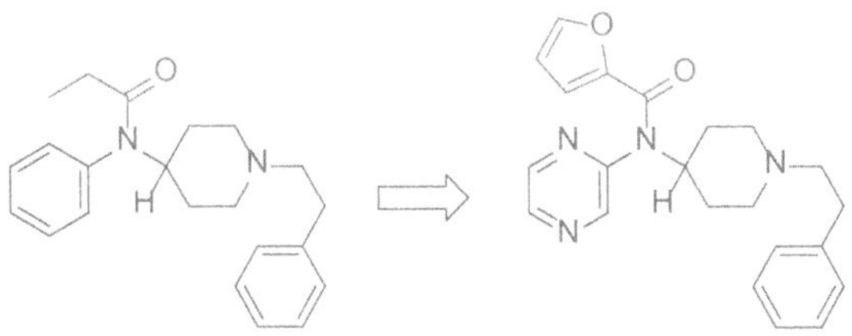

fentanyl (Duragesic, **9**)
Janssen, 1968
μ opioid agonist

mirfentanil (**10**)
μ opioid agonist

In drug discovery, pyrazine has served as a bioisostere of phenyl and heteroaryl moieties. For instance, mirfentanil (**10**), a μ opioid agonist, is an analog of fentanyl (Duragesic, **9**). Here, pyrazine on mirfentanil (**10**) replaced the fentanyl (**9**)'s phenyl group.[1] One of the factors that contributed to today's opioid epidemic is the fact that so many fentanyl analogs exist, which makes even their detection a challenge.

Bayer's sorafenib (Nexavar, **11**) is a dual platelet-derived growth factor receptor (PDGF) and vascular endothelial growth factor receptor (VEGF) inhibitor. An exercise of replacing the pyridine ring on sorafenib (**11**) with pyrazine led to a series of analogs. Their biochemical activities suggested that the substituents on urea are essential for interaction with c-Raf. One of the pyrazine derivatives **12** exerted cytostatic activities that surpassed sorafenib (**11**) in inhibitory effects on proliferation of cancer cell lines including Hela, A549, and HepG2.[2]

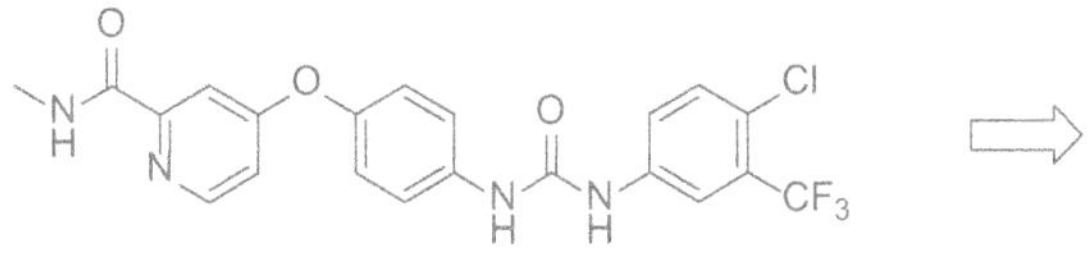

sorafenib (Nexavar, **11**)
Bayer/Onyx, 2007
PDGF and VEGF inhibitor
HeLa cell , IC_{50} = 7.1 nM
A549 cell , IC_{50} = 6.1 nM
HepG2 cell , IC_{50} = 6.2 nM

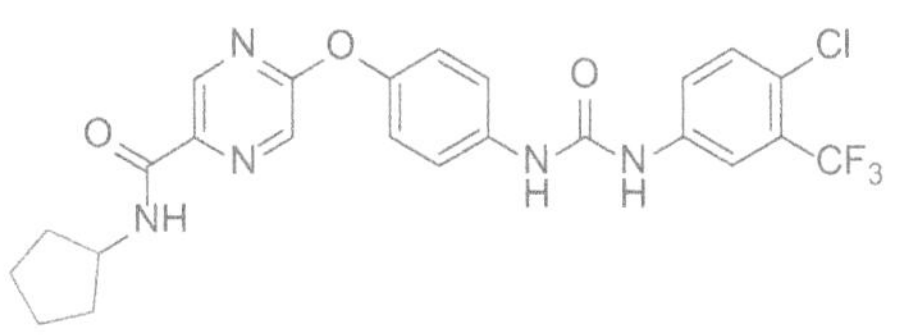

pyrazine analog **12**
HeLa cell , IC_{50} = 0.6 nM
A549 cell , IC_{50} = 0.8 nM
HepG2 cell , IC_{50} = 0.7 nM

Point mutations in isocitrate dehydrogenase (IDH) 1 and 2 are found in multiple tumors, including glioma, cholangiocarcinoma, chondrosarcoma, and acute myeloid leukemia (AML). Agios's ivosidenib (Tibsovo) is a potent, selective, and, more importantly, metabolically stable, IDH1 inhibitor that was approved by the FDA for the treatment of IDH1-mutant cancers.

Novartis reported a novel class of mutant IDH1 (mIDH1) inhibitors with a 3-pyrimidin-4-yl-oxalidin-2-one motif. Among them, NI-1 (**13**) inhibited both biochemical and cellular production of oncometabolite D-2-hydroxyglutatrate (D2HG) with an IC_{50} value of 94 nM (reduced D2HG level by 24.7% at 50 μM) for IDH1 R132H. Mutant R132H is the predominant mutant of the IDH1 enzyme where Arg_{132} is substituted by His. Employing pyrazine as a bioisostere of NI-1 (**13**)'s pyridine gave rise to 3-pyrazine-2-yl-oxazolidin-2-one (**14**). It too effectively suppressed the D2HG production level in cells transfected with IDH1-R132H mutation. Pyrazine analog **14** also showed a good ability to penetrate the blood-brain barrier (BBB) using a parallel artificial membrane permeability assay (PAMPA).[3]

Docking analysis of both NI-1 (**13**) and pyrazine **14** revealed that the carbonyl of their oxazolidinone establishes a hydrogen bond with the amine of Leu_{120}. While pyrazine **14**'s C2-amine group forms a donor–acceptor polar interaction (a salt bridge) with Ile_{128}, the 4-N atom of its pyrazine cannot interact with Ile_{128} although NI-1 (**13**) does. This may be the reason why pyrazine **14** showed decreased inhibitory activity compared to NI-1 (**13**).[3]

NI-1 (**13**)
D2HG level (IDH1 R132H)
@50 μM = 24.7%
LogP = 3.71

pyrazine analog **14**
D2HG level (IDH1 R132H)
@50 μM = 24.6%
LogP = 4.40

Checkpoint kinase 1 (CHK1) is an intracellular, serine/threonine kinase that plays a central role in the DNA damage response pathway. CHK1 inhibitors, meanwhile, are of current interest as potential antitumor agents although many advanced CHK1 inhibitors are not orally bioavailable. From the hybridization of two lead scaffolds derived from the fragment-based drug design (FBDD) and subsequent optimization, CCT244747 (**15**) was obtained using a cellular-based assay cascade. Although the compound showed high biochemical kinase potency and selectivity for CHK1-dependent mechanism of action (MOA) in human cancer cells, its micromolar human ether-a-go-go-related gene (hERG) potassium channel activity inhibition was a source of concern.[4a]

Generally speaking, hERG inhibition is dependent on the lipophilicity and basicity of the compounds. During further optimization of **15**, its 5-aminopyrazine-2-carbonitrile group was retained because it was shown, through extensive structural biology, to optimally interact with a unique *protein-bound water molecule* in CHK1, conferring high selectivity over other kinases. Eventually, extensive SAR produced CCT245737 (**16**) as a potent and selective CHK1 inhibitor. It also has low predicted doses and exposure in humans which mitigated the residual weak *in vitro* human ether-a-go-go (hERG) potassium channel activity inhibition. It is now a clinical candidate as an oral treatment of RAS-mutant non-small cell lung cancer (NSCLC) and Eμ-MYC driven B-cell lymphoma.[4b]

The crystal structure of CCT245737 (**16**) bound to CHK1 revealed that the cyanopyrazine was positioned close to the side-chain of Lys$_{38}$. Importantly, both the nitrile and *N*-4 of the pyrazine ring were positioned to interact with *protein-bound water* at the entrance to the pocket beyond the gatekeeper residue, one of a network of conserved water molecules resulting from the presence of the unique polar residue Asn$_{59}$ in this pocket in CHK1 instead of the more common lipophilic side chains. Interactions with these bound waters are a CHK1 selectivity determinant for this and other series of CHK1 inhibitors including other pyrazines.[4b]

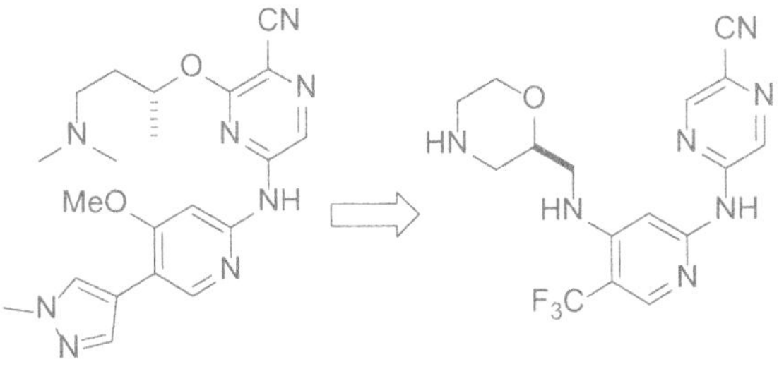

CCT244747 (**15**)
CHK1 IC$_{50}$ = 7.7 nM
hERG IC$_{50}$ = 5.0 μM

CCT245737 (**16**)
CHK1 IC$_{50}$ = 1.3 nM
CHK2 IC$_{50}$ = 2,440 nM

Even though electron-deficient, pyrazine is still an aromatic ring. Like many aromatic rings, pyrazine has made frequent appearance as a fragment of receptor tyrosine kinase (RTK) inhibitors. One of the two nitrogen atoms on pyrazine may serve as a hydrogen bond acceptor (HBA) to form a hydrogen bond with target kinase protein.

Tropomyosin receptor kinase (Trk) was first discovered as an oncogene that is activated through chromosomal rearrangement in human colon carcinoma. A computational Trk hit **17** was generated from a kinase-directed virtual library screen. The crystal structure of compound **17** and TrkA indicated that it forms a hydrogen bond at the kinase hinge through a weak hydrogen bond (~4 Å) with Met$_{620}$. Extensive SAR investigations identified compound **18** as a potent and selective TrkA inhibitor (there are three Trk isoforms: TrkA, TrkB, and TrkC).[5]

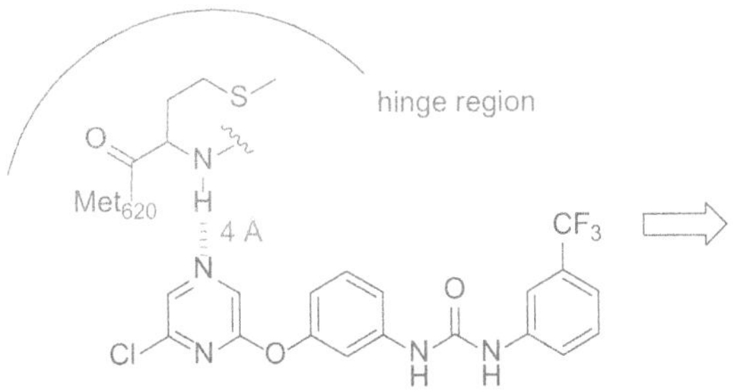

Trk computational hit **17**
IC$_{50}$ = 3.5 μM

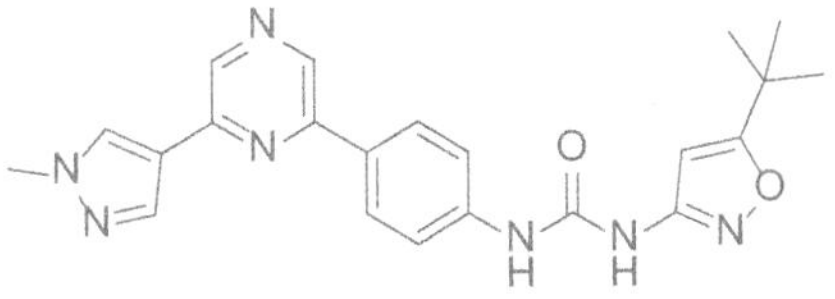

TrkA inhibitor **18**
IC$_{50}$ = 5 nM

Spleen tyrosine kinase (Syk) is a cytosolic non-receptor protein tyrosine kinase that plays an essential role in immune-receptor signaling, mainly in B cell receptors. X-Ray structure of a Syk inhibitor **19** bound to Syk indicated that its aminopyrazine motif forms that key hydrogen bonding with Glu$_{449}$ and Ala$_{451}$ in the hinge region.[6]

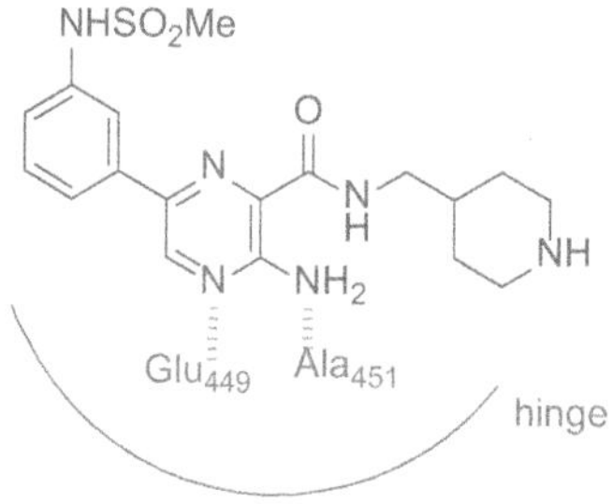

Syk kinase inhibitor **19**
IC$_{50}$ = 670 nM

Deregulation of the well-known phosphoinositide-3-kinase (PI3K) pathway has been implicated in numerous pathologies such as cancer, diabetes, thrombosis, rheumatoid arthritis (RA), and asthma. Two PI3K inhibitors have been approved by the FDA. One is Gilead's idelalisib (Zydelig), which is a PI3Kδ selective inhibitor. The other is Bayer's copanlisib (Aliqopa), which is a pan-PI3K inhibitor.

A novel series of potent, selective (against PI3Kα, PI3Kβ, and PI3Kδ), and orally bioavailable PI3Kγ inhibitors, as represented by inhibitor **20**, have been reported. The cocrystal structure of **20** bound to PI3Kγ indicated that its aminopyrazine moiety forms key hydrogen bonding with the kinase hinge residue Val$_{882}$.[7]

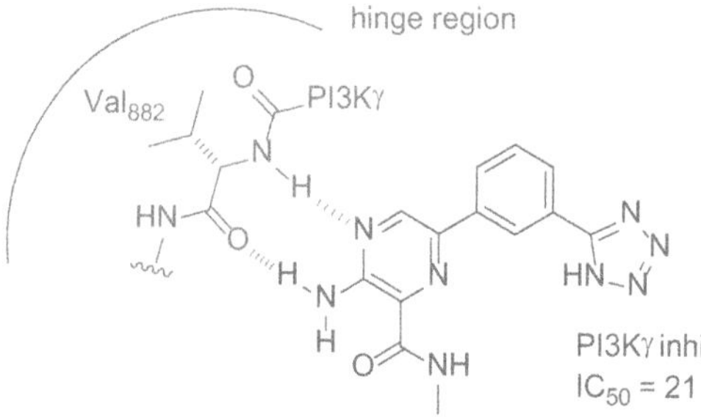

PI3Kγ inhibitor **20**
IC$_{50}$ = 21 nM

SYNTHESIS OF SOME PYRAZINE-CONTAINING DRUGS

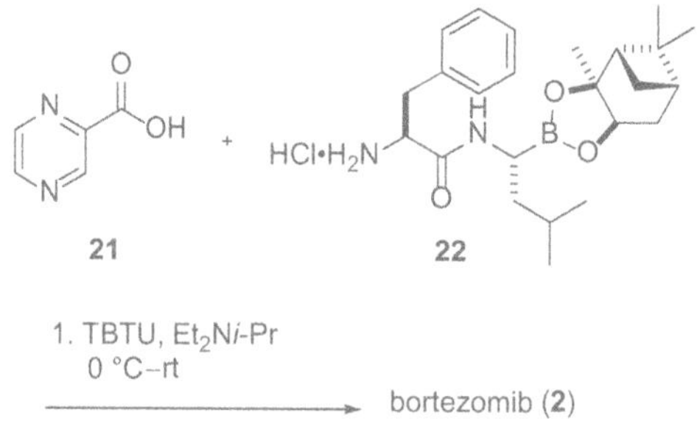

21   **22**

1. TBTU, Et$_2$Ni-Pr
   0 °C–rt
   $\longrightarrow$ bortezomib (**2**)
2. i-BuB(OH)$_2$, 1 N HCl
   MeOH/pentane

For the preparation of Vertex's proteasome inhibitor bortezomib (**2**), its pyrazine motif was installed at the end of the synthesis. Amide formation between pyrazine-carboxylic acid (**21**) and pinanediol-boronate intermediate **22** was facilitated by 2-(1*H*-benzotriazole-1-yl)-1,1,3,3-tetramethylaminium tetrafluoroborate (TBTU) as the coupling agent. Conversion of pinanediol-boronate to the corresponding boronic acid was carried out using 1 N HCl and isobutylboronic acid to deliver bortezomib (**2**).[8]

Rhone–Poulenc's synthesis of racemic zopiclone (**3**) commenced with the treatment of pyrazine anhydride (**23**) with 2-amino-5-chloropyridine (**24**) to give amide **25** in good yield. Refluxing amide **25** with thionyl chloride produced imide **26**. Mono-reduction of the imide with KBH$_4$ afforded alcohol **27**. Acylation of alcohol **27** with 4-methylpiperazine-1-carbonyl chloride (**28**) using NaH in DMF then produced racemic zopiclone (**3**).[9]

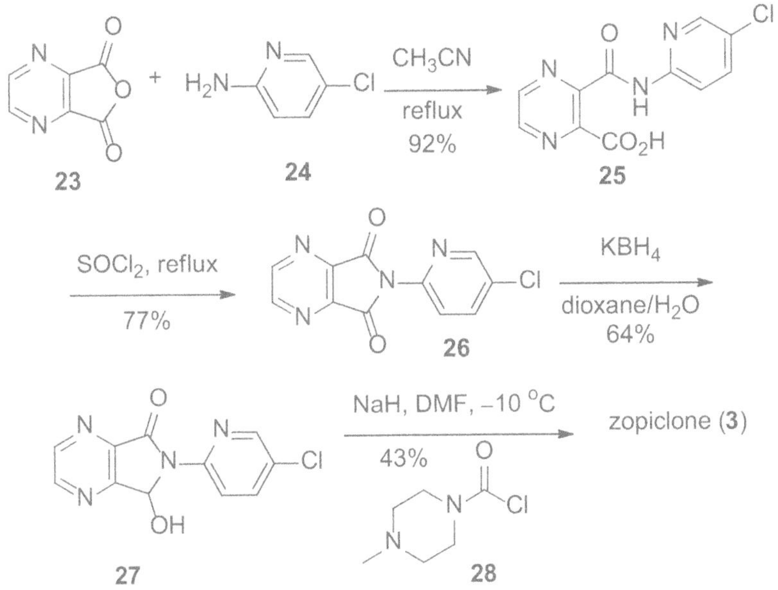

Several methods have been described for the preparation of the eszopiclone (**4**). Chiral resolution of **3** by diastereomeric salt formation and recrystallization has been described with both malic acid and *O,O'*-dibenzoyltartaric acid. Another conceptually different approach involving an enzymatic resolution of carbonate intermediates has also been described.[10] An immobilized form of lipase B from *Candida Antarctica* (Chirazyme-L2) catalyzed the hydrolysis of carbonate **29** to give optically active carbonate **30** with the correct absolute (*S*)-stereochemistry required for the synthesis of eszopiclone (**4**) and racemic **27**. The enzyme preparation can be recycled ten times without any loss to the activity of the catalyst or enantioselectivity of the reaction. Furthermore, racemic **27** can be recycled to improve the efficiency of the asymmetric synthesis.[10]

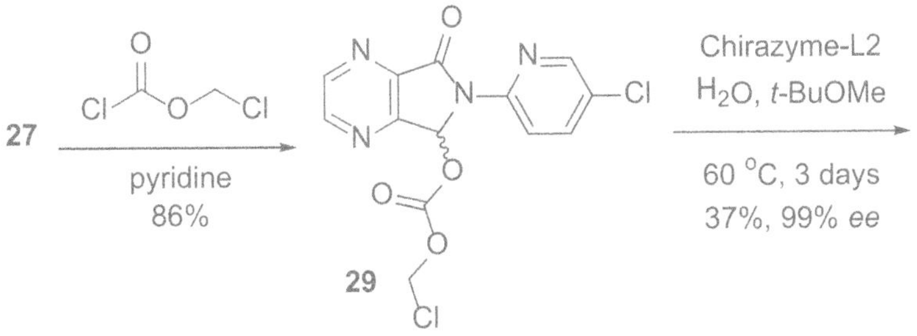

Many synthetic routes have been published for the preparation of varenicline (**5**). In one of Pfizer's process routes, bicyclic intermediate [3,2,1]-benzazepine (**31**) was assembled before installation of the pyrazine ring. Trifluoroacetamide protection was specifically introduced to insulate the nitrogen by removing electron density to avoid the formation of doubly charged cationic intermediates. This protection allowed nitration to proceed. With an excess of two equivalents of nitronium triflate, dinitrated product **32** was obtained in >75% yield. Regioselectivity for this conversion likely derives from steric and electronic factors driven by the bicyclic core, leading to 7:1–11:1 preference for the desired *ortho*-dinitro over the *meta*-dinitro regioisomer.[11]

Dinitro intermediate **32** is readily reduced to dianiline **33** via palladium-catalyzed hydrogenation. Although dianiline **33** is stable in solid form, it decomposes after prolonged standing in solution, thus processes were set up without isolation of **33** using direct addition of aqueous glyoxal to form the desired quinoxaline **34**. Added bicarbonate controlled the pH to avoid unwanted side products in this step. The process was completed with rapid and quantitative trifluoroacetamide hydrolysis via treatment with sodium hydroxide, thus combining steps via non-isolated intermediates and telescoping directly into the formation of the *L*-tartrate salt in methanol to deliver varenicline (**5**)-tartarate. The reaction control of the tartrate salt polymorph to the desired form B has been described by Rose and co-workers.[12]

A synthetic route of PGI-2 receptor agonist selexipag (**7**) began with condensation of benzil (**35**) with glycinamide to assemble hydroxypyrazine **36**. Chlorination of **36** in refluxing $POCl_3$ was facilitated by a catalytic amount of H_2SO_4 to afford chloride **37**. S_NAr displacement of the chloride by amino-alcohol **38** gave rise to adduct **39**. A phase-transfer catalyzed S_N2 reaction between **39** and *tert*-butyl bromoacetate provided ether **40**. After basic hydrolysis of **40** to make acid **41**, it was converted to selexipag (**7**) by coupling acid **41** with methanesulfonamide.[13]

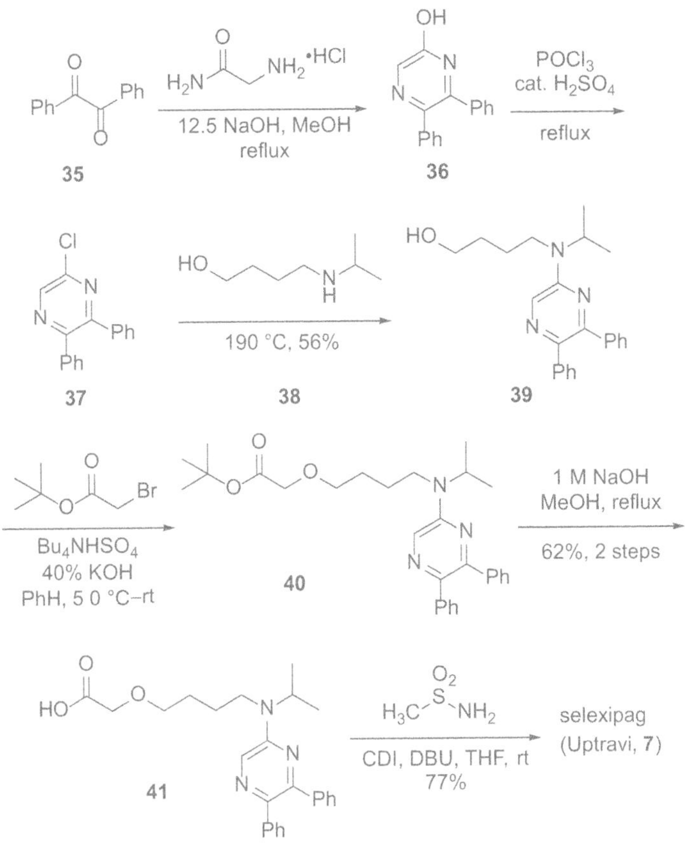

In summary, at least eight pyrazine-containing drugs have been approved by the FDA. In medicinal chemistry, pyrazine has been employed as a bioisostere of benzene, pyridine, and pyrimidine. For pyrazine-containing kinase inhibitors, the pyrazine nitrogen atom frequently serves as a hydrogen bond acceptor to interact with an amino acid in the hinge region of the kinase protein.

REFERENCES

1. Dolezal, M.; Zitko, J. *Expert Opin. Ther. Pat.* **2015**, *25*, 33–47.
2. Džolić, Z. R.; Perković, I.; Pavelic, S. K.; Sedić, M.; Ilić, N.; Schols, D.; Zorc, B. *Med. Chem. Res.* **2016**, *25*, 2729–2741.
3. Ma, T.; Zou, F.; Pusch, S.; Yang, L.; Zhu, Q.; Xu, Y.; Gu, Y.; von Deimling, A.; Zha, X. *Bioorg. Med. Chem.* **2017**, *25*, 6379–6387.
4. (a) Lainchbury, M.; Matthews, T. P.; McHardy, T.; Boxall, K. J.; Walton, M. I.; Eve, P. D.; Hayes, A.; Valenti, M. R.; de Haven Brandon, A. K.; Box, G.; et al. *J. Med. Chem.* **2012**, *55*, 10229–10240. (b) Osborne, J. D.; Matthews, T. P.; McHardy, T.; Proisy, N.; Cheung, K.-M. J.; Lainchbury, M.; Brown, N.; Walton, M. I.; Eve, P. D.; Boxall, K. J.; et al. *J. Med. Chem.* **2016**, *59*, 5221–5237.
5. Frett, B.; McConnell, N.; Wang, Y.; Xu, Z.; Ambrose, A.; Li, H.-Y. *MedChemComm* **2014**, *5*, 1507–1514.
6. Forns, P.; Esteve, C.; Taboada, L.; Alonso, J. A.; Orellana, A.; Maldonado, M.; Carreno, C.; Ramis, I.; Lopez, M.; Miralpeix, M.; et al. *Bioorg. Med. Chem. Lett.* **2012**, *22*, 2784–2788.
7. Leahy, J. W.; Buhr, C. A.; Johnson, H. W. B.; Kim, B. G.; Baik, T.; Cannoy, J.; Forsyth, T. P.; Jeong, J. W.; Lee, M. S.; Ma, S.; et al. *J. Med. Chem.* **2012**, *55*, 5467–5482.
8. Adams, J.; Behnke, M.; Chen, S.; Cruickshank, A. A.; Dick, L. R.; Grenier, L.; Klunder, J. M.; Ma, Y.-T.; Plamondon, L.; Stein, R. L. *Bioorg. Med. Chem. Lett.* **1998**, *8*, 333–338.
9. Racemic zopiclone synthesis: (a) Cotrel, C.; Jeanmart, C.; Messer, M. N. US Patent 3,862,149. (1975 to Rhone-Poulenc S.A.). (b) Cotrel, C.; Roussel, G. EP 495,717 (1992 to Rhone-Poulenc Rorer SA).

10. (a) Solares, L. F.; Diaz, M.; Brieva, R.; Sánchez, V. M.; Bayod, M.; Gotor, V. *Tetrahedron: Asymmetry* **2002**, *13*, 2577–2582. (b) Palomo, J. M.; Mateo, C.; Fernández-Lorente, G.; Solares, L. F.; Diaz, M.; Sánchez, V. M.; Bayod, M.; Gotor, V.; Guisan, J. M.; Fernandez-Lafuente, R. *Tetrahedron: Asymmetry* **2003**, *14*, 429–438.

11. (a) Handfield, Jr., R. E.; Watson, T. J. N.; Johnson, P. J.; Rose, P. R. U.S. Patent US2007/7285686 (2007). (b) Busch, F. R.; Withbroe, G. J.; Watson, T. J.; Sinay, T. G.; Hawkins, J. M.; Mustakis, I. G. U.S. Patent US2008/0275051 (2008).

12. Bogle, D. E.; Rose, P, R.; Williams, G. R. U.S. USPatent 6890927 (2005).

13. Asaki, T.; Hamamoto, T.; Sugiyama, Y.; Kuwano, K.; Kuwabara, K. *Bioorg. Med. Chem. Lett.* **2007**, *15*, 6692–6704.

21 Pyrazoles

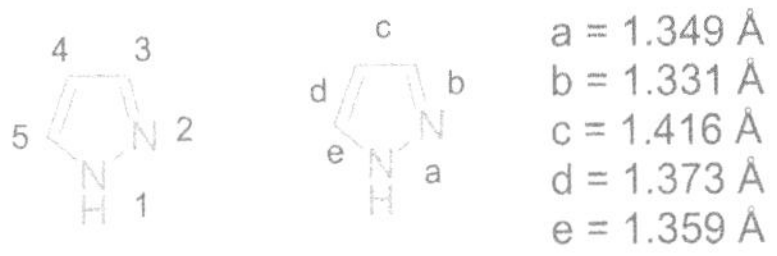

a = 1.349 Å	
b = 1.331 Å	
c = 1.416 Å	
d = 1.373 Å	
e = 1.359 Å	

Pyrazole's numbering and bond lengths

Pyrazole is a five-membered aromatic heterocycle with two N heteroatoms. Its N-1 is similar to the NH of pyrrole, and its N-2 behaves closely to the nitrogen atom of pyridine. Due to different bonding environments, all five bonds a–e have different bond lengths. Pyrazole's aromaticity lies somewhere in the middle of the scale among other aromatic heterocycles:

furan 12 pyrrole 31 imidazole 43 thiophene 45 pyrazole 61

pyridazine 65 pyrimidine 67 triazole 71 tetrazole 80 pyridine 82 benzene 100

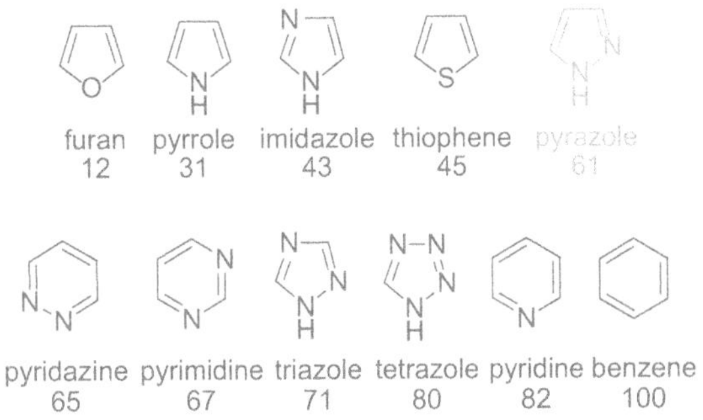

Relative aromaticity of aromatic heterocycles

With a pKa of 2.5, pyrazole is significantly less basic than imidazole, whose pKa is 7.1. In fact, having an adjacent heteroatom near the N atom always has the effect of lowering the basicity of the N because of its inductive effect. Nonetheless, pyrazole is basic enough to be protonated with the most strong inorganic acids. When pyrazole is unsymmetrically substituted, it may exist as a mixture of two tautomers. For instance, 5-methylpyrazole and 3-methylpyrazole coexist in a solution.

The ramification of the tautomerization is that alkylation of unsymmetrically substituted pyrazoles often gives rise to a mixture of two isomers, one is the *N*-1 alkylation and the other is the *N*-2 alkylation. The ratio depends on the nature of the substrate and the electrophile as well as on the solvent and base.

PYRAZOLE-CONTAINING DRUGS

Not many pyrazoles exist in nature. However, over a dozen pyrazole-containing synthetic medicines are on the market. For instance, an anti-inflammatory cyclooxygenase-2 (COX-2) selective inhibitor celecoxib (Celebrex, **1**) has a tri-substituted pyrazole core structure.

During the last few years, more and more pyrazole-containing drugs have gained regulatory approval, most conspicuously in the field of kinase inhibitors. Nearly all kinase inhibitors are competitive inhibitors, occupying the adenine triphosphate (ATP) binding pocket and replacing ATP. As a consequence, all of the ATP-competitive kinase inhibitors possess a flat aromatic core structure, mimicking the adenine portion of ATP. Thanks to pyrazole's aromaticity, it is making frequent appearances in the realm of kinase inhibitors. Pfizer's first-in-class dual anaplastic lymphoma kinase (ALK) and mesenchymal–epithelial transition factor (c-MET) inhibitor, crizotinib (Xalkori, 2), contains a di-substituted pyrazole. It was discovered using sunitinib-like 3-substituted indolin-2-ones as the starting point. The co-crystal structure of crizotinib (2) and the c-MET protein indicated that the 5-pyrazol-4-yl group is bound through the narrow lipophilic tunnel surrounded by Ile-1084 and Tyr-1159.[1] Since crizotinib (2) has little CNS exposure, Pfizer, guided by a structure-based drug design (SBDD), lipophilicity efficiency (LipE), and physical property-based optimization, converted the crizotinib (2) to macrocycles. Among them, lorlatinib (Lorbrena, 3) emerged with good absorption, distribution, metabolism, and excretion (ADME), low propensity for p-glycoprotein (pgp) 1-mediated efflux, and good passive permeability with significant CNS exposure. It is thus suitable for treating metastasized brain tumors.[2]

celecoxib (Celebrex, 1)
Pfizer, 1998
COX-2 inhibitor

crizotinib (Xalkori, 2)
Pfizer, 2011
ALK inhibitor

Janus kinases (JAK) recruit signal transducers and activators of transcription (STATS) to cytokine receptors, leading to modulation of gene expression. Incyte was one of the pioneers in the discovery of JAK inhibitors. Their pyrazole-containing ATP-competitive JAK1/2 dual inhibitor ruxolitinib (Jakafi, 4) was approved by the FDA in 2011 for the treatment of patients with myelofibrosis (MF, a bone marrow disorder). Incyte's *encore* JAK1/2 dual inhibitor baricitinib (Olumiant, 5) was marketed in 2017 for treating rheumatoid arthritis (RA).

lorlatinib (Lorbrena, 3)
Pfizer, 2018
ALK inhibitor

ruxolitinib (Jakafi, 4)
Incyte, 2010
JAK1/2 inhibitor

baricitinib (Olumiant, 5)
Incyte/Lilly, 2017 (for RA)
JAK1/2 inhibitor

Approximately ten dipeptidyl peptidase-4 (DPP4) inhibitors are currently available for the treatment of type II diabetes mellitus (T2DM). One of them, Mitsubishi's teneligliptin (Tenelia, 6) has a tri-substituted pyrazole core structure. Lexicon's tryptophan hydroxylase inhibitor telotristat ethyl

(Xermelo, **7**) was approved in 2017. In 2018, Array/Novartis's B-raf kinase inhibitor, encorafenib (Braftovi, **8**), gained the nod from the FDA for treating melanoma and colorectal cancers. It is a tri-substituted pyrazole. The latest entry of pyrazole-containing drug is Bayer's androgen receptor antagonist darolutamide (Nubeqa, **9**), which contains, not one, but two pyrazoles.

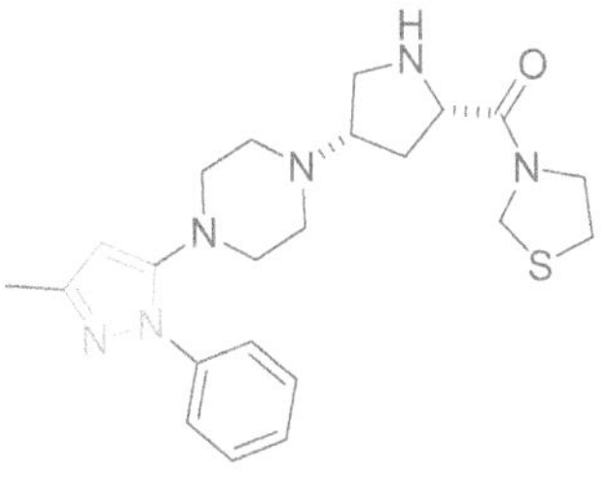

teneligliptin (Tenelia, **6**)
Mitsubishi Tanabe, 2012 (Japan)
DPP-4 inhibitor

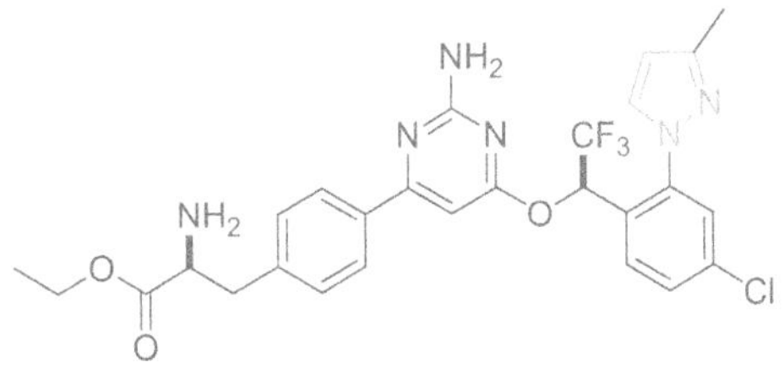

telotristat ethyl (Xermelo, **7**)
Lexicon, 2017
tryptophan hydroxylase inhibitor

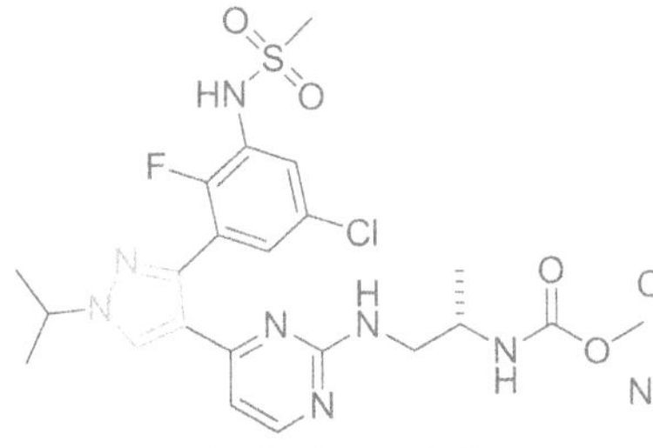

encorafenib (Braftovi, **8**)
Novartis, 2018
B-raf inhibitor

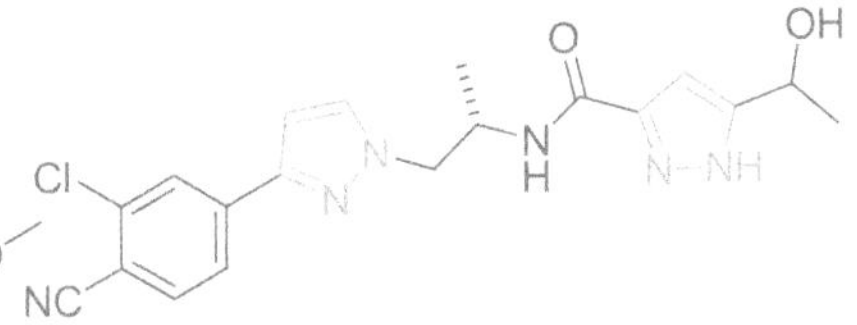

darolutamide (Nubeqa, **9**)
Bayer, 2019
Anti-androgen

Several pyrazole-fused drugs are also on the market. In addition to the well-known phosphodiesterase V (PDE5) inhibitor sildenafil (Viagra), they also include BMS's factor Xa inhibitor apixaban (Eliquis, **10**) and Merck's DPP-4 inhibitor omarigliptin (Marizev, **11**).

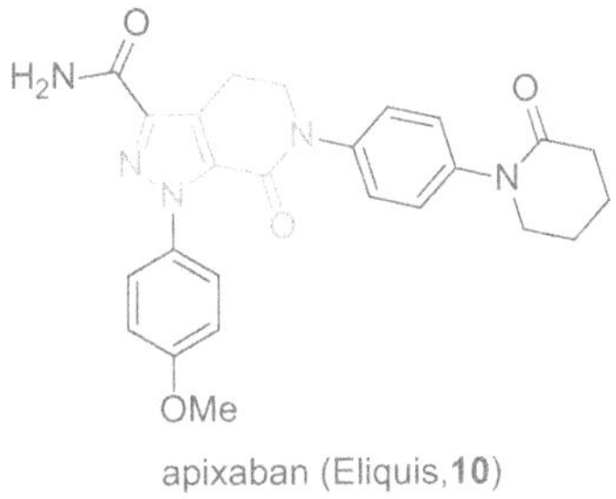

apixaban (Eliquis, **10**)
BMS, 2010
Factor Xa inhibitor

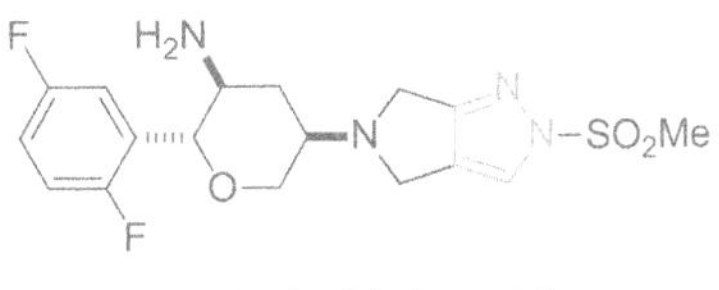

omarigliptin (Marizev, **11**)
Merck, 2015 (Japan)
DPP-4 inhibitor

PYRAZOLES IN DRUG DISCOVERY

PYRAZOLE AS A BIOISOSTERE OF ARENES AND HETARENES

With regard to aromaticity, pyrazole (65%) is not the closest to benzene (100%). However, the former is significantly less lipophilic (ClogP=0.241) than the latter (ClogP=2.142). Therefore, pyrazole has been employed as a bioisostere of benzene and other arenes, resulting in improved potency and physiochemical properties, such as lipophilicity and aqueous solubility.

pyrazole
CLogP = 0.241

benzene
CLogP = 2.142

While 2-aryl quinolone **12** demonstrated good antimalarial activities against various strains of *Plasmodium falciparum* by inhibiting two mitochondrial enzymes in the electron transport chain, the cytochrome bc_1 complex and type-II NADH:ubiquinone oxidoreductase (PfNDH2), its physiochemical properties needed optimization. Replacing the central phenyl ring with a pyrazole group resulted in a series of 2-pyrazolyl quinolones with improved antimalarial potency and various *in vitro* drug metabolism and pharmacokinetics (DMPK) features. In particular, 2-pyrazolyl quinolone **13** displayed no cross-resistance with multidrug-resistant parasite strains (W2) compared to drug-sensitive strains (3D7), with IC_{50} values in the range of 15–33 nM. Furthermore, 2-pyrazolyl quinolone **13** also retained moderate activity against the atovaquone-resistant parasite isolate (TM90C2B). It also displayed improved DMPK properties, including improved aqueous solubility compared to previously reported quinolone series as represented by 2-aryl quinolone **12**, as well as acceptable safety margin through *in vitro* cytotoxicity assessment.[3]

2-aryl quinolone **12**
IC_{50} (3D7) = 117 nM
IC_{50} (W2) = 26 nM
IC_{50} (TM90C2B) = 122 nM
aq. sol. (pH7.4) = 0.03 μM
cLogP = 5.67

2-pyrazolyl quinolone **13**
IC_{50} (3D7) = 33 nM
IC_{50} (W2) = 15 nM
IC_{50} (TM90C2B) = 500 nM
aq. sol. (pH7.4) = 0.3 μM
cLogP = 3.70
TI = 333

Inhibition of kynurenine aminotransferase (KAT) II may offer a new therapeutic approach for schizophrenia and other CNS diseases. Pfizer's irreversible KAT II inhibitor **14** was found to be very potent, forming a covalent bond with the enzyme's cofactor pyridoxal phosphate (PLP) in the enzyme active site. In order to reduce phenyl **14**'s lipophilicity, Dounay and colleagues chose pyrazole as a replacement for the phenyl core structure, arriving at a new series of pyrazoles that possessed superior physiochemical properties. A representative, pyrazolyl **15**, had a ten-fold drop of calculated shake flask (cSF) distribution coefficient (log D) at pH 7.4 in comparison to **14**. Its lipophilic efficiency (LipE) was 8.53, a nearly 2-log boost from its prototype **14**.[4]

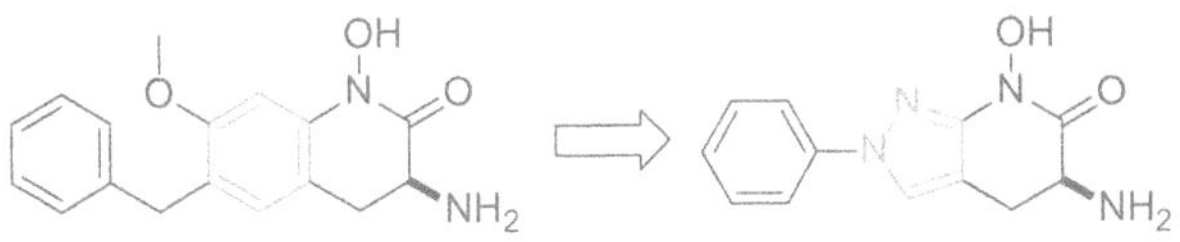

phenyl 14
KAT II IC$_{50}$, 37 nM
k$_{inact}$/K$_i$, 129,000 M^{-1}S^{-1}
cSF log D, 2.07
LipE, 6.73

pyrazolyl 15
KAT II IC$_{50}$, 25nM
k$_{inact}$/K$_i$, 112,000 M^{-1}S^{-1}
cSF log D, 0.20
LipE, 8.53

Du Pont's losartan (Cozaar, **16**) was the first selective nonpeptide angiotensin II receptor antagonist on the market for treating hypertension. It was soon discovered that its carboxylic acid metabolite, imidazole acid **17**, was significantly more potent than losartan (**16**) with a similar or longer duration of action. Merck initially explored triazole analogs in place of losartan (**16**)'s imidazole, but triazoles were generically covered by Du Pont's patent application. To circumvent the intellectual property (IP) issue, Merck switched to pyrazole as the bioisostere of losartan's imidazole moiety. To that end, they discovered a series of pyrazole compounds as represented by pyrazole **18**, which had similar potency as the imidazole derivatives.[5]

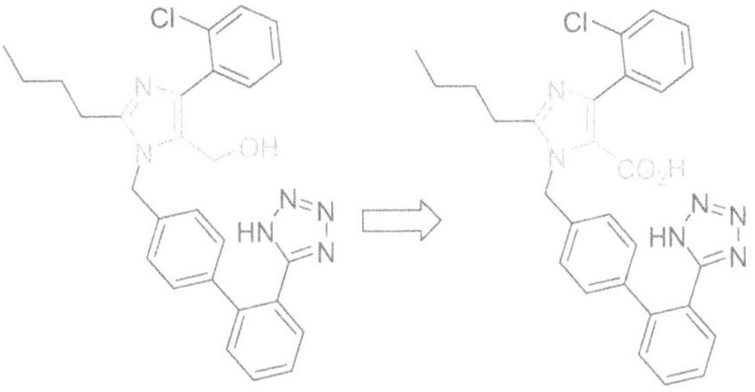

losartan (Cozarr, **16**)
rabbit aorta AT$_1$ IC$_{50}$, 40 nM

imidazole **17**
rabbit aorta AT$_1$ IC$_{50}$, 0.55 nM

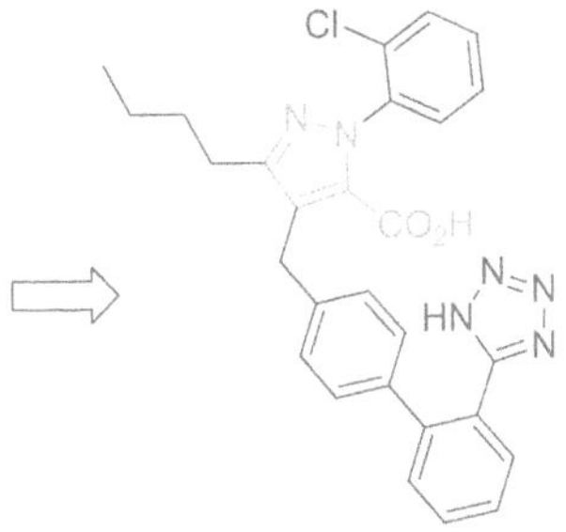

pyrazole **18**
rabbit aorta
AT$_1$
IC$_{50}$, 0.35 nM

The enormous success of apixaban (Eliquis, **10**) proved that factor Xa inhibition is a valid approach for discovering anticoagulants. Factor IXa inhibitors, in the same vein, have also shown promise in modulating the intrinsic pathway of the blood clotting cascade. Celera initially obtained phenol **19** with an IC$_{50}$ value of 99 nM as a reasonable lead. However, the potency was not translated into observed efficacy in *ex vivo* clotting efficacy assays, possibly because of **19**'s unfavorable physiochemical properties. Replacing phenol with hydroxypyrazole as a bioisostere led to compound **20**, which showed similar potency without significantly increasing the molecular weight. Furthermore, the more polar pyrazole analog **20** had improved physiochemical properties, which translated into better *ex vivo* efficacy.[6]

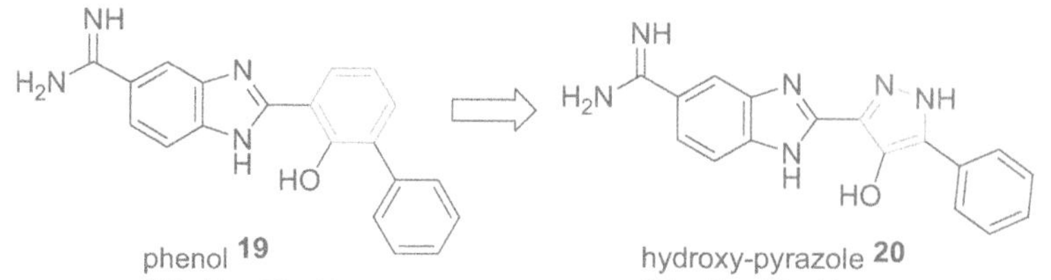

phenol **19**
flXa K_i = 99 nM
CLogP = 3.89

hydroxy-pyrazole **20**
flXa K_i = 50 nM
CLogP = 2.64

PYRAZOLE AS HYDROGEN BOND DONOR AND ACCEPTOR

Back in the 1980s, Schering–Plough reported the discovery of the first high-affinity and selective D_1/D_5 antagonist SCH 23390 (**21**). However, benzazepine **21** was inactive in rhesus monkeys with a very short duration of action. The pharmacokinetic evaluation revealed that extensive *O*-glucuronidation of the phenol and *N*-dealkylation of the N–Me group *in vivo* may contribute to the poor PK profile. Taking a page from Lilly's[7a] and Purdue's[7b] exploits of the pyrazole bioisostere of phenol in the dopaminergic field, Schering–Plough chose to replace the metabolically problematic phenol with pyrazole and other heterocycles with a hydrogen bond donating NH functionality. Generally speaking, phenol's heterocyclic bioisosteres tend to be more lipophilic and less vulnerable to phase I and II metabolisms in comparison to phenol. Indazole **22** was tested about ten-fold less potent than SCH 23390 (**21**) but began to show an improved PK profile. As a testimony to the importance of the hydrogen bond donating NH functionality, methylated indazole **23** had significantly decreased affinity for the D_1 receptor.[7c]

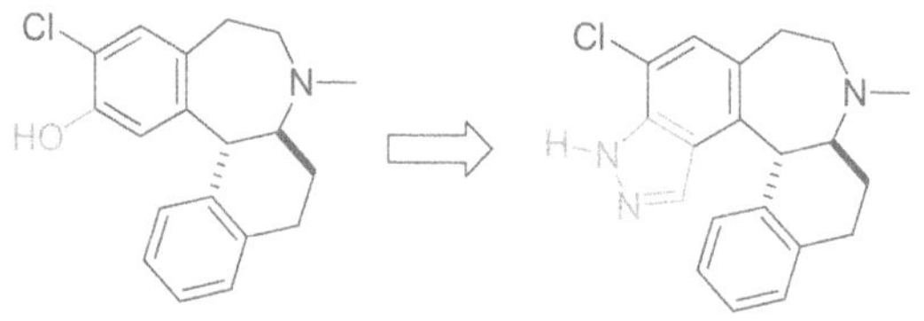

SCH 23390 (**21**)
D_1 K_i = 1.4 nM
D_5 K_i = 2.8 nM

pyrazole **22**
D_1 K_i = 14 nM
D_5 K_i = 30 nM

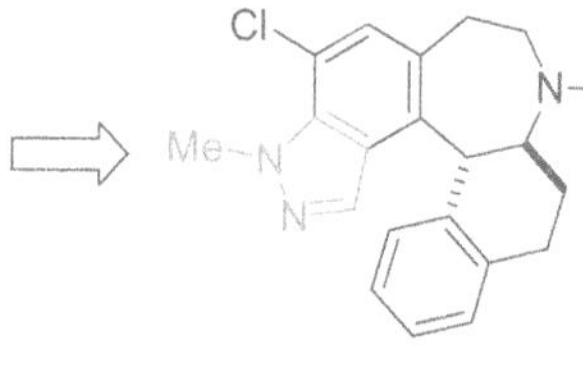

methylated pyrazole **23**
D_1 K_i = 183 nM

17β-Hydroxysteroid dehydrogenase (17β-HSD) catalyzes the reduction of estrone (**24**) to estradiol (**25**). 17β-HSD inhibitors have shown promise in several therapeutic areas including oncology, CNS, and osteoporosis. It was speculated that hydrogen bonds between estone (**24**) [or estradiol (**25**)] and the catalytically active His_{198} and His_{201} of the 17β-HSD protein stabilize substrate complexes as shown on the bottom left in the figure below.

The aforementioned hypothesis was supported by the affinity shown by the pyrazole analog **26**. While estrone (**24**) had a K_i value of 9.50 µM, inhibitor **26** had a K_i value of 4.08 µM, heterocycle such as isoxazole without a hydrogen bond donor was significantly less active. For pyrazole **26**, its NH donating an H-bond to His_{198}, with His_{201} adopting a tautomeric configuration in order to donate an H-bond to the pyrazole nitrogen atom as depicted by the structure at the bottom right corner. The potency difference between estrone (**24**) and inhibitor **26** was attributed to an additional H-bond donor interaction to His[198] by the pyrazole NH.[8]

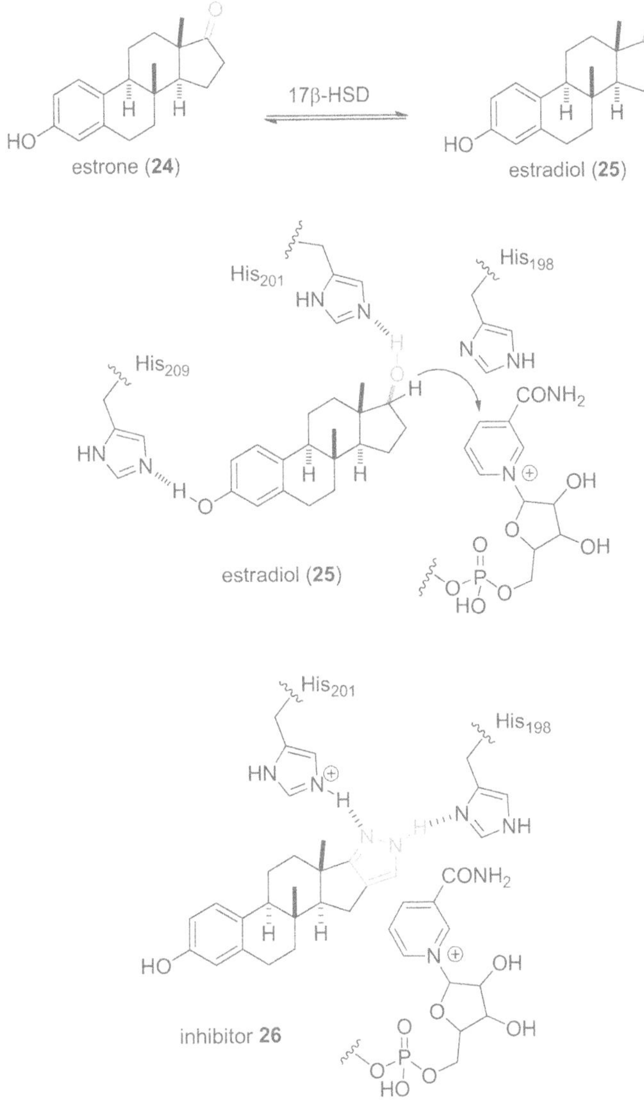

The serine/threonine kinase glycogen synthase kinase 3 (GSK3) inhibitors have been investigated as potential therapies for T2DM and Alzheimer's disease (AD). Vertex's quinazoline series GSK3 inhibitors highlighted the importance of H-bonds in kinase ligand design. Although the isoxazole analog **27** was virtually inactive, pyrazole **28** was potent and exhibited inhibition kinetics consistent with the compound acting as a competitive inhibitor of ATP binding. An X-ray co-crystal structure confirmed the mode of inhibition and revealed three H-bonding interactions between the GSK3 enzyme and pyrazole **28**, which adopted an overall *planar* topography in the active site, as shown below.[9]

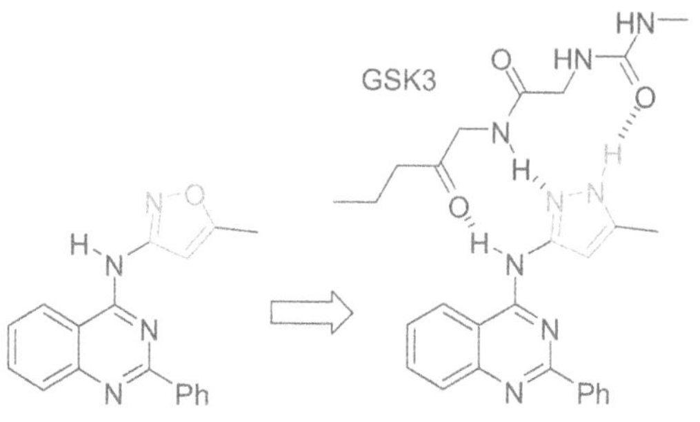

isoxazole **27**
GSK K_i > 2.0 μM

pyrazole **28**
GSK K_i = 0.024 μM

SYNTHESIS OF SOME PYRAZOLE-CONTAINING DRUGS

Production of celecoxib (**1**) is very straightforward. It entails simple condensation between dione **29** and 4-sulfonamidophenylhydrazine hydrochloride (**30**) to deliver the desired diarylpyrazole **1**.[10]

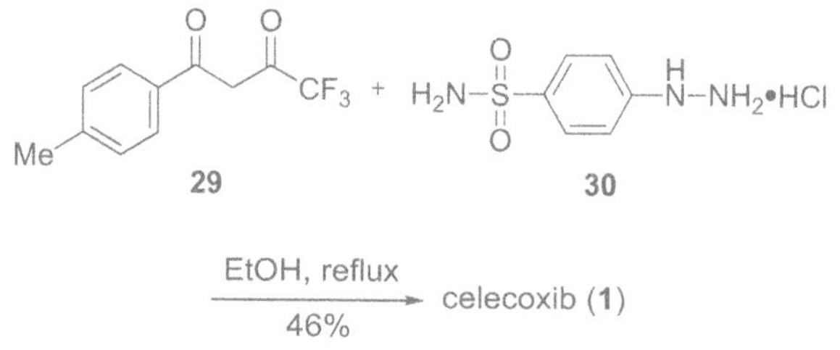

29   **30**

EtOH, reflux
————————→ celecoxib (**1**)
46%

Pfizer's synthesis of its ALK inhibitor crizotinib (**2**) began with an S_N2 reaction between 4-iodo-1*H*-pyrazole (**31**) and mesylate **32**. The resulting adduct **33** underwent a halogen–metal exchange, followed by quenching with borolane reagent **34** to afford arylboronate **35**. Finally, a Suzuki coupling with between **35** and iodide **36** delivered crizotinib (**2**).[1]

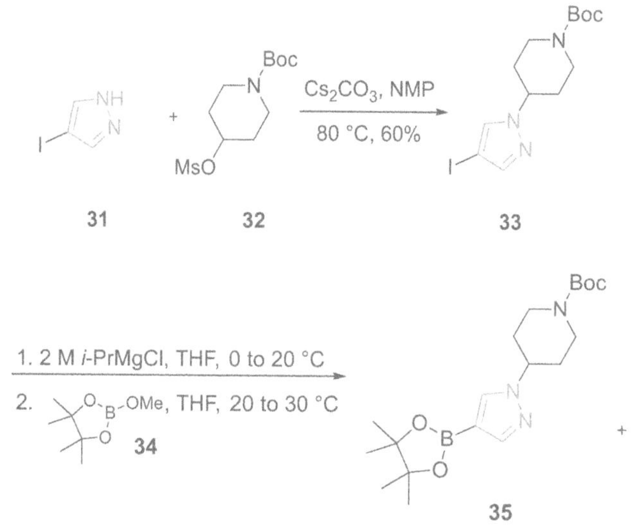

Cs$_2$CO$_3$, NMP
————————→
80 °C, 60%

31   **32**   **33**

1. 2 M *i*-PrMgCl, THF, 0 to 20 °C

2. O–B–OMe, THF, 20 to 30 °C
        34

35

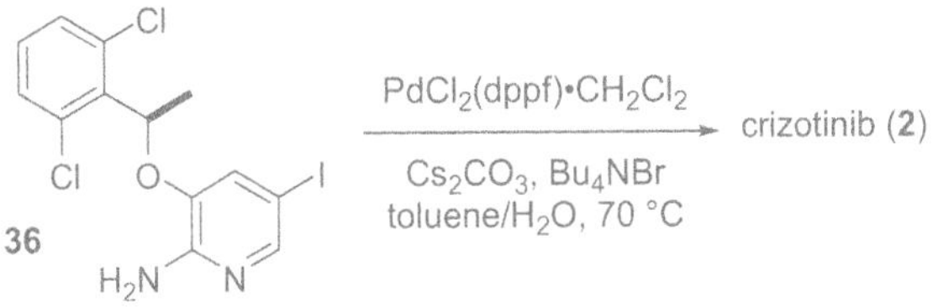

For the preparation of Pfizer's macrocyclic ALK inhibitor lorlatinib (**3**), a palladium-catalyzed aminocarbonylation between aryl iodide **37** and pyrazole amine **38** assembled the linear precursor **39**. After regioselective bromination of **39** with NBS, the resulting bromide **40** was protected as its bis-acetamide **41** otherwise the palladium-catalyzed cyclization did not work. Eventually, an intramolecular arylation of **41** was followed by de-protection to offer lorlatinib (**3**).[2]

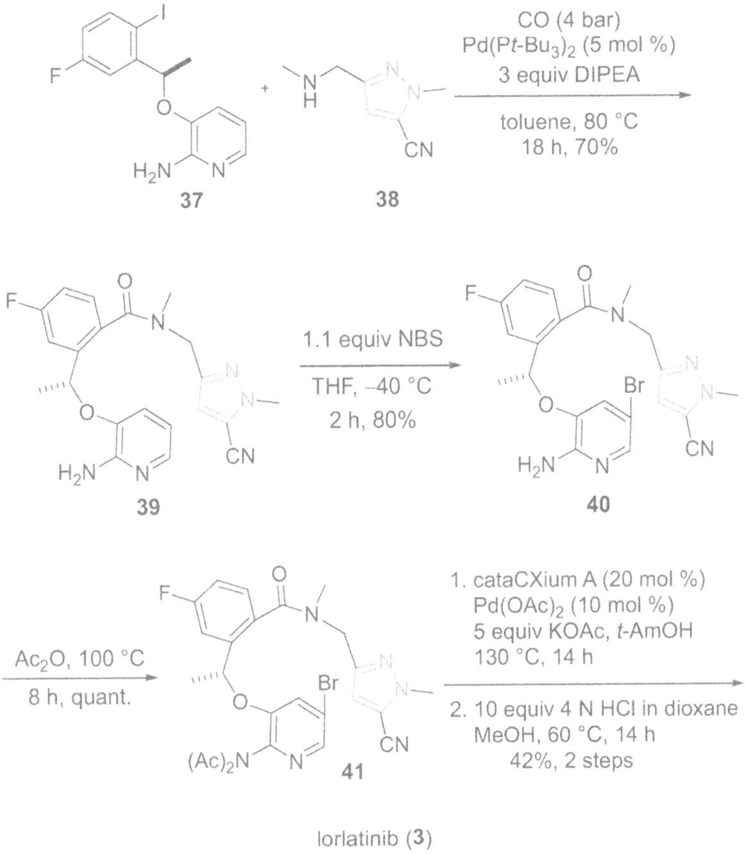

lorlatinib (**3**)

Incyte's synthesis of their JAK1/2 inhibitor ruxolitinib (**4**) commenced with a Suzuki coupling between chloropyrrolopyrimidine **42** and pyrazole pinacolatoboronate **43** to assemble pyrazole **44**. *aza*-Michael addition of pyrazole **44** to 3-cyclopentylacrylonitrile (**45**) was promoted by DBU to arrive at SEM-protected ruxolitinib **46** in excellent yield. Chiral separation of the two resultant enantiomers was followed by deprotection, salt formation, and recrystallization to produce ruxolitinib (**4**).[11]

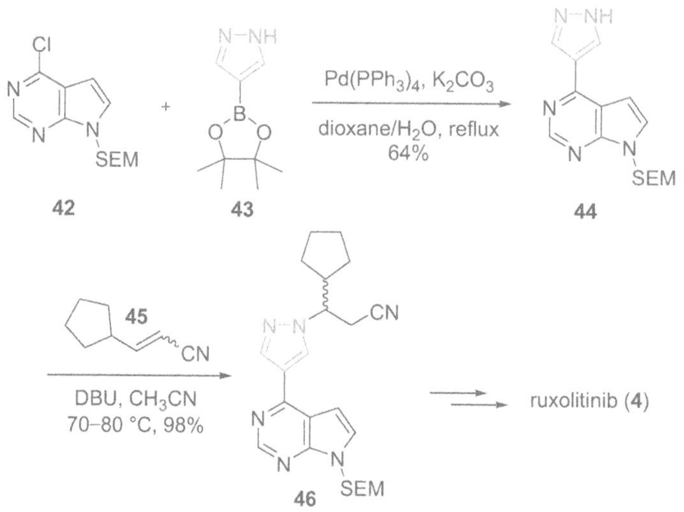

The aforementioned tactic of *aza*-Michael addition of pyrazole to acrylonitrile was also employed in one of the syntheses of Incyte's second JAK1/2 inhibitor baricitinib (**5**). To that end, *aza*-Michael addition of the same pyrazole **43** to cyanomethylazetidine **47** gave rise to pinacolatoboronate **48**. The final Suzuki coupling between **48** and chloropyrrolopyrimidine **49** proceeded smoothly *without* the protection of the pyrrole NH to deliver baricitinib (**5**).[12]

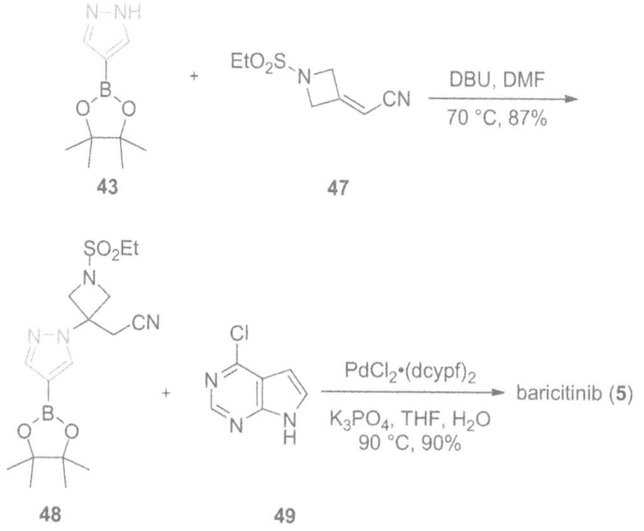

The only reported synthetic approach of DPP4 inhibitor teneligliptin (**6**) began with condensation of phenylhydrazine with acetoacetamide **50**, followed by cyclodehydration, to produce pyrazole **51**. After removal of the Boc protection, the resultant piperazine **52** underwent a reductive amination with amide-ketone **53** to assemble adduct **54**. Deprotection of the second Boc group and HBr salt formation then give rise to teneligliptin (**6**).[13]

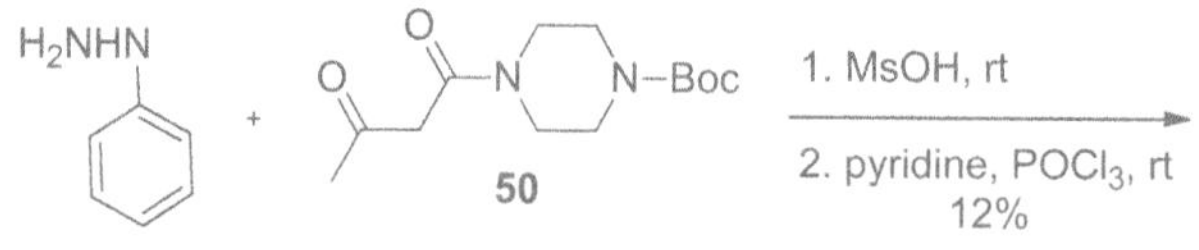

51 52 53

54

1. TFA, CH_2Cl_2, rt, 93%
2. 48% HBr, reflux EtOH, 90%

teneligliptin (**6**)

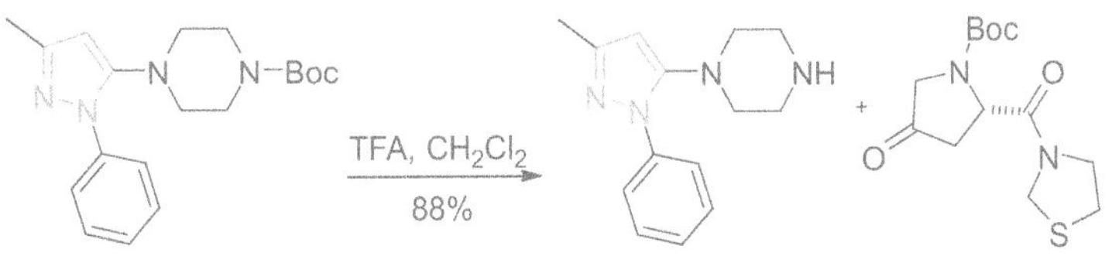

Preparation of the pyrazole fragment on Lexicon's tryptophan hydroxylase inhibitor telotristat ethyl (**7**) involves an interesting asymmetric reduction of a ketone. Thus, the halogen–metal exchange intermediate from bromide **55** was quenched by ethyl trifluoroacetate to afford ketone **56**. Asymmetric reduction of **56** was accomplished via an iridium-catalyzed hydride transfer with the aid of a chiral ligand **57** to give trifluoroalcohol **58** without resorting to any silica gel chromatography thus far. Subsequently, an S_NAr reaction between **58** and pyrimidylchloride **59** gave rise to adduct **60**. Since the reaction conditions were harsh enough to partially remove the Boc protection, it was put back on to help purification. Adjustment of the reaction to pH ~ 4 using 6 N HCl was key to ensure a good overall yield. After purification, the Boc protective group was removed and the ethyl ester was installed to provide telotristat ethyl (**7**). The API was then prepared as the hippurate salt.[14]

55 56

58 59

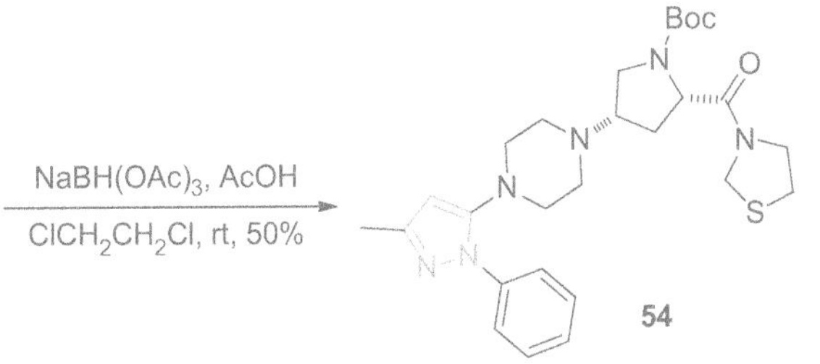

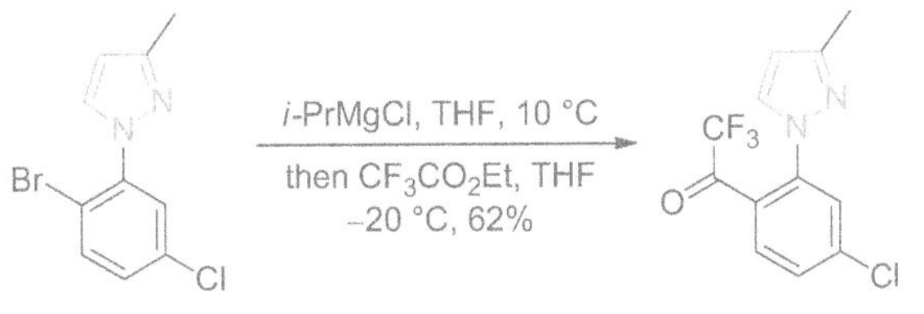

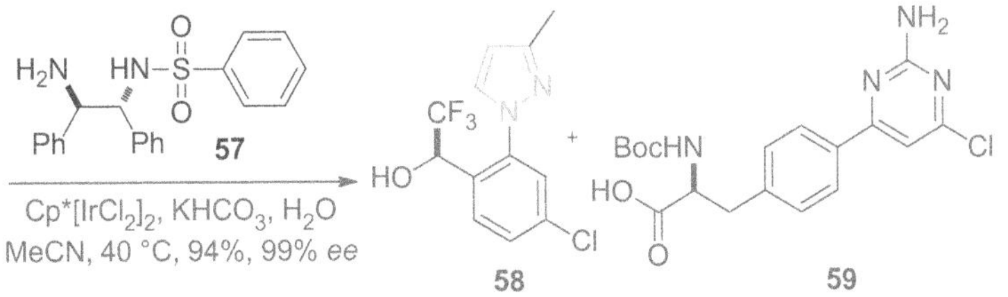

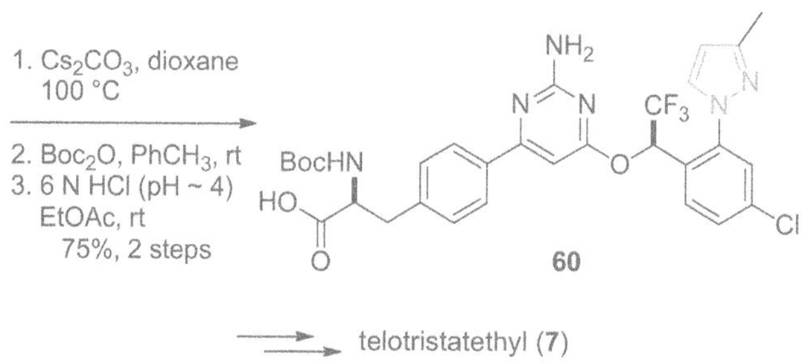

Novartis's synthesis of the B-raf inhibitor encorafenib (**8**) resorted conversion of pyrazole-amine **61** to the corresponding iodide **62** via the intermediacy of its diazonium salt. A two-step sequence involving *N,N*-dimethylformamide dimethyl acetal (DMF–DMA), and guanidine hydrochloride constructed the aminopyrimidine **63**. The amine functionality was converted to the alcohol, which was subsequently chlorinated to offer chloropyrimidine **64**. An S_NAr reaction between **64** and primary amine **65** assembled **66**, which underwent a Suzuki coupling with pinacolatoboronate **67** to deliver encorafenib (**8**).[15]

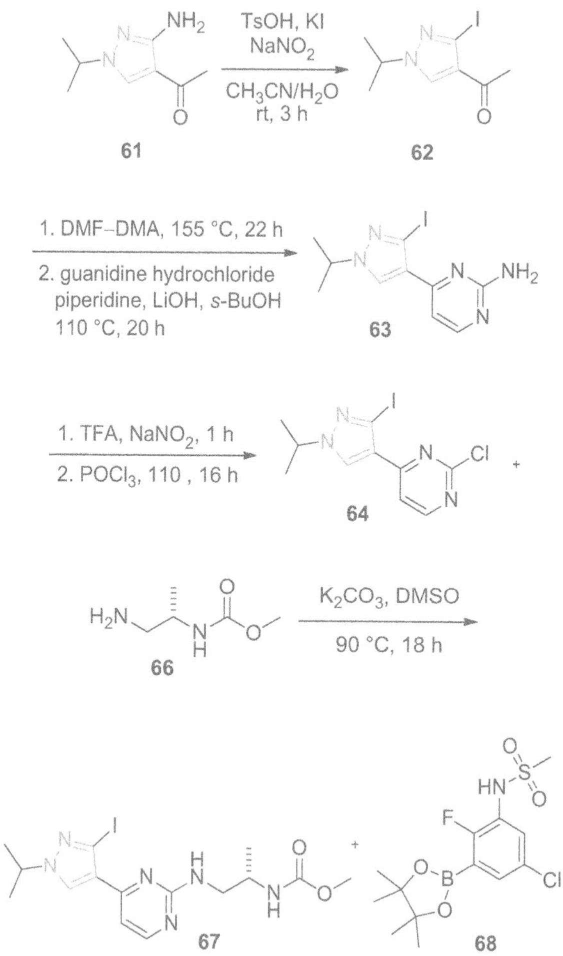

$$\xrightarrow[\substack{\text{toluene/EtOH, 85 °C} \\ \text{16 h}}]{\text{Pd(Ph}_3\text{P)}_4\text{, aq. K}_2\text{CO}_3} \text{encorafenib (8)}$$

In summary, the pyrazole fragment may serve as a bioisostere to replace an arene or a hetarene with improved potency and physiochemical properties, such as lipophilicity and aqueous solubility. In addition, as an H-bond donating heterocycle, pyrazole has been employed as a more lipophilic and more metabolically stable bioisostere of phenol. Recent trend shows that more and more pyrazole-containing drugs are in drug pipelines.

REFERENCES

1. Cui, J. J.; Tran-Dube, M.; Shen, H.; Nambu, M.; Kung, P.-P.; Pairish, M.; Jia, L.; Meng, J.; Funk, L.; Botrous, I.; et al. *J. Med. Chem.* **2011**, *54*, 6342–6363.
2. Johnson, T. W.; Richardson, P. F.; Bailey, S.; Brooun, A.; Burke, B. J.; Collins, M. R.; Cui, J. J.; Deal, J. G.; Deng, Y.-L.; Dinh, D.; et al. *J. Med. Chem.* **2014**, *57*, 4720–4744.
3. Hong, W. D.; Leung, S. C.; Amporndanai, K.; Davies, J.; Priestley, R. S.; Nixon, G. L.; Berry, N. G.; Hasnain, S. S.; Antonyuk, S.; Ward, S. A.; et al. *ACS Med. Chem. Lett.* **2018**, *9*, 1205–1210.
4. Dounay, A. B.; Anderson, M.; Bechle, B. M.; Evrard, E.; Gan, X.; Kim, J.-Y.; McAllister, L. A.; Pandit, J.; Rong, S. B.; Salafia, M. A.; et al. *Bioorg. Med. Chem. Lett.* **2013**, *23*, 1961–1966.
5. Ashton, W. T.; Hutchins, S. M.; Greenlee, W. J.; Doss, G. A.; Chang, R. S. L.; Lotti, V. J.; Faust, K. A.; Chen, T. B.; Zingaro, G. J.; et al. *J. Med. Chem.* **1993**, *36*, 3595–3605.
6. Vijaykumar, D.; Sprengeler, P. A.; Shaghafi, M.; Spencer, J. R.; Katz, B. A.; Yu, C.; Rai, R.; Young, W. B.; Schultz, B.; Janc, J. *Bioorg. Med. Chem. Lett.* **2006**, *16*, 2796–2799.
7. (a) Bach, N. J.; Kornfeld, E. C.; Clemens, J. A.; Smalstig, E. B. *J. Med. Chem.* **1980**, *48*, 812–814. (b) Doll, M. K.-H.; Nichols, D. E.; Kilts, J. D.; Prioleau, C.; Lawler, C. P.; Lewis, M. M.; Mailman, R. B. *J. Med. Chem.* **1999**, *42*, 935–940. (c) Wu, W.-L.; Burnett, D. A.; Spring, R.; Greenlee, W. J.; Smith, M.; Favreau, L.; Fawzi, A.; Zhang, H.; Lachowicz, J. E. *J. Med. Chem.* **2005**, *48*, 680–693.
8. Sweet, F.; Boyd, J.; Medina, O.; Konderski, L.; Murdock, G. L. *Biochem. Biophys. Res. Commun.* **1991**, *180*, 1057–1063.
9. Pierce, A. C.; ter Haar, E.; Binch, H. M.; Kay, D. P.; Patel, S. R.; Li, P. *J. Med. Chem.* **2005**, *48*, 1278–1281.
10. Penning, T. D.; Talley, J. J.; Bertenshaw, S. R.; Carter, J. S.; Collins, P. W.; Docter, S.; Graneto, M. J.; Lee, L. F.; Malecha, J. W.; Miyahiro, J. M.; et al. *J. Med. Chem.* **1997**, *40*, 1347–1365.
11. Zhou, J.; Liu, P.; Lin, Q.; Metcalf, B. W.; Meloni, D.; Pan, Y.; Xia, M.; Li, M.; Yue, T.-Y.; Rodgers, J. D.; et al. WO2010083283A2 (2010).
12. Xu, J.; Cai, J.; Chen, J.; Zong, X.; Wu, X.; Ji, M.; Wang, P. *J. Chem. Res.* **2016**, *40*, 205–208.
13. Yoshida, T.; Akahoshi, F.; Sakashita, H.; Kitajima, H.; Nakamura, M.; Sonda, S; Takeuchi, M.; Tanaka, Y.; Ueda, N.; Sekiguchi, S.; et al. *Bioorg. Med. Chem.* **2012**, *20*, 5705–5719.
14. (a) Bednarz, M. S.; Burgoon, H. A., Jr.; Iimura, S.; Kanamarlapudi, R. C.; Song, Q.; Wu, W.; Yan, J.; Zhang, H. WO2009029499A1 (2009). (b) Bednarz, M. S.; De Paul, S.; Kanamarlapudi, R. C.; Perlberg, A.; Zhang, H. WO2009042733A1 (2009).
15. Huang, S.; Jin, X.; Liu, Z.; Poon, D.; Tellew, J.; Wan, Y.; Wang, X.; Xie, Y. WO2011025927 (2011).

22 Pyridazines

Although not too many pyridazine-containing drugs are on the market, it has been argued that pyridazines should be considered privileged structures.[1] As a phenyl bioisostere, pyridazine's two nitrogen atoms are capable of forming hydrogen bonds with target proteins. Pyridazine has also been frequently employed as an isostere of pyridine, pyrimidine, and pyrazine in drug discovery. In addition, aminopyridazines can be used as carboxamide and amine surrogates. Among the three diazines shown below, pyridazine is the most polar due to the concentration of electronegative nitrogen lone pairs on one side of the molecule. As a consequence, pyridazine could help to boost the aqueous solubility of a drug. It often aids the formation of crystalline, water-soluble salts as well.[2]

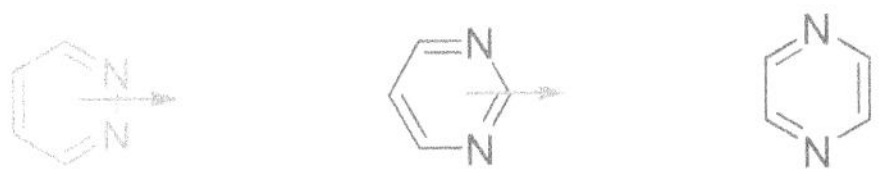

pyridazine, 3.9 D   pyrimidine, 2.42 D   pyrazine, ~0 D
pK$_a$, 2.3   1.3   0·6

Dipole moments (D=Debye) and pK$_a$ values of diazines

PYRIDAZINE-CONTAINING DRUGS

Only a few pyridazine-containing drugs are currently on the market. Sulfamethoxy-pyridazine (**1**) is an old sulfa drug. Ciba's hydralazine (Apresoline, **2**) is an old antihypertensive, but its safety profile is sub-ideal because of several side effects. Since its "naked" hydrazine was considered the culprit of toxicity and is a conspicuous structural alert under today's standards, it was masked by as a hydrazine-carboxylate to afford cadralazine (**3**).

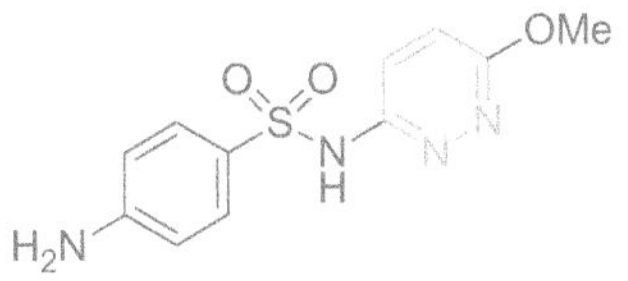

sulfamethoxypyridazine (**1**)

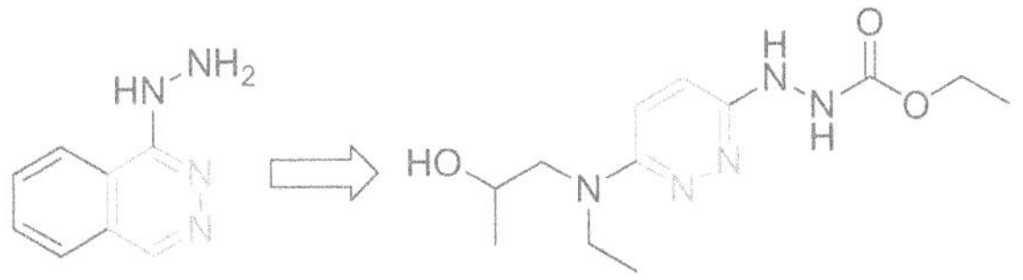

hydralazine (Apresoline, **2**)
Ciba, 1949
anti-hypertensive

cadralazine (**3**)
anti-hypertensive

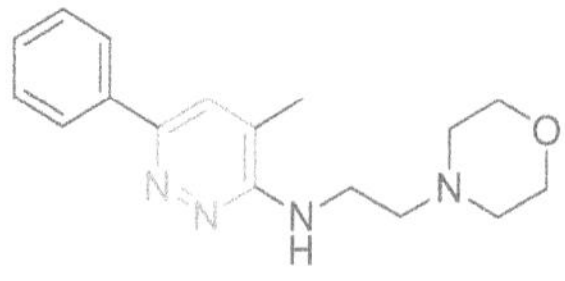

minaprine (Cantor, **4**)
anti-depressant (France)
reversible MAO-A inhibitor
withdrawn in 1996

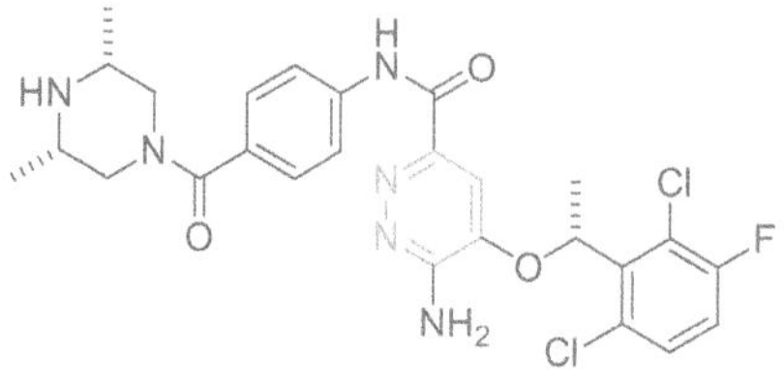

ensartinib (X-396, **5**)
Xcovery
ALK inhibitor

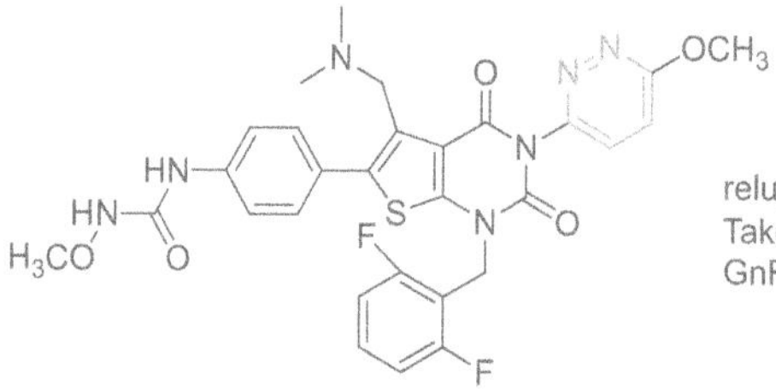

relugolix (Relumina, **6**)
Takeda, 2019
GnRH receptor antagonist

Minaprine (Cantor, **4**) is an old antidepressant only marketed in France. Like most reversible mono-amine oxidase (MAO) inhibitors, minaprine (**4**) has many side effects and was withdrawn in 1996.

Considering so many aromatic heterocycles have made appearances in kinase inhibitors, it is surprising that no pyridazine-containing kinase inhibitor has made its way to market. Xcovery's ana-plastic lymphoma kinase (ALK) inhibitor ensartinib (X-396, **5**) is currently in phase II clinical trials.

Takeda's relugolix (Relumina, **6**) is the latest entry of pyridazine-containing drugs just approved in 2019 for the treatment of uterine fibroids. It is a gonadotropin-releasing hormone (GnRH) receptor antagonist.

A pyridazine derivative, pyridazone [pyridazin-3(2*H*)-one], has made appearances in several drugs. Orion's levosimendan (Simdax, **7**) was approved in 2000 in Sweden for the treatment of congestive heart failure (CHF). It functions as a calcium sensitizer. In addition, BioMarin's phthalazin-1(2*H*)-one-containing talazoparib (Talzenna, **8**) is a poly(ADP-ribosyl) polymerase (PARP)-1/2 inhibitor approved by the FDA in 2018 for the treatment of germline BRCA-mutated, HER2-negative, locally advanced or metastatic breast cancer.

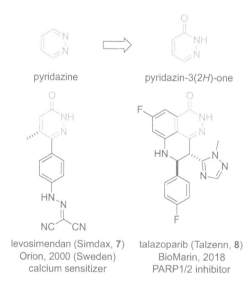

pyridazine

pyridazin-3(2*H*)-one

levosimendan (Simdax, **7**)
Orion, 2000 (Sweden)
calcium sensitizer

talazoparib (Talzenn, **8**)
BioMarin, 2018
PARP1/2 inhibitor

Similar to that of talazoparib (**8**), AstraZeneca's PARP-1 inhibitor olaparib (Lyparza, **9**) also contains a phthalazin-1(2*H*)-one motif.

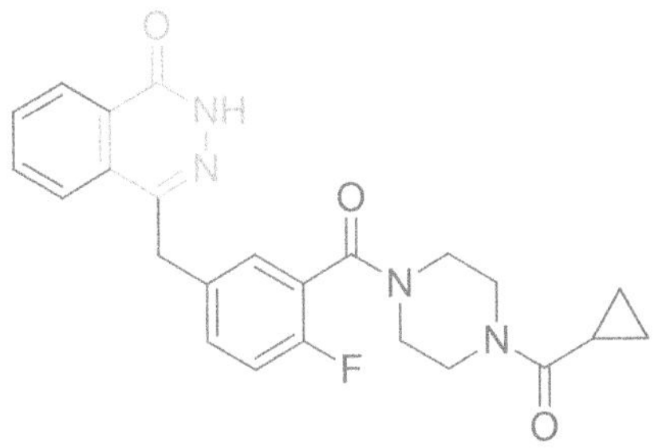

olaparib (Lyparza, **9**)
AstraZeneca, 2018
PARP1 inhibitor

On the other hand, bicyclic fused pyridazines have shown up as fragments of many drugs. Since it is not the focus of this review, only one example is given here. Ariad's ponatinib (Iclusig, **10**) was approved by the FDA in 2012 for the treatment of chronic myeloid leukemia (CML) and Philadelphia chromosome-positive (Ph+) acute lymphoblastic leukemia (ALL). It is a dual Bcr-abl/vascular endothelial growth factor receptor (VEGFR) inhibitor. CML with the T315I mutation is resistant to imatinib (Gleevec). Ponatinib (**10**), along with several second-generation Bcr-Abl inhibitors such as dasatinib (Sprycel, BMS, 2006), nilotinib (Tasigna, Novartis, 2007), and bosutinib (Bosulif, Pfizer, 2012), has been designed to be effective against T315I-mutant CML.

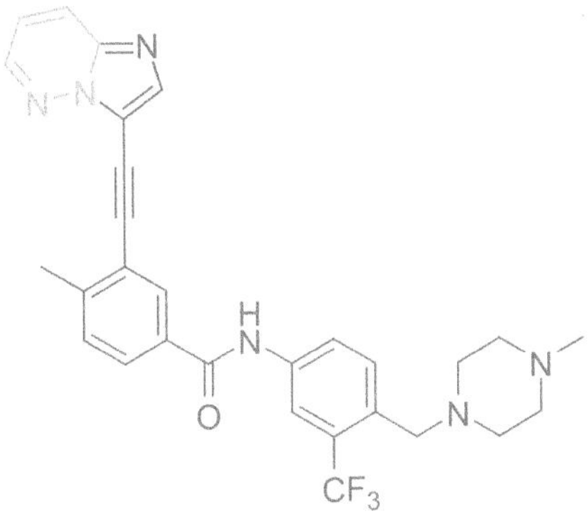

ponatinib (Iclusig, **10**)
Ariad, 2012
Bcr-abl/VEGFR inhibitor

PYRIDAZINES IN DRUG DISCOVERY

Pyridazine motif has been employed to boost a drug's aqueous solubility and facilitate the formation of crystalline, water-soluble salts. Furthermore, pyridazine may serve as a bioisostere of many heterocycles such as pyridine, pyrazine, and pyrimidine.

Boost a Drug's Aqueous Solubility

As an isostere of the phenyl ring (ClogP=2.14), the pyridazine fragment (ClogP=−0.58) is significantly less lipophilic than the former as indicated by their respective ClogP values. Diazepam (Valium, **11**), a γ-aminobutyric acid (GABA)$_A$ receptor modulator, was one of the most successful minor tranquilizers, even garnered the nickname of "Mother's Little Helper". With a log P value of 2.84, it is quite lipophilic with low aqueous solubility. In contrast, its corresponding pyridazine analog **12** has a log P value of 0.96, which translates to boosted aqueous solubility.

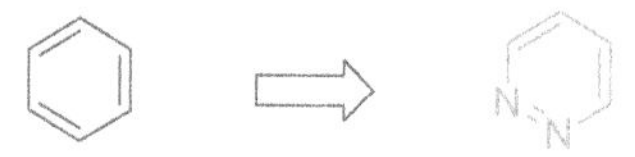

benzene
CLog P = 2.14

pyridazine
CLog P = −0.58

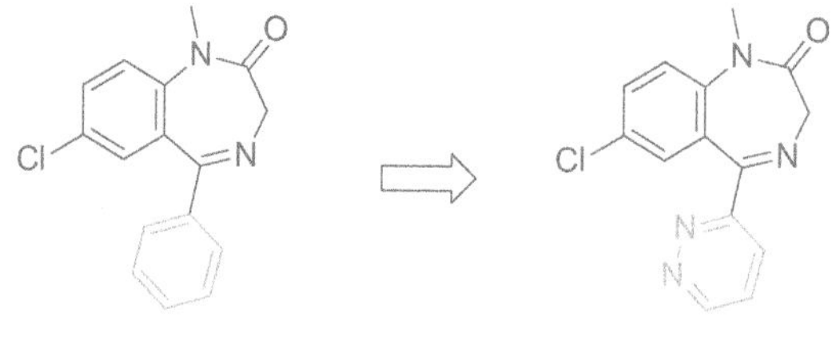

diazepam (Valium, **11**)
Log P = 2.84

pyridazine analog **12**
Log P = 0.96

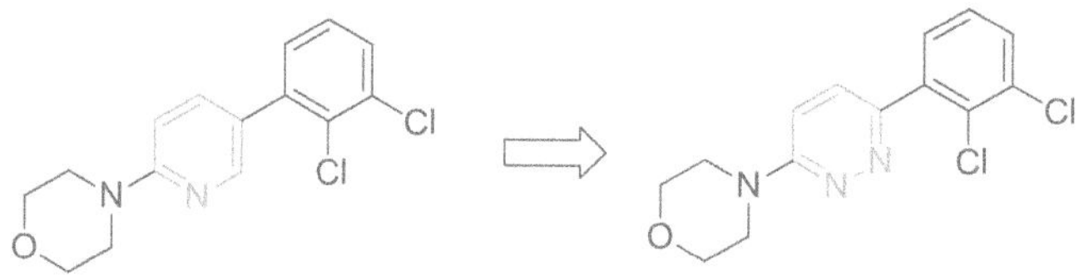

pyridine **13**
CB_2, pEC_{50}, 7.1 (efficacy, 88%)
aqueous solubility, 0.5 μM/mL
rat clearance, 9.5 mL/min/g

pyridazine **14**
CB_2, pEC_{50}, 7.1 (efficacy, 76%)
aqueous solubility, 6 μM/mL
rat clearance, 2.8 mL/min/g

Naturally occurring cannabinoids, such as Δ^9- tetrahydrocannabinol (THC), act as agonists of three G-protein coupled receptors (GPCRs): cannabinoid receptor type-1 (CB_1R) and type-2 (CB_2R), and GPR55. A series of 3-amino-6-aryl-pyridazines have been identified as CB_2R agonists with high efficacy and selectivity against the CB_1R. While pyridine **13** had an abysmal aqueous solubility of 0.5 μM/mL with high *in vitro* metabolism, its diazine analog, pyridazine **14**, saw a 12-fold boost of aqueous solubility (6 μM/mL). Diazines, including pyridazine **14**, also showed more favorable metabolic stability in comparison to the pyridine prototype **13**. Moreover, in the presence of two basic nitrogen atoms on the pyridazine ring, the formation of salt becomes possible, which may further enhance the drug's aqueous solubility. For instance, the HCl salt of pyridazine **14** has a solubility of 120 μg/mL.[3]

FACILITATE FORMATION OF CRYSTALLINE, WATER-SOLUBLE SOLID SALTS

An old antidepressant, minaprine (**4**), acts on both forms of monoamine oxidase: MAO-A and MAO-B. The free base of minaprine (**4**) is an oily, water-insoluble liquid. In stark contrast, its di-hydrochloride salt **4'** is a white, crystalline, water-soluble solid power.[4]

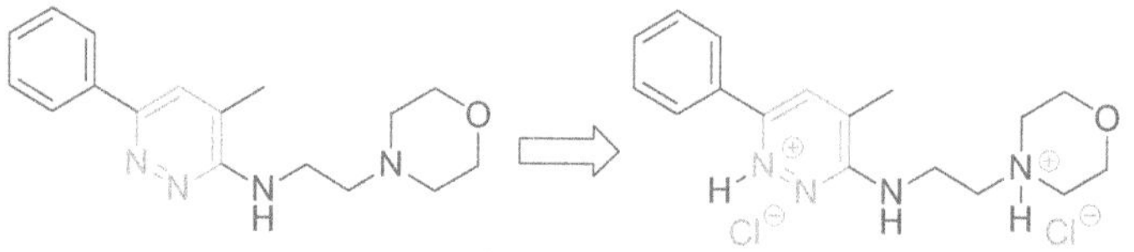

minaprine (**4**), free base
oily, water-insoluble liquid

minaprine-di-HCl **4'**, white,
water-soluble crystalline power

AS A BIOISOSTERE OF HETARYL RINGS

The pyridazine motif has been employed as a bioisostere of phenyl and hetaryl rings including pyridine, pyrazine, and pyrimidine.

Ridogrel (15) is an antiplatelet that functions as a combined thromboxane A_2 receptor (TXA$_2$R) antagonist and thromboxane synthase (TxS) inhibitor. Bioisosteric replacement of the 3-pyridyl moiety (15, IC$_{50}$, 7 nM) with 2-pyrazinyl (IC$_{50}$, 6 nM), 4-pyridazinyl (16, IC$_{50}$, 16 nM), and 5-pyrimidyl (IC$_{50}$, 39 nM) analogs inhibited TxS with comparable potency in gel-filtered human platelets. Meanwhile, the three diazines 2-pyrazinyl (IC$_{50}$, 11 μM), 4-pyridazinyl (16, IC$_{50}$, 6 μM), and 5-pyrimidyl (IC$_{50}$, 1.5 μM) analogs tested to have comparable potency in blocking the TXA$_2$R in comparison to the prototype 15 (IC$_{50}$, 1.7 μM). In the end, testing of inhibition of collagen-induced platelet aggregation in human platelet-rich plasma with 2-pyrazinyl, 4-pyridazinyl (16), or 5-pyrimidinyl analogs of ridogrel (15) indicated that these heteroaromatic moieties may serve as bioisosteric substitutes of a 3-pyridyl group in dual-acting antiplatelet agents.[5]

ridogrel (15)
TxS, IC$_{50}$, 7 nM
TXA$_2$R, C$_{50}$, 1.7 μM

4-pyridazinyl derivative 16
TxS, IC$_{50}$, 16 nM
TXA$_2$R, C$_{50}$, 6 μM

Janus kinases (JAKs) are intracellular tyrosine kinases that mediate the signaling of numerous cytokines and growth factors involved in the regulation of immunity, inflammation, and hematopoiesis. There are four members of the Janus kinase family: JAK1, JAK2, JAK3, and tyrosine kinase 2 (TYK2). In response to the stimulation of these receptors, the Janus kinases phosphorylate signal transducer and activator of transcription (STAT) proteins, which then dimerize, translocate to the nucleus, and activate gene transcription. As a member of the JAK family of nonreceptor tyrosine kinases, TYK2 plays an important role in mediating the signaling of pro-inflammatory cytokines including IL-12, IL-23, and type 1 interferon. Therefore, selective TYK2 inhibitors may be the treatment of autoimmune diseases.

In an effort to identify a new *allosteric* TYK2 JH2-binding scaffold, BMS screened their corporate compound collection by utilizing a scintillation proximity assay (SPA) and identified nicotinamide 17 as a potent TYK2 JH2 inhibitor. Unfortunately, in the IL-23 and IFNα-stimulated reporter assays, 17 was not selective over the catalytic (JH1) domains of the four JAK family members. Furthermore, 17 was found to be fairly promiscuous, inhibiting 85 of 261 kinases tested by at least 50% (at 1 μM concentration) in the homogeneous time-resolved fluorescence (HTRF) assay. Extensive structure–activity relationship (SAR) investigations eventually led to the discovery of pyridazinyl derivative 18 as the clinical candidate. Allosteric inhibitor 18 provided robust inhibition in a mouse IL-12-induced IFNγ pharmacodynamic model as well as efficacy in an IL-23 and IL-12-dependent mouse colitis model. These results demonstrated the ability of TYK2 JH2 domain binders to provide a highly selective alternative to conventional TYK2 *orthosteric* inhibitors.[6]

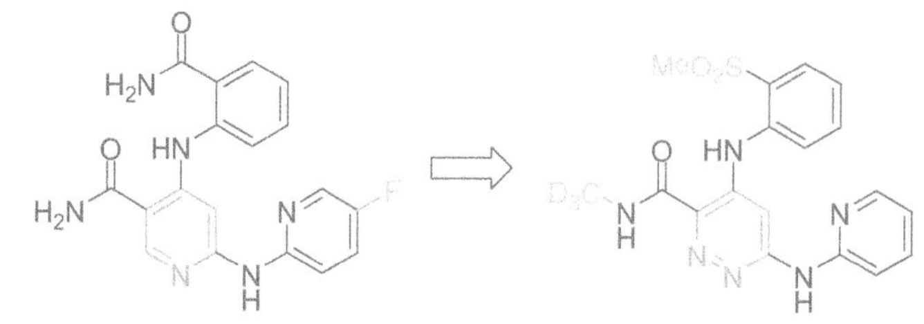

pyridyl derivative **17**
$K_{i,app}$ = 0.06 nM
TYK2 JH2 IC$_{50}$ = 0.46 nM
TYK2/JAK1/JAK2 kinase domain
IC$_{50}$ = 15/26/24 nM
IFNα IC$_{50}$ = 37 nM

pyridazinyl derivative **18**
$K_{i,app}$ = 0.07 nM
TYK2 JH2 IC$_{50}$ = 0.53 nM
TYK2/JAK1/JAK2 kinase domain
IC$_{50}$ = >50/>40/>50 nM
IFNα IC$_{50}$ = 27 nM

Perhaps a more "exotic" form of isostere of amide, aminopyridazine successfully served to expand the SAR of sulpiride (Dogmatil, **19**), an antipsychotic drug. *Wermuth and coworkers replaced* sulpiride (**19**)'s carboxamide with aminopyridazine and arrived at Do-754 (**20**, K_i=3.9 nM), which showed a higher affinity toward target protein than its prototype **19** (K_i=7.6 nM).[7]

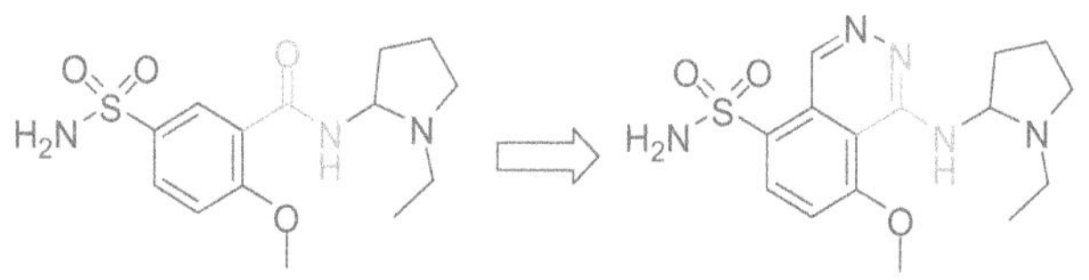

sulpiride (Dogmatil, **19**)
K_i = 7.6 nM

Do-754 (**20**)
K_i = 3.9 nM

SYNTHESIS OF SOME PYRIDAZINE-CONTAINING DRUGS

Fatty acid amide hydrolase (FAAH) is an integral membrane serine hydrolase responsible for the degradation of fatty acid amide signaling molecules such as endocannabinoid anandamide (AEA), which has been shown to possess cannabinoid-like analgesic properties. FAAH belongs to the amidase class of enzymes, a subclass of serine hydrolases that have an unusual Ser$_{241}$–Ser$_{217}$–Lys$_{142}$ catalytic triad since the Ser–His–Asp catalytic triad is more common among hydrolyses. Pfizer discovered PF-04457845 (**21**), a highly potent and selective FAAH inhibitor that reduces inflammatory and noninflammatory pain. Mechanistic and pharmacological characterization of PF-04457845 (**21**) revealed that it is an irreversible covalent inhibitor involved in carbamylation of FAAH's catalytic Ser$_{241}$ nucleophile, which results in four products including inactive covalently modified FAAH, a serine, a lysine, and by-product pyridazin-3-amine as the leaving group.[8]

Preparation of PF-04457845 (**21**) is quite straightforward. The reaction between benzylidenepiperidine hydrochloride salt **22** and phenyl pyridazin-3-ylcarbamate (**23**) in the presence of the Hünig base provided urea **21**.[8]

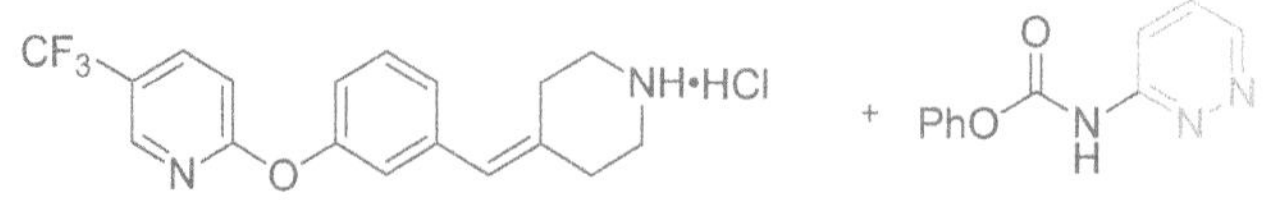

22

23

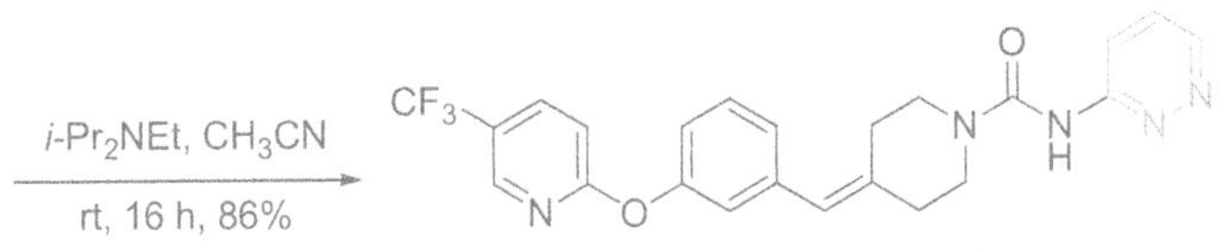

PF-04457845 (21)

Pyridazine amide **30** is an orally efficacious inhibitor of cell spleen tyrosine kinase (Syk). Its synthesis began with an S_NAr displacement of dichloropyridazine ester **24** with pyridyl aniline **25** to assemble adduct **26**. Conversion of ester on **26** to amide **27** was accomplished with ease by using 7 N ammonia in methanol. Finally, another S_NAr reaction between **27** and *tert*-butyl ((1*S*,2*R*)-2-aminocyclo-hexyl)-carbamate (**28**) produced adduct **29**, which was deprotected to offer pyridazine amide **30** as a potent Syk inhibitor.[9]

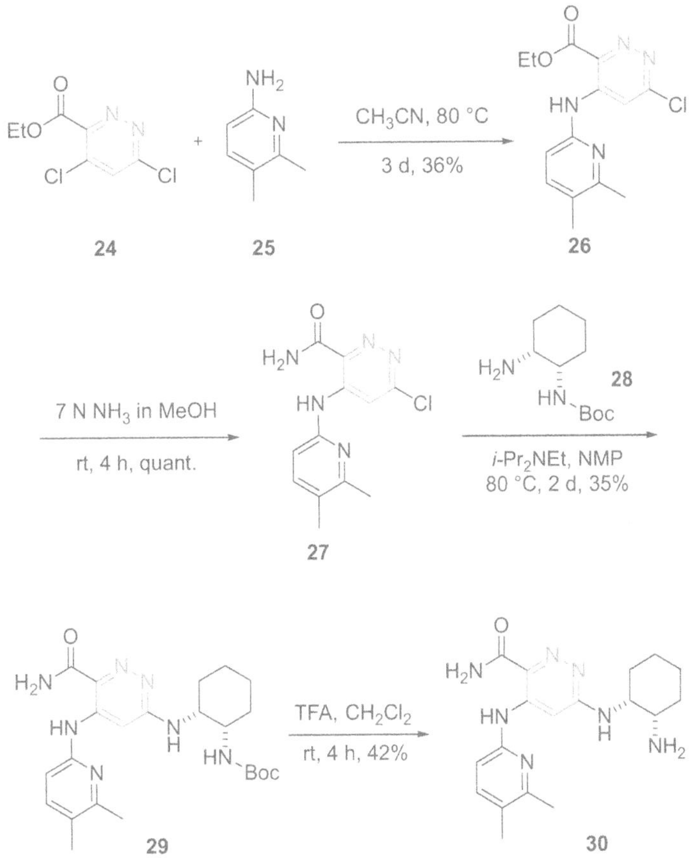

3-Pyridazinyl-coumarin **36** is a selective and reversible monoamine oxidase B (MAO-B) inhibitor, a potential drug for treating Parkinson's disease (PD) and Alzheimer's disease (AD). Its pyridazinone intermediate **32** may be prepared by condensation of ethyl (*E*)-2-(5-oxofuran-2(5*H*)-ylidene) acetate (**31**) with hydrazine hydrate. Another condensation between ester **32** and α-hydroxyl-phenol **33** in the presence of piperidine then assembled the pyridazinyl-coumarin hybrid **34** in good yield. Refluxing pyridazinone **34** with excess POCl₃ then gave rise to pyridazinyl chloride **35**, which was then converted to the final MAO-B inhibitor **36** via an S_NAr reaction with sodium methoxide.[10]

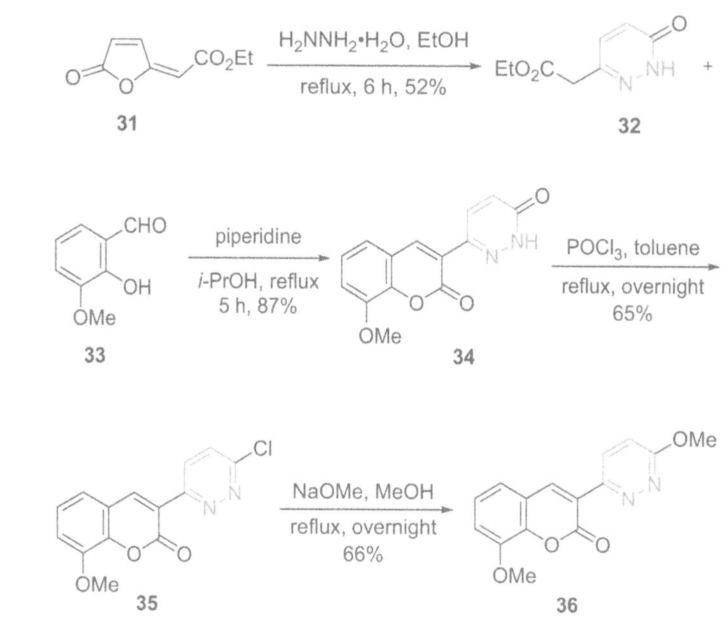

As a potential treatment of spinal muscular atrophy (SMA), branaplam (**41**) is a splicing modulator of survival motor neuron-2 (SMN2). *En route* to its synthesis, intermediate **39** was obtained from the Suzuki coupling between pinacol boronic ester **37** and pyridazinyl chloride **38**. Subsequently, another Suzuki coupling between **39** and pinacol boronic ester **40** was followed by an HCl-promoted deprotection to deliver SMN2 modulator **41**.[11]

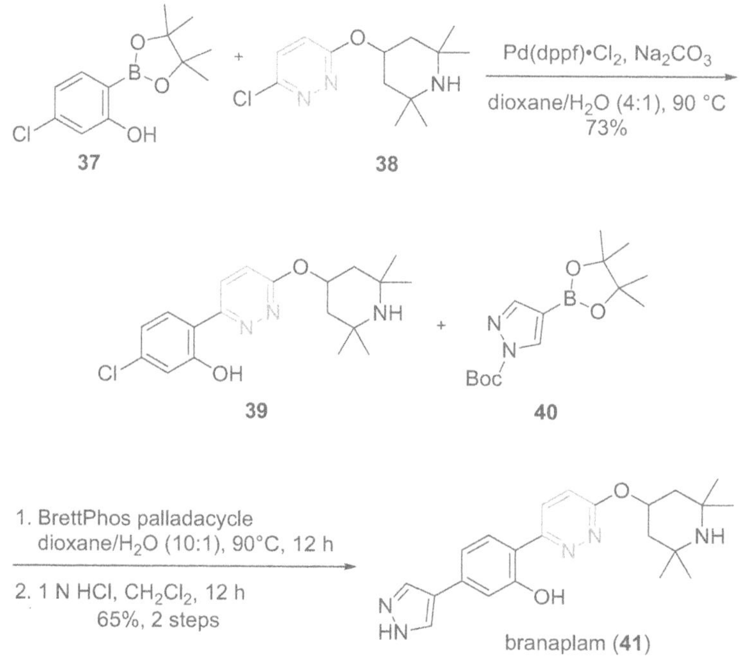

In summary, pyridazine motif has been employed in drug discovery to (a) serve as a bioisostere of hetaryl rings such as pyridine, pyrimidine, and pyrazine. Furthermore, aminopyridazines can be used as carboxamide and amine surrogates; (b) boost a drug's aqueous solubility; and (c) facilitate the formation of crystalline, water-soluble solid salts.

REFERENCES

1. Wermuth, C. G. *MedChemComm* **2011**, *2*, 935–941.
2. Jaballah, M. Y.; Serya, R. T.; Abouzid, K. *Drug Res.* **2017**, *67*, 138–148.
3. Gleave, R. J.; Beswick, P. J.; Brown, A. J.; Giblin, G. M.; Goldsmith, P.; Haslam, C. P.; Mitchell, W. L.; Nicholson, N. H.; Page, L. W.; Patel, S.; et al. *Bioorg. Med. Chem. Lett.* **2010**, *20*, 465–468.
4. Kan, J. P.; Mouget-Goniot, C.; Worms, P.; Biziere, K. *Biochem. Pharmacol.* **1986**, *35*, 973–978.
5. Heinisch, G.; Holzer, W.; Kuntz, F.; Langer, T.; Lukavski, P.; Pechlaner, P.; Weissenberger, H. *J. Med. Chem.* **1996**, *39*, 4058–4064.
6. Moslin, R.; Zhang, Y.; Wrobleski, S. T.; Lin, S.; Mertzman, M.; Spergel, S.; Tokarski, J. S.; Strnad, J.; Gillooly, K.; McIntyre, K. W.; et al. *J. Med. Chem.* **2019**, *62*, 8953–8972.
7. Rognan, D.; Sokokoff, P.; Mann, A.; Martres, M.-P.; Schwartz, J.-C.; Costentin, J.; Wermuth, C. G. *Eur. J. Pharmacol.* **1990**, *189*, 59–70.
8. Johnson, D. S.; Stiff, C.; Lazerwith, S. E.; Kesten, S. R.; Fay, L. K.; Morris, M.; Beidler, D.; Liimatta, M. B.; Smith, S. E.; Dudley, D. T.; et al. *ACS Med. Chem. Lett.* **2011**, *2*, 91–96.
9. Lucas, M. C.; Bhagirath, N.; Chiao, E.; Goldstein, D. M.; Hermann, J. C.; Hsu, P.-Y.; Kirchner, S.; Kennedy-Smith, J. J.; Kuglstatter, A.; Lukacs, C.; et al. *J. Med. Chem.* **2014**, *57*, 2683–2691.
10. Costas-Lago, M. C.; Besada, P.; Rodriguez-Enriquez, F.; Vina, D.; Vilar, S.; Uriarte, E.; Borges, F.; Teran, C. *Eur. J. Med. Chem.* **2017**, *139*, 1–11.
11. Cheung, A. K.; Hurley, B.; Kerrigan, R.; Shu, L.; Chin, D. N.; Shen, Y.; O'Brien, G.; Sung, M. J.; Hou, Y.; Axford, J.; et al. *J. Med. Chem.* **2018**, *61*, 11021–11036.

23 Pyridines—The Magic of Phenyl–Pyridyl Switch

With 6π-electrons, pyridine is an electron-deficient aromatic heterocycle containing a ring nitrogen atom. The aromatic π-electron system does not require the participation of the lone pair of electrons on the nitrogen atom. The ring nitrogen is more electronegative than the ring carbons, making the two α-ring carbons and the γ-ring carbon more electropositive than otherwise would be expected from benzene.

benzene   CH → N   pyridine

In the context of medicinal chemistry, replacement of a CH group with an N atom on a phenyl ring may influence a drug's molecular and physicochemical properties, as well as its intramolecular and intermolecular interactions that can translate to improved pharmaceutical profiles.

PYRIDINE-CONTAINING DRUGS

The FDA has approved over 60 pyridine-containing drugs, making pyridine a privileged scaffold, only eclipsed by (72) piperidine-containing drugs. We showcase only eight representative examples here for brevity.

loratadine (Claritin, **1**)
Schering–Plough, 1993
H$_1$ receptor antagonist

nevirapine (Viramune, **2**)
Boehringer Ingelheim,1996
NNRTI

Schering–Plough's antihistamine franchise dominated the allergy drug market for decades. Its non-sedating histamine receptor-1 (H$_1$) antagonist loratadine (Claritin, **1**), containing one pyridine ring, was one of the most popular treatments for allergy without central nervous system (CNS) side effects. Its carbamate moiety is key to retard its brain penetration and the 8-chlorine atom blocks a probable site of metabolism and offers a longer duration of action[1]—since pyridine is electron-deficient, the phenyl ring is oxidized more readily by cytochrome P450 (CYP) in the liver. Nevirapine (Viramune, **2**), Boehringer Ingelheim's non-nucleoside human immunodeficiency virus type 1 (HIV-1) reverse transcriptase inhibitor (NNRTI) for treating AIDS, contains two pyridine rings. It adopts a "butterfly-like" conformation when binding to the allosteric site of the reverse transcriptase enzyme. Although the corresponding N11-ethyl derivative was more potent

(IC_{50}, 35 nM) than nevirapine (**2**, IC_{50}, 84 nM) in both enzymatic and cellular assays and more soluble, the N11-cyclopropyl analog nevirapine (**2**) was selected as the drug candidate because it was more bioavailable due to the fact that cyclopropane was more resistant to metabolism while the ethyl group was more prone to undergo dealkylation.[2]

The pyridine motif is indispensable to the biological activities of AstraZeneca's proton pump (H^+, K^+-ATPase) inhibitor omeprazole (Prilosec, **3**) as a treatment of peptic ulcer. The lone pair electrons on pyridine are the "engine" to start the "omeprazole cycle". Indeed, the "omeprazole cycle" begins by nucleophilic attack of protonated benzimidazole by the N atom of pyridine to form a benzimidazoline intermediate.[3] In a sense, omeprazole (**2**) is a pro-drug, only becoming active when protonated. On the other hand, Roche's netupitant (**4**, Akynzeo when combined with palonosetron, a $5HT_3$ receptor antagonist) is a nerokinin-1 (NK-1) receptor antagonist for the treatment of nausea and vomiting caused by chemotherapy or surgery.[4]

omeprazole (Prilosec, **3**)
AstraZeneca, 1998
proton pump inhibitor

netupitant (Akynzeo, **4**)
Roche, 2015
NK-1 receptor antagonist

Kinase inhibitors are the most fruitful class of medicines as targeted cancer drugs during the last three decades. Half a dozen of them possess the pyridine fragment, which includes Lilly's cyclin-dependent kinase (CDK)4/6 inhibitor abemaciclib (Verzenio, **5**)[5] and Pfizer's anaplastic lymphoma kinase (ALK) inhibitor lorlatinib (Lorbrena, **6**). The case of lorlatinib (**6**) serves as a clever medicinal chemistry maneuver that converting linear ALK inhibitor crizotinib (Xalkori) to macrocycle may offer superior physicochemical properties. As the first-in-class ALK inhibitor, crizotinib (Xalkori) has little CNS exposure. Guided by a structure-based drug design (SBDD), lipophilicity efficiency (LipE), and physical property-based optimization, Pfizer discovered macrocyclic lorlatinib (**6**), which has good absorption, distribution, metabolism, and excretion (ADME), low propensity for p-glycoprotein (pgp) 1-mediated efflux, and good passive permeability with significant CNS exposure. It is thus suitable for treating metastasized brain tumors.[6]

abemaciclib (Verzenio, **5**)
Lilly, 2017
CDK4/6 inhibitor

lorlatinib (Lorbrena, **6**)
Pfizer, 2018
ALK inhibitor

The latest entries of pyridine-containing drugs on the market may be exemplified by Jassen's androgen receptor (AR) antagonist apalutamide (Erleada, **7**)[7] for treating castration-resistant prostate cancer (CRPC) and Agios' isocitrate dehydrogenase 1 (IDH1) inhibitor ivosidenib (Tibsovo, **8**) for treating IDH1-mutant cancers such as acute myeloid leukemia (AML).[8] The structure of apalutamide (**7**), discovered by Jung's group at UCLA in the 2000s, is similar to that of enzalutamide

(Xtandi, developed by Medivation and approved in 2012), also discovered by Jung. However, in murine xenograft models of metastasized-CRPC (mCRPC), apalutamide (**7**) demonstrated greater antitumor activity than enzalutamide. Furthermore, apalutamide (**7**) penetrates less effectively the BBB (blood–brain barrier) than enzalutamide, suggesting that the chance of developing seizures may be less than with enzalutamide. In the end, the fact that both Janssen and Medivation were able to secure intellectual properties for their respective AR antagonists also speaks volume of the power of the phenyl–pyridyl switch. Finally, Agios' ivosidenib (**8**) is the first-in-class IDH1 inhibitor, approved hot on the heel of FDA approval of their IDH2 inhibitor enasidenib (Idhifa, incidentally, also contains two pyridine rings), another first-in-class cancer drug.

apalutamide (Erleada, **7**)
Janssen, 2018
androgen receptor antagonist

ivosidenib (Tibsovo, **8**)
Agios, 2018
IDH1 inhibitor

PYRIDINES IN DRUG DISCOVERY

In 2017, Pennington and Moustakas published an excellent review on "The Necessary Nitrogen Atom" covering the influence of pyridine as a bioisostere of the phenyl fragment (dubbed as the N scan SAR strategy).[9] The phenyl–pyridyl switch may have a profound impact on a drug's *in vitro* binding affinity; *in vitro* functional affinity; *in vitro* PK/ADME profile; *in vitro* safety profile; and *in vivo* pharmacological profile. Here only a few interesting examples are highlighted.

Boosting Biochemical Potency

Replacing a phenyl ring with a pyridine offers a nitrogen atom as a hydrogen bond acceptor, which may make contact with the target, forming a hydrogen bond in addition to a subtle change in π-stacking. The consequence may be a higher binding affinity and a better biochemical potency.

Phenyl-linked sulfonamide **9** as a dual mammalian target of rapamycin (mTOR) and phosphoinositide 3-kinase α (PI3Kα) inhibitor was not very potent (mTOR, IC$_{50}$ = 1,500 nM; PI3Kα, IC$_{50}$ = 48 nM). A partial N-scan on the phenyl scaffold of **9** identified pyridyl-linked sulfonamide **10** displaying 140- and 30-fold improved potency toward mTOR and PI3Kα, respectively (mTOR, IC$_{50}$ = 11 nM; PI3Kα, IC$_{50}$ = 1.6 nM). An X-ray cocrystal structure revealed that the pyridyl N atom engaging in a hydrogen bond with an ordered water molecule located between Tyr867 and Asp841 in the affinity pocket of the enzyme.[10]

phenyl-linked sulfonamide **9**
mTOR IC$_{50}$ = 1,500 nM
PI3Kα K_i = 48 nM

140-fold
biochemical potency

pyridyl-linked sulfonamide **10**
mTOR IC$_{50}$ = 11 nM
PI3Kα K_i = 1.6 nM

FIXING CYP450 LIABILITY

One of Agios' IDH1 inhibitors AGI-14100 (11) looked largely promising as a drug candidate with a good profile of single-digit nM potency in enzyme and cell-based assays and desired metabolic stability. However, assessment in the human pregnane X receptor (hPXR) screen indicated that it was potentially a CYP 3A4 inducer. hPXP activation by AGI-14100 (11) was approximately 70% of rifampicin, a known strong CYP 3A4 inducer. Miraculously, changing one of the two C–F bonds with an N atom embedded in the ring led to ivosidenib (Tibsovo, 12) with two pyridine rings and a balance of desirable properties: good enzyme and cellular potency, good stability in human liver microsomes (HLM), reduced hPXPR activation, good permeability, and low efflux ratio (E_h).[8] It is a small wonder that it eventually became an FDA-approved drug.

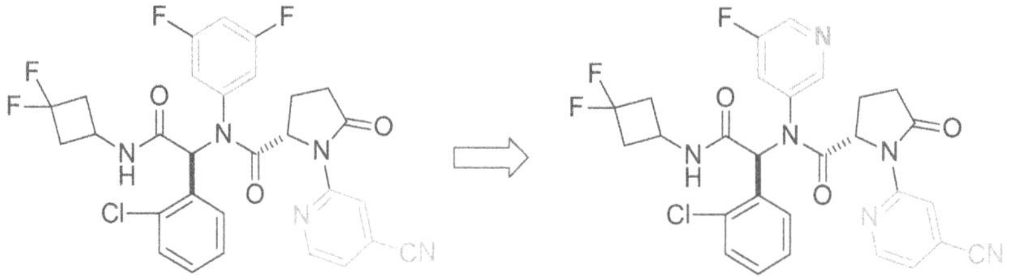

AGI-14100 (11)
enzyme IC_{50}, 6 nM
cellular IC_{50}, 1 nM
E_h 0.23
hPXR activation(% of rifampicin at 1/10 μM):
69/70

ivosidenib (Tibsovo, 12)
12 nM
8 nM
0.15
2/21

ELEVATING PERMEABILITY

Pfizer's CP-533,536 (13) is a selective and non-prostanoid EP2 receptor agonist (EP2 stands for prostaglandin E2). Switching one CH to N led to omidenepag (OMD, 14), which was 15-fold more potent than its progenitor 13. However, OMD (14)'s cell membrane permeability was insufficient and the rate measured with the parallel artificial membrane permeability assay (PAMPA) was 0.9×10^{-6} cm/s. The isopropyl ester prodrug, omidenepag isopropyl (OMDI, 15), gratifyingly, had adequate cell membrane permeability with a rate of 2.8×10^{-5} cm/s. After showing efficacy in lowering intraocular pressure (IOP) following ocular administration in ocular normotensive monkeys, omidenepag isopropyl (15) was selected as a clinical candidate for the treatment of glaucoma.[11]

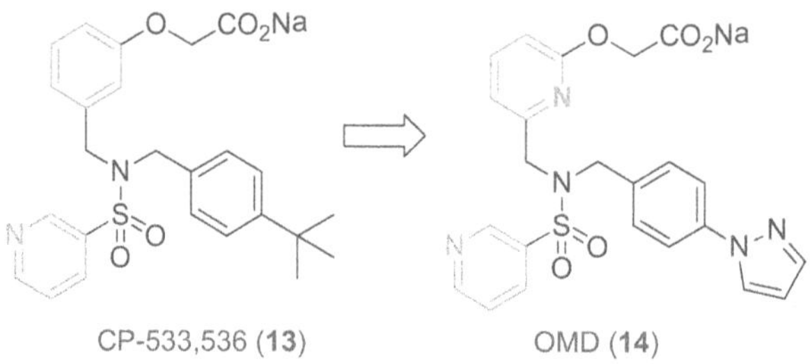

CP-533,536 (13)
h-EP2 EC_{50}: 17 nM

OMD (14)
h-EP2 EC_{50}: 1.1 nM
PAMPA: 0.9×10^{-6} cm/s

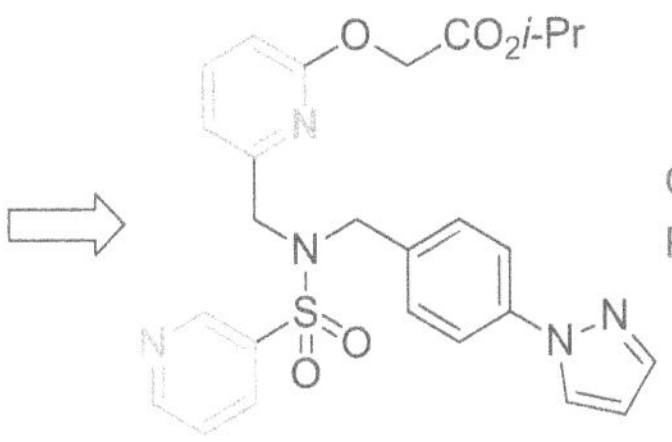

OMDI (15)
PAMPA: 2.8×10^{-5} cm/s

Some may consider "cheating" to use ester prodrug **15** to further boost the drug's cellular permeability. A *bona fide* example of permeability improvement via the phenyl–pyridyl switch may be found in Lundbeck's endeavor in their tricyclic thiazolopyrazole derivatives as metabotropic glutamate receptor 4 (mGluR4) positive allosteric modulators (PAMs). Phenyl aniline **16** displayed weak PAM activity and poor permeability in PAMPA. Nitrogen scan afforded 3-pyridyl aniline **17** with >190-fold increase of permeability although at the price of losing potency. The 2-pyridyl aniline saw >30-fold improvement in permeability and a ten-fold increase in potency, but its kinetic solubility was low. Eventually, the corresponding 2-pyrimidyl aniline derivative had a good balance of potency, permeability, and solubility.[12] In this case, one more nitrogen atom is good, two more are even better!

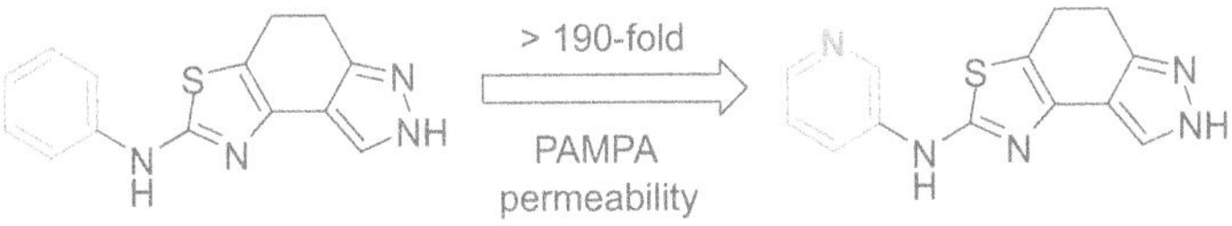

> 190-fold
PAMPA
permeability

phenyl aniline **16**
mGlu4 EC_{50} = 2.2 μM
$P_{app}(A \rightarrow B) < 0.10 \times 10^{-6}$ cm/s

3-pyridyl aniline **17**
mGlu4 EC_{50} > 10 μM
$P_{app}(A \rightarrow B) = 19 \times 10^{-6}$ cm/s

ADDRESSING PROTEIN BINDING ISSUE

The phenyl–pyridyl switch has been employed to fix protein binding (plasma protein shift) issues. For example, olefin **18** as a selective estrogen receptor degrader (SERD) had an excellent potency toward lowering steady-state ERα levels but was highly protein-bound in diluted mouse plasma (f_u=0.30%). The phenyl–pyridyl switch offered several pyridyl analogs. One of them, 2-pyridyl **19** exhibited an 11-fold lower protein binding (f_u=3.2%). Apparently, reduction of the molecule's lipophilicity was beneficial in reducing protein binding.[13]

Again, many useful examples on the merits of the phenyl–pyridyl switch may be found in Pennington's scholastic 2017 JMC review.[9]

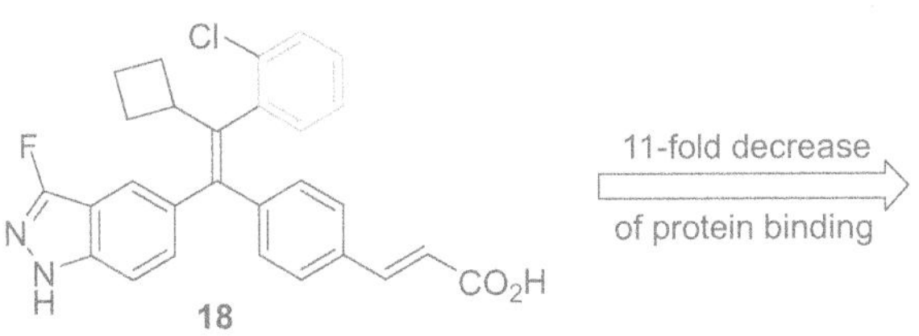

11-fold decrease
of protein binding

18
MCF-7 ERα EC_{50} = 0.40 nM
E_{max} = 90%
f_u = 0.30%

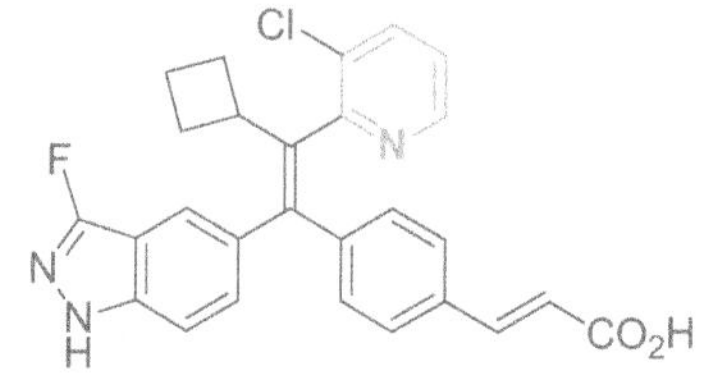

19
MCF-7 ERα EC$_{50}$ = 0.30 nM
E$_{max}$ = 90%
f_u = 3.2%

SYNTHESIS OF SOME PYRIDINE-CONTAINING DRUGS

LORATADINE (CLARITIN, 1)

Schering–Plough's synthesis of loratadine (**1**) commenced with a Ritter reaction of 3-methylpico-linonitrile in *t*-butanol with the aid of concentrated H_2SO_4 as the catalyst to prepare *N*-(*tert*-butyl)-3-methylpicolinamide (**20**) as a means of masking the nitrile group. After treating **20** with two equivalents of BuLi, the resulting intensely purple-colored dianion underwent an S_N2 reaction with *m*-chlorobenzyl chloride to assemble adduct **21**. Refluxing **21** in POCl$_3$ restored the original nitrile functionality on **22**, which was then treated with *N*-methyl-piperidinyl Grignard reagent to produce ketone **23** after acidic hydrolysis of the imine intermediate. Superacid cyclodehydration of ketone **23** employing HF/BF$_3$ delivered the C8-chloro-tricyclic **24**. Superacid HF/BF$_3$ was key to the success of regioselectivity with minimal C10-chloro isomer formation. The penultimate derivative **24** was then treated with 3 equivalents of ethyl chloroformate to deliver antihistamine **1**.[14]

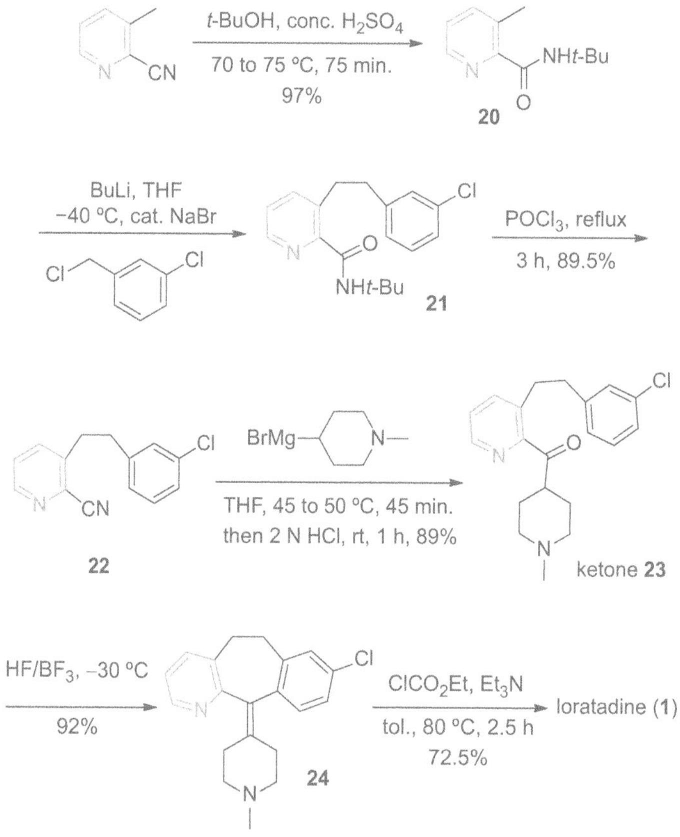

NEVIRAPINE (VIRAMUNE, 2)

The precursor for one of the syntheses of nevirapine (2) was 3-aminopyridine 25, which was prepared from the reduction of the corresponding 3-nitropyridine analog. Coupling 25 with acid chloride 26 resulted in bis-pyridyl-amide 27. Subsequent S$_N$Ar reaction between 27 and cyclopropylamine gave adduct 28 in 83% yield. Such an excellent regioselectivity favoring the chloride on the right reflects its attachment to an electron-withdrawing carbonyl group. Finally, adduct 28, on cyclization under basic conditions at 160°C, afforded nevirapine (2) in 67% yield.[15]

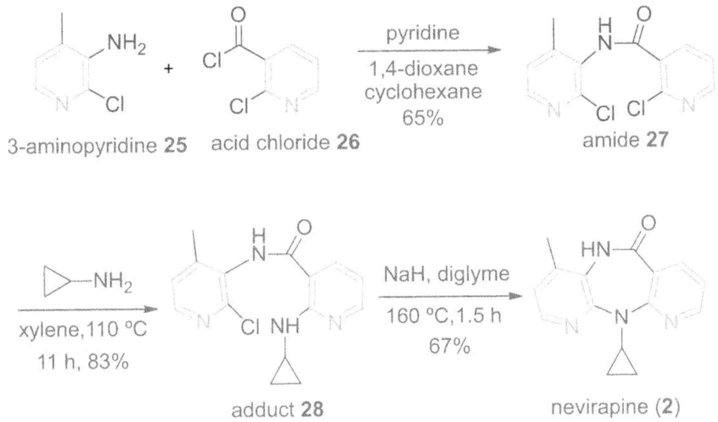

3-aminopyridine 25   acid chloride 26   amide 27

adduct 28   nevirapine (2)

OMEPRAZOLE (PRILOSEC, 3)

One of the syntheses of omeprazole (Prilosec, 3) employed 2,3,5-trimethylpyridine as its starting material, which underwent an oxidation and subsequent nitration to produce 4-nitro-pyridine N-oxide 29. S$_N$Ar replacement of the nitro group with NaOMe afforded 4-methoxy-pyridine N-oxide 30. The Boekelheide reaction entailing treatment of N-oxide 30 with acetic anhydride offered hydroxymethyl-pyridine 31 after basic hydrolysis of the acetate intermediate. Chlorination of alcohol 31 led to chloride 32, which was coupled with thiol 33 to assemble pymetazole (34). Oxidation of sulfide 34 then delivered omeprazole (3).[16]

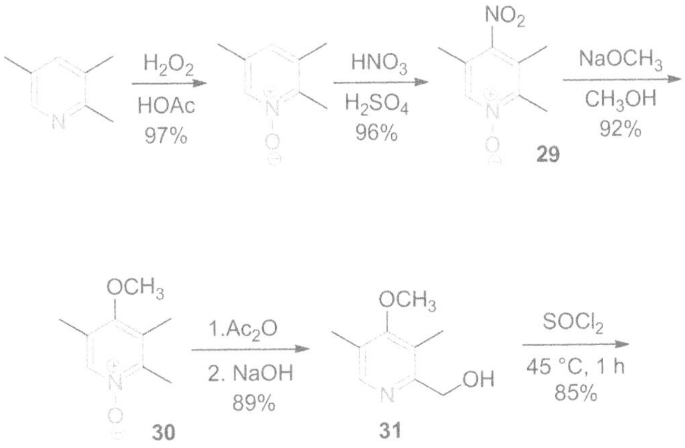

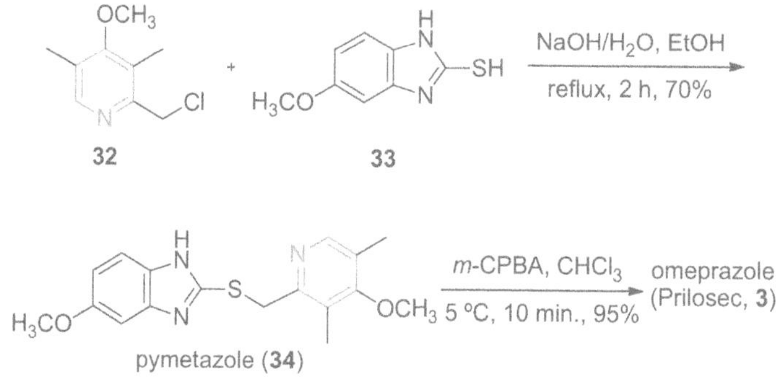

NETUPITANT (4)

Roche's preparation of netupitant (4) was initiated by a conjugate addition of *o*-tolylmagnesium chloride to 6-chloronicotinic acid to afford the corresponding dihydropyridine intermediate, which was then oxidized to phenyl-pyridine **35**. Converting its carboxylic acid to the corresponding primary carboxamide was followed by an S_NAr displacement of the chlorine with 1-methylpiperazine to give tri-substituted pyridine **36**. The Hoffmann rearrangement of **36** was initiated by NBS to produce carbamate **37**, which was then reduced to the corresponding methylaniline and treated with the requisite acid chloride to deliver netupitant (4).[17]

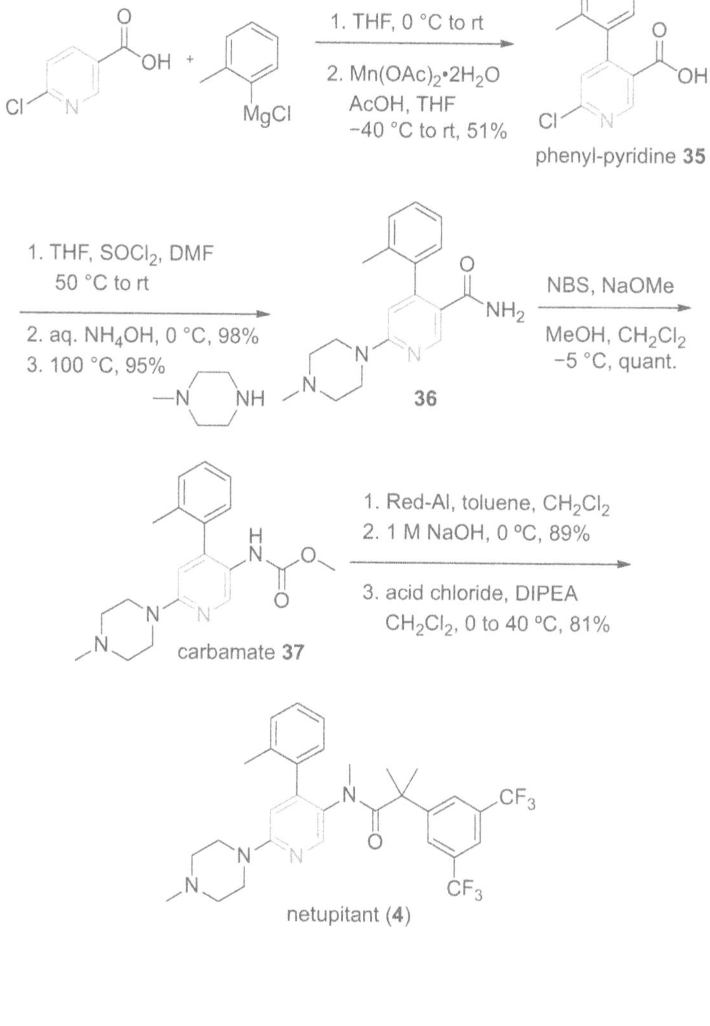

ABEMACICLIB (VERZENIO, 5)

As the first step for a route to Lilly's CDK4/6 inhibitor abemaciclib (**5**), a reductive amination between 6-bromonicotinaldehyde and 1-ethylpiperazine led to adduct **38**. Conversion of the bromide to the corresponding amine employing Hartwig's procedure—a combination of LiHMDS and Pd$_2$(dba)$_3$ with the aid of CyJohnphos as the ligand—led to aminopyridine **39** after acidic hydrolysis. A Buchwald–Hartwig coupling of **39** with chloropyrimidine **40** then produced abemaciclib (**5**).[18]

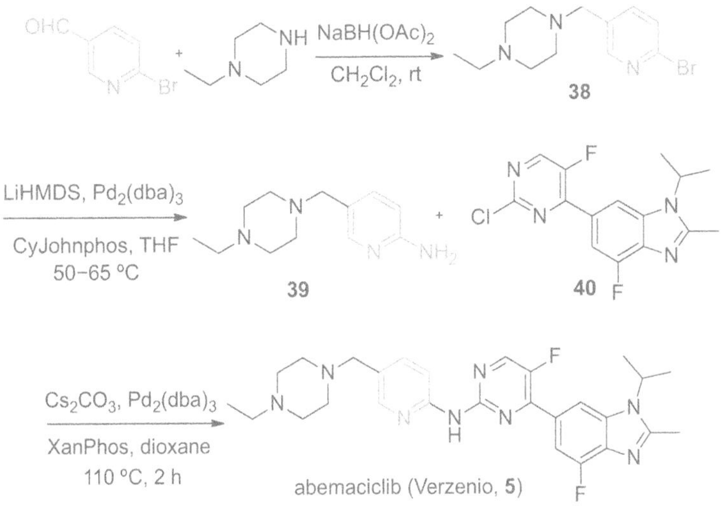

LORLATINIB (LORBRENA, 6)

Pfizer's synthesis of lorlatinib (**6**) started with an S$_N$2 displacement of mesylate **41** with 2-aminopyridin-3-ol to prepare ether **42**, which coupled with amine **43** under palladium-catalyzed carbonylation conditions to assemble amide **44**. Regioselective bromination was readily achieved to install 5-bromopyridine **45** since the pyridine ring became quite electron-rich under the influence of two strongly electron-donating substituents. Because the intramolecular Heck reaction did not work well in the presence of a "naked" C2-amine group, it had to be protected as bis-actamide **46**, which was then cyclized to deliver lorlatinib (**6**) after acidic deprotection.[6]

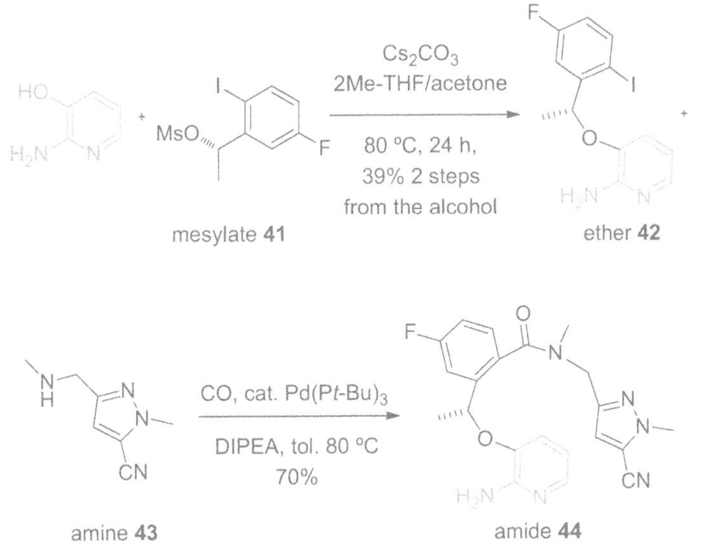

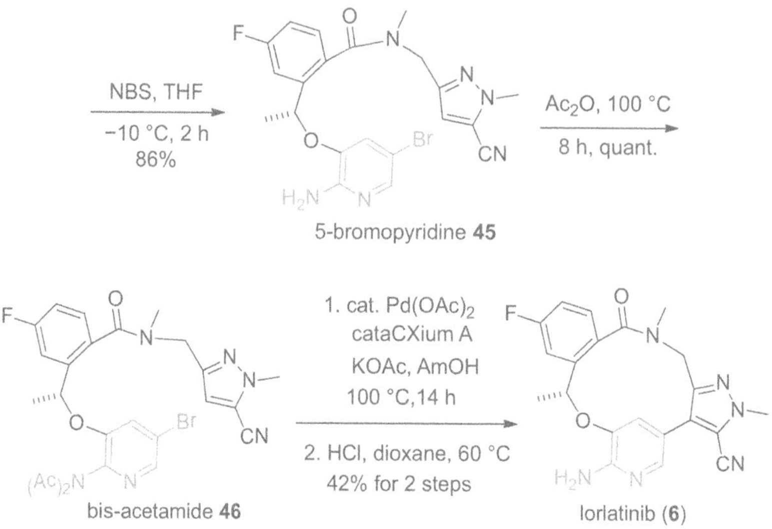

5-bromopyridine **45**

bis-acetamide **46**   lorlatinib (**6**)

APALUTAMIDE (ERLEADA, 7)

Hydrolysis of commercially available 2-chloro-3-(trifluoromethyl)pyridine (**47**) was Jung's first step toward apalutamide (**7**) to afford pyridone **48**, which offered requisite reactivity and regioselectivity for nitration to produce 5-nitropyridone **49**. Refluxing **49** with a mixture of POCl₃ and PCl₅ restored the chloropyridine functionality on **50**, which was then reduced to 6-chloro-5-(trifluoromethyl) pyridin-3-amine (**51**). The choice of Raney nickel as the catalyst for hydrogenation was a wise one because a palladium-based catalyst would have caused concurrent dechlorination. After Boc protection of the amine as **52**, an S$_N$Ar reaction took place to install 6-cyano-pyridine on **53**, which underwent an acidic deprotection to unmask the amine on **54**. Transformation of **54** to isothiocyanate **55** was accomplished by treating **54** with thiophosgene. Coupling between isothiocyanate **55** and aniline **56** then delivered apalutamide (**7**) after acidic hydrolysis.[19]

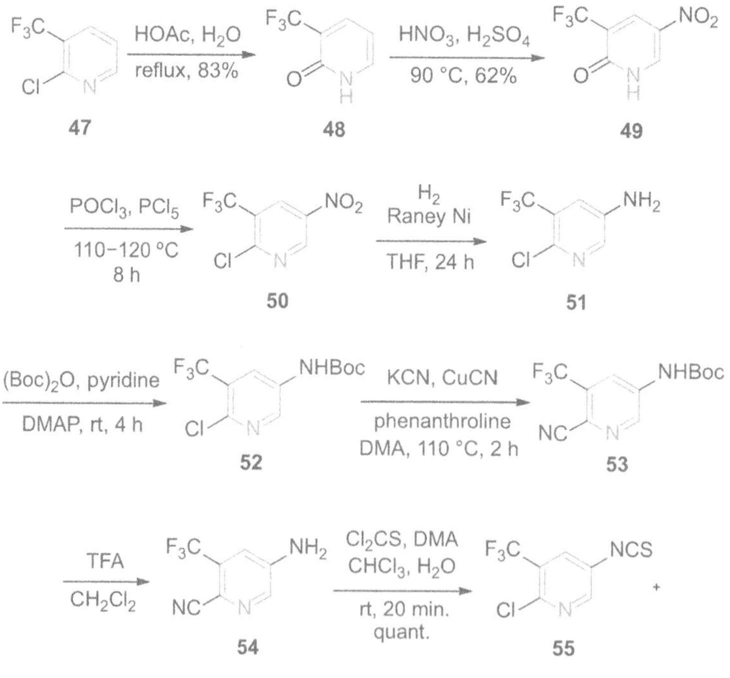

47   **48**   **49**

50   **51**

52   **53**

54   **55**

1. DMF, 80 °C, MW, 20 h

apalutamide (**7**)

2. HCl, MeOH, 2 h

87%, 2 steps

To conclude, the nitrogen atom on pyridine has a profound impact on its physiochemical properties, making pyridine a good bioisostere of the phenyl fragment. The phenyl–pyridyl switch may
improve a drug's *in vitro* binding affinity; *in vitro* functional affinity; *in vitro* PK/ADME profile;
in vitro safety profile; and *in vivo* pharmacological profile.

REFERENCES

1. Barnett, A.; Green, M. J. *Loratadine*, In *Chronicles of Drug Discovery III*, Lednicer, D., ed., American Chemical Society: Washington, DC, **1993**, *vol 3*, pp 83–99.
2. Lindberg, P.; Bränström, A.; Wallmark, B.; Mattsson, H.; Rikner, L.; Hoffman, K.-J. *Med. Res. Rev.* **1990**, *10*, 1–54.
3. Hargrave, K. D.; Proudfoot, J. R.; Grozinger, K. G.; Cullen, E.; Kapadia, S. R.; Patel, U. R.; Fuchs, V. U.; Mauldin, S. C.; Vitous, J.; et al. *J. Med. Chem.* **1991**, *34*, 2231–2241.
4. Keating, G. M. *Drugs* **2015**, *75*, 2131–2141.
5. Kim, E. S. *Drugs* **2017**, *77*, 2063–2070.
6. Johnson, T. W.; Richardson, P. F.; Bailey, S.; Brooun, A.; Burke, B. J.; Collins, M. R.; Cui, J. J.; Deal, J. G.; Deng, Y.-L.; Dinh, D.; et al. *J. Med. Chem.* **2014**, *57*, 4720–4744.
7. Jung, M. E.; Ouk, S.; Yoo, D.; Sawyers, C. L.; Chen, C.; Tran, C.; Wongvipat, J. *J. Med. Chem.* **2010**, *53*, 2779–2796.
8. Popovici-Muller, J.; Lemieux, R. M.; Artin, E.; Saunders, J. O.; Salituro, F. G.; Travins, J.; Cianchetta, G.; Cai, Z.; Zhou, D.; Cui, D.; et al. *ACS Med. Chem. Lett.* **2018**, *9*, 300–305.
9. Pennington, L. D.; Moustakas, D. T. *J. Med. Chem.* **2018**, *61*, 4386–4396.
10. D'Angelo, N. D.; Kim, T.-S.; Andrews, K.; Booker, S. K.; Caenepeel, S.; Chen, K.; D'Amico, D.; Freeman, D.; Jiang, J.; Liu, L.; et al. *J. Med. Chem.* **2011**, *54*, 1789–1811.
11. Iwamura, R.; Tanaka, M.; Okanari, E.; Kirihara, T.; Odani-Kawabata, N.; Shams, N.; Yoneda, K. *J. Med. Chem.* **2018**, *61*, 6869–6891.
12. Hong, S.-P.; Liu, K. G.; Ma, G.; Sabio, M.; Uberti, M. A.; Bacolod, M. D.; Peterson, J.; Zou, Z. Z.; Robichaud, A. J.; Doller, D. *J. Med. Chem.* **2011**, *54*, 5070–5081.
13. Govek, S. P.; Nagasawa, J. Y.; Douglas, K. L.; Lai, A. G.; Kahraman, M.; Bonnefous, C.; Aparicio, A. M.; Darimont, B. D.; Grillot, K. L.; Joseph, J. D.; et al. *Bioorg. Med. Chem. Lett.* **2015**, *25*, 5163–5167.
14. Schumacher, D. P.; Murphy, B. L.; Clark, J. E.; Tahbaz, P.; Mann, T. A. *J. Org. Chem.* **1989**, *54*, 2242–2244.
15. Grozinger, K. G.; Fuchs, V.; Hargrave, K. D.; Mauldin, S.; Vitous, J.; Campbell, S.; Adams, J. *J. Heterocycl. Chem.* **1995**, *32*, 259–263.
16. Braendstroem, A. E. WO911895 (1991).
17. (a) Hoffmann-Emery, F.; Hilpert, H.; Scalone, M.; Waldmeier, P. *J. Org. Chem.* **2006**, *71*, 2000–2008. (b) Fadini, L.; Manni, P.; Pietra, C.; Guiliano, C.; Lovati, E.; Cannella, R.; Venturini, A.; Stella, V. J. U.S. Patent US20150011510A1 (2015). (c) For an alternative synthesis, see: Hoffmann, T.; Bos, M.; Stadler, H.; Schnider, P.; Hunkeler, W.; Godel, T.; Galley, G.; Ballard, T. M.; Higgins, G. A.; Poli, S. M.; Sleight, A. J. *Bioorg. Med. Chem. Lett.* **2006**, *16*, 1362–1365.
18. Coates, D. A.; De Dios Magana, A.; De Prado Gonzales, A.; Del Prado Catalina, M. F.; Garcia Paredes, M. C.; Gelbert, L. M.; Knobeloch, J. M.; Martin De La Nava, E. M.; Martin Ortega Finger, M. D.; Martinez Perez, J. A.; et al. WO2010075074A1 (2010).
19. Jung, M. E.; Sawyers, C. L.; Ouk, S.; Tran, C.; Wongvipat, J. WO2007126765 (2007).

24 Pyrimidines

Electronically, pyrimidine is a deactivated arene comparable to 3-nitropyridine or dinitrobenzene. As far as aromaticity is concerned, pyrimidine's aromaticity is 67% relative to benzene (100%). As a privileged structure, pyrimidine has appeared in more than 19 drugs, with the majority of them as kinase inhibitors.

pyrimidine   3-nitropyridine   dinitrobenzene

PYRIMIDINE-CONTAINING DRUGS

More than 19 pyrimidine-containing drugs are currently on the market. One of the better-known pyrimidine-containing drugs is AstraZeneca's 3-hydroxy-3-methylglutaryl coenzyme A (HMG-CoA) reductase inhibitor rosuvastatin (Crestor, **1**) for lowering cholesterol. In place of the isopropyl group of many other statins, rosuvastatin (**1**)'s methylsulfonamide group renders the drug a non-substrate of CYP3A4, thus lowering the potential for drug–drug interactions (DDIs) and bestows the molecule with novel intellectual properties.

Both of Tibotec's non-nucleoside reverse transcriptase inhibitors (NNRTI), etravirine (Intelence, **2**) and rilpivirine (Edurant, **3**) contain the diamino-pyrimidine core structures.

The majority of pyrimidine-containing drugs are protein kinase inhibitors. This has not come as a surprise because pyrimidine closely mimics the adenine fragment of adenosine triphosphate (ATP), which is critical to the phosphorylation process, the key function of kinases. The first kinase inhibitor on the market, Novartis's Brc-abl inhibitor imatinib (Gleevec, **4**) contains a pyrimidine ring in addition to three additional aromatic rings. BMS's dual Bcr–abl and Src inhibitor dasatinib (Sprycel, **5**) have a diaminopyrimidine fragment. Novartis's follow-up Brc-abl inhibitor nilotinib (Tasigna, **6**) retained imatinib (**4**)'s pyridyl-pyrimidine motif. The difference between imatinib (**4**)'s and nilotinib (**6**)'s central methylphenyl structures is that the former has two amine groups and thus higher chances of forming reactive metabolites via CYP oxidation.

Apparently, diaminopyrimidine is a good pharmacophore for kinase inhibitors. GSK's dual vascular endothelial growth factor receptor (VEGFR) and c-Kit inhibitor pazopanib (Votrient, **7**) also have a diaminopyrimidine core structure.

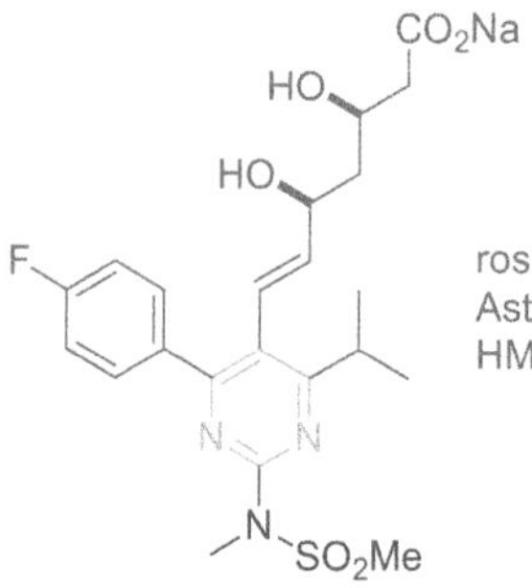

rosuvastatin (Crestor, **1**)
AstraZeneca, 2002
HMG-CoA reductase inhibitor

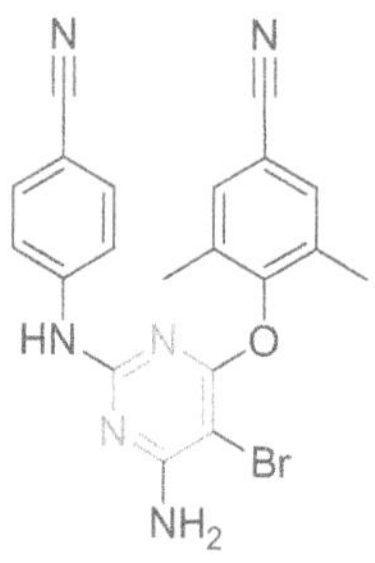

etravirine (Intelence, **2**)
Tibotec/J&J, 2008
NNRTI

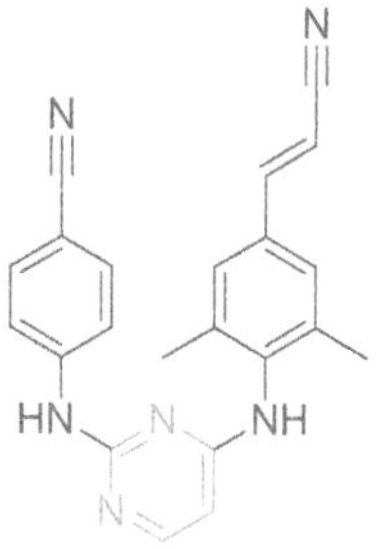

rilpivirine (Edurant, **3**)
Tibotec/J&J, 2011
NNRTI

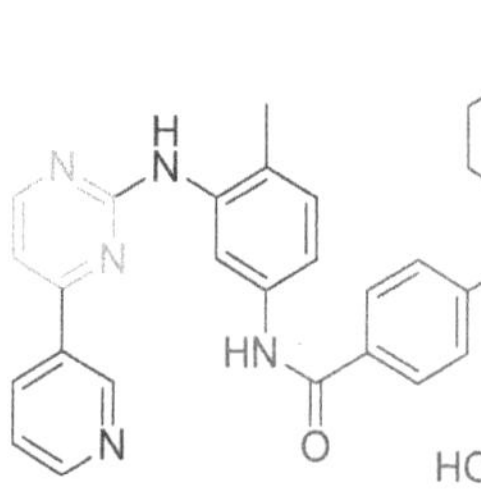

imatinib (Gleevec, **4**)
Novartis, 2001
Brc-abl inhibitor

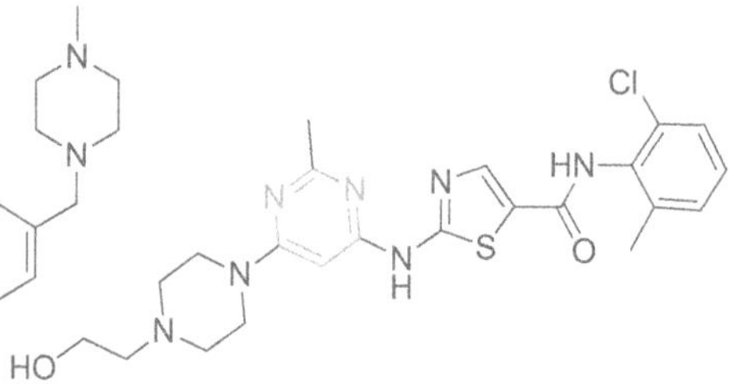

dasatinib (Sprycel, **5**)
BMS, 2006
dual Bcr-abl and Src inhibitor

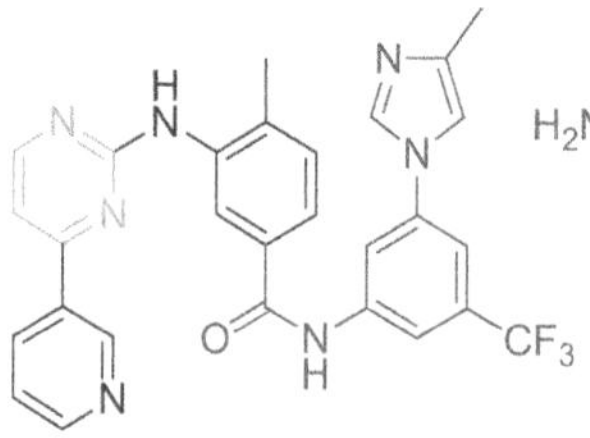

nilotinib (Tasigna, **6**)
Novartis, 2007
Bcr-Abl inhibitor

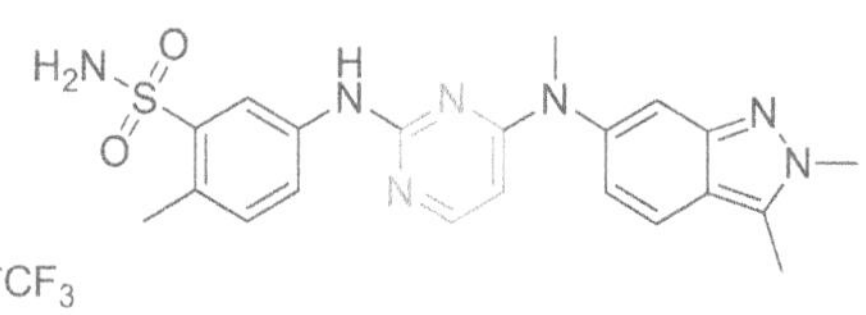

pazopanib (Votrient, **7**)
GSK, 2009
VEGFR, c-Kit inhibitor

Epidermal growth factor receptor (EGFR) inhibitors are among the earliest kinase inhibitors on the market. But resistance invariably developed and covalent inhibitors have been invented to combat the L858R and T790M mutations by taking advantage of Cys-797 at EGFR's active site. AstraZeneca's third-generation EGFR inhibitor osimertinib (Tagrisso, **8**) is a covalent inhibitor expressly designed to overcome the T790M mutation. Lilly's aminopyrimidine-containing abemaciclib (Verzanio, **9**) is a cyclin-dependent kinase (CDK)4/6 inhibitor.

In addition, Novartis's anaplastic lymphoma kinase (ALK) inhibitor ceritinib (Zykadia, **10**) contains a chloro-diaminopyrimidine core, and Bayer's copanlisib (Aliqopa, **11**) is a pan-phosphoinositide-3-kinase (PI3K) inhibitor with an aminopyrimidine fragment.

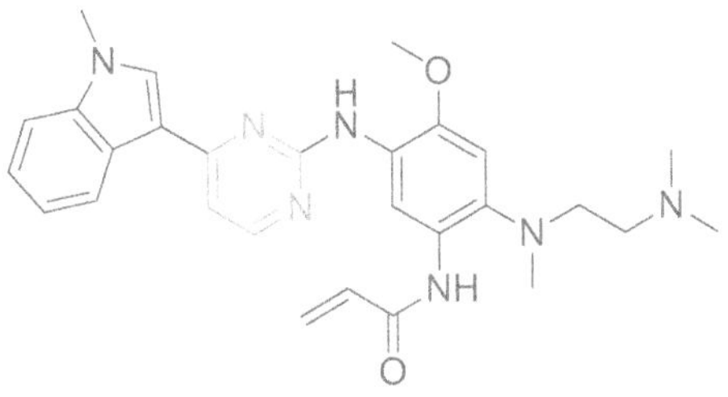

osimertinib (Tagrisso, 8)
AZ, 2015
EGFR inhibitor (3rd generation)
C797S mutation

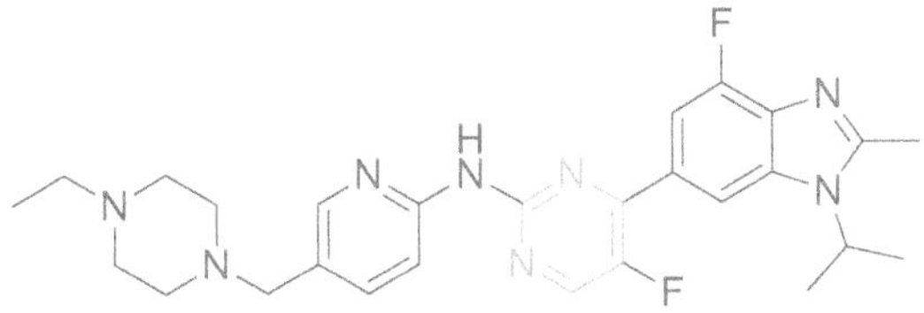

abemaciclib (Verzanio, 9)
Lilly, 2017
CDK4/6 inhibitor

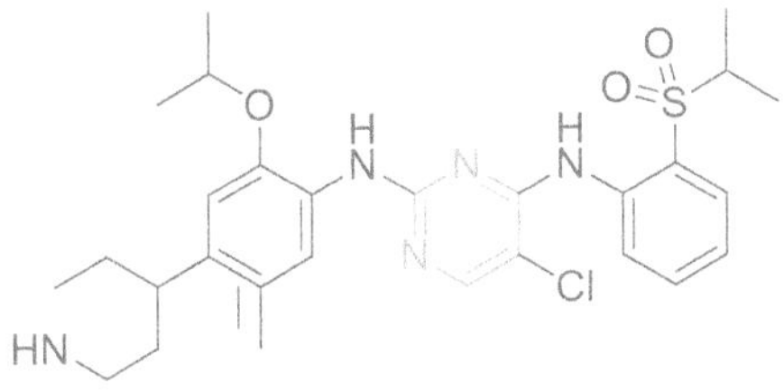

ceritinib (Zykadia, 10)
Novartis, 2014
ALK inhibitor

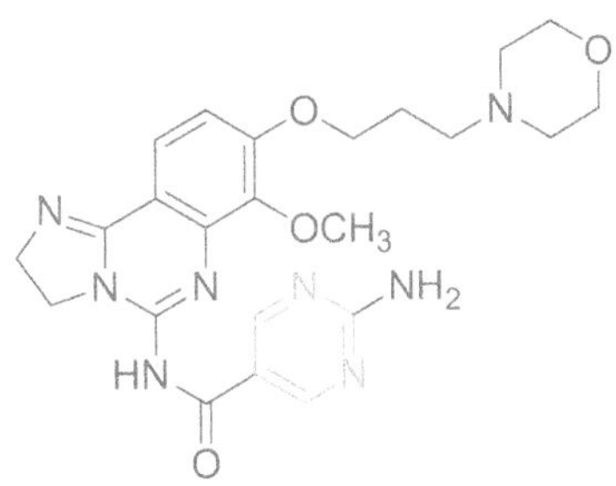

copanlisib (Aliqopa, 11)
Bayer, 2017
pan-PI3K inhibitor

The Ras–MAP kinase pathway has been implicated in tumor progression for a variety of human cancers. The Raf kinases, which are components of this cascade, are serine/threonine kinases that activate mitogen-activated protein kinase (MEK)1/2. Mutant B-Raf containing a V600E substitution (where B-Raf protein's 600th amino acid valine is replaced by glutamic acid) causes aberrant constitutive activation of this pathway and has a high occurrence in several human cancers. Two of the three B-Raf kinase inhibitors approved by the FDA, dabrafenib (Tafinlar, **12**) and encorafenib (Braftovi, **13**), contain an aminopyrimidine moiety.

The chloro-diaminopyrimidine core seems to be a favored structure for ALK inhibitors. It appeared on Novartis's ALK inhibitor ceritinib (**10**) and it also made an appearance on Ariad's ALK and EGFR inhibitor brigatinib (Alunbrig, **14**).

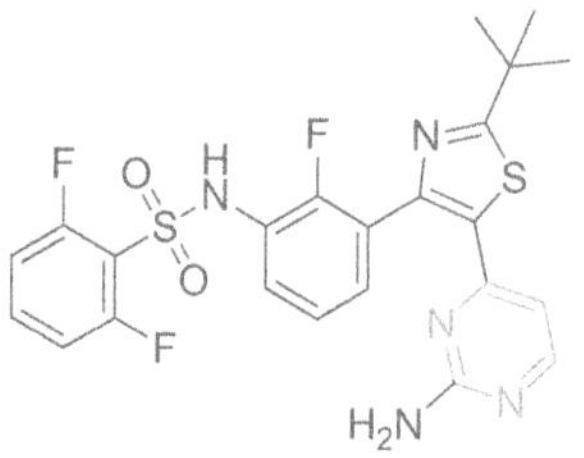

dabrafenib (Tafinlar, **12**)
GSK, 2013
B-raf inhibitor

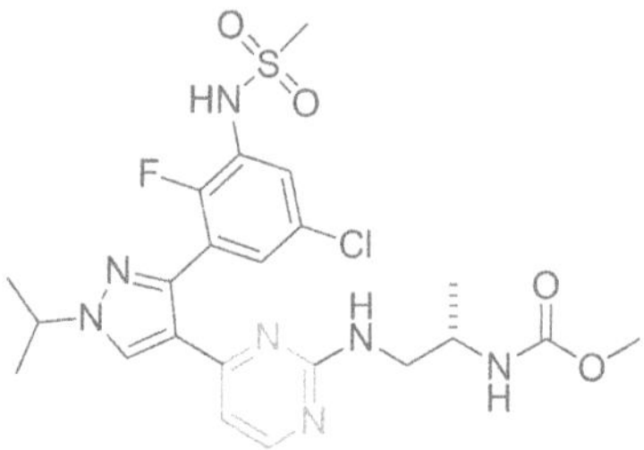

encorafenib (Braftovi, **13**)
Array BioPharma/Novartis, 2018
B-raf inhibitor

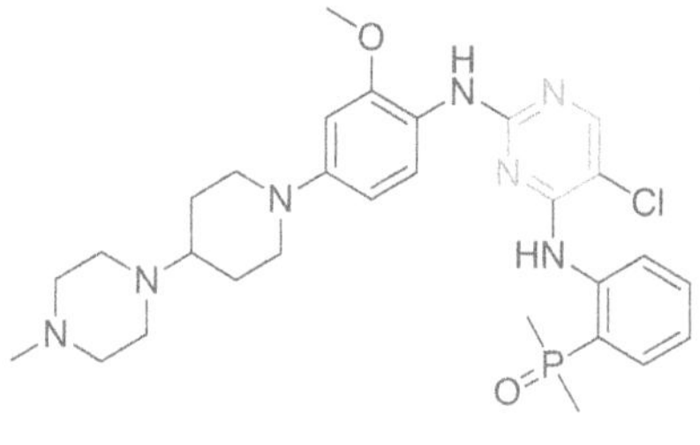

brigatinib (Alunbrig, **14**)
Ariad, 2017
ALK and EGFR inhibitor

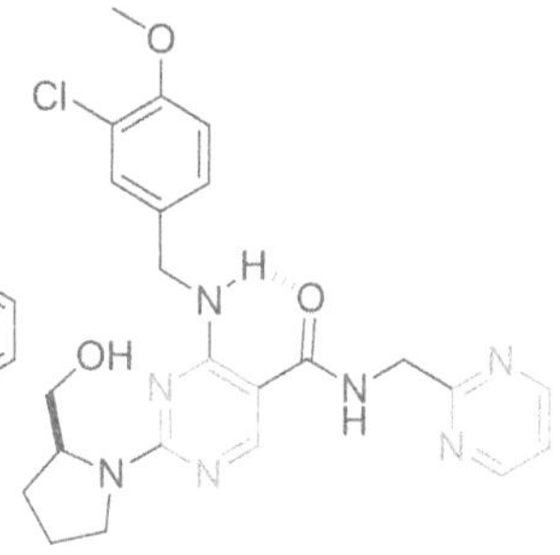

avanafil (Stendra, **15**)
Vivus, 2012
PDE5 inhibitor

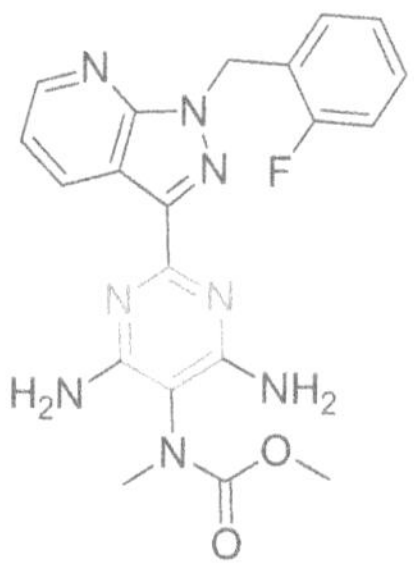

riociguat (Adempas, **16**)
Bayer, 2013
sGC stimulator

Phosphodiesterase 5 (PDE5) inhibitors include well-known names such as sildenafil (Viagra), vardenafil (Levitra), and tadalafil (Cialis) for the treatment of erectile dysfunction (ED). Vivus's PDE5 inhibitor avanafil (Stendra, **15**) possesses two pyrimidine rings. An intramolecular hydrogen bond forms a pseudo ring and renders the core pyrimidine "behave" almost like a bicycle. Triaminopyrimidine riociguat (Adempas, **16**) by Bayer is a stimulator of soluble guanylate cyclase (sGC) approved for the treatment of adults with persistent/recurrent chronic thromboembolic pulmonary hypertension (CTEPH).

Three pyrimidine-containing drugs were approved in 2019. Karyopharm's selinexor (Xpovio, **17**) is a first-in-class selective inhibitor of nuclear export (SINE) XPO1 antagonist for the treatment of adult patients with relapsed or refractory multiple myeloma (RRMM). Eisai's lemborexant (Dayvigo, **18**) is a dual antagonist of the orexin OX1 and OX2 receptors approved for the treatment of insomnia. Impact's fedratinib (Inrebic, **19**) is a Janus kinase (JAK)-2-selective inhibitor for treating high-risk myelofibrosis.

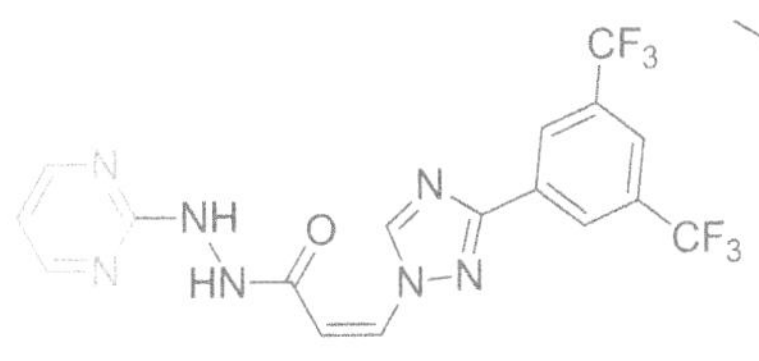

selinexor (Xpovio, **17**)
Karyopharm, 2019
XPO1 antagonist

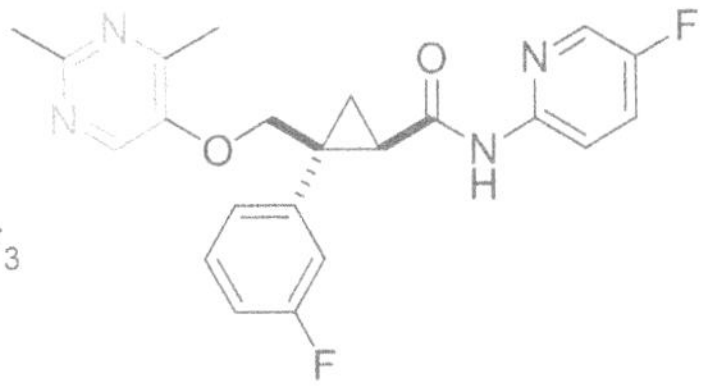

lemborexant (Dayvigo, **18**)
Eisai, 2019
dual OX1 and OX2 antagonist

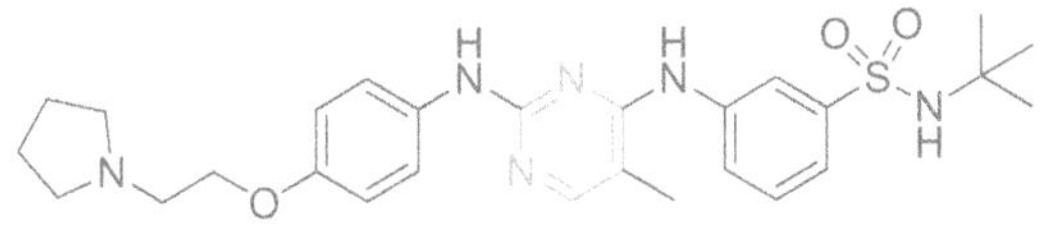

fedratinib
(Inrebic, **19**)
Impact, 2019
JAK2 inhibitor

PYRIMIDINES IN DRUG DISCOVERY

PYRIMIDINE AS A BIOISOSTERE

In theory, pyrimidine may serve as an isostere for all aromatic and heteroaryl rings. Historically, pyrimidine as a bioisostere had mixed results. A case in point was the pyrimidine isostere **21** for SDZ EAB 515 (**20**), a potent *N*-methyl-D-aspartate (NMDA) receptor antagonist. To overcome **20**'s high polarity, Sandoz scientists in 1994 prepared its pyrimidine analog **21**, which displayed a pK_i of 4.3 and a pA_2 value of 4.6 was measured in the cortical wedge test for functional activity. For the original drug **20**, these values were 6.7 and 6.94, respectively. Thus, the pyrimidine isostere **21** was more than two orders of magnitude less active than the potent NMDA antagonist **20**. The results are quite astonishing in light of the seemingly minor structural differences.[1]

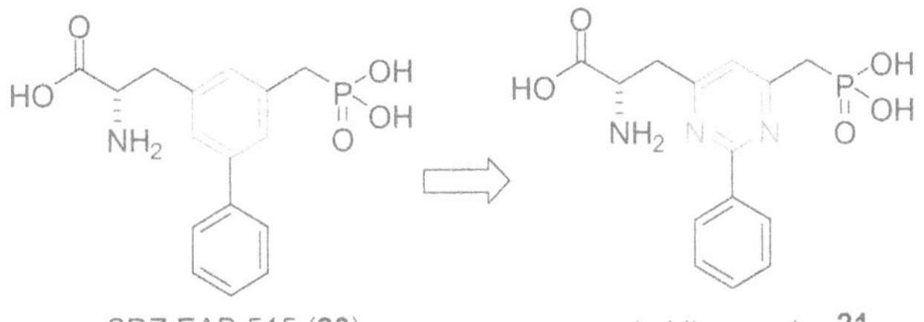

SDZ EAB 515 (**20**)
$pK_i = 6.7$, $pA_2 = 6.94$

pyrimidine analog **21**
$pK_i = 4.3$, $pA_2 = 4.6$

The next example of isostere had a more positive outcome. Pyrrolopyridine **22** as a type I inhibitor of troponin I-interacting kinase (TNNI3K) is not very potent with an IC_{50} value of 8,000 nM. Switching the pyridine moiety to the pyrimidine isostere gave rise to pyrrolopyrimidine **23**, which is 100-fold more potent in binding affinity with an IC_{50} value of 80 nM. Scrutiny of the X-ray crystal structure of a similar TNNI3K inhibitor bound to the ATP binding site of the kinase revealed that the steric clash of H1 and H2′ in the coplanar aryl orientation of **22** prevented it from adopting the correct (more flat) conformation required for effective binding. On the other hand, pyrrolopyrimidine **23** with an extra strategically placed nitrogen atom allows free rotation to adopt a more planar conformation and therefore more effective binding.[2]

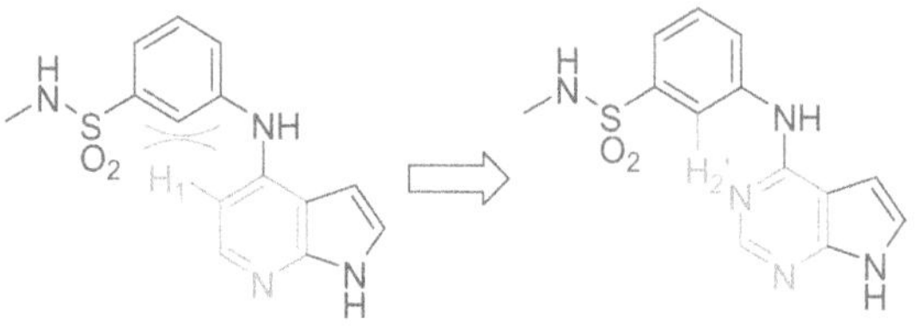

pyrrolopyridine **22**
TNNI3K IC$_{50}$ = 8,000 nM

pyrrolopyrimidine **23**
TNNI3K IC$_{50}$ = 80 nM

Antagonists of retinol-binding protein 4 (RBP4) have potential as a treatment of atrophic age-related macular degeneration (AMD) and Stargardt disease. An RBP4 antagonist **24** with the anthranilic acid appendage had a reasonable *in vitro* and *in vivo* PK and PD profile already. To further enhance the *in vitro* and *in vivo* potency, 6-methyl-pyrimidine-4-carboxylic acid **25** was prepared because it could provide a suitable isostere for the amide of **24** while still presenting the acid group as a favorable interaction. Pyrimidine here reduces *the number of rotatable bonds* and was expected to lead improved RBP4 binding affinity. Pyrimidine-acid **25** was tested indeed more potent in an *in vitro* binding assays (RBP4 SPA IC$_{50}$ = 12.8 nM, SPA = scintillation proximity assay) in comparison to the parent compound **24** (IC$_{50}$ = 72.7 nM), a more than fivefold improvement. Meanwhile, pyrimidine-acid **25** showed boosted functional RBP4–TTR (TTR = transthyretin) interaction antagonist (HTRF) activity (RBP4 HTRF IC$_{50}$ = 43.6 nM, HTRF = homogeneous time-resolved fluorescence), a nearly 7-fold enhancement against the parent compound **24** (IC$_{50}$ = 294 nM).[3]

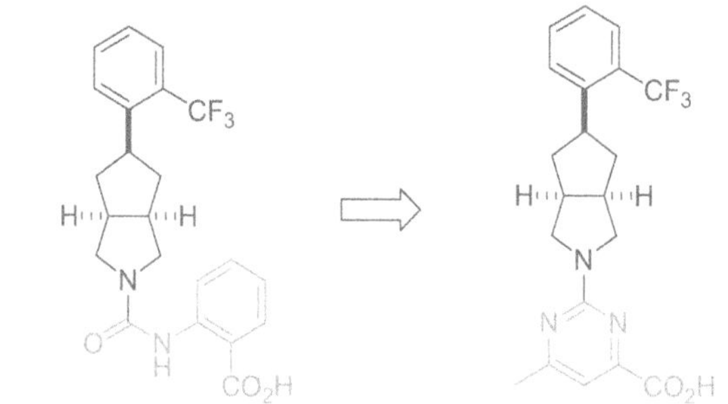

anthranilic acid **24**
RBP4 SPA IC$_{50}$ = 72.7 nM
RBP4 HTRF IC$_{50}$ = 294 nM

pyrimidine-4-carboxylic acid **25**
RBP4 SPA IC$_{50}$ = 12.8 nM
RBP4 HTRF IC$_{50}$ = 43.6 nM

INTERACTIONS WITH TARGET PROTEINS

There are two nitrogen atoms on pyrimidine, providing two possible contact points with target proteins in the form of hydrogen bond acceptors. A quick glance of pyrimidine-containing kinase inhibitors revealed that aminopyrimidines are prevalent. As shown below, since aminopyrimidine closely mimics the left portion of adenine, an important fragment of ATP, which is critical to the phosphorylation process, the key function of kinases. The amino group on aminopyrimidines not only offers its electron-donating properties to the electron-deficient pyrimidine ring but also provides interacting points with target kinase proteins as a hydrogen bond donor.

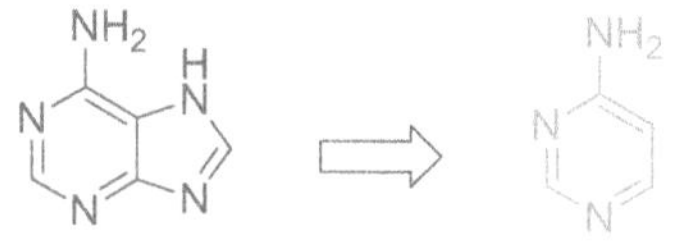

adenine

aminopyrimidine

The first marketed kinase inhibitor, Brc-abl inhibitor imatinib (**4**) contains a pyridyl-aminopyrimidine motif, as does Novartis's follow-up Brc-abl inhibitor nilotinib (**6**). The crystal structure of imatinib (**4**)–abelson (Abl) tyrosine kinase complex indicated that the amine sandwiched between the pyrimidine and phenyl rings serves as a hydrogen bond donor to interact with the side chain hydroxyl group of Thr315 of the target protein. The pyridyl nitrogen accepts a hydrogen bond from the amide of Met318, which is normally hydrogen-bonded to the nitrogen N1 in ATP. These two interactions contribute to imatinib (**4**)'s selectivity. One of the two nitrogen atoms on the pyrimidine ring serves as a hydrogen bond acceptor with a water molecule nearby.[4]

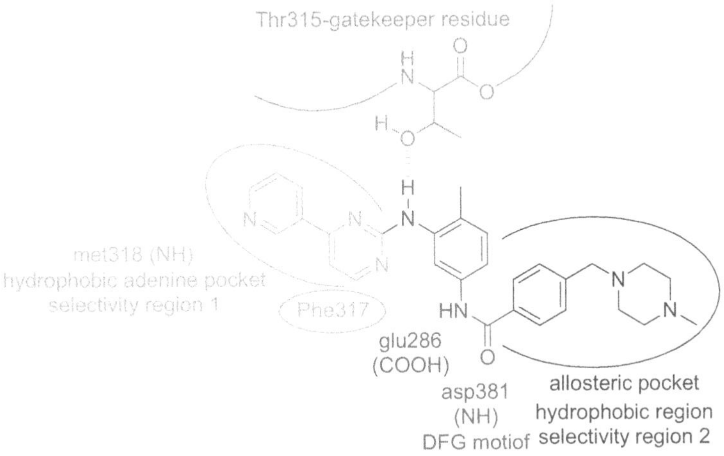

interactions between imatinib (**4**) and Brc-abl enzyme

Protein–ligand interactions between ceritinib (**10**) and its target protein, wild-type ALK, are shown below. Ceritinib (**10**) forms two hydrogen bonds at the hinge area via the pyrimidine and amino nitrogen atoms with the backbone nitrogen and oxygen of M1199. The central pyrimidine ring of the ceritinib (**10**) is sandwiched between residues A1148 and L1256, probably interacting via π-stacking. The chlorine substituent on the pyrimidine ring is involved in a hydrophobic interaction with the gatekeeper residue L1196. The piperidine ring extends to the solvent and the isopropyl group adjacent to the sulfonyl bends down into the cavity lined by residues L1256 and D1270.[5]

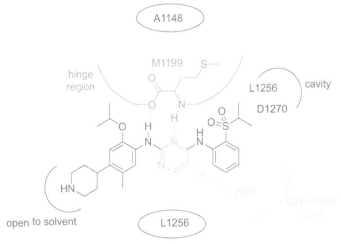

interactions between ceritinib (**10**) and wild-type ALK enzyme

SYNTHESIS OF SOME PYRIMIDINE-CONTAINING DRUGS

The 12-step synthesis of imatinib mesylate (**4**) in the manufacturing process was accomplished by Novartis in an astonishingly short time. The synthesis began with a condensation between 3-acetylpyridine (**26**) and ethyl formate. Deprotonation of the methyl group on **26** using freshly prepared sodium methoxide afforded an enolate. Condensation of the enolate with ethyl formate was followed by an exchange with dimethylamine to produce 3-dimethylamino-1-(3-pyridyl)-2-propen-1-one (**27**). Alternatively, **27** could be prepared from the condensation of **26** and *N,N*-dimethylformamide dimethylacetal [(MeO)$_2$CHN(Me)$_2$]. Meanwhile, nitration of 2-amino-toluene (**28**) gave 2-amino-4-nitrotoluene nitrate, with the nitro group serving as a masked amine group. Refluxing 2-amino-4-nitrotoluene nitrate with cyamide furnished 2-amino-5-nitrophenylguanidine (**29**). Subsequently, condensation of 3-dimethylamino-enone **27** and guanidine **29** was achieved in the presence of NaOH in refluxing isopropanol to assemble pyrimidine **30**. Palladium-catalyzed hydrogenation of nitrophenylpyrimidine **30** unmasked the nitro group to provide aminophenyl-pyrimidine. Amide formation was accomplished by treatment of the aminophenyl-pyrimidine with 4-(4-methylpiperazinomethyl)benzoyl chloride to deliver **4**. Finally, the mesylate of **4** was readily accessed by the addition of one equivalent of methanesulfonic acid.[6]

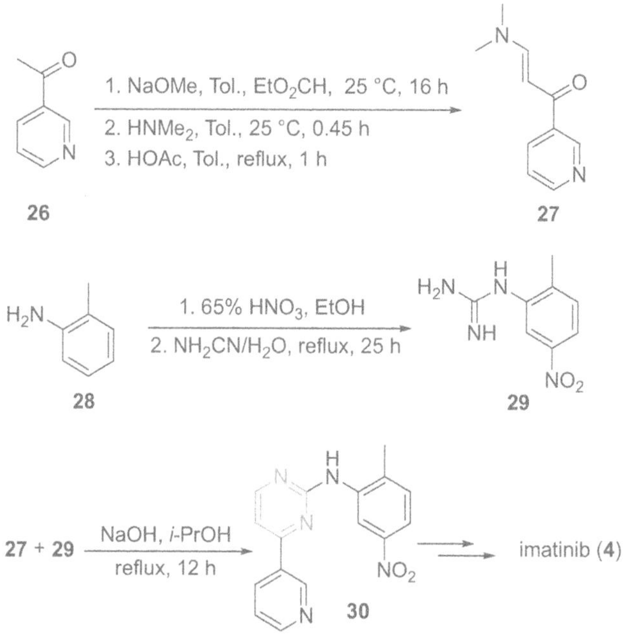

The synthesis of pazopanib (**7**) involves the amination of 2,4-dichloropyrimidine (**31**) with 6-amino-2,3-dimethylindazole (**32**) in the presence of sodium bicarbonate in ethanol/THF to produce **33**. Subsequent *N*-methylation of **33** with iodomethane and cesium carbonate produced **34**. The 2-chloro group of pyrimidine on **34** was then allowed to react with 5-amino-2-methylbenzene-sulfonamide (**35**) in catalytic HCl/isopropanol at reflux to deliver pazopanib hydrochloride (**7**) in good yield.[7]

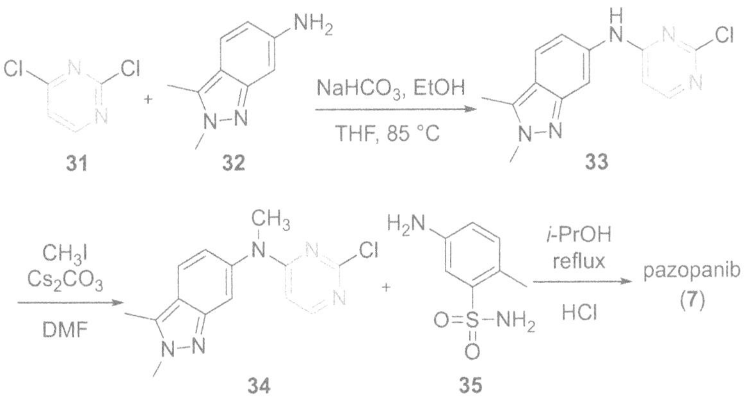

AstraZeneca's synthesis of osimertinib (**8**) also employed 2,4-dichloro-pyrimidine (**31**) as its starting material. Its coupling with N-methylindole assembled adduct **36** with the help of FeCl$_3$. Further coupling of **36** with aniline **37** afforded 2-phenylamino-3-indolylpyrimidine **38**, which was converted to the desired osimertinib (**8**) in three additional steps.[8]

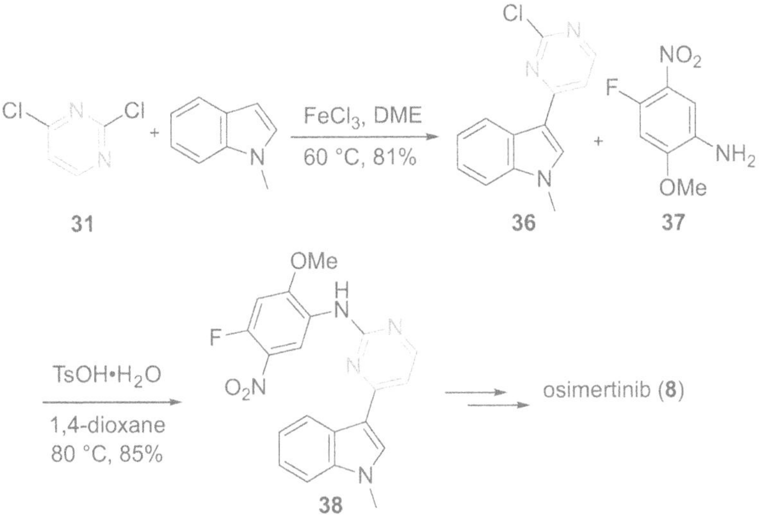

For the process synthesis of riociguat (**16**), condensation of amidine **39** with malononitrile **40** assembled pyrimidine **41**. The phenyldiazine group on **41** was cleaved via palladium-catalyzed hydrogenation to afford tri-aminopyrimidine **42**. Treatment of **42** with dimethyl dicarbonate selectively protected the amine at the 4′-position in the middle to offer carbamate **43** in 95% yield. Eventually, carbamate **43** was methylated and the resulting compound was recrystallized with DMSO and EtOAc sequentially to deliver the desired active pharmaceutical ingredient (API) riociguat (**16**).[9]

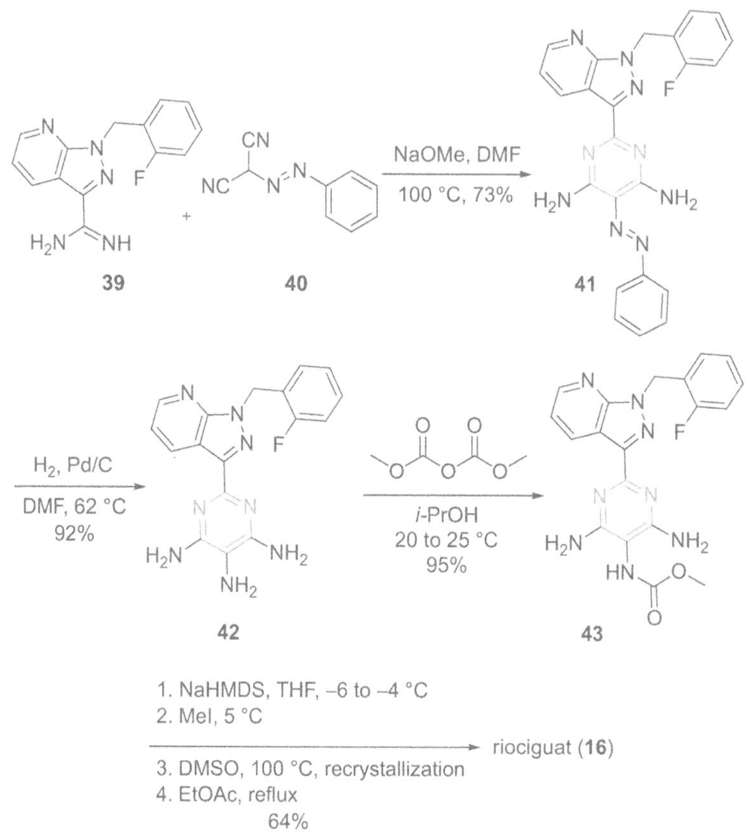

In summary, pyrimidine motif is a privileged fragment in drug discovery, boasting more than 19 marketed pyrimidine-containing drugs, with the majority of them as kinase inhibitors. In particular, amino- and diamino-pyrimidine fragments are especially prevalent.

REFERENCES

1. Cai, C.; Buechler, D.; Lowe, D.; Urwyler, S.; Shapiro, G. *Tetrahedron* **1994**, *50*, 5735–5740.
2. Lawhorn, B. G.; Philp, J.; Zhao, Y.; Louer, C.; Hammond, M.; Cheung, M.; Fries, H.; Graves, A. P.; Shewchuk, L.; Wang, L.; et al. *J. Med. Chem.* **2015**, *58*, 7431–7448.
3. Cioffi, C. L.; Racz, B.; Freeman, E. E.; Conlon, M. P.; Chen, P.; Stafford, D. G.; Schwarz, D. M. C.; Zhu, L.; Kitchen, D. B.; Barnes, K. D.; et al. *J. Med. Chem.* **2015**, *58*, 5863–5888.
4. Schindler, T.; Bornmann, W.; Pellicena, P, Miller, W. T.; Clarkson, B.; Kuriyan, J. *Science* **2000**, *289*, 1938–1942.
5. Ni, Z.; Zhang, T.-C. *J. Mol. Mod.* **2015**, *21*, 1–10.
6. (a) Zimmermann, J.; Buchdunger, E.; Mett, H.; Meyer, T.; Lydon, N. B. *Bioorg. Med. Chem. Lett.* **1997**, *7*, 187–192. (b) Zimmermann, J. U.S. Patent US5521184 (1996). (c) Torley, L. W.; Johnson, B. B.; Dusza, J. P. EP0233461 (1987). (d) Dusza, J. P.; Albright, J. D. U.S. Patent US4281000 (1981).
7. Harris, P. A.; Boloor, A.; Cheung, M.; Kumar, R.; Crosby, R. M.; Davis-Ward, R. G.; Epperly, A. H.; Hinkle, K. W.; Hunter III, R. N.; Johnson, J. H.; et al. *J. Med. Chem.* **2008**, *51*, 4632–4640.
8. (a) Finlay, M. R. V.; Anderton, M.; Ashton, S.; Ballard, P.; Bethel, P. A.; Box, M. R.; Bradbury, R. H.; Brown, S. J.; Butterworth, S.; Campbell, A.; et al. *J. Med. Chem.* **2014**, *517*, 8249–8267. (b) Butterworth, S.; Finlay, M. R.; Verschoyle; Ward, R. A.; Kadambar, V. K.; Chandrashekar, R. C.; Murugan, A.; Redfearn, H. M. WO2013014448 (2013).
9. Mais, F.-J.; Rehse, J.; Joentgen, W.; Siegel, K. U.S. Patent US20110130410A1 (2011).

25 Pyrrolidines

Pyrrolidine is a part of proline (Pro, P), a natural amino acid. Not surprisingly, pyrrolidine plays an important role in drug discovery. The NH group may serve as a hydrogen bond donor and, when its NH is masked as in the case of nicotine, the N atom may serve as a hydrogen bond acceptor to its target protein. Meanwhile, the pyrrolidine motif may offer enhanced aqueous solubility and improve other physiochemical properties in addition to being part of the pharmacophore.

pyrrolidine   proline 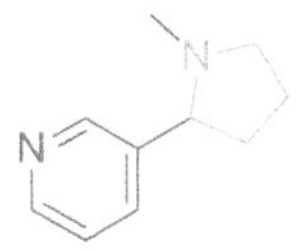   nicotine

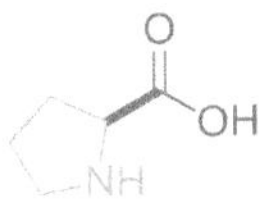

PYRROLIDINE-CONTAINING DRUGS

More than two dozens of pyrrolidine-containing drugs are currently on the market. Hoechst's high-ceiling sulfonamide diuretic piretanide (Arelix, **1**) was one of the first *synthetic* pyrrolidine-containing drugs. This was partially because the synthetic methodology for installing the pyrrolidine ring to benzene was not available until then. Angiotensin-converting enzyme (ACE) inhibitors are a class of antihypertensive drugs that emerged after sulfonamide diuretics. The prototype was BMS's captopril (Capoten, **2**) approved in 1980 and its second-generation ACE inhibitor fosinopril (Fozitec, **3**) appeared on the market in 1991. To overcome captopril (**2**)'s trio of shortcomings (short half-life, rashes, and loss of taste perception), Merck took advantage of a previously unappreciated S_1' hydrophobic pocket on the ACE protein and arrived at enalapril (Vasotec, **4**) and subsequently lisinopril (Zestril, **5**).

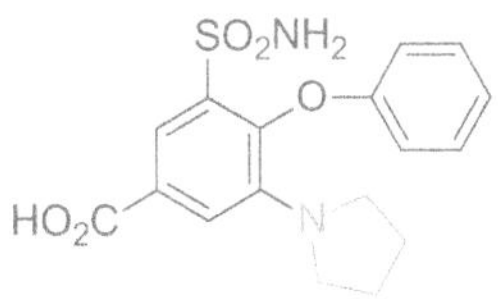

piretanide (Arelix, **1**)
Hoechst/Aventis, 1981
high-ceiling diuretic

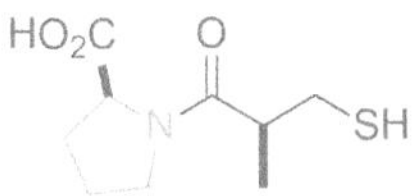

captopril (Capoten, **2**)
BMS, 1980
ACE inhibitor

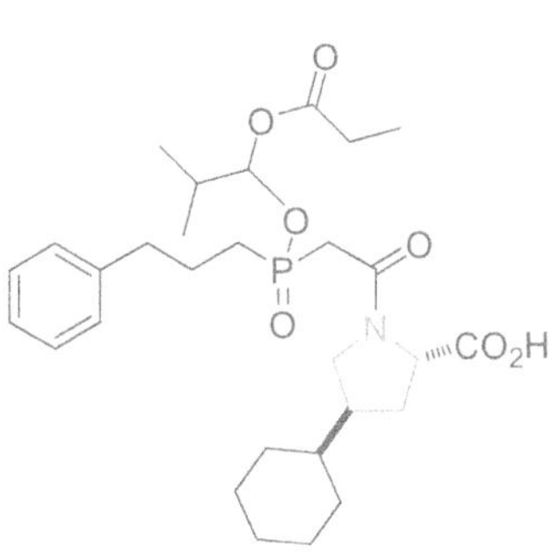

fosinopril (Fozitec, **3**)
BMS, 1991
ACE inhibitor

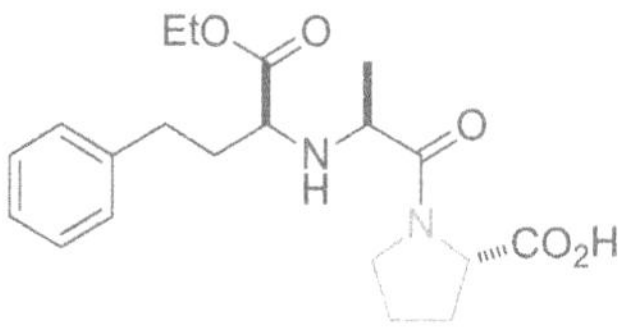

enalapril (Vasotec, **4**)
Merck, 1984
ACE inhibitor

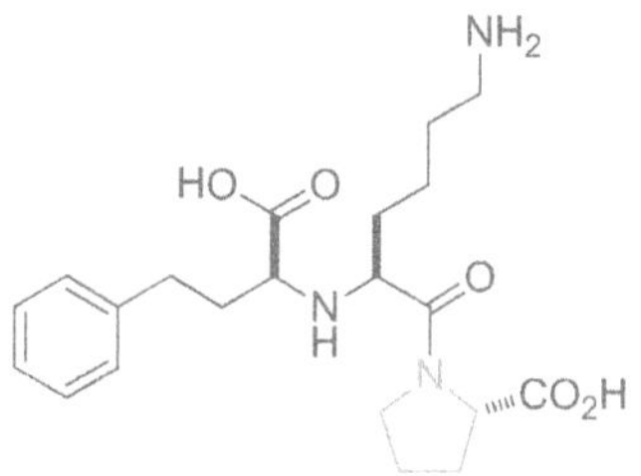

lisinopril (Zestril, **5**)
Merck, 1987
ACE inhibitor

In the realm of drugs to treat type II diabetes mellitus (T2DM), four of approximately 10 dipeptidyl peptidase-IV (DPP-4) inhibitors on the market contain the pyrrolidine fragment. They are Novartis's vildagliptin (Galvus, **6**), BMS's saxagliptin (Onglyza, **7**), Kenkyusho's anagliptin (Suiny, **8**), and Mitsubishi Tanabe's teneligliptin (Tenelia, **9**). The last two DPP-4 inhibitors **8** and **9** are only available in Japan. In terms of mechanism of action (MOA), the nitrile group on compounds **6–8** forms a reversible covalent bond to Ser_{630} in coordination with Tyr_{547} in the S_1 pocket of the DPP-4 enzyme.

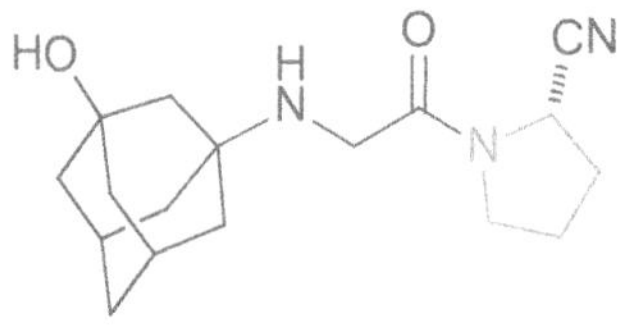

vildagliptin (Galvus, **6**)
Novartis, 2007
DPP-4 inhibitor

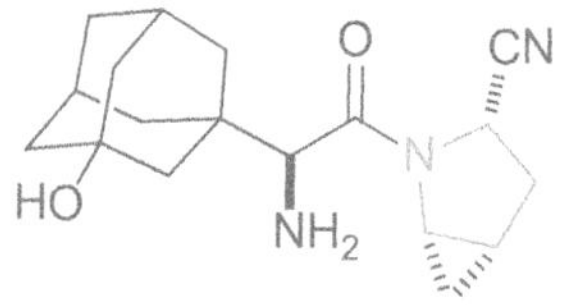

saxagliptin (Onglyza, **7**)
Bristol-Myers Squibb, 2009
DPP-4 inhibitor

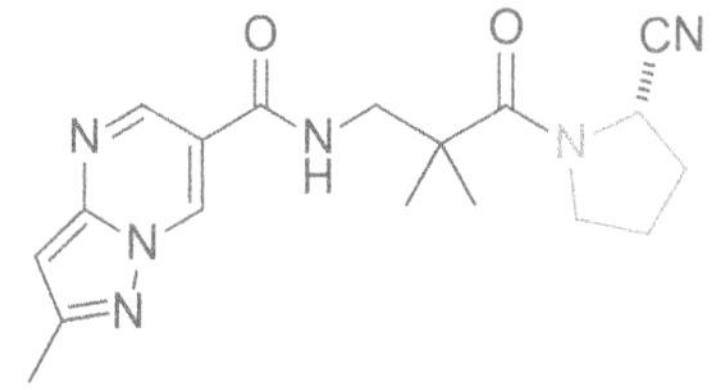

anagliptin (Suiny, **8**)
Sanwa Kagaku Kenkyusho, 2012
DPP-4 inhibitor

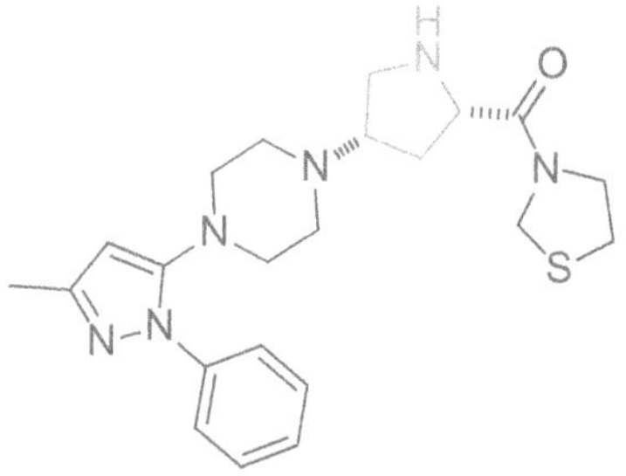

teneligliptin (Tenelia, **9**)
Mitsubishi Tanabe, 2012 (Japan)
DPP-4 inhibitor

As far as antibacterial medicines are concerned, the third generation quinolone tosufloxacin (Ozex, **10**) is only sold in Japan due to its controversial safety profile. Pyrrolidine-containing cefepime (Maxipime, **11**) is a fourth-generation cephalosporin antibiotic, which has a greater resistance to β-lactamases than third-generation cephalosporins.

The first marketed carbapenem antibiotic in the US was imipenem (Primaxin), which does not contain a pyrrolidine group. Rather, it has an amidine tail. Pyrrolidine-containing meropenem (Merrem, **12**) was the second. Incorporation of the hydrophobic pyrrolidinyl sidechain enhanced its potency against *Pseudomonas aeruginosa* and other Gram-negative pathogens. Merck's ertapenem (Invanz, **13**) was the third marketed carbapenem and doripenem (Doribax, **14**), approved by the FDA in 2007, was the fourth.

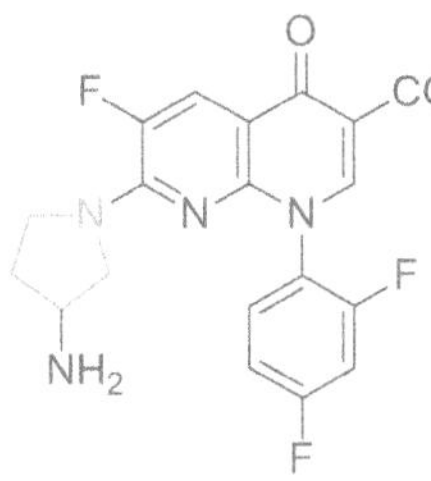

tosufloxacin (Ozex, **10**)
third generation
quinolone antibiotic

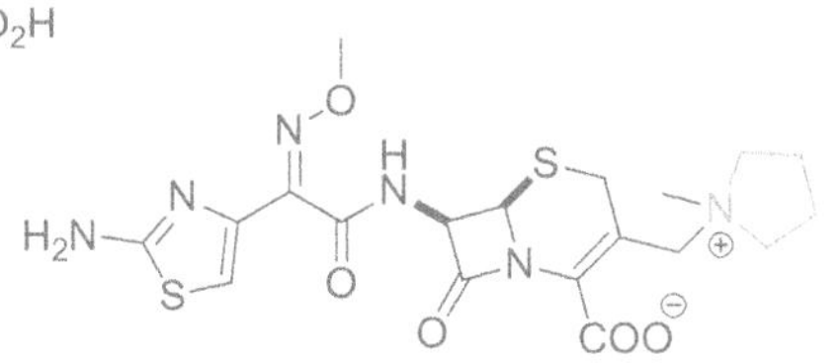

cefepime (Maxipime, **11**)
BMS, 1994
cephalosporin antibiotic

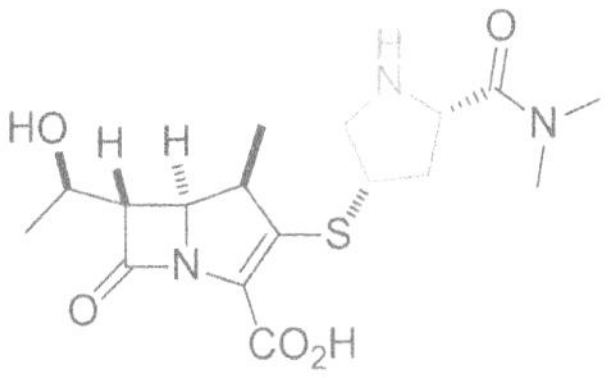

meropenem (Merrem, **12**)
AstraZeneca, 1996
carbapenem antibiotic

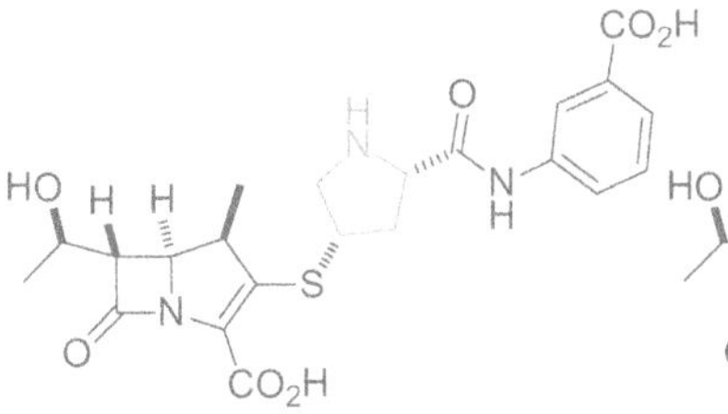

ertapenem (Invanz, **13**)
Merck, 2001
carbapenem antibiotic

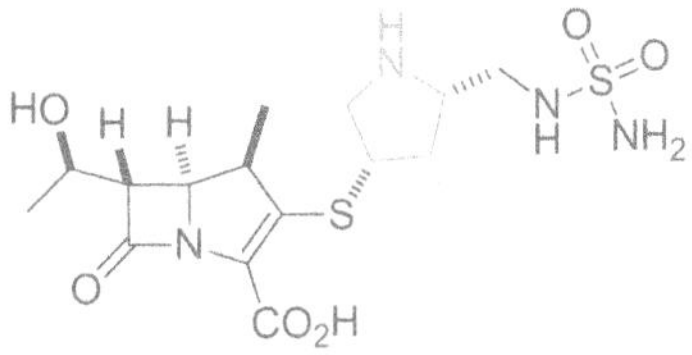

doripenem (Doribax, **14**)
Shionogi, 2007
carbapenem antibiotic

With the exception of the three MEK inhibitors, nearly all kinase inhibitors on the market are competitive inhibitors occupying the adenine triphosphate (ATP) binding pocket. As a consequence, they possess a flat aromatic core structure, mimicking the adenine portion of ATP. However, flat aromatic compounds tend to be highly crystalline and have low aqueous solubility. Imatinib (Gleevec) has a piperazine group specifically installed to boost its solubility. For AstraZeneca/Acerta's covalent Bruton's tyrosine kinase (BTK) inhibitor acalabrutinib (Calquence, **15**), its ynamide "warhead" is attached to a pyrrolidinyl ring to improve its solubility.

Not all kinase inhibitors are cancer drugs. Abbvie's Janus kinase (JAK)1/2 inhibitor upadacitinib (Rinvoq, **16**) is approved for treating rheumatoid arthritis (RA). Its pyrrolidinyl fragment, albeit as a substituent of urea, helps in boosting the drug's solubility and improving other physiochemical properties.

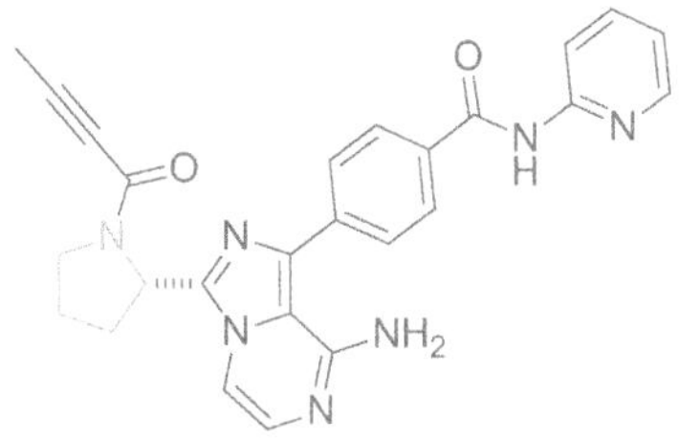

acalabrutinib (Calquence, **15**)
AZ/Acerta, 2017
BTK inhibitor

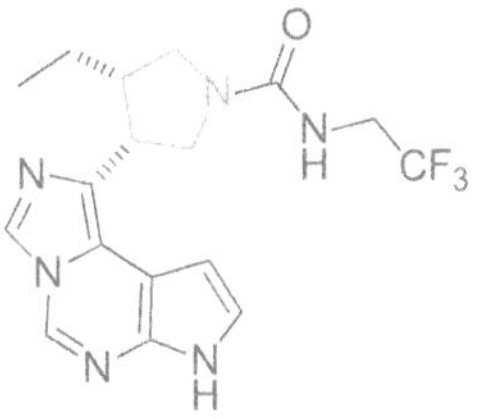

upadacitinib (Rinvoq, **16**)
Abbvie, 2019 (for RA)
JAK1/2 inhibitor

The most abundant pyrrolidine-containing drugs are found in the field of hepatitis C virus (HCV) drugs. HCV has a positive-sense single-stranded RNA genome, which includes nonstructural proteins NS2, NS3, NS4A, NS4B, NS5A, and NS5B. Schering–Plough/Merck's boceprevir (Victrelis, **17**) and Vertex's telaprevir (Incivek, **18**) are the first two NS3/4A serine protease inhibitors approved in 2011. Their keto-amide "warheads" form reversible covalent tetrahedral hemiacetal intermediates with the protein's serine$_{139}$ in concert with Asp$_{81}$. Both boceprevir (**17**) and telaprevir (**18**) have fused pyrrolidine backbones.

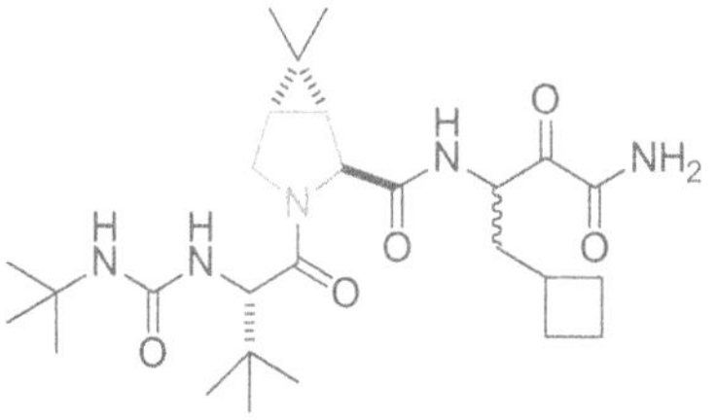

boceprevir (Victrelis, **17**)
Schering-Plough/Merck, 2011
HCV NS3/4A
Serine Protease Inhibitor

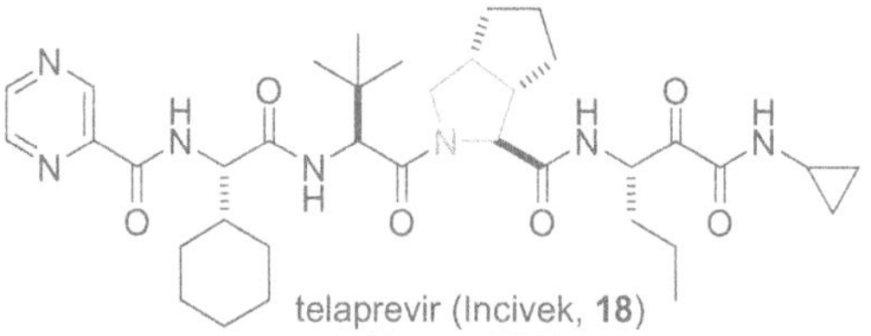

telaprevir (Incivek, **18**)
Vertex, 2011
HCV NS3/4A
Serine Protease Inhibitor

Four additional NS3/4A serine protease inhibitors gained the FDA approval: glecaprevir (Mavyret, **19**), voxilaprevir (Vosevi, **20**), paritaprevir, and simeprevir. The first two contain a pyrrolidine group.

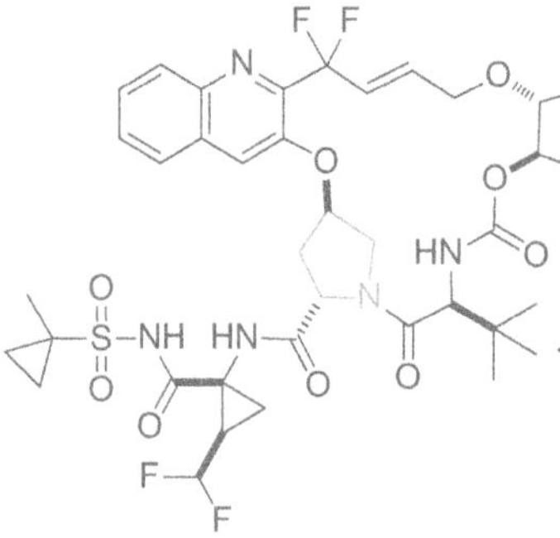

glecaprevir (Mavyret, **19**)
Gilead, 2017
HCV NS3/4A inhibitor

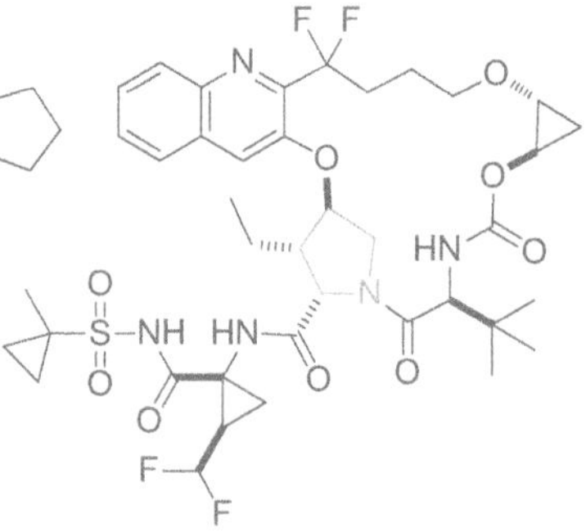

voxilaprevir (Vosevi, **20**)
Gilead, 2017
HCV NS3/4A inhibitor

Six HCV NS5A polymerase inhibitors daclatasvir (Daklinza, **21**), elbasvir, ledipasvir, ombitasvir, velpatasvir (Vosevi, **22**), and pibrentasvir (Mavyret, **23**) are currently on the market. The two NS5B polymerase inhibitors dasabuvir (Exviera) and sofosbuvir (Sovaldi), however, do not contain the pyrrolidine moiety.

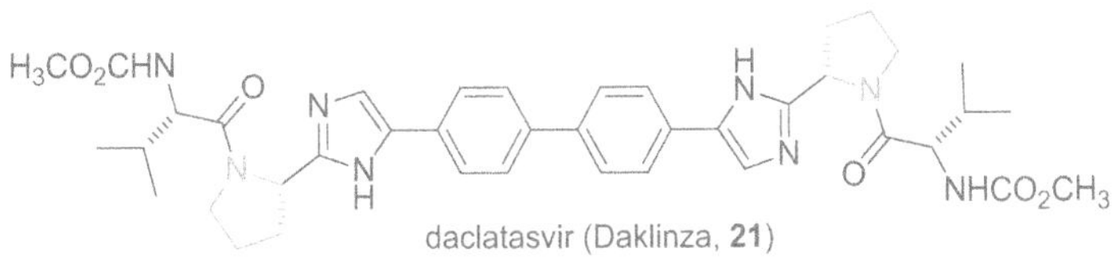

daclatasvir (Daklinza, **21**)
BMS, 2015
HCV NS5A Inhibitor

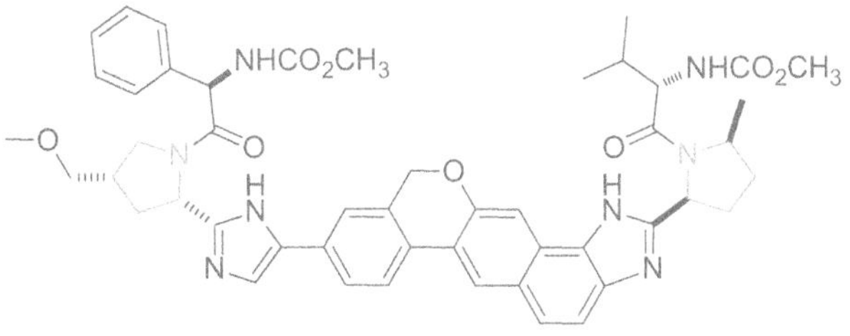

velpatasvir (Vosevi, **22**)
Gilead, 2016
HCV NS5A inhibitor

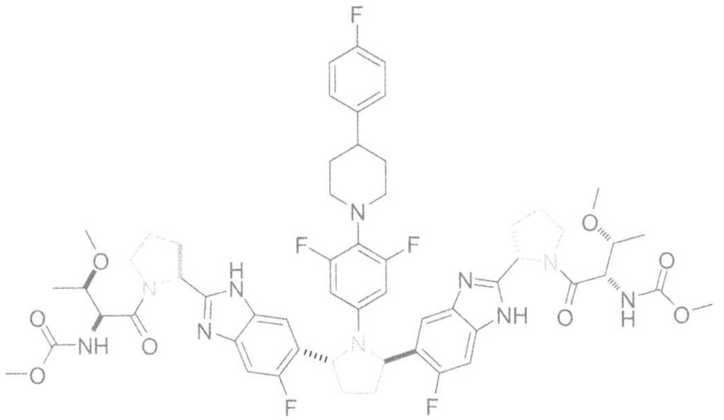

pibrentasvir (Mavyret, **23**)
Gilead, 2017
HCV NS5A inhibitor

Tracing their roots to procainamide, pyrrolidine-containing *o*-anisamides sultopiride (Barnetil, **24**, a sulfone), remoxipride (Roxiam, **25**), and sulpiride (dogmatil, **26**, a sulfonamide) have been used as treatments of psychosis. They are not remarkable drugs in terms of either efficacy or safety. But an important lesson may be learned here. Thanks to their intramolecular hydrogen bonds, all three very polar drugs can cross the cell membrane and blood-brain barrier (BBB) with ease.

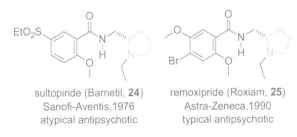

sultopiride (Barnetil, **24**)
Sanofi-Aventis,1976
atypical antipsychotic

remoxipride (Roxiam, **25**)
Astra-Zeneca,1990
typical antipsychotic

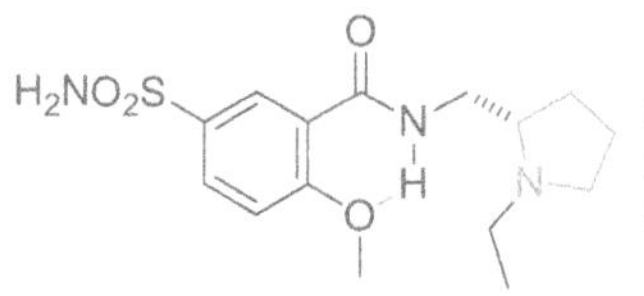

sulpiride (dogmatil, **26**)
Sanofi,1993
atypical antipsychotic

PYRROLIDINES IN DRUG DISCOVERY

Pyrrolidine motif has been employed to improve a drug's potency, selectivity, and pharmacokinetic profile.

As a natural amino acid, proline is ubiquitous as a building block in enzymes, receptors, ion channels, and endogenous ligands. Not surprisingly, some drugs derived from nature contain the proline moiety as well.

In the early 1970s, Squibb isolated teprotide (**27**, a nonapeptide: 9 amino acids) by degrading dried extract of the venom of the poisonous Brazilian pit viper, *Bothrops jararaca*. It was shown to reduce blood pressure in healthy volunteers and confirmed that it was a selective ACE inhibitor in humans. Cushman and Ondetti at Squibb truncated teprotide (**27**) and obtained succinoyl-1-proline **28** with an IC_{50} value of 330 nM for ACE inhibition. The activity of **28** demonstrated that a small molecular weight drug could be a potent inhibitor of ACE by occupying only a small fraction of the extended active site cavities because the chemistry of peptide bond hydrolysis is typically dependent on a small number of critical amino acids (sometimes merely a triad of amino acids). Compound **28** was chosen as the starting point because the C-terminal amino acid occurs at the free C-terminus of all the naturally occurring peptidic inhibitors.

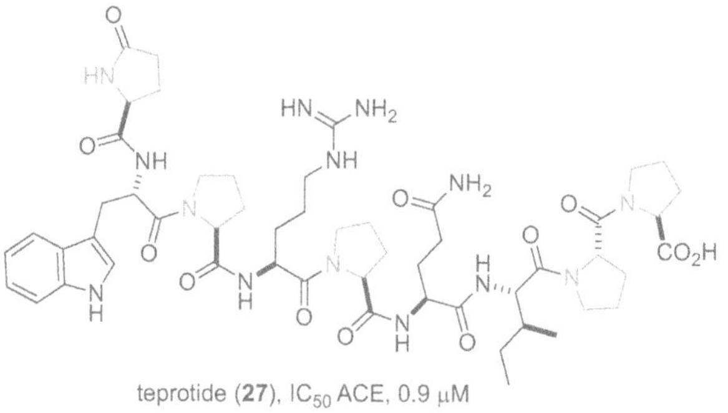

teprotide (**27**), IC_{50} ACE, 0.9 μM

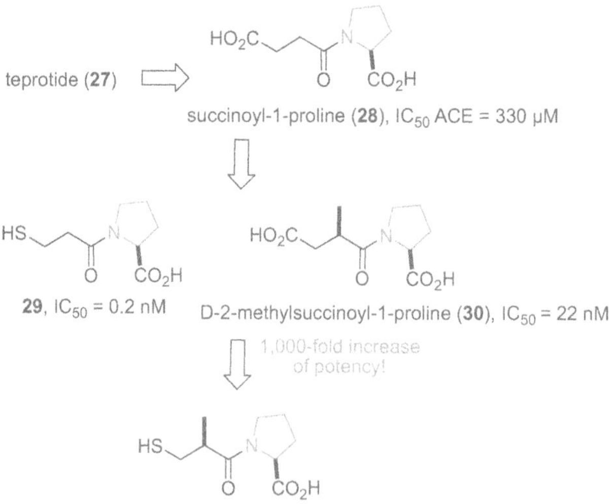

teprotide (**27**) ⟹ succinoyl-1-proline (**28**), IC_{50} ACE = 330 μM

29, IC_{50} = 0.2 nM   D-2-methylsuccinoyl-1-proline (**30**), IC_{50} = 22 nM

1,000-fold increase of potency!

captopril (Capoten, **31**), IC_{50} = 23 nM

Later on, two important observations were made stemming from succinoyl-1-proline **28**. (a) A moderate boost of activity (1.6-fold) was observed when the carboxylic acid was replaced with a sulfhydryl as in **29**. (b) A significant increase of the binding activity was obtained when an extra methyl group was installed at the α-position of the amide bond, giving rise to D-(*R*)-2-methylsuccinoyl-1-proline (**30**) with an IC_{50} of 22 nM. Interestingly, the L-(*S*)-enantiomer of **30** was found to be much less active. A breakthrough came when both features of **29** and **30** were incorporate into one molecule (**31**). Thus, they replaced the carboxylate group with a sulfhydryl (−SH) and installed an additional methyl group to achieve a 1,000-fold improvement in potency for ACE inhibition. The drug became the first oral ACE inhibitor, captopril (Capoten, **31**), which was approved by the FDA in 1978 and significantly contributed to the management of hypertension and hypertension-related target-organ damage. Squibb arrived at captopril (**31**) from only 60 compounds logically synthesized and tested.[1]

KRAS p.G12C mutation is prevalent among the top most deadly cancer types in the United States: it has a causal role in 14% of lung adenocarcinomas (non-small cell lung cancer, NSCLC) and 5% of colorectal adenocarcinomas (colorectal cancer, CRC). A breakthrough came in 2013 when Shokat et al. at UCSF reported the identification of compounds that covalently bound to a previously unappreciated pocket near the KRAS Switch-II effector region. The compounds form an irreversibly covalent bond to the mutant cysteine-12, locking the protein in its inactive GDP-bound state. Targeting the Switch-II binding site was a clear advance in the field.[2]

Building upon Shokat's ground-breaking discovery, Mirati and Array BioPharma pursued their own KRAS[G12C] inhibitors. Initial screening of "warheads" and SAR investigations led to their discovery of compound **32a**. Its percent of control modification (POC Mod.) value was low (8%), indicating a low percentage of protein and adduct formation. This was reflected by a weaker potency in the cellular assay using an H358 *KRAS*[G12C]-driven cell line (IC_{50}, 7.6 μM). They hypothesized that substitution at the C-2 position of the pyrimidine ring would afford access to the carboxylate of Glu_{62} on the KRAS[G12C] enzyme. The potential ionic interactions (salt bridge) with Glu_{62} were tested with compounds **32b** and **32c** by varying the chain length to place a basic amine near the carboxylate. Both of these compounds were more active in the protein modification assay under the 15 min/3 μM conditions. The rigidified α-methyl analog **32d** showed a further increase in cell potency but only the (*R*)-enantiomer was seen in the cocrystal structure with KRAS[G12C]. On the basis of this X-ray structure and the observed amine interactions with Glu_{62} and His_{95}, Mirati designed compound **32e** with the aim of eliminating one rotatable bond and introducing more hydrophobic contacts with His_{95}. Docking studies of compound **32e** suggested that the pyrrolidine ring amine would make the same salt bridge interaction with Glu_{62} and cation−π interaction with His_{95}.

32
KRAS[G12C] covalent inhibitor

MRTX1257 (**33**)

	R =	POC Mod. (15 min/3 µM)	H358 IC$_{50}$ (µM)
32a	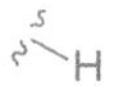	8%	7.6
32b	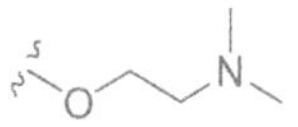	52%	1.9
32c	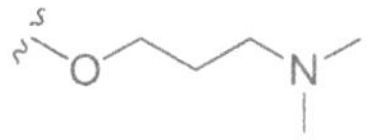	22%	1.5
32d	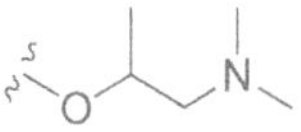	21%	0.54
32e	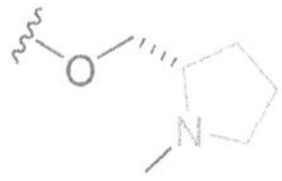	84%	0.070
33	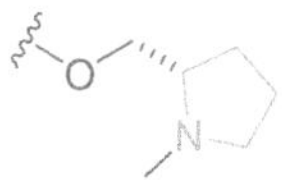	59%	0.0009!

Furthermore, the B-subunit of crystal structure with **32e** has pyrrolidine interacting with Asp$_{92}$ via a salt bridge and a water-mediated hydrogen bond. Indeed, compound **32e** with the N-methyl pyrrolidine C-2 amine substituent displayed a dramatic boost in biochemical and cellular potency (cell IC$_{50}$=0.070 µM) and protein modification (POC=84%). This potency increase was attributed to the aforementioned interactions and removal of one rotatable bond.[3] Additional efforts eventually led to orally bioavailable KRAS^{G12C} covalent inhibitor MRTX1257 (**33**). Its close analog MRTX849 is currently undergoing phase I clinical trials in 2019.

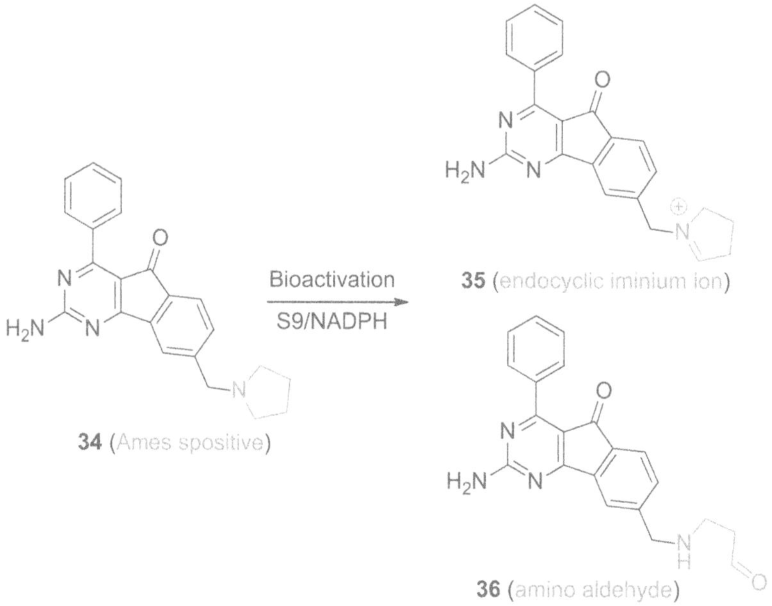

34 (Ames spositive) → Bioactivation S9/NADPH → **35** (endocyclic iminium ion)

36 (amino aldehyde)

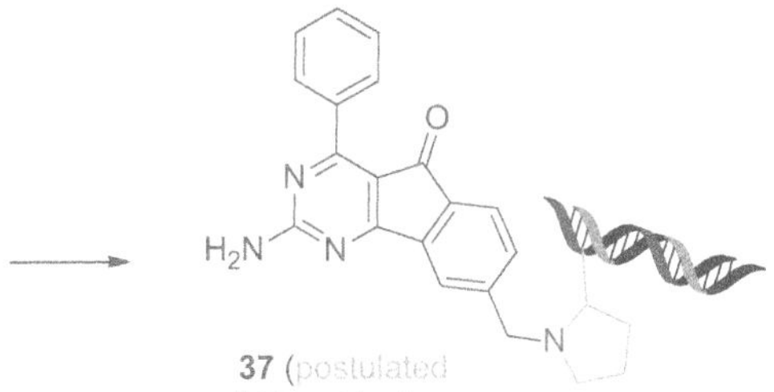

37 (postulated
DNA adduction:
Positive Ames Test!)

There is a potential liability associated with pyrrolidine-containing drugs since pyrrolidine may be oxidized by CYP450 enzymes to reactive metabolites that form "hard" electrophiles with endogenous proteins. For instance, pyrrolidine-substituted arylindenopyrimidine **34** as a potent and selective dual adenosine A (2A)/A1 antagonist was determined to be genotoxic in both the bacterial *Salmonella* Ames reverse mutation and mouse lymphoma L5178Y assays (in an Aroclor 1254-induced rat liver S9/NADPH-dependent fashion).[4] It was proposed and experimentally confirmed that endocyclic iminium ion **35** and amino aldehyde **36** were the reactive metabolites after bioactivation by CYP450 enzymes. In addition to genotoxicity, it was concluded that **34** also had the potential to be mutagenic in humans based on observing the endocyclic iminium ion following incubation with a human liver S9 preparation and the commensurate detection of DNA adducts **37**. It was suggested and experimentally confirmed that the corresponding dimethyl-substituted pyrrolidine and pyridine analogs were devoid of genotoxicity since those two fragments minimized their bio-activation to the corresponding iminium ion as a reactive intermediate.[4]

In fact, compound **34**'s genotoxicity is not unique. Cases linking the generation of iminium ion metabolites with protein covalent binding have been reported with xenobiotics such as phencyclidine (contains a piperidine ring) and nicotine (contains a pyrrolidine ring).[5]

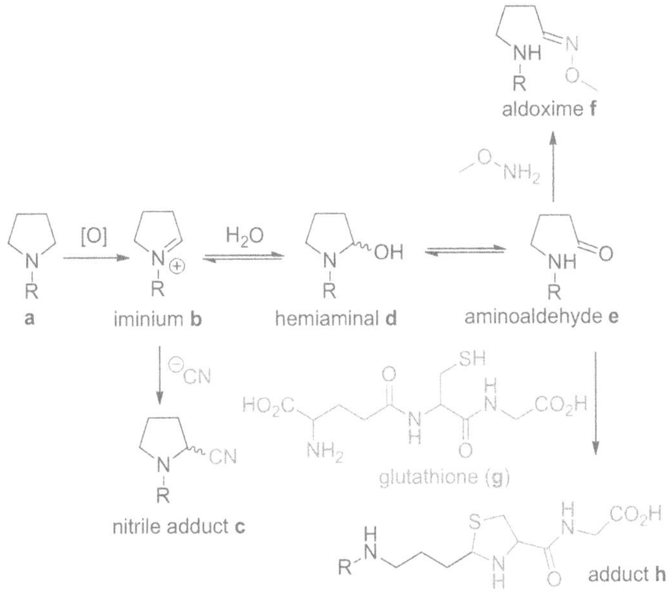

Kalgutkar summarized the metabolic pathway of cyclic amines such as pyrrolidines and piperidines. As shown below, a cyclic amine as generically represented by pyrrolidine **a** is oxidized to iminium **b** by CYP450 enzymes. If cyanide was added to the experiment, monocyano adduct **c** would be obtained although the addition of two cyanide groups was also possible to give the bis-cyano adduct. Under *in vivo* conditions, water addition to iminium **b** would give rise to hemiaminal **d**, which could be subsequently hydrolyzed to aminoaldehyde **e**. Again, if methoxyamine was added to the experiment, aldoxime **f** would be obtained. Otherwise, aminoaldehyde **e** would be captured by endogenous glutathione (**g**) to offer adduct **h**.[6]

SYNTHESIS OF SOME PYRROLIDINE-CONTAINING DRUGS

Merck's vernakalant (Kynapid, **41**) is an atrial potassium channel blocker. In one of the synthetic routes leading to vernakalant (**41**), racemic cyclohexyl epoxide (**38**) was opened with protected prolinol **39** as the nucleophile in hot water. The resulting mixture of diastereomers was separated by classical resolution of the corresponding tartrate salt to afford *cis*-isomer **40**. Subsequent ether formation from **40** was followed by debenzylation to deliver the desired active pharmaceutical ingredient (API) **41**.[7]

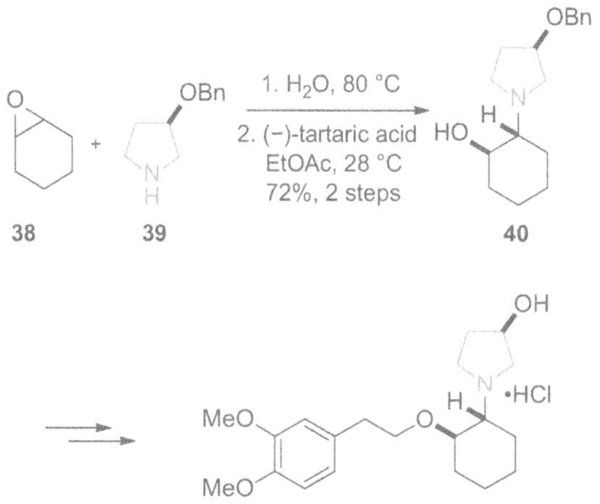

vernakalant (Kynapid, **41**)

The only reported synthetic approach of DPP-4 inhibitor teneligliptin (**9**) resorted reductive amination of piperazine **42** and pyrrolidinone **43** to assemble adduct **44**. Deprotection of the Boc group and HBr salt formation then gave rise to teneligliptin (**9**).[8]

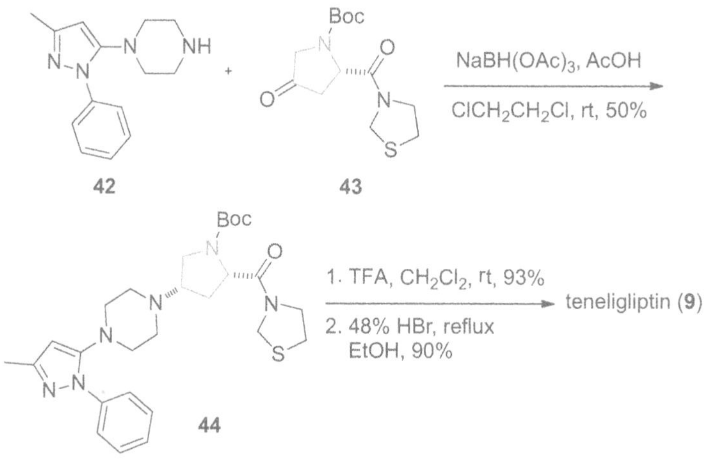

Synthesis of BMS's HCV NS5A replication complex inhibitor daclatasvir (**21**) commenced with bromination of bis-ketone **45** to make bis-bromide **46**. S_N2 displacement of the two bromide leaving groups on **46** by *N*-Boc-L-proline (**47**) assembled bis-ester **48**, which was transformed to bis-imidazole **49** by condensation of **48** with ammonium acetate. The two NH groups on the pyrrolidine rings were exposed by treating **49** with HCl and the resultant "naked" bis-pyrrolidine **50** was coupled with *N*-(methoxycarbonyl)-L-valine (**51**) to produce daclatasvir (**21**). The final API was prepared as its di-hydrochloride salt after purification with 3M's Cuno Zeta Carbon™ 55SP.[9]

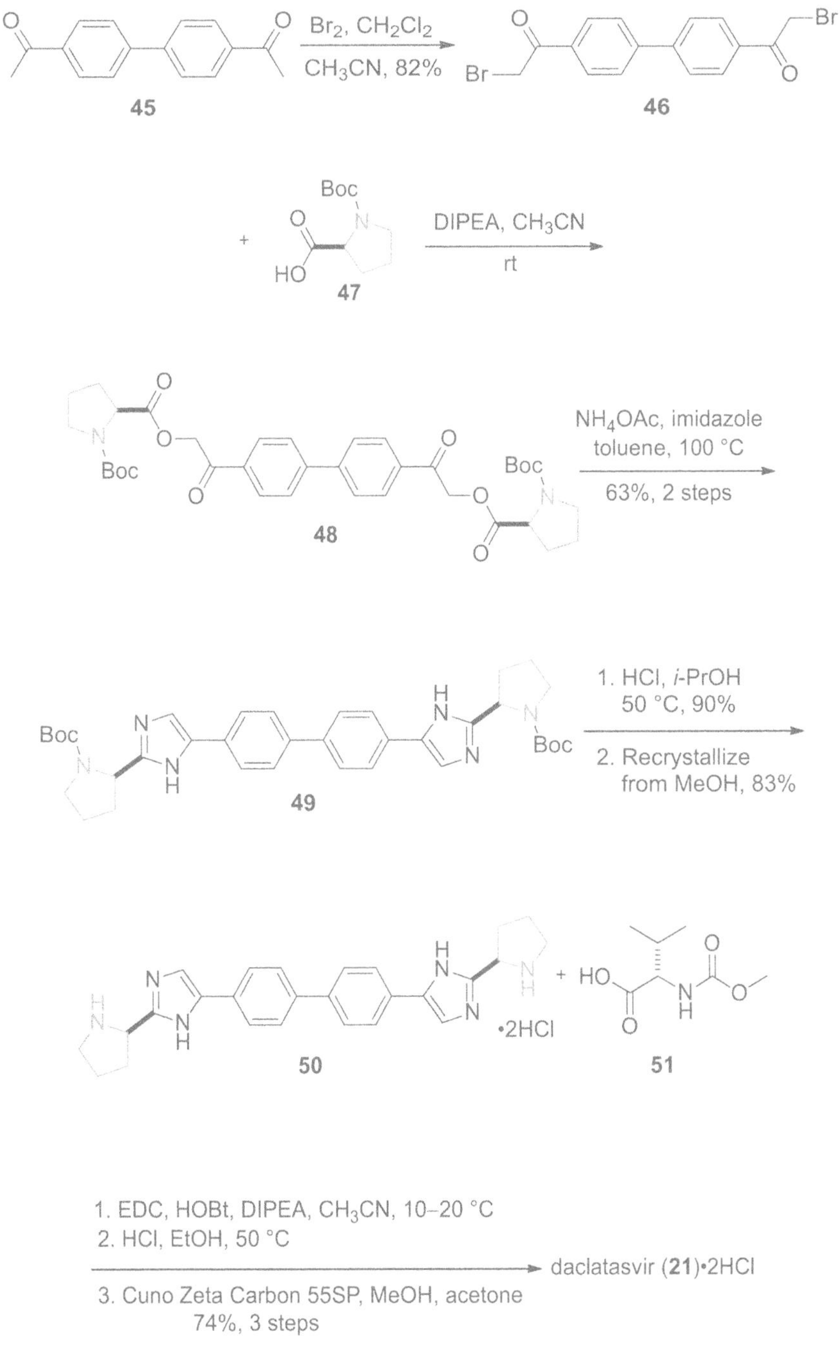

Gilead's synthesis of pibrentasvir (**23**) employed an interesting Buchwald amidation. Thus, bis-phenyl chloride **52** was coupled with Boc-proline-amide **53** to assemble bis-amide **54** in 54% yield using Pd$_2$(dba)$_3$ as the catalyst, Xantphos as the ligand, and Cs$_2$CO$_3$ as the base.[10]

Not all pyrrolidine building blocks are commercially available. For more complicated pyrrolidines with many substitutions, they have to be prepared from commercially available starting materials. For instance, both pyrrolidine fragments on Gilead's HCV NS5A polymerase inhibitor velpatasvir (**22**) had to be "pre-fabricated".

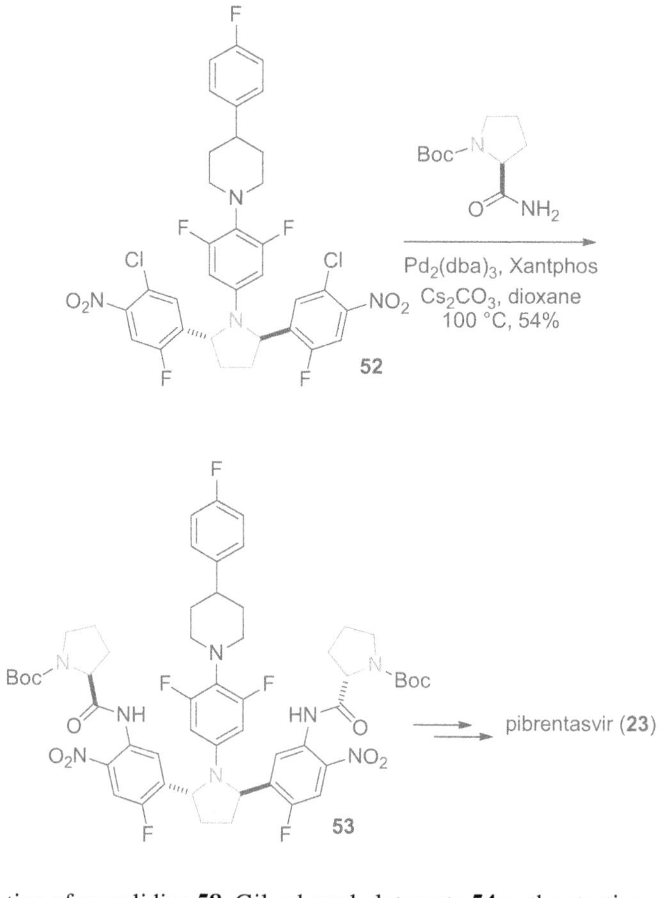

For the preparation of pyrrolidine **58**, Gilead used glutamate **54** as the starting material. An intramolecular condensation of **54** was facilitated by NaHMDS and further *careful* exposure to TFA afforded dihydropyrrole **55** (there was some ambiguity on the patent with regard to the survival of the two *t*-butyl groups).[11a,b] Double reduction of **55** produced pyrrolidine alcohol **56**. After removal of both Boc and *t*-butyl ester on **56**, Boc was put back on to prepare **57**. Methylation of **57** was followed by salt formation with dicyclohexylamine [HN(Cy)$_2$, a common, useful tactic to purify carboxylic acids because dicyclohexylamine tends to form crystalline salts with acids], which allowed isolation of the desired *cis*-isomer. A subsequent "salt break" then delivered pyrrolidine **58**.[11]

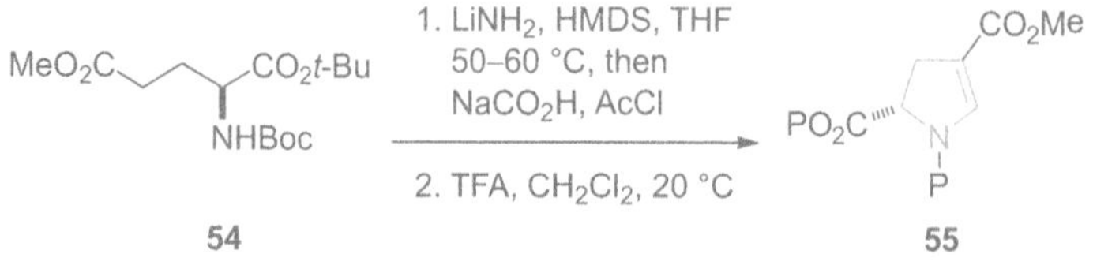

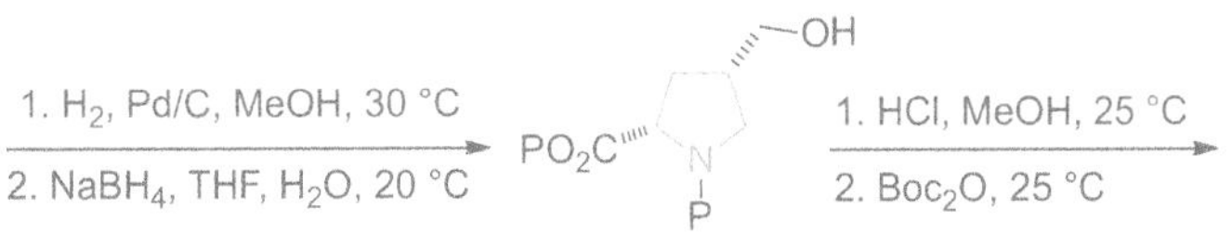

The other pyrrolidine fragment **61** was prepared from pyrrolidinone **59**, which was treated with methyl Grignard reagent to give ketone **60**. A one-pot Boc deprotection and subsequent ring-closing reductive amination assembled pyrrolidine fragment **61**. Installation of pyrrolidines **58** and **61** onto the corer structure was followed by further manipulations to deliver velpatasvir (**22**).[11]

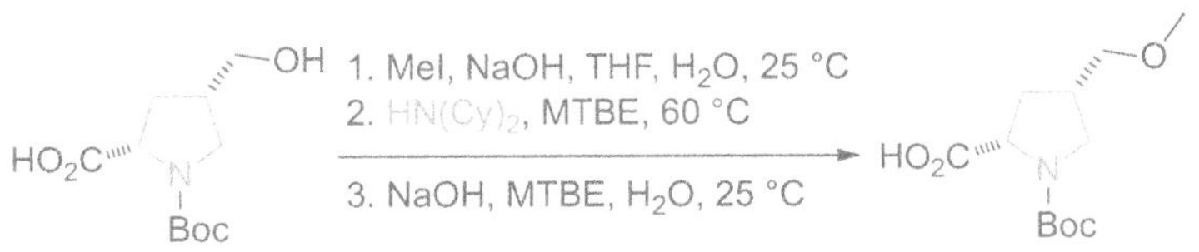

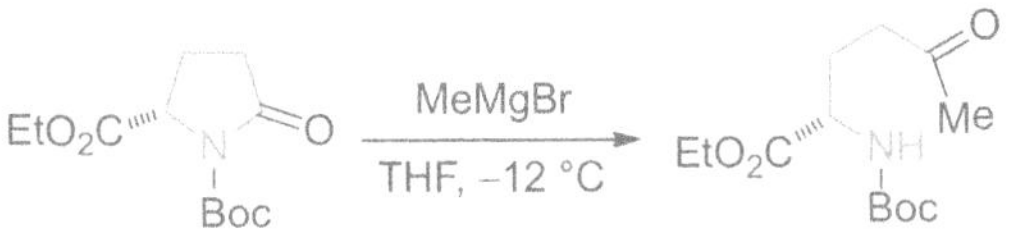

In summary, pyrrolidine motif on a drug may offer enhanced aqueous solubility and improve other physiochemical properties in addition to being part of the pharmacophore. The NH may serve as a hydrogen bond donor and, when its NH is masked, the N atom may serve as a hydrogen bond acceptor to its target protein. While the pyrrolidine moiety appears frequently in drugs, it may have a potential liability of being bioactivated to the corresponding iminium ion and aminoaldehyde. These reactive metabolites have potential genotoxicity and mutagenicity. While this is not prevalent, it is always a good idea to be vigilant.

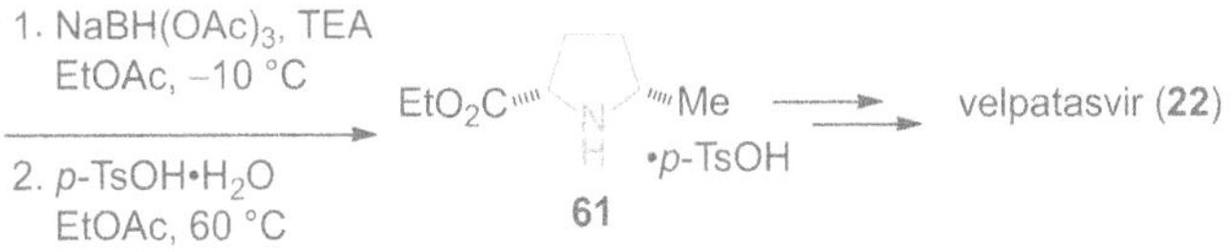

REFERENCES

1. Ondetti, M. A.; Williams, N. J.; Sabo, E. F.; Pluscec, J.; Weaver, E. R.; Kocy, O. *Biochem.* **1971**, *10*, 4033–4039.
2. Ostrem, J. M.; Peters, U.; Sos, M. L.; Wells, J. A.; Shokat, K. M. *Nature* **2013**, *503*, 548–551.
3. Fell, J. B.; Fischer, J. P.; Baer, B. R.; Ballard, J.; Blake, J. F.; Bouhana, K.; Brandhuber, B. J.; Briere, D. M.; Burgess, L. E.; Burkard, M. R.; et al. *ACS Med. Chem. Lett.* **2018**, *9*, 1230–1234.
4. Lim, H. K.; Chen, J.; Sensenhauser, C.; Cook, K.; Preston, R.; Thomas, T.; Shook, B.; Jackson, P. F.; Rassnick, S.; Rhodes, K.; et al. *Chem. Res. Toxicol.* **2011**, *24*, 1012–1030.
5. Shigenaga, M. K.; Trevor, A. J.; Castagnoli, N., Jr. *Drug Metab. Dispos.* **1988**, *16*, 397–402.
6. Kalgutkar, A. S. *Chem. Res. Toxicol.* **2017**, *30*, 220–238.
7. Chuo, D. T. H.; Jung, G.; Plouvier, B.; Yee, J. G. K. WO2006138673A2 (2006).

8. Yoshida, T.; Akahoshi, F.; Sakashita, H.; Kitajima, H.; Nakamura, M.; Sonda, S.; Takeuchi, M.; Tanaka, Y.; Ueda, N.; Sekiguchi, S.; et al. *Bioorg. Med. Chem.* **2012**, *20*, 5705–5719.

9. Pack, S. K.; Geng, P.; Smith, M. J.; Hamm, J. WO2009020825A1 (2009).

10. (a) Rodgers, J. D.; Shepard, S.; Li, Y.-L.; Zhou, J.; Liu, P.; Meloni, D.; Xia, M. WO2009114512 (2009). (b) Kobierski, M. E.; Kopach, M. E.; Martinelli, J. R.; Varie, D. L.; Wilson, T. M. WO2016205487 (2016).

11. (a) Allan, K. M.; Fujimori, S.; Heumann, L. V.; Huynh, G. M.; Keaton, K. A.; Levins, C. M.; Pamulapati, G. R.; Roberts, B. J.; Sarma, K.; Teresk, M. G.; et al. WO2015191437A1 (2015). (b) Hughes, D. L. *Org. Process Res. Dev.* **2016**, *20*, 1404–1415. (c) Flick, A. C.; Ding, H. X.; Leverett, C. A.; Fink, S. J.; O'Donnell, C. J. *J. Med. Chem.* **2018**, *61*, 7004–7031.

26 Pyrrolotriazines

Pyrrolotriazine-containing drugs have attracted intense attention recently since Gilead's nucleoside antiviral drug remdesivir (Veklury) has been catapulted to the limelight when, on May 1st, 2020, the FDA approved it for emergency use to treat COVID-19 patients.

Pyrrolotriazines and their aza-analogs imidazolotriazines have found widespread applications in nucleoside antiviral drugs and kinase inhibitors.

Bioisosterism is a powerful tool for drug discovery. Many old nucleoside antiviral drugs were discovered by mimicking the structures of DNA bases. There are two pairs of DNA bases: adenine, guanine, thymine, and cytosine. Another base, uracil, is only seen in RNAs. Pyrrolotriazines and their aza-analogs imidazolotriazines, non-natural bases, bear a striking resemblance to both adenine and guanine. Therefore, they are ideal bioisosteres of those two DNA bases. On the other hand, protein kinases phosphorylate amino acids such as tyrosine, serine, or threonine amino acids on target proteins by taking a phosphate from adenine triphosphate (ATP). Nearly all kinase inhibitors on the market are ATP competitive inhibitors by occupying the ATP-binding pockets of the kinases. Not surprisingly, pyrrolotriazines and imidazolotriazines are good bioisosteres of the adenine portion of ATP.

PYRROLOTRIAZINE-CONTAINING DRUGS

Only one imidazolotriazine-containing drug is currently on the market, to the best of our knowledge. It is Bayer's vardenafil (Levitra, **2**),[1] a phosphodiesterase V (PDE5) inhibitor for the treatment of erectile dysfunction (ED). It is a "me-too" drug of Pfizer's blockbuster drug and social sensation, sildenafil (Viagra, **1**). Bayer discovered vardenafil (**2**) employing a classic "scaffold-hopping" strategy. Although vardenafil (**2**)'s ethyl group on piperazine bestowed the molecule with a superior PDE5/PDE1 selectivity over the prototype drug sildenafil (**1**), the minor change would not have created any novel intellectual properties (IP) for Bayer. What "brought home the bacon" was vardenafil (**2**)'s imidazolotriazinone fragment to replace sildenafil (**1**)'s pyrrolopyrimidinone core structure. This approach is also known as the "analog-based drug discovery" strategy because imidazolotriazinone is an analog of pyrrolopyrimidinone.

sildenafil (Viagra, **1**)
Pfizer, 1998
PDE5 inhibitor

vardenafil (Levitra, **2**)
Bayer, 2003
PDE5 inhibitor

Several pyrrolotriazine- and imidazolotriazine-containing drugs have entered clinical trials, mostly as nucleoside antiviral drugs and kinase inhibitors. The most conspicuous of all is, of course, Gilead's nucleoside antiviral drug remdesivir (Veklury, **4**), initially discovered as a treatment of Ebola virus (EBOV) and entered phase III clinical trials.[2] It evolved from GS-6620 (**3**), the first *C-nucleoside* HCV polymerase inhibitor with demonstrated antiviral response in HCV infected patients.[3] Unlike normal nucleosides where the base and the sugar are tied together by an N–C bond, the pyrrolotriazine fragment connects the sugar motif via a C–C bond for C-nucleosides. *C-Nucleosides have the potential for improved metabolism and pharmacokinetic properties over their N-nucleoside counterparts due to the presence of a strong carbon–carbon glycosidic bond and a non-natural heterocyclic base.*

GS-6620 (**3**)

remdesivir (Veklury, **4**)

The field of kinase inhibitors is a fertile ground for flat heterocycles in general, and pyrrolotriazines in particular, because they mimic the adenine motif of the ATP molecule, the key player of the kinase functions. Bayer's rogaratinib (**5**) is a pan-fibroblast growth factor receptor (FGFR) inhibitor, inhibiting all four subtypes of this receptor: FGFR1–4. The drug has good drug metabolism and pharmacokinetic (DMPK) properties. Since rogaratinib (**5**) demonstrated tumor growth reduction in preclinical models bearing different FGFR-alterations in mono- and combination-therapy, it was brought to phase I clinical trials in 2018.[4]

Similarly, BMS's pyrrolotriazine-containing drug brivanib alaninate (**6**) is a dual inhibitor of FGFR-1 and vascular endothelial growth factor-2 (VEGFR2). It is a pro-drug of BMS-540215 in an attempt to boost solubility and bioavailability. Brivanib alaninate (**6**) has completed phase III clinical trials as a treatment of malignant tumors and is probably in the process of seeking governmental approval.[5]

BMS also employed the pro-drug tactic to arrive at BMS-751324 (**8**), a clinical p38α mitogen-activating protein (MAP) kinase inhibitor. It is a carbamoylmethylene-linked pro-drug of the parent drug, BMS-582949 (**7**), which was advanced to phase II clinical trials for the treatment of rheumatoid arthritis (RA), but suffered low solubility and low exposure at higher pH. The unique pro-drug BMS-751324 (**8**) is not only stable but also water-soluble under both acidic and neutral conditions. It is effectively biotransformed into the parent drug BMS-582949 (**7**) *in vivo* by alkaline phosphatase and esterase in a stepwise manner.[6]

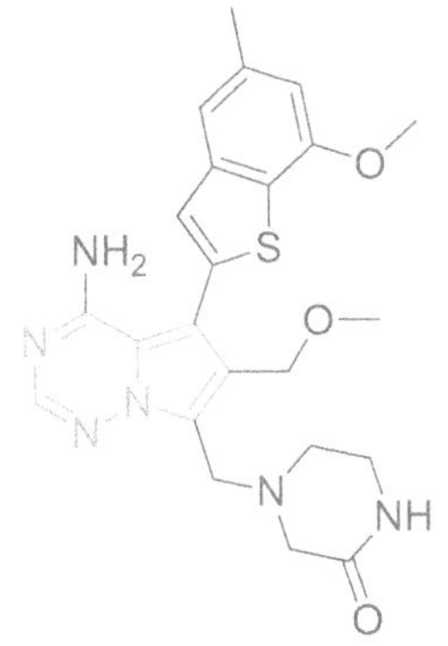

rogaratinib (**5**)
Bayer
pan-FGFR inhibitor

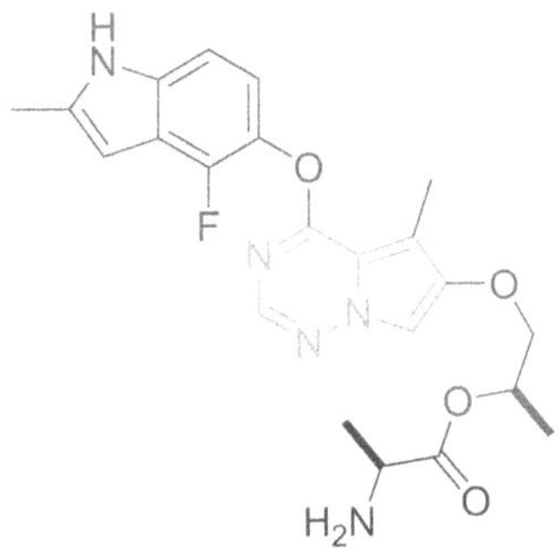

brivanib alaninate (**6**)
BMS, dual FGFR-1
and VEGFR-2 inhibitor

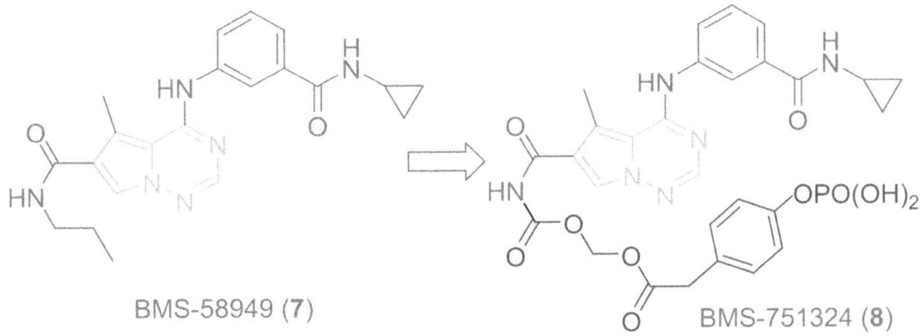

BMS-58949 (**7**)

BMS-751324 (**8**)

PYRROLOTRIAZINES IN DRUG DISCOVERY

How does remdesivir (**4**) work as a nucleoside antiviral drug?

Nucleosides have long been recognized as direct-acting antivirals (DAAs). In terms of mechanism of action (MoA), remdesivir (**4**) belongs to a class of nucleoside antiviral drugs known as antimetabolites. By definition, antimetabolites are drugs that interfere with normal cellular function, particularly the synthesis of DNA that is required for replication. In another word, remdesivir (**4**) "pretends" to be a nucleoside building block and participates in the viral DNA synthesis. But since it is not a *bona fide* DNA building block, DNA replication is interrupted and the virus is killed.

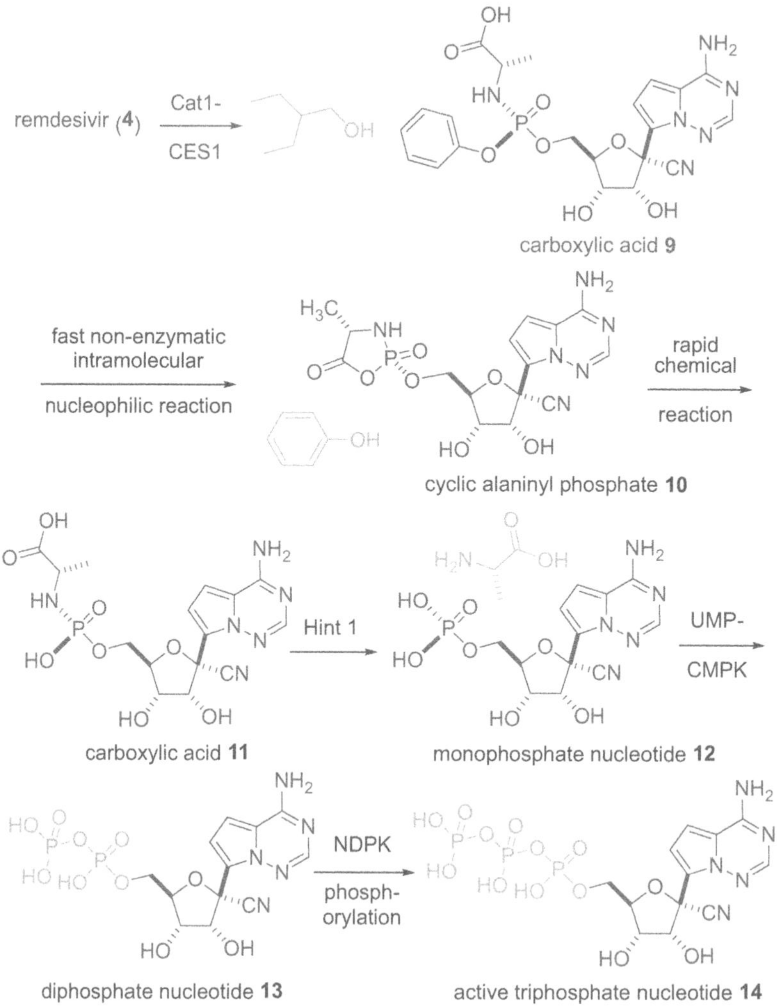

carboxylic acid 9

cyclic alaninyl phosphate 10

carboxylic acid 11

monophosphate nucleotide 12

diphosphate nucleotide 13

active triphosphate nucleotide 14

But the reality is not that simple. Because triphosphate nucleotide 14 (once a nucleoside is attached to a phosphate, it becomes a nucleotide) as the active antimetabolite has low solubility and low bioavailability and could even be toxic. To overcome these shortcomings, the pro-drug tactic is resorted here, but with a twist. To achieve an optimal pharmacokinetic and pharmacodynamic outcome, several forms of pro-drugs are installed onto the triphosphate nucleotide 14. Therefore, remdesivir (4) may be considered as a pro-drug of a pro-drug of a pro-drug of triphosphate nucleotide 14.

As shown in the scheme, once remdesivir (4) is injected via IV, the key enzymes initially involved in the metabolism of remdesivir (4) are human cathepsin A 1 (CatA1) and carboxylesterase 1 (CES1), which are responsible for the hydrolysis of the carboxyl ester between the alaninyl moiety and the isopropyl alcohol. This stereospecific reaction gives rise to the corresponding carboxylic acid 9. A nonenzymatic intramolecular nucleophilic attack then results in the formation of an alaninyl phosphate intermediate 10, which undergoes a rapid chemical reaction to hydrolyze the cyclic phosphate to a linear phosphate as carboxylic acid 11. The next step is speculated to involve the histidine triad nucleotide-binding protein 1 (Hint 1) enzyme in which the alaninyl phosphate intermediate is deaminated to form a monophosphate nucleotide 12. The final two steps involve consecutive phosphorylation reactions mediated by cellular kinases, uridine monophosphate–cytidine monophosphate kinase (UMP–CMPK) and nucleoside diphosphate kinase (NDPK), producing the diphosphate nucleotide **13** and subsequently the active triphosphate nucleotide **14**.[7] Triphosphate nucleotide **14** is then incorporated to the virus DNA to stop its replication.

In addition to remdesivir (**4**), there are several additional series of pyrrolotriazine-containing drugs in the areas of nucleotide antiviral drugs and protein kinase inhibitors as anti-cancer drugs.

Biota and Boehringer Ingelheim collaborated to discover a series of pyrrolotriazine-containing C-nucleosides as potential novel anti-HCV agents. The adenosine analog **15** was found to inhibit HCV NS5B polymerase (its triphosphate to be exact) and have excellent pharmacokinetic properties. *In vitro*, C-nucleoside **15** was tested efficacious across all genotypes with minimal cytotoxicity ($CC_{50} > 100\,\mu M$). Unfortunately, it was not pursued as a drug candidate (DC) because adverse effects were observed in its rat safety studies due to cytotoxicities.[8] Prodrugs of pyrrolotriazine C-nucleoside **15** were prepared to overcome the toxicity issues. Very similar to Gilead's GS-6620 (**3**) structurally, pro-drug **16** was found to increase anti-HCV activity and enhance nucleotide triphosphate concentrations *in vitro*.[9]

In the realm of kinase inhibitors, both pyrrolotriazines and imidazolotriazines have been used as they are bioisosteres of adenine moiety of ATP, a key player of kinase functions.

HCV NS5B polymerase inhibitor **15**
EC_{50} = (cross genotype) = 0.4–0.6 μM
$CC_{50} > 100\,\mu M$
Dog PK: $t_{1/2}$= 7.6 h, *F*% = 96

pro-drug **16**

Interleukin-1 receptor-associated kinase 4 (IRAK4) is an enzyme important in innate immunity, and its inhibition is predicted to be beneficial in treating inflammatory diseases. The fact that PF-06650833, a selective IRAK4 inhibitor, was advanced to phase I clinical trials for the treatment of RA and inflammatory bowel disorder (IBD). This has boosted the confidence in rationale (CIR) for this drug target.

pyrrolopyrimidine **17**
IRAK4, IC_{50} = 5 nM
Caco2 A2B
P_{app} = 14 x 10^{-6} cm/s
Caco2 efflux ratio, 2.8

pyrrolotriazine **18**
IRAK4, IC_{50} = 22 nM
Caco2 A2B
P_{app} = 38 x 10^{-6} cm/s
Caco2 efflux ratio, 0.5

AstraZeneca initially obtained a series of pyrrolopyrimidines as represented by **17** as IRAK4 inhibitors. While the pyrrolopyrimidines were potent in both enzymatic and cellular assays, they suffered from low permeability and high efflux. A scaffold-hopping strategy led to the discovery of a series of pyrrolotriazines as analogous IRAK4 inhibitors. *In comparison to the pyrrolopyrimidines, the pyrrolotriazines contain one fewer formal hydrogen bond donor and are intrinsically more lipophilic.* Cell permeability increases as the number of hydrogen bond donors decreases. Optimization of the series culminated the discovery of pyrrolotriazine **18** with higher permeability and lower efflux in Caco2 assays in comparison to the prototype **17**. Pyrrolotriazine **18** was a promising *in vivo* probe to assess the potential of IRAK4 inhibition in cancer treatment. It demonstrated tumor regression in combination with ibrutinib, a covalent Bruton's tyrosine kinase (Btk) inhibitor.[10]

Just like pyrrolotriazines, imidazolotriazines are also used as fragments in nucleoside antivirals and kinase inhibitors. Furthermore, imidazolotriazines also found applications in the fields of PDE and histone deacetylase (HDAC) inhibitors.

(*R*)-Roscovitine (seliciclib) is a cyclin-dependent kinase (CDK) inhibitor in phase II clinical trials. In a scaffold-hopping exercise, the purine core structure on (*R*)-roscovitine was replaced with its imidazolotriazine isostere, giving rise to an imidazolotriazine analog as a potent CDK inhibitor.[11]

PDE 2A inhibitors may have potential as central nervous system (CNS) drugs. Employing a late-stage microsomal oxidation of an existing PDE 2A inhibitor, Pfizer identified PF-06815189, an imidazolotriazinone derivative, with reduced clearance by cytochrome 450 (CYP450) enzymes, minimizing the risk of drug–drug interactions (DDIs) and improving physiochemical properties.[12]

Pfizer identified pyrazolopyrimidine **19** as a potent, selective, and brain penetrant PDE 2A inhibitor. Its estimated human dose, 108 mg/day, is high. A design strategy to reduce its clearance in human liver microsomes (HLM C_{lint}) by reducing the drugs' lipophilicity led to the discovery of imidazolotriazine **20**. It is a potent, highly selective, and brain penetrant PDE 2A inhibitor clinical candidate with a significantly lower human dose, 30 mg/day, qd. It is likely that the core structure changes boosted hydrogen bonding strength, thus reducing lipophilicity and clearance while elevating its lipophilic ligand efficiency (LipE).[13]

SYNTHESIS OF SOME PYRROLOTRIAZINE-CONTAINING DRUGS

In Gilead's process-scale synthesis of remdesivir (**4**), aminopyrrolotriazine **21** was brominated with 1,3-dibromo-5,5-dimethylhydantoin (DBDMH, **22**) to prepare bromide **23**. After protecting the aniline with two TMS groups, halogen–metal exchange prepared the lithium intermediate **24**, which was added to a cold solution of lactone **25** to assemble lactol **26** as a 1:1 mixture of two anomers after quenching the reaction with a weakly acidic aqueous solution. Separation of the anomers and installation of the pro-drug motif then delivered remdesivir (**4**) after a few minor functional group manipulations.

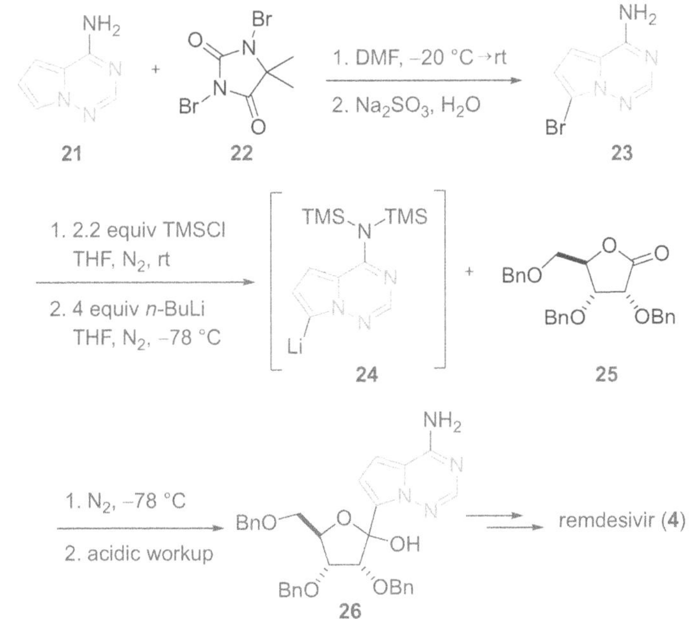

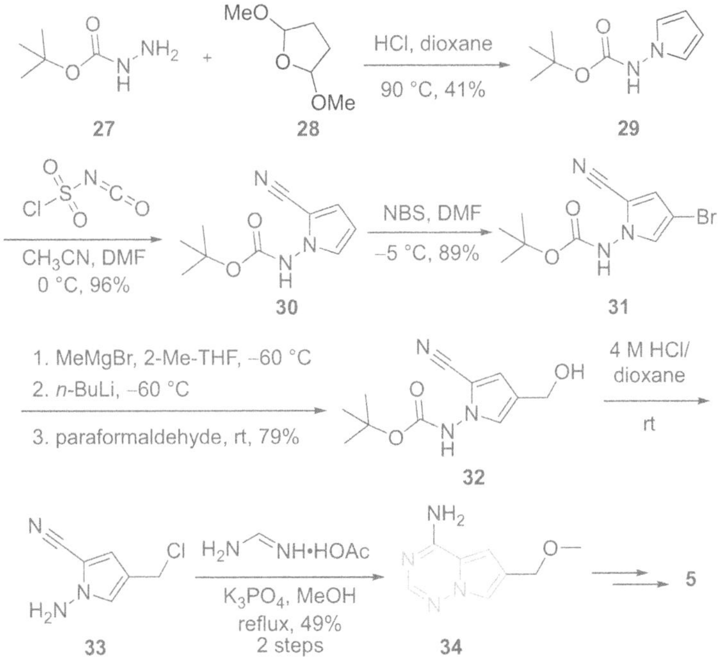

Preparation of the aminopyrrolotriazine fragment **30** of Bayer's pan-FGFR inhibitor rogaratinib (**5**) is accomplished in six linear steps. At first, an acid-catalyzed condensation of *tert*-butyl carbazate (**27**) and 2,5-dimethoxytetrahydro-furan (**28**) afforded the protected aminopyrrole **29** in 41% yield after crystallization. Subsequent C1 electrophilic substitution was achieved by treating **29** with chlorosulfonylisocyanate followed by an intramolecular elimination to install cyanopyrrole

30. Bromination of **30** at the C4 position was carried out at low temperature to boost the desired regioselectivity and minimize the overreaction (to dibromopyrrole). The resulting bromide **31** then underwent a halogen–metal exchange reaction and was quenched by paraformaldehyde to afford the hydroxymethylpyrrole **32**. Treating **32** with 4 M HCl not only removed the Boc protective group but also converted the alcohol to chloride to produce **33**. The very reactive chloromethyl **33** was directly converted to methoxymethyl-pyrrolotriazine **34** in a one-pot, two-step reaction sequence by treating **33** with methanol, followed by formamidine acetate and potassium phosphate at reflux. The key intermediate **34** was employed as a building block to construct rogaratinib (**5**).[4]

For the synthesis of imidazolotriazine-containing drugs, Pfizer's PDE 2A inhibitor clinical candidate imidazolotriazine **20** is showcased here as an example. As shown underneath, 4-bromo-1-methylpyrazole (**35**) was lithiated and trapped with *N*-methoxy-*N*-methylacetamide to give the methyl ketone, which was selectively brominated to afford α-bromoketone **36**. Condensation of **36** with N-aminoamidine **37** led to imidazole **38**. Treatment of **38** with TFA was followed by reaction with formamidine, and the resulting intermediate underwent an intramolecular cyclization with the aid of CDI to produce imidazolotriazinone **39**. Chlorination of **39** was followed by an S_NAr reaction with azetidine to assemble the key intermediate **40**, which was transformed to imidazolotriazine **20** in several additional steps.[13,14]

In summary, pyrrolotriazines and imidazolotriazines are bioisosteres of bases such as adenine, guanine, purine, and so on. As a consequence, they have found utility in the fields of nucleoside antiviral drugs, kinase inhibitors, PDE, and HDAC inhibitors.

REFERENCES

1. Haning, H.; Niewohner, U.; Schenke, T.; Es-Sayed, M.; Schmidt, G.; Lampe, T.; Bischoff, E. *Bioorg. Med. Chem. Lett.* **2002**, *12*, 865–868.
2. Siegel, D.; Hui, H. C.; Doerffler, E.; Clarke, M. O.; Chun, K.; Zhang, L.; Neville, S.; Carra, E.; Lew, W.; Ross, B.; et al. *J. Med. Chem.* **2017**, *60*, 1648–1661.
3. Cho, A.; Zhang, L.; Xu, J.; Lee, R.; Butler, T.; Metobo, S.; Aktoudianakis, V.; Lew, W.; Ye, H.; Clarke, M.; et al. *J. Med. Chem.* **2014**, *57*, 1812–1825.
4. Collin, M.-P.; Lobell, M.; Huebsch, W.; Brohm, D.; Schirok, H.; Jautelat, R.; Lustig, K.; Boemer, U.; Voehringer, V.; Heroult, M.; et al. *ChemMedChem* **2018**, *13*, 437–445.

5. Cai, Z.-W.; Zhang, Y.; Borzilleri, R. M.; Qian, L.; Barbosa, S.; Wei, D.; Zheng, X.; Wu, L.; Fan, J.; Shi, Z.; et al. *J. Med. Chem.* **2008**, *51*, 1976–1980.

6. (a) Liu, C.; Lin, J.; Hynes, J.; Wu, H.; Wrobleski, S. T.; Lin, S.; Dhar, T. G. M.; Vrudhula, V. M.; Sun, J.-H.; Chao, S.; et al. *J. Med. Chem.* **2015**, *58*, 7775–7784. (b) Liu, C.; Lin, J.; Everlof, G.; Gesenberg, C.; Zhang, H.; Marathe, P. H.; Malley, M.; Galella, M. A.; Mckinnon, M.; Dodd, J. H.; et al. *Bioorg. Med. Chem. Lett.* **2013**, *23*, 3028–3033.

7. (a) Murakami, E.; Tolstykh, T.; Bao, H.; Niu, C.; Steuer, H. M.; Bao, D.; Chang, W.; Espiritu, C.; Bansal, S.; Lam, A. M.; et al. *J. Biol. Chem.* **2010**, *285*, 34337–34347. (b) Ma, H.; Jiang, W.-R.; Robledo, N.; Leveque, V.; Ali, S.; Lara-Jaime, T.; Masjedizadeh, M.; Smith, D. B.; Cammack, N.; Klumpp, K.; et al. *J. Biol. Chem.* **2007**, *282*, 29812–29820.

8. Draffan, A. G.; Frey, B.; Pool, B.; Gannon, C.; Tyndall, E. M.; Lilly, M.; Francom, P.; Hufton, R.; Halim, R.; Jahangiri, S.; et al. *ACS Med. Chem. Lett.* **2014**, *5*, 679–684.

9. Tyndall, E. M.; Draffan, A. G.; Frey, B.; Pool, B.; Halim, R.; Jahangiri, S.; Bond, S.; Wirth, V.; Luttick, A.; Tilmanis, D.; et al. *Bioorg. Med. Chem. Lett.* **2015**, *25*, 869–873.

10. Degorce, S. L.; Anjum, R.; Dillman, K. S.; Drew, L.; Groombridge, S. D.; Halsall, C. T.; Lenz, E. M.; Lindsay, N. A.; Mayo, M. F.; Pink, J. H.; et al.. *Bioorg. Med. Chem.* **2018**, *26*, 913–924.

11. Popowycz, F.; Fournet, G.; Schneider, C.; Bettayeb, K.; Ferandin, Y.; Lamigeon, C.; Tirado, O. M.; Mateo-Lozano, S.; Notario, V.; Colas, P.; et al. *J. Med. Chem.* **2009**, *52*, 655–663.

12. Stepan, A. F.; Tran, T. P.; Helal, C. J.; Brown, M. S.; Chang, C.; O'Connor, R. E.; De Vivo, M.; Doran, S. D.; Fisher, E. L.; Jenkinson, S.; et al. *ACS Med. Chem. Lett.* **2018**, *9*, 68–72.

13. Helal, C. J.; Arnold, E.; Boyden, T.; Chang, C.; Chappie, T. A.; Fisher, E.; Hajos, M.; Harms, J. F.; Hoffman, W. E.; Humphrey, J. M.; et al. *J. Med. Chem.* **2018**, *61*, 1001–1018.

14. Clarke, M.; O'Neil, H.; Feng, J. Y.; Jordan, R.; Mackman, R, L.; Ray, A. S.; Siegel, D. U.S. Patent US20170071964 (2017).

27 Spiroazetidines

Spirocycles are making more and more frequent appearances in drugs as both core structures and peripheries of drug molecules. They have at least three advantages as drug fragments:

a. Spirocyclic scaffolds are inherently three-dimensional and can project functionalities in all three dimensions to interact more extensively with the protein target of interest and lower off-target effects in comparison to their two-dimensional counterparts.
b. They are sp^3-carbon-rich with a high value of a *fraction of saturated carbon* (Fsp^3) and are correlated with favorable physiochemical properties such as higher aqueous solubility.[1]
c. Spirocycles often offer new chemical space to create novel intellectual properties, as amply documented by Zheng et al.[2]

    With regard to spiroazetidines, the other ring may be fused to azetidine at either the α- or the β-position of the nitrogen atom. Overwhelming examples in the literature are the β-fused azetidine spirocycles.

β-fused azetidines   α-fused azetidines

SPIROAZETIDINES IN DRUG DISCOVERY

No spiroazetidine-containing drug has been approved for marketing yet, although two spiroazetidine-containing drugs have entered clinical trials: spiroazetidine **6** and azetidine–piperidine **12** (*vide infra*). More and more spiroazetidines are appearing in the literature for drug discovery.

SPIROAZETIDINES AS BIOISOSTERES

Spiroazetidines, like all spirocyclic scaffolds, are inherently three-dimensional and offer structural novelty. In terms of isosterism, 2,6-diazaspiro[3.3]heptane can mimic piperazine and 2-oxa-6-aza-spiro[3.3]heptane bears striking resemblance to morpholine. In the same vein, 2-azaspiro[3.3]heptane has served as an excellent isostere of piperidine and has found several important applications in drug discovery. Therefore, these spiroazetidines have found utility in drug discovery as isosteres of piperazine, morpholine, and piperidine, respectively.

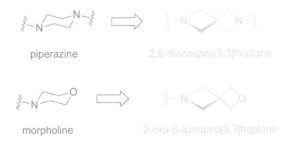

piperazine   2,6-diazaspiro[3.3]heptane

morpholine   2-oxa-6-azaspiro[3.3]heptane

piperidine   2-azaspiro[3.3]heptane

Inhibition of ketohexokinase (KHK, also known as fructokinase) promotes fructose metabolism and thus KHK inhibitors have potential in the diabetes treatment. From high-throughput screening (HTS) and further optimization guided by a structure-based drug design (SBDD), Maryanoff and coworkers arrived at pyrimidinopyrimidine piperazine 1 as a potent, selective human hepatic KHK (KHK-C isoform) inhibitor. Since piperazine is considered a conformationally constrained isostere of ethyldiamine, 2,6-diazaspiro[3.3]heptane is even further constrained than piperazine. Along this line of speculation, spiroazetidine analog 2 was prepared and tested to have similar enzymatic potency in inhibiting recombinant human hepatic KHK-C as piperazine 1, indicating that conformational rigidity is tolerated for the structure–activity relationship (SAR) for this pyrimidinopyrimidine series of KHK-C selective inhibitors. Furthermore, both piperazine 1 and spiroazetidine 2 exhibited reasonably potent cellular KHK inhibition (400 and 360 nM, respectively), which relates to their intrinsic potency versus KHK and their ability to enter cells.[3]

piperazine 1
KHK-C, IC_{50} = 12 nM
cellular, IC_{50} = 400 nM

spiroazetidine 2
KHK-C, IC_{50} = 8 nM
cellular, IC_{50} = 360 nM

Like KHK inhibitors, stearoyl-CoA desaturase (SCD) inhibitors are also potential treatments for metabolic disorders such as diabetes and obesity. To minimize adverse events associated with eyes and skin, liver-targeted SCD inhibitors are preferred. Merck-Frost arrived at piperazine 3 as a liver-selective SCD inhibitor with an enzymatic potency of 28 nM against rat SCD. The strategic presence of polar acid moiety was important because carboxylic acids and tetrazoles are recognized organic anionic transporter proteins (OATPs), thus providing the desired in vivo properties, i.e., a high liver concentration (target organ for efficacy) and a low systemic concentration to minimize exposures in off-target tissues and cell-associated adverse events (eyes and skin).[4]

Spiroazetidine isostere 4 has a similar enzymatic potency as its prototype 3. Moreover, both of them are also virtually inactive in a whole cell assay in a human hepatocellular carcinoma line (HepG2), indicating that both spiroazetidine 4 and piperazine 3 do not enter off-target cells via passive diffusion since HepG2 cells are devoid of active OATPs. Gratifyingly, both spiroazetidine 4 and piperazine 3 do cross cell membranes through active transporters when assessed in a rat hepacyte (Rat Hep, which contains functional OATPs) assay.[4]

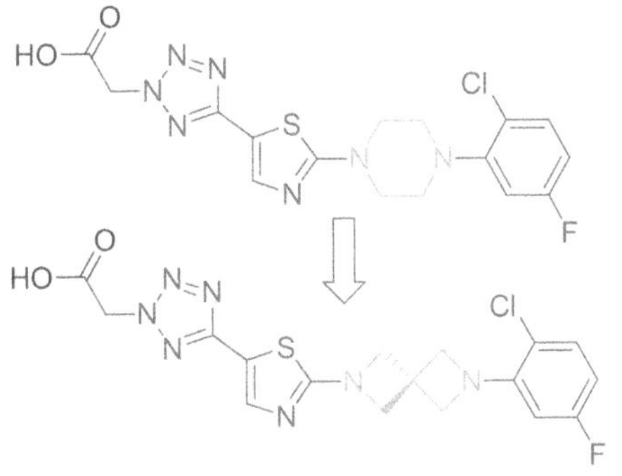

piperazine **3**
Rat SCD IC$_{50}$ = 28 nM
HepG2 IC$_{50}$ = 4,200 nM
Rat Hep IC$_{50}$ = 134 nM

spiroazetidine **4**
Rat SCD IC$_{50}$ = 43 nM
HepG2 IC$_{50}$ = 50,900 nM
Rat Hep IC$_{50}$ = 103 nM

Continuing the theme of diabetes treatments, melanin-concentrating hormone receptor 1 (MCHr1) antagonists have been explored for weight control. AstraZeneca discovered a series of oxadiazole-containing compounds as represented by morpholine **5**. Their efforts culminated in clinical candidate **6** with a novel peripheral spiroazetidine moiety. Spiroazetidine **6** displayed appropriate lipophilicity for a CNS indication, showed excellent permeability with no efflux, and possessed good off-target selectivity, including hERG. Preclinical good-laboratory practice (GLP) toxicology and safety pharmacology studies were without findings and spiroazetidine **6** (AZD1979) was taken into clinical trials.[5]

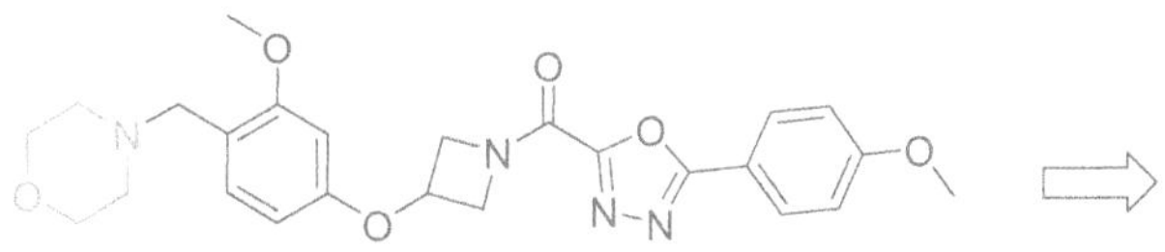

morpholine **5**
MCH GTPγS, IC$_{50}$ = 27 nM
MCH bind Mu, IC$_{50}$ = 16 nM
hERG, IC$_{50}$ = 9.8 μM

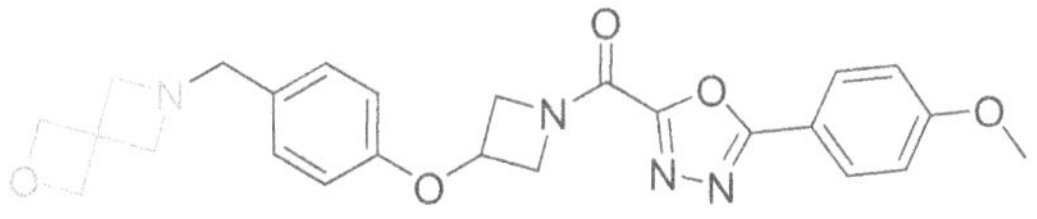

2-oxa-6-azaspiro[3.3]heptane **6**
MCH GTPγS, IC$_{50}$ = 27 nM
MCH bind Mu, IC$_{50}$ = 11 nM
hERG, IC$_{50}$ = 22 μM

Similar to Schering–Plough's boceprevir (Victrelis), Vertex's telaprevir (Incivek, **5**) is a hepatitis C virus (HCV) NS3/4A *serine* protease inhibitor. They are also competitive, covalent reversible inhibitors. Working in concert with His$_{57}$ and Asp$_{81}$, the Ser$_{139}$ on HCV NS3/4A serine protease enzyme adds to the ketone warhead on telaprevir (**5**) to form a covalent tetrahedral hemiacetal intermediate **6**, which closely mimics the transition state of the hydrolysis processes by the serine protease.[6] Introduction of spiroazetidine moieties at the P2 unit of telaprevir (**5**) resulted in inhibitors with good potency as measured using inhibition of HCV RNA replication in Huh7 cells in a subgenomic HCV replicon system. In particular, spiroazetidine **7** displayed a potency (EC$_{50}$=0.8 μM) similar to that of telaprevir (**5**, EC$_{50}$=0.4 μM).[7]

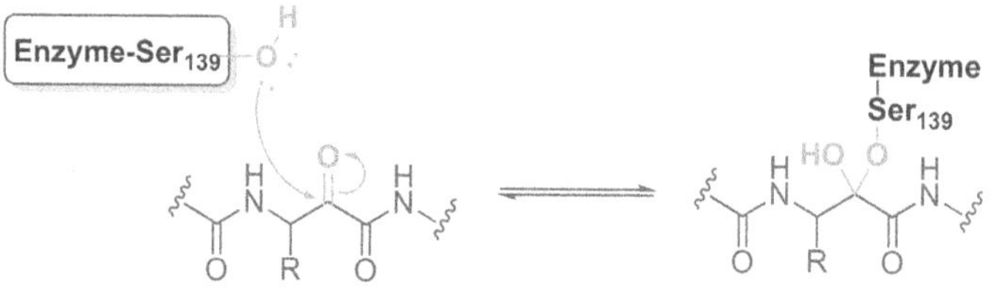

telaprevir (Incivek, **5**)   covalent intermediate **6**

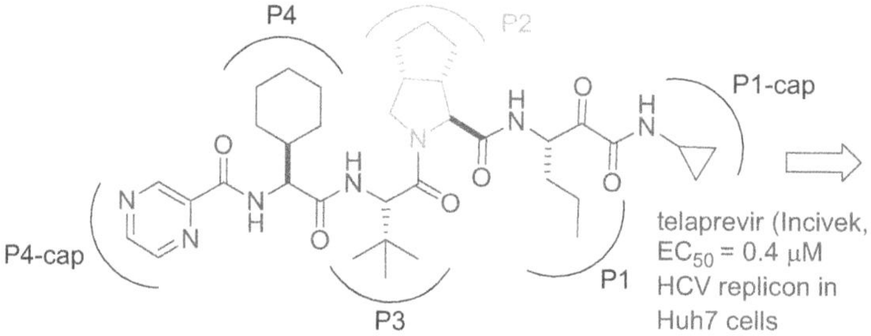

telaprevir (Incivek, **5**)
EC_{50} = 0.4 µM
HCV replicon in
Huh7 cells

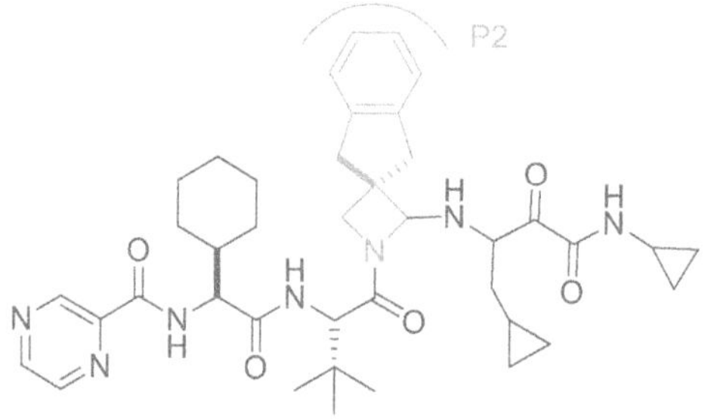

spiroazetidine **7**
EC_{50} = 0.8 µM
HCV replicon in
Huh7 cells

Not surprisingly, not all spiroazetidine isosteres worked as a panacea.

Leucine-rich repeat kinase 2 (LRRK2) inhibitors are potentially useful in the treatment of Parkinson's disease (PD). From HTS and lead optimization, Pfizer arrived at morpholine **8** (PF-06447475) as a highly potent (LRRRK2 enzymatic assay: IC_{50}=3 nM), selective, brain penetrant, and *in vivo* active LRRK2 inhibitor. Attempt to replace the morpholino group with the 2-oxa-6-aza-spiro[3.3]heptane isostere led to spiroazetidine **9**, which was virtually inactive toward LRRK2, regrettably.[8]

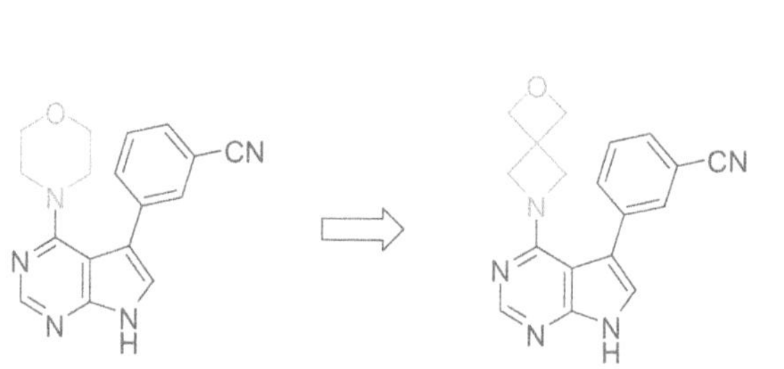

morpholine **8**
LRRK2 IC_{50} = 3 nM

2-oxa-6-azaspiro[3.3]heptane **9**
LRRK2 IC_{50} > 2,300 nM

UTILITY OF SPIROAZETIDINES IN MEDICINAL CHEMISTRY

The spirocyclic azetidine–piperidine (2,7-diazaspiro[3,5]nonane) fragment served as the backbone of a series of inverse agonists of the ghrelin receptor (GR) discovered by Pfizer.[9–11]

GR is a G-protein coupled receptor (GPCR) that plays a role in obesity and glucose homeostasis. Starting from an HTS hit, azetidine–piperidine **10** was discovered as a potent GR *inverse agonist* (hGR IC$_{50}$=4.6 nM, K_i=7.0 nM, an *inverse agonist* is a drug that binds to the same receptor as an agonist but induces a pharmacological response opposite to that of the agonist) with 43% bioavailability in Sprague–Dawley rat. But it suffered from an undesired off-target effect; namely, it displayed muscarinic acetylcholine receptor (mAChR) M2 activity (K$_i$=269 nM).[9]

Conformational restriction of the right-hand portion of compound **10** led to chiral indane **11**, which maintained its potency (hGR pK$_i$= 8.2, i.e., K$_i$=6.3 nM) with improved selectivity over mAChR M2 (pK$_i$=4.85, i.e., K$_i$=14.1 µM) as well as high receptor occupancy.[10]

Eventually, meticulous fine-tuning of the SAR culminated in PF-5190457 (**12**, pK$_i$=8.36, i.e., K$_i$=4.4 nM), which showed a better balance of receptor activity and off-target selectivity (M2 K$_b$/ GR K$_i$ ratio=266). As a potent, selective, and orally bioavailable GR inverse agonist, azetidine–piperidine **12** was advanced to clinical trials for the treatment of diabetes on the basis of its promising pharmacological and safety profile.[11]

azetidine-piperidine **10**
hGR IC$_{50}$ = 4.6 nM
hGR K_i = 7.0 nM
LE = 0.33

indane **11**
hGR pK$_i$ = 8.2
mAChR M2 pK$_i$ = 4.85

azetidine-piperidine **12**
hGR pK$_i$ = 8.36
GTPγS, pK$_i$ = 8.18
(inverse agonist)

The azetidine–piperidine (2,7-diazaspiro[3,5]nonane) fragment also found success in the fatty acid amide hydrolase (FAAH) inhibitors program as well. FAAH is an integral membrane serine hydrolase responsible for the degradation of fatty acid amide signaling molecules such as endocannabinoid anandamide (AEA), which has been shown to possess cannabinoid-like analgesic properties. Therefore, FAHH inhibitors are explored as a treatment of pain. A group of chemists at Janssen prepared heteroarylurea FAHH inhibitors with dozens of diamine linkers. One of them, 2,7-diazaspiro[3,5]nonane **13**, was found to be a potent FAHH inhibitor (hFAHH IC$_{50}$=8 nM). In addition, it was found to inhibit FAAH centrally, elevate the brain levels of three fatty acid ethanolamides [FAAs: AEA, oleoyl ethanolamide (OEA), and palmitoyl ethanolamide (PEA)], and was moderately efficacious in a rat model of neuropathic pain.[12]

2,7-diazaspiro[3,5]nonane **13**
hFAHH IC$_{50}$ = 8 nM

Respiratory syncytial virus (RSV) is a major cause of pneumonia and bronchiolitis in young children, immunocompromised adults, and the elderly. Unfortunately, there is no effective treatment except an expensive monoclonal antibody (palivizumab) with a dubious safety profile. BMS's pyridinoimidazolone **14** is an efficacious RSV F fusion glycoprotein inhibitor. Using **14** as a starting point, a "patent-busting" or "scaffold-hopping" exercise led to the discovery of a series of novel spiroazetidine 2-oxo-indoline derivatives. The lead compound **15** exhibited excellent *in vitro* potency with an EC_{50} value of 0.8 nM and demonstrated 71% oral bioavailability in mice.[13]

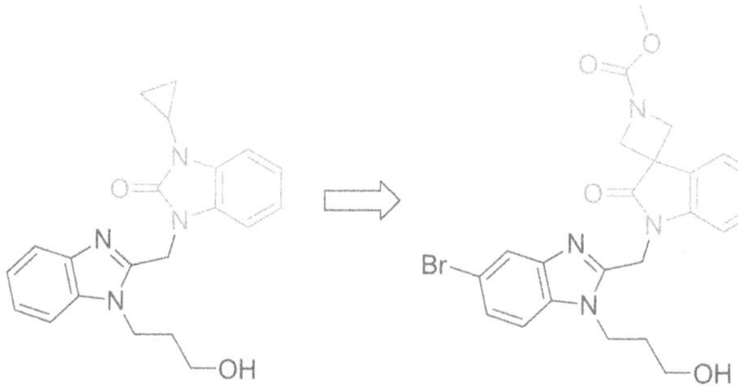

pyridinoimidazolone **14**
RSV EC_{50} = 22 nM
CC_{50} > 100 μM

spiroazetidine **15**
RSV EC_{50} = 0.8 nM
CC_{50} > 100 μM
%F = 71%

SYNTHESIS OF SOME SPIROAZETIDINES

Synthesis of Merck-Frost's liver-selective SCD inhibitor spiroazetidine **4** commenced with the preparation of bromothiazole-nitrile **17** from bromothiazole-ester **16** in two steps. Tetrazole **18** was then elaborated by condensation of bromothiazole-nitrile **17** with sodium azide. S_N2 alkylation of tetrazole **18** with *t*-butyl bromoacetate gave rise to adduct **19**, which was followed by an S_NAr reaction with aryl-2,6-diazaspiro[3.3]heptane **20** to assemble the adduct **21**. TFA-promoted deprotection of the ester then delivered spiroazetidine **4**.[4]

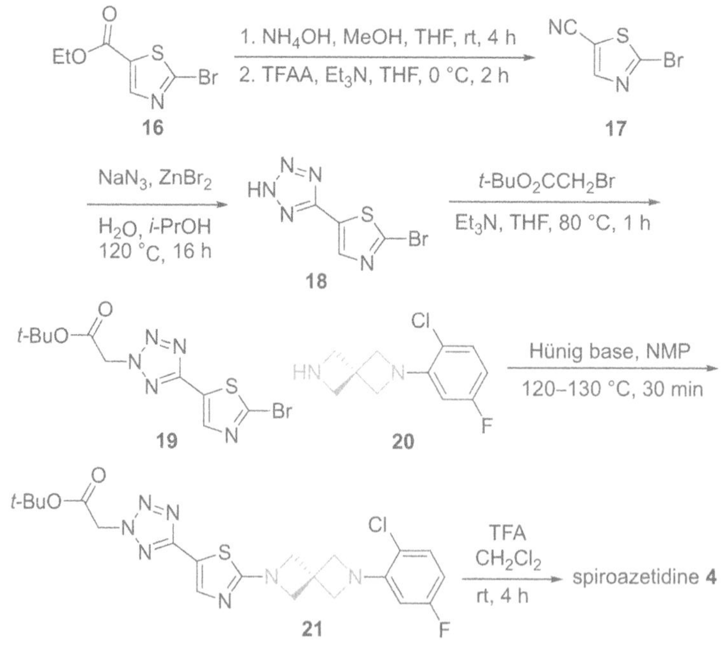

AstraZeneca began the synthesis of their MCHr1 antagonist clinical candidate AZD1979 (**6**) with a reductive amination of *p*-anisaldehyde with 2-oxa-6-azaspiro[3.3]heptane to prepare benzylamine **22**. Subsequently, a Mitsunobu reaction between benzylamine **22** and Cbz-protected azetidine **23** provided azetidine ether **24**. After deprotection, the resulting "naked" azetidine **25** was coupled with oxadiazole **26** with the aid of a catalytic amount of sodium cyanide to deliver the desired 2-oxa-6-azaspiro[3.3]heptane **6**.[5]

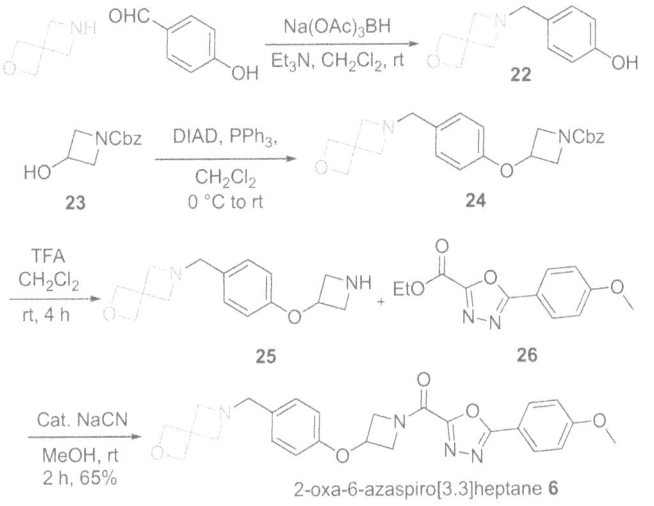

To produce GR inverse agonist PF-5190457 (**12**), Pfizer first carried out a reductive amination of chloroaldehyde **27** with indane-amine **28** to assemble azetidine–piperidine **29**. Scaffold **29** served as a linchpin-like modular framework for further functionalizations. Palladium-catalyzed Miyaura reaction between bromide **29** and bis(pinacolato)diboron produced boronate intermediate **30**, which underwent a Suzuki coupling with 4-bromo-6-methylpyrimidine to provide adduct **31**. After removal of the Boc protective group on **31**, the revealed azetidine–piperidine was coupled with the requisite acid to deliver PF-5190457 (**12**).[11]

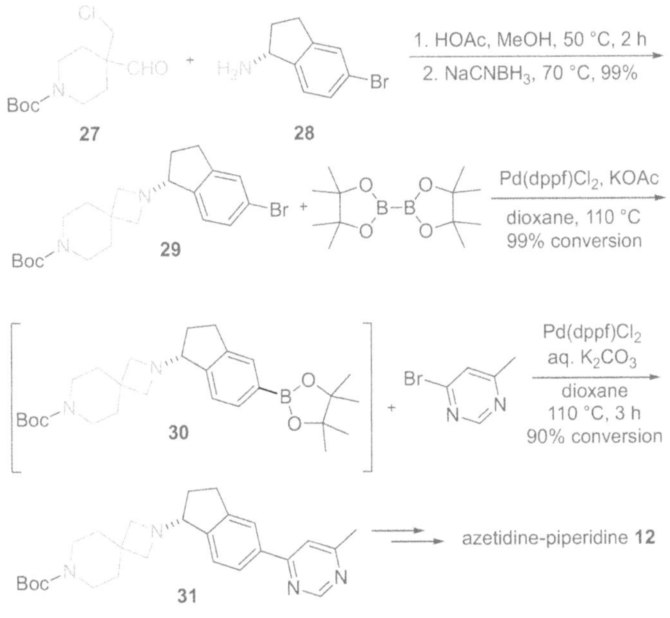

For spiroazetidine building blocks that are not commercially available, Carreira[14,15] and Mykhailik[16,17] have published excellent works on their preparations.

To conclude, spiroazetidines, like all spirocyclic scaffolds, are bestowed with three advantages over their flat counterparts: inherent three-dimensional structures may offer more interactions with target proteins; spiroazetidines may provide superior physiochemical properties and thus are more drug like; and novel structures may offer fresh intellectual properties. Meanwhile, although no spiroazetidine-containing drugs are currently approved for marketing, at least two of them have been advanced to clinical trials. Their applications in medicinal chemistry are destined to grow, especially since many of them are now commercially available.

REFERENCES

1. (a) Lovering, F.; Bikker, J.; Humblet, C. *J. Med. Chem.* **2009**, *52*, 6752–6756. (b) Lovering, F. *MedChemComm* **2013**, *4*, 515–519.
2. (a) Zheng, Y. J.; Tice, C. M. *Exp. Opin. Drug Discov.* **2016**, *11*, 831–834. (b) Zheng, Y. J.; Tice, C. M.; Singh, S. B. *Bioorg. Med. Chem. Lett.* **2014**, *24*, 3673–3682.
3. Maryanoff, B. E.; O'Neill, J. C.; McComsey, D. F.; Yabut, S. C.; Luci, D. K.; Jordan, A. D.; Masucci, J. A.; Jones, W. J.; Abad, M. C.; Gibbs, A. C.; et al. *ACS Med. Chem. Lett.* **2011**, *2*, 538–543.
4. Lachance, N.; Gareau, Y.; Guiral, S.; Huang, Z.; Isabel, E.; Leclerc, J.-P.; Leger, S.; Martins, E.; Nadeau, C.; Oballa, R. M.; et al. *Bioorg. Med. Chem. Lett.* **2012**, *22*, 980–984.
5. Johansson, A.; Löfberg, C.; Antonsson, M.; von Unge, S.; Hayes, M. A.; Judkins, R.; Ploj, K.; Benthem, L.; Linden, D.; Brodin, P.; et al. *J. Med. Chem.* **2016**, *59*, 2497–2511.
6. Kwong, A. D.; Kauffman, R. S.; Hurter, P.; Mueller, P. *Nat. Biotechnol.* **2011**, *29*, 993–1003.
7. Bondada, L.; Rondla, R.; Pradere, U.; Liu, P.; Li, C.; Bobeck, D.; McBrayer, T.; Tharnish, P.; Courcambeck, J.; Halfon, P.; et al. *Bioorg. Med. Chem. Lett.* **2013**, *23*, 6325–6330.
8. Henderson, J. L.; Kormos, B. L.; Hayward, M. M.; Coffman, K. J.; Jasti, J.; Kurumbail, R. G.; Wager, T. T.; Verhoest, P. R.; Noell, G. S.; Chen, Y.; et al. *J. Med. Chem.* **2015**, *58*, 419–432.
9. Kung, D. W.; Coffey, S. B.; Jones, R. M.; Cabral, S.; Jiao, W.; Fichtner, M.; Carpino, P. A.; Rose, C. R.; Hank, R. F.; Lopaze, M.G.; et al. *Bioorg. Med. Chem. Lett.* **2012**, *22*, 4281–4287.
10. McClure, K. F.; Jackson, M.; Cameron, K. O.; Kung, D. W.; Perry, D. A.; Orr, S. T.; Zhang, Y.; Kohrt, J.; Tu, M.; Gao, H.; et al. *Bioorg. Med. Chem. Lett.* **2013**, *23*, 5410–5414.
11. Bhattacharya, S. K.; Andrews, K.; Beveridge, R.; Cameron, K. O.; Chen, C.; Dunn, M.; Fernando, D.; Gao, H.; Hepworth, D.; V. Jackson, M.; et al. *ACS Med. Chem. Lett.* **2014**, *5*, 474–479.
12. Keith, J. M.; Jones, W. M.; Pierce, J. M.; Seierstad, M.; Palmer, J. A.; Webb, M.; Karbarz, M. J.; Scott, B. P.; Wilson, S. J.; Wennerholm, M. L.; et al. *Bioorg. Med. Chem. Lett.* **2014**, *24*, 737–741.
13. Shi, W.; Jiang, Z.; He, H.; Xiao, F.; Lin, F.; Sun, Y.; Hou, L.; Shen, L.; Han, L.; Zeng, M.; et al. *ACS Med. Chem. Lett.* **2018**, *9*, 94–97.
14. Guérot, C.; Tchitchanov, B. H. T.; Knust, H.; Carreira, E. M. *Org. Lett.* **2011**, *11*, 780–783.
15. Carreira, E. M.; Fessard, T. C. *Chem. Rev.* **2014**, *114*, 8257–8322.
16. Kirichok, A. A.; Shton, I.; Kliachyna, M.; Pishel, I.; Mykhaiiuk, P. K. *Angew. Chem. Int. Ed.* **2017**, *56*, 8865–8869.
17. Kirichok, A. A.; Shton, I.; Pishel, I.; Zozulya, S. A.; Borysko, P. O.; Kubyshkin, V.; Zaporozhets, O. A.; Tolmachev, A. A.; Mykhaiiuk, P. K. *Chem. Eur. J.* **2018**, *24*, 5444–5449.

28 Spirocyclic Piperidines

Spirocycles have made an appearance in drugs as both core structures and peripheries of the molecules. They are advantageous in several aspects. First and foremost, spirocyclic scaffolds are inherently three-dimensional and can project functionalities in all three dimensions to interact more extensively with the protein target of interest and lower off-target effects. In addition, they are sp^3-carbon-rich with a high fraction of saturated carbon (Fsp^3, defined as equation 28.1) and correlated with favorable physiochemical properties such as higher aqueous solubility.[1] Last but not the least, spirocycles often offer new chemical space to create novel intellectual properties.[2] Recently, spirocyclic piperidines have found impressive utility in Novartis's allosteric SHP2 inhibitors program (*vide infra*).

$$Fsp^3 = \left(\text{number of } sp^3\text{-hybridized carbon}\right)\big/\left(\text{total carbon count}\right) \qquad (28.1)$$

SPIROCYCLIC PIPERIDINE-CONTAINING DRUGS

Only one spirocyclic piperidine-containing drug is currently on the market: Tesaro/Schering–Plough's NK_1 receptor antagonist rolapitant (Varubi, **1**). It was approved by the FDA in 2016, in combination with other antiemetic agents, for the prevention of delayed chemotherapy-induced nausea and vomiting (CINV) associated with initial and repeat courses of emetogenic cancer therapy.[3]

An old antipsychotic fluspirilene (Redeptin, **2**) also possesses a spirocyclic piperidine motif.[4] Fenspiride (Eurespal, **3**), an old antitussive with a spirocyclic piperidine fragment approved in Russia and France, was withdrawn from the market due to QT elongation-associated adverse effects such as torsades de pointes.[5]

rolapitant (Varubi, **1**)
Tesaro/Schering–Plough, 2016
NK_1 receptor antagonist

fluspirilene (Redeptin, **2**)
Janssen, 1963
antipsychotic

fenspiride (Eurespal, **3**)
antitussive

SPIROCYCLIC PIPERIDINES IN DRUG DISCOVERY

The G-protein-coupled receptor 40 (GPR40) is also known as free fatty acid receptor 1 (FFA1). Since endogenous ligand binding to FFA1 regulates the secretion of insulin in pancreatic β-cells. FFA1 agonists have the potential as a treatment of type 2 diabetes mellitus (T2DM). Lilly's FFA1 agonist LY2881835 (**4**), now in phase I clinical trials, has been shown to increase insulin and glucagon-like protein-1 (GLP-1) secretion in preclinical animal models.[6] It contains a spirocyclic piperidine substituent in addition to the 3-phenylpropanoic acid pharmacophore. Toxicity has been associated with the high lipophilicity of earlier FFA1 agonists. Conspicuously, Takeda's greasy fasiglifam (TAK-875) failed phase III clinical trials in 2013 due to idiosyncratic liver toxicity. Krasavin et al. sought to lower the drugs' lipophilicity by adding an oxygen atom to Lilly's spirocyclic piperidine periphery. In due course, they arrived at 1-oxa-9-spiro[5.5]undecane **5**, which not only had similar efficacy to that of LY2881835 (**4**), it also demonstrated excellent aqueous solubility and Caco-2 permeability although it showed somewhat rapid metabolism in the presence of human liver microsomes (HLMs).[7]

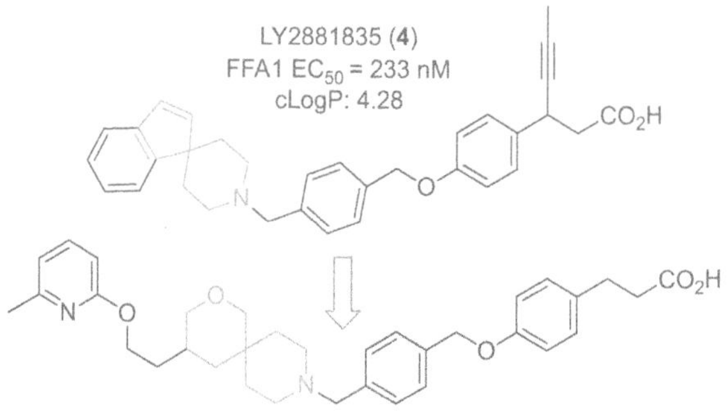

LY2881835 (**4**)
FFA1 EC_{50} = 233 nM
cLogP: 4.28

1-oxa-9-spiro[5,5]undecane **5**
FFA1 EC_{50} = 260 nM
cLogP: 3.73

The same spirocyclic piperidine core structure found utility in soluble epoxide hydrolase (sEH) inhibitors, which may serve as a potential treatment of cardiovascular disease, inflammation, and pain. Employing two existing sEH inhibitors 2,8-diazaspiro[4.5]decane **6** and 1-oxa-4,9-diaza-spiro[5.5]undecane **7**, Lukin and colleagues arrived at a polar spirocyclic scaffold 1-oxa-9-spiro[5.5] undecane-4-amine **8** in an attempt to improve the drug's bioavailability.[8]

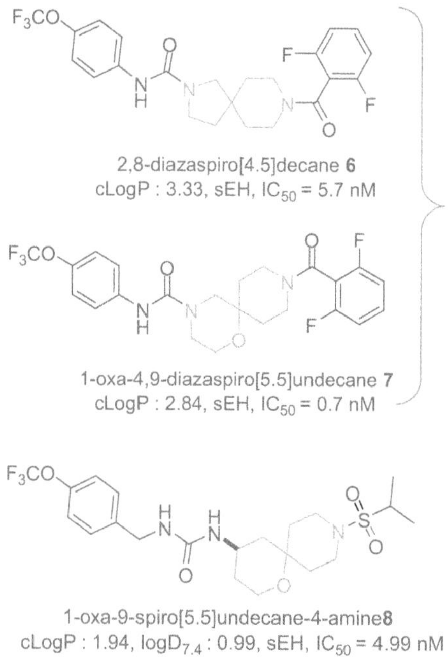

2,8-diazaspiro[4.5]decane **6**
cLogP : 3.33, sEH, IC_{50} = 5.7 nM

1-oxa-4,9-diazaspiro[5.5]undecane **7**
cLogP : 2.84, sEH, IC_{50} = 0.7 nM

1-oxa-9-spiro[5.5]undecane-4-amine **8**
cLogP : 1.94, $logD_{7.4}$: 0.99, sEH, IC_{50} = 4.99 nM

Spirocyclic piperidine amides served as potent, non-peptide inhibitors of human mast cell tryptase. Transforming spirocyclic piperidine-indane **9** to the more polar spirocyclic piperidine-dihydro-benzofuran **10** reduced lipophilicity by nearly one log as measured by cLogP (from 4.36 to 3.40) while still maintaining their potency for tryptase. Both **9** and **10** also have excellent selectivity over trypsin (5,000× and 2,500×, respectively).[9]

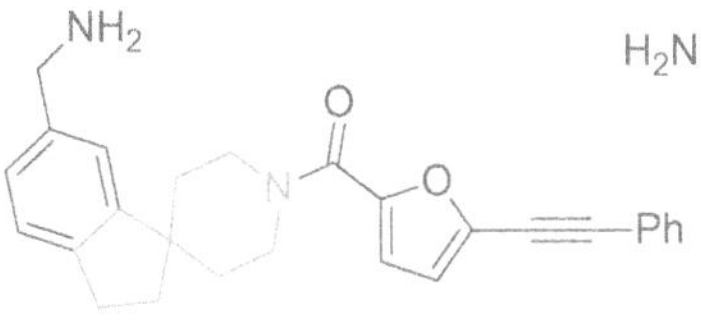

spirocyclic piperidine-indane **9**   spirocyclic piperidine-dihydrobenzofuran **10**
tryptase IC$_{50}$ = 3.1 nM   tryptase IC$_{50}$ = 3.6 nM
cLogP: 4.36   cLogP: 3.40

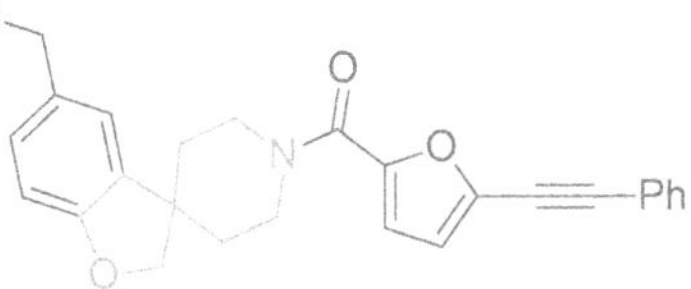

SHP2 (Src homology region 2-containing protein tyrosine phosphatase), an oncogenic, non-receptor protein tyrosine phosphatase and scaffold protein, operates downstream of multiple receptor tyrosine kinases as a positive transducer in numerous oncogenic signaling cascades (e.g., RAS-ERK, PI3K-AKT, and JAK-STAT). Due to its significant potential in cancer treatment, SHP2 inhibitors have been pursued with rigor during the last two decades. Regrettably, earlier SHP2 inhibitors displayed low selectivity, cell permeability, and oral bioavailability, largely due to the highly solvated nature of the catalytic pocket and the polar substituents needed for enhanced binding potency. Novartis made a breakthrough with a new pyrazinyl class of SHP2 allosteric inhibitors exemplified by SHP099 (**11**). It functions as a molecular glue, stabilizing an inactive conformation of concurrent binding to the interface of the N-terminal SH2, C-terminal SH2, and protein tyrosine phosphatase domains. More importantly, SHP099 (**11**) is potent (SHP2 IC$_{50}$ = 70 nM, p-ERK IC$_{50}$ = 250 nM), selective, orally bioavailable, and efficacious in KYSE-520 tumor-bearing nude mouse model.[10a]

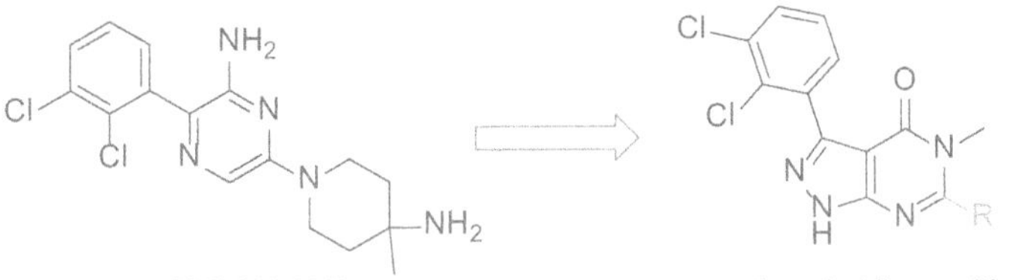

SHP099 (**11**)   pyrrolopyrimidinone **12**

R	SHP2 IC$_{50}$ (nM)	p-ERK IC$_{50}$ (nM)	Antiproliferation IC$_{50}$ (μM)	cLogP/LipE	hERG IC$_{50}$ (nM)
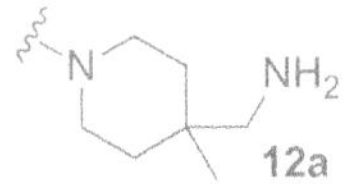 **12a**	34	355	13.49	3.1/3.1	980
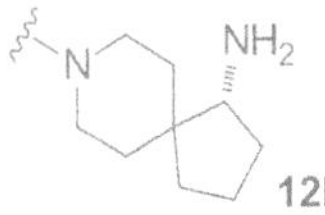 **12b**	31	30	0.465	3.4/3.2	250
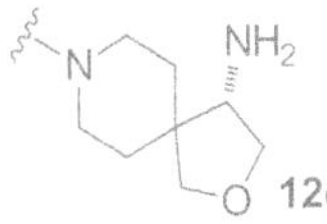 **12c**	50	123	1.73	2.3/4.7	930
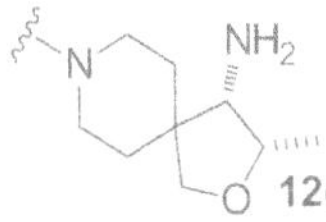 **12d**	28	12	0.167	2.9/4.0	290

Improving upon SHP099 (**11**) via scaffold morphing resulted in many diverse series. Among them, pyrrolopyrimidinone **12** was explored with regard to the SAR of the piperidine-amine moiety. Extending **11**'s basic piperidine-amine by one methylene to displace the proposed water gave **12a** with a slight improvement in biochemical assay and human ether-a-go-go (hERG) potassium channel activity selectivity. Yet, little improvement was observed for its *p*-ERK or KYSE proliferation cellular assays. Cyclization to spiro[4.5]-amine led to **12b** with no boost of biochemical potency but provided >10-fold improvement in the *p*-ERK or KYSE proliferation cellular assays. Tetrahydrofuran-fused piperidine was designed to attenuate the basicity of the amine, the resulting **12c** lost certain potency for all three assays but its lipophilicity efficiency (LipE) was improved since it had a one log reduction in lipophilicity. The expected lipophilicity and basicity offered by **12c** produced only a minor benefit to hERG inhibition. Finally, in an attempt to restore cellular potency by rebalancing lipophilicity, the spirocyclic ether was methylated and the resulting **12d** was bestowed with improved cellular potency in both *p*-ERK and KYSE proliferation cellular assays.[10b]

Although not thoroughly profiled, the following spirocyclic amines were investigated for their biochemical activities by the Novartis team. Analogs **12e** with a spirocyclic pipridine-cyclobutylamine and **12f** with a spirocyclic pipridine-cyclohexylamine were both somewhat less potent than **12d**.[11]

pyrrolopyrimidinone **12e**
SHP2 IC_{50} = 44 nM

pyrrolopyrimidinone **12f**
SHP2 IC_{50} = 72 nM

SYNTHESIS OF SOME SPIROCYCLIC PIPERIDINES

Since spirocyclic piperidines can be readily installed onto the core structure via S_NAr reactions to make **12b**–**12f**, only their syntheses are summarized here.

Preparation of spirocyclic piperidine-cyclobutylamine **16** began with condensation between cyclobutanone **13** and the (*R*)-2-*t*-butyl-2-sulfinamide (**14**) with the aid of titanium isopropoxide. The resulting imine was reduced with lithium borohydride to afford (*S*)-spirocyclic amine **15** and its Boc protective group was readily removed using TFA in methylene chloride to produce **16**. The chiral auxiliary on **16** was then removed under alcoholic acidic conditions to expose the free spirocyclic bis-amine **17**.[11]

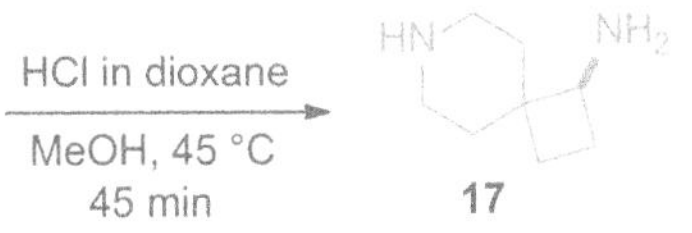

The spirocyclic piperidine fragment of **12f** is 3-azaspiro[5.5]undecan-7-amine (**22**). Since it is racemic, the auxiliary does not have to be chiral. Thus, condensation of cyclohexanone **18** with 2-*t*-butyl-2-sulfinamide (**19**) was followed by a reduction to provide **20**. Removal of the Boc group on **20** gave **21**, which was treated with HCl in methanol to deliver **22**.[11]

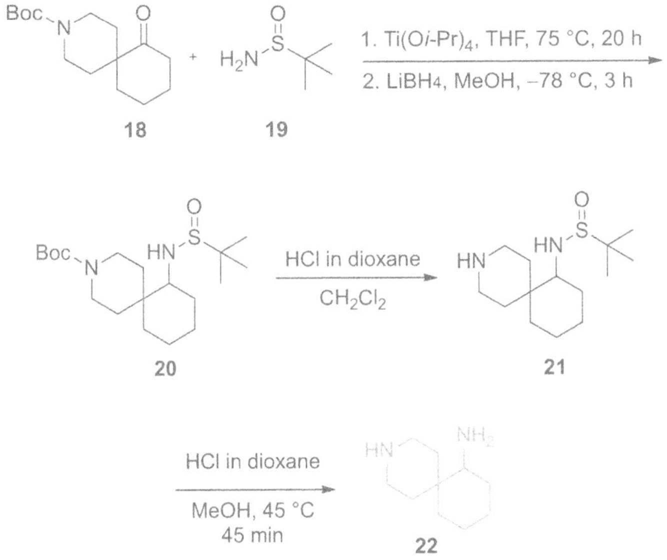

Preparation of the spirocyclic amine **26** on **12c** commenced with the Dess–Martin oxidation of alcohol **23** to afford ketone **24**. Installation of Ellman's chiral auxiliary (*R*)-2-*t*-butyl-2-sulfinamide (**14**) was followed by reduction of the resulting imine with LiBH₄ in methanol to give intermediate **25**. Acidic removal of the sulfonamide chiral auxiliary then led to spirocyclic bis-amine **26**.[11]

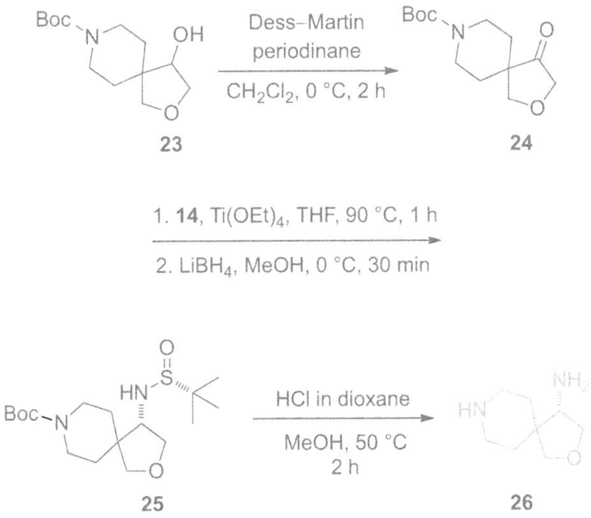

The synthesis of spirocyclic bis-amine **32** is relatively lengthy since there are two chiral centers. LDA lithiation of piperidine ester **27** was followed by quenching the resulting enolate with chiral aldehyde **28** to assemble adduct **29**. Although the hydroxyl group was generated without selectivity, it was inconsequential since it would be oxidized later on. Thus, compound **29** was reduced to give a diol. After removal of the silyl protective group, the new alcohol was converted to the corresponding tosylate and a concurrent S_N1 displacement reaction led to alcohol **31**. The Dess–Martin oxidation of **31** was followed by installation of the Ellman's chiral auxiliary (**14**), reduction, and removal of the sulfonamide to deliver **32**.[11]

In summary, spirocyclic scaffolds in general, spirocyclic piperidines in particular, have several advantages over the sp^2-carbon-rich, flat structures. Their 3-D trajectory offers more points of contact with the protein target of interest and can give rise to more potent and selective drugs. The sp^3-carbon-rich molecules are also blessed with superior physiochemical properties such as higher aqueous solubility. Finally, unexplored new spirocyclic structures can create novel intellectual property spaces. The success of Novartis's allosteric SHP2 inhibitors is a testimony to the utility of spirocyclic scaffolds. Their applications in medicinal chemistry are destined to grow rapidly, especially since many of them are now commercially available.

REFERENCES

1. (a) Lovering, F.; Bikker, J.; Humblet, C. *J. Med. Chem.* **2009**, *52*, 6752–6756. (b) Lovering, F. *MedChemComm* **2013**, *4*, 515–519.
2. (a) Zheng, Y. J.; Tice, C. M. *Exp. Opin. Drug Discov.* **2016**, *11*, 831–834. (b) Zheng, Y. J.; Tice, C. M.; Singh, S. B. *Bioorg. Med. Chem. Lett.* **2014**, *24*, 3673–3682.
3. Syed, Y. Y. *Drugs* **2016**, *75*, 1941–1945.
4. Awouters, F. H. L.; Lewi, P. J. *Arzneimittel-Forsch* **2007**, *57*, 625–632.
5. Shmelev, E. I.; Kunicina, Y. L. *Clin. Drug Investig.* **2006**, *26*, 151–159.
6. Hamdouchi, C.; Kahl, S. D.; Patel Lewis, A.; Cardona, G. R.; Zink, R. W.; Chen, K.; Eessalu, T. E.; Ficorilli, J. V.; Marcelo, M. C.; Otto, K. A.; et al. *J. Med. Chem.* **2016**, *59*, 10891–10916.
7. Krasavin, M.; Lukin, A.; Bagnyukova, D.; Zhurilo, N.; Golovanov, A.; Zozulya, S.; Zahanich, I.; Moore, D.; Tikhonova, I. G. *Eur. J. Med. Chem.* **2017**, *127*, 357–368.

8. Kato, Y.; Fuchi, N.; Saburi, H.; Nishimura, Y.; Watanabe, A.; Yagi, M.; Nakadera, Y.; Higashi, E.; Yamada, M.; Aoki, T. *Bioorg. Med. Chem. Lett.* **2013**, *23*, 5975–5979. (b) Kato, Y.; Fuchi, N.; Nishimura, Y.; Watanabe, A.; Yagi, M.; Nakadera, Y.; Higashi, E.; Yamada, M.; Aoki, T.; Kigoshi, H. *Bioorg. Med. Chem. Lett.* **2014**, *24*, 565–570. (c) Lukin, A.; Kramer, J.; Hartmann, M.; Weizel, L.; Hernandez-Olmos, V.; Falahati, K.; Burghardt, I.; Kalinchenkova, N.; Bagnyukova, D.; Zhurilo, N.; et al. *Bioorg. Chem.* **2018**, *80*, 655–667.

9. Costanzo, M. J.; Yabut, S. C.; Zhang, H.-C.; White, K. B.; de Garavilla, L.; Wang, Y.; Minor, L. K.; Tounge, B. A.; Barnakov, A. N.; Lewandowski, F.; et al. *Bioorg. Med. Chem. Lett.* **2008**, *18*, 2114–2121.

10. Bagdanoff, J. T. Garcia Fortanet, J.; Chen, C. H.-T.; Chen, Y.-N. P.; Chen, Z.; Deng, Z.; Firestone, B.; Fekkes, P.; Fodor, M.; Fortin, P. D.; et al. *J. Med. Chem.* **2016**, *59*, 7773–7782. (b) Acker, M.; Chen, Y.-N.; Chan, H.; Dore, M.; Firestone, B.; Fodor, M.; Fortanet, J.; Hentemann, M.; Kato, L. *J. Med. Chem.* **2019**, *62*, 1781–1792. (c) Sarver, P.; Acker, M.; Bagdanoff, J. T.; Chen, Z.; Chen, Y.-N.; Chan, H.; Firestone, B.; Fodor, M.; Fortanet, J.; Hao, H.; et al. *J. Med. Chem.* **2019**, *62*, 1793–1802.

11. Bagdanoff, J. T.; Chen, Z.; Dore, M.; Fortanet, J. G.; Kato, M.; Lamarche, M. J.; Sarver, P. J.; Shultz, M.; Smith, T. D.; Williams, S. WO2016203404A1 (2016).

29 Spirocyclic Pyrrolidines

Spirocyclic pyrrolidines combine the characteristics of both pyrrolidines and spirocycles. For pyrrolidine itself, its N atom may serve as a hydrogen bond acceptor to interact with the drug target protein. Meanwhile, it may offer enhanced aqueous solubility and other physiochemical properties.

Spirocyclization has been taken advantage of to provide rigidification of floppy molecules. Spirocyclic pyrrolidines may be considered a bioisostere of pyrrolidines, but with a three-dimensional (3-D) geometry. Like most 3-D structures, spirocyclic pyrrolidines have three common benefits in comparison to their flat counterparts. (a) Spirocyclic scaffolds project functionalities in all three dimensions to interact more extensively with the drug target proteins of interest, leading to higher potency and fewer off-target effects; (b) they may be associated with more favorable physiochemical properties such as higher aqueous solubility; and (c) novel intellectual properties may be obtained.

SPIROCYCLIC PYRROLIDINE-CONTAINING DRUGS

Three spirocyclic pyrrolidine-containing drugs are currently on the market. One of the early ones is Sandoz's angiotensin-converting enzyme (ACE) inhibitor spirapril (Renormax, **1**), approved by the FDA in 1995 for the treatment of hypertension.[1]

Daiichi Sankyo's sitafloxacin (Gracevit, **2**) contains a cyclopropane-fused spiropyrrolidine fragment. It is a fluoroquinolone antibiotic only sold in Japan.[2]

Gilead has been a leader in the hepatitis C virus (HCV) therapeutics field. Its HCV non-structural protein 5A (NS 5A) inhibitor ledipasvir (**3**) is combined with sofosbuvir and was approved for marketing in 2014 with the brand name Harvoni.[3]

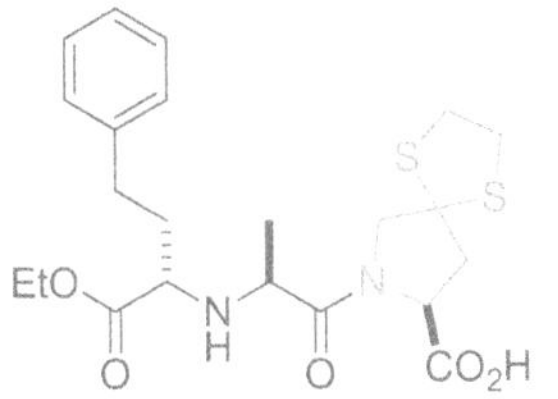

spirapril (Renormax, **1**)
Sandoz, 1995
ACE inhibitor

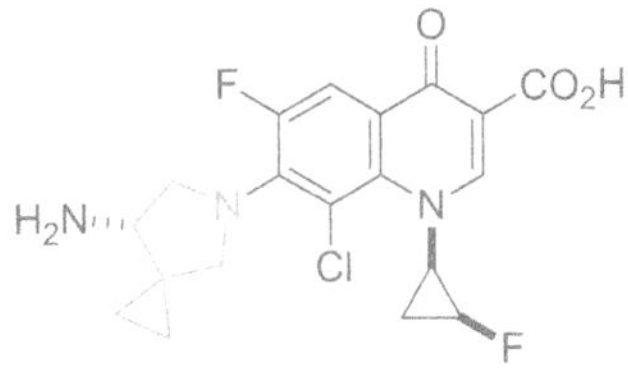

sitafloxacin (Gracevit, **2**)
Daiichi Sankyo, 2008 in Japan
fluoroquinolone antibiotic

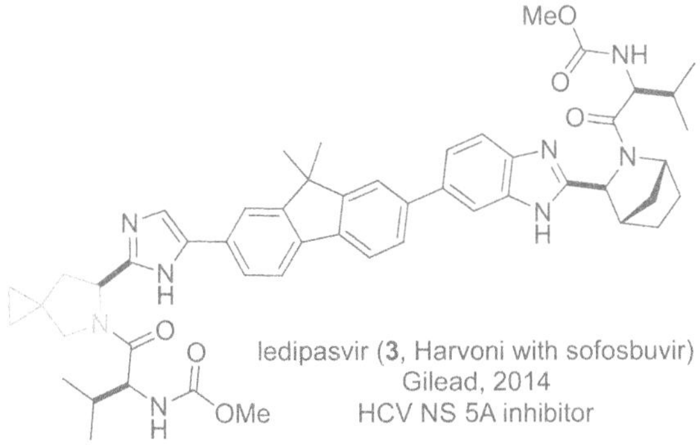

ledipasvir (**3**, Harvoni with sofosbuvir)
Gilead, 2014
HCV NS 5A inhibitor

SPIROCYCLIC PYRROLIDINES IN DRUG DISCOVERY

Some spirocyclic pyrrolidines have been employed as three-dimensionally extended pyrrolidines. Merck's ACE inhibitor lisinopril (Zestril, **4**, FDA approval, 1987) was significantly more potent than Squibb's first-in-class ACE inhibitor captopril (Capten, FDA approval, 1980). Lisinopril (**4**) occupies the S_1 hydrophobic pocket on the ACE enzyme not taken advantage of by the captopril. Sandoz scientists took one step further to explore additional hydrophobic binding pockets. They achieved success with spirapril (**1**) by installing a spirocyclic proline where the dithioketal extended deeper to the S_2' pocket. Apparently, it offered more extensive binding to the enzyme because spirapril (**1**)'s *in vitro* inhibition was 80-fold more potent than lisinopril (**4**). This has also translated to *in vivo* efficacy where spirapril (**1**) was tested 3.5-fold more efficacious than lisinopril (**4**) in rat and anesthetized dog models.[2]

lisinopril (Zestril, **4**)
ACE *in vitro*, IC_{50}, 1273 nM
in vivo, ID_{50}, 57 µg/kg

spirapril (Renormax, **1**)
ACE *in vitro*, IC_{50}, 16 nM
in vivo, ID_{50}, 16 µg/kg

ciprofloxacin (Cipro, **5**)
Bayer, 1987
fluoroquinolone antibiotic

sitafloxacin (Gracevit, **2**)
bacterial DNA gyrase and
topoisomerase IV inhibitor

tosufloxacin (Ozex, **6**)
Nichi-iko, 2000, Japan only
fluoroquinolone antibiotic

Drug discovery is extremely challenging for first-in-class drugs. It is relatively easier when it comes to "me-too" drugs, especially if you know what you are doing.

The genesis of quinolone antibiotics traces back to Sterling–Winthrop's nalidixic acid (Nevigramon). Bayer's ciprofloxacin (Cipro, **5**), a second-generation quinolone antibiotic, is likely the "best-in-class" drug, possibly because its cyclopropyl motif helps boosting its bioavailability and subsequently efficacy. The third-generation quinolone antibiotic tosufloxacin (Ozex, **6**) contains a 3-amino-pyrrolidine moiety. Sandoz's sitafloxacin (**2**) incorporates features of both ciprofloxacin

(**5**)'s cyclopropane and tosufloxacin (**6**)'s 3-amino-pyrrolidine. If one cyclopropyl group is good, two is probably even better. Sandoz decided to add a spirocyclic cyclopropyl fragment to the 3-amino-pyrrolidine group and arrived at sitafloxacin (**2**). Like all fluoroquinolones, sitafloxacin (**2**)'s mechanism of action (MoA) is through inhibiting bacterial DNA gyrase and topoisomerase IV.[2]

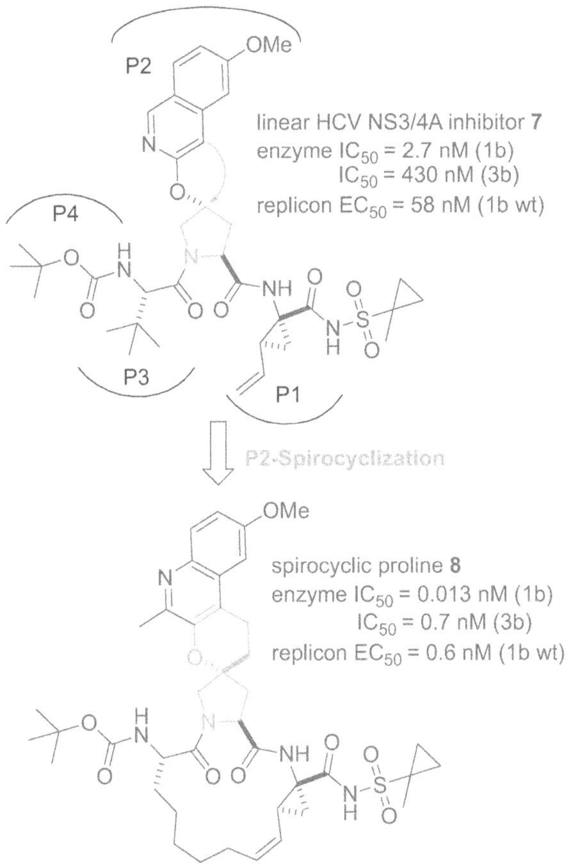

Two HCV NS3/4A serine protease inhibitors are on the market. One is Schering–Plough/Merck's boceprevir (Victrelis, 2011), and the other is Vertex's telaprevir (Incivek, 2011). Building on the success of boceprevir, Schering–Plough/Merck scientists pursued superior backup HCV NS3/4A inhibitors. Linear molecule **7** is a reasonably good HCV NS3/4A inhibitor. However, it can adopt many conformations as an open-chain molecule. Restraining the molecule would significantly rigidify the molecule and limit the number of conformations binding to the enzyme.[4]

Through molecular modeling around the P2 region, it was observed that a spirocyclization of quinolone moiety of **7** onto the proline could make favorable van der Waals contact with histidine (H57), an invariant catalytic residue, which may result in an improved mutant profile. *Spirocyclization would also impart greater conformational rigidity to the molecule and may provide an advantage with reduced entropic cost of binding by biasing it toward the bioactive conformation.* Indeed, the resulting spirocyclic proline **8** was tested significantly more potent in both enzymatic assay and replicon efficacy assay. More important, minor optimization led to the discovery of MK-8831 (**44**, *vide infra*), a novel spiro-proline macrocycle as a pan-HCV NS3/4A protease inhibitor with a good pharmacokinetic profile and excellent safety. MK-8831 (**44**) was carried to phase I clinical trials in 2015.[4] This tactic has been employed to discover additional P2–P4 macrocycles containing this unique spirocyclic proline core structure.[5]

spiroindole **9**
Sky IC_{50} = 495 nM
solubility = 5.3 μM
$t_{1/2}$ = 3.1 h
Cl = 60 mL/min/kg
V_{dss} = 13 L/kg
F% = 14%
HLM $t_{1/2}$ = 1 h

A group of my former colleagues at Pfizer in Ann Arbor discovered a series of novel and selective spiroindole-based inhibitors of Sky, a tyrosine kinase receptor. For example, spiroindole **9** binds in the ATP-binding site and exhibits a high level of kinome selectivity through filling the Ala571-subpocket. In addition, the nitrogen on the pyrrolidine sandwiched between indoline and the top pyrrolidine also forms two hydrogen bonds with Asp663 on the target Sky protein. In all, the spirocyclic pyrrolidine motif provides at least two advantages here: (a) The 3-D geometry offers selectivity for the target Sky protein; (b) The nitrogen atom on pyrrolidine serves as a hydrogen bond acceptor. Spiroindole **9** exhibits moderate bioavailability in the rat due to low absorption across the gut wall.[6]

Another Pfizer project also involved spirocyclic pyrrolidines: it was the β-secretase (BACE1) inhibitors project employing the fragment-based drug design (FBDD) technique. An X-ray-based fragment screen of Pfizer's proprietary fragment library of 340 compounds identified spiropyrrolidine **10** as the only BACE binder. The nitrogen atom on pyrrolidine forms two hydrogen bonds with Asp32 and Asp228, respectively, on the BACE enzyme. Despite exhibiting only weak inhibitory activity against the BACE enzyme, spiropyrrolidine hit **10** was verified by biophysical and NMR-based methods as a genuine BACE inhibitor.[7]

Identification of a viable vector led to extension at the 5′ position of the pyrrolidine fragment. Guided by structure-based drug design (SBDD), computational prediction of physiochemical properties, installation of a series of esters gave rise to the optimized spiropyrrolidine inhibitor **11**, which was an approximately 1000-fold improvement in potency over fragment **10**. It also has a high ligand efficiency (LE) and properties predictive of good permeability and low P-gp liability.[7]

spiropyrrolidine hit **10**
BACE1, K_D = 1.4 mM

spiropyrrolidine inhibitor **11**
BACE1, IC_{50} = 4 μM

P2Y$_1$ antagonist **12**
K_i = 4.3 nM
IC_{50} = 4.9 μM
HLM, 89%
Cl = 5.3 mL/min/kg
V_{dss} = 3.1 L/kg
$t_{1/2}$ = 9.7 h
F% = 55%

Blockbuster antiplatelet drug clopidogrel (Plavix) works as a P2Y$_{12}$ antagonist. Similarly, P2Y$_1$ antagonists have the potential as blood thinners as well. BMS scientists used a high throughput screen (HTS) hit, a urea-containing compound, as their starting point and carried out an extensive drug discovery campaign. The fruit of their labor was spirocyclic pyrrolidine **12** as a potent P2Y$_1$ antagonist with good *in vitro* binding and functional activities. Moreover, spirocyclic pyrrolidine **12** demonstrated a favorable PK profile and was the first P2Y$_1$ antagonist that inhibited arterial and venous thrombosis at doses that produced a limited prolongation of bleeding time via oral as well as IV dosing.[8]

Proteolysis targeting chimera (PROTAC) functions as a protein degrader. It uses hetero-bifunctional small molecules to remove specific proteins from cells to achieve targeted protein degradation. Wang and coworkers recently published a spirocyclic pyrrolidine PROTAC MD-224 (**13**). It links a murine double minute 2 (MDM2)–p53 inhibitor on the left and immunomodulatory drug (IMiD) lenalidomide as E3 ligase binders on the right, as a result of careful linker optimization. The CRL4–CRBN E3 ubiquitin ligase on **13** degraded MDM2, but co-treatment of **13** with lenalidomide, a cereblon (CRBN) binder, effectively blocked MDM2 degradation via competitive displacement of cereblon from the ternary complex, confirming the drug was on target. PROTAC **13** was tested as a nanomolar drug in cells and efficacious in RS4;11 xenograft animal models when given multiple IV-dosing at 25 mg/kg every second day (Q2D).[9]

MD-224 (**13**)
RS4;11 cell growth inhibit. IC$_{50}$ = 1.5 nM
RS4;11 xenograft = 50% regression
at 25 mpk, iv Q2D dosing

SYNTHESIS OF SOME SPIROCYCLIC PYRROLIDINE-CONTAINING DRUGS

For the synthesis of Daiichi's sitafloxacin (**2**), the key intermediate is a cyclopropane-fused spirocyclic pyrrolidine **22**. Initially, fragment **22** was prepared via a chiral auxiliary [(*R*)-1-phenylethan-1-amine]-aided separation. Later on, a more efficient and greener route was developed via asymmetric microbial reduction. As shown beneath, N-benzyl glycine (**14**) was protected as its Boc analog **15**. After treating **15** with carbonyldiimidazole, the intermediate was exposed to the magnesium enolate of hydrogen ethyl malonate to afford the corresponding β-keto-γ-amino ester **16** in 84% yield. Cyclopropanation of **16** with 1,2-dibromoethane and potassium carbonate in acetone gave **17** in 72% yield. Treatment of **17** with TFA was followed by refluxing in toluene produced the key intermediate **18** in 72% yield.[10]

With substrate keto-lactam **18** in hand, optimization of the microbial reduction conditions led to the use of *Phaeocreopsis* sp. JSM 1880, giving rise to chiral alcohol **19**. Subsequently, a Mitsunobu reaction using DPPA afforded the corresponding azide, which was reduced to amine **20**. After protection, the resulting Boc analog **21** was subjected to hydrogenolysis to deliver cyclopropyl-pyrrolidine **22**, which was used to prepare sitafloxacin (**2**).[10]

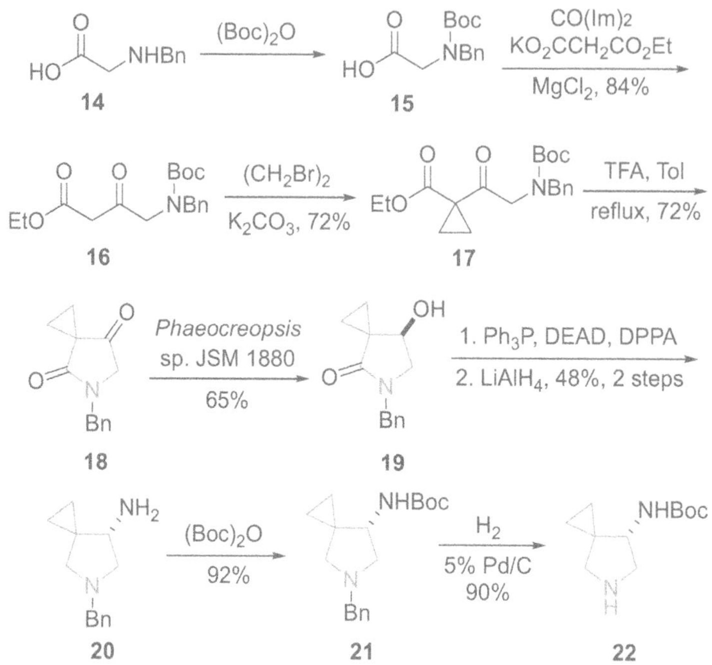

For the synthesis of Gilead's HCV NS 5A inhibitor ledipasvir (**3**), here we only focus on the synthesis of the closely related cyclopropyl-proline portion **28**. Thus, diol **23** was iodinated using the routine combination of iodine, triphenylphosphine, and imidazole to prepare di-iodide **24**. Spirocyclopropyl-proline **26** was assembled as a racemate treating N-Boc-glycine ethyl ester (**25**) with sodium hydride followed by di-iodide **24**. Saponification of ester **26** followed by a classical resolution with (1*S*,2*R*)-aminoindanol provided enantiomerically pure salt **27** after recrystallization in 2-methyltetrahydrofuran. After the liberation of the free carboxylic acid, treatment with potassium *tert*-butoxide produced enantiopure potassium **28**, ready to be used as a building block to synthesize ledipasvir (**3**).[11]

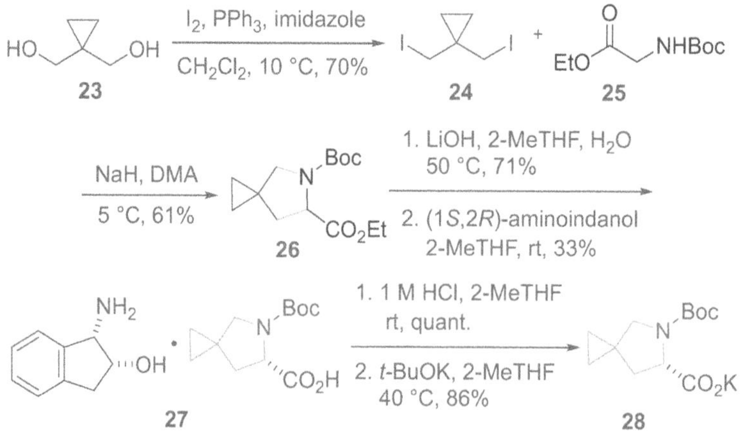

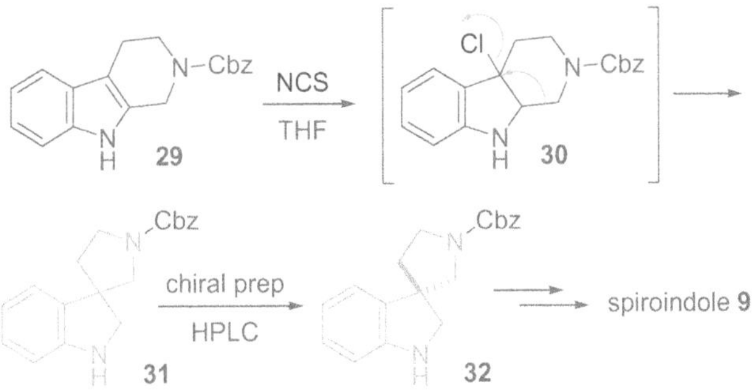

Synthesis of spiroindole **9**, Pfizer's Sky kinase inhibitor, resorted to an interesting rearrangement reaction. When treated with NCS, tetrahydrocarboline **29** was chlorinated at its 3-position to give chloride **30**. Subsequently, unstable chloride **30** underwent a rearrangement to afford spirocyclic pyrrolidine **31** as a racemate, which was separated by chiral SFC to give enantiomerically pure **32**.[6]

Oxindoles are privileged structures in drug discovery. Pfizer's BACE1 inhibitor spiropyrrolidine **11** contains a spirocyclic oxindolyl-pyrrolidine moiety. Preparation **11** also resorted the fascinating rearrangement reaction employed to make spirocyclic pyrrolidine **35** for another Pfizer project on Sky kinase inhibitors. Substrate tetrahydrocarboline **33** was treated with NBS in acetic acid. The NBS-induced oxidative rearrangement, via the intermediacy of **34**, furnished the desired spirocyclic pyrrolidine intermediate as a mixture of diastereomers **35** and **35′**. The diastereoselectivity favored the desired diastereomer **35**, which was readily separated via normal flash chromatography and eventually converted to the desired spiropyrrolidine **11** as a potent and bioavailable BACE1 inhibitor.[7]

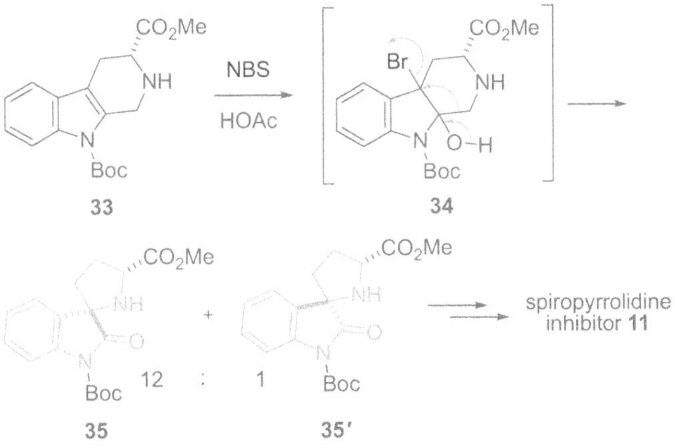

BMS's P2Y$_1$ antagonist **12** contains a spirocyclic piperidinyl indoline core structure **40**. Assembly of the key intermediate **40** began with treating 2-fluorobenzeneacetonitrile (**36**) with bis-chloride **37** to prepare piperidine **38**. After deprotection of the Boc group, the resulting nitrile **39** was reduced by LiAlH(OEt)$_3$, freshly prepared by LiAlH$_4$ and HOEt in ethylene glycol and dimethyl ether, to effect spirocyclization, giving rise to spirocyclic piperidinyl indoline **40**.[8]

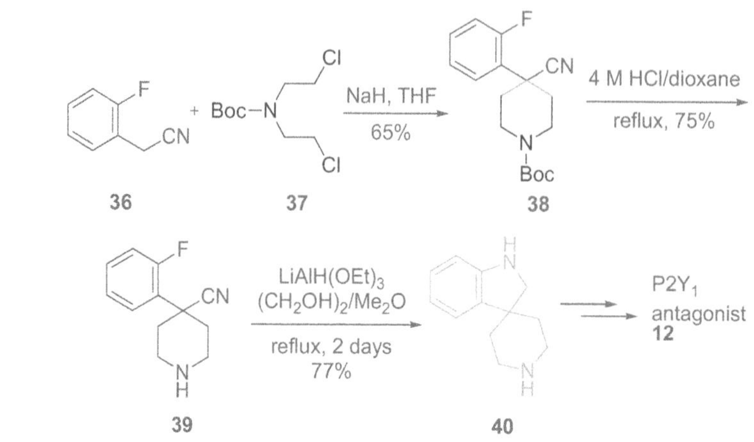

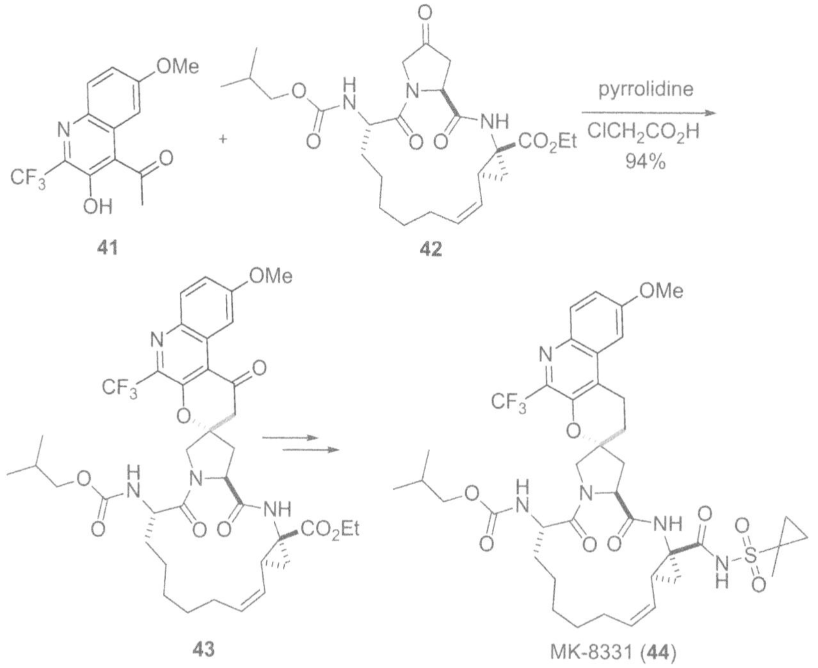

MK-8831 (**44**), a novel spiro-proline macrocycle as a pan-HCV NS3/4A protease inhibitor, evolved from spiro-proline **8**. Spirocyclization was carried out by condensing keto-phenol **41** with keto-proline derivative **42** to assemble the spirocyclic pyranone-proline **43**. Several steps of further manipulations then delivered MK-8831 (**44**).[12]

In summary, spirocyclic pyrrolidines combine the characteristics of both pyrrolidines and spirocycles. The nitrogen atom on the pyrrolidine may serve as a hydrogen bond acceptor, interacting with the target protein. The pyrrolidine motif on a drug may also offer enhanced aqueous solubility and other physiochemical properties. Meanwhile, spirocyclic pyrrolidines, like most spirocycles, are bestowed with the three advantages over their flat counterparts: (a) inherent three-dimensional geometry offers tighter interactions with the target protein and thus they are more potent and selective with fewer off-target effects; (b) potential superior physiochemical properties; and (c) possible novel intellectual properties.

REFERENCES

1. Smith, E. M.; Swiss, G. F.; Neustadt, B. R.; McNamara, P.; Gold, E. H.; Sybertz, E. J.; Baum, T. *J. Med. Chem.* **1989**, *32*, 1600–1666.
2. Kimura, Y.; Atarashi, S.; Kawakami, K.; Sato, K.; Hayakawa, I. *J. Med. Chem.* **1994**, *37*, 3344–3352.
3. Link, J. O.; Taylor, J. G.; Xu, L.; Mitchell, M.; Guo, H.; Liu, H.; Kato, D.; Kirschberg, T.; Sun, J.; Squires, N.; et al. *J. Med. Chem.* **2014**, *57*, 2033–2046.
4. Neelamkavil, S. F.; Agrawal, S.; Bara, T.; Bennett, C.; Bhat, S.; Biswas, D.; Brockunier, L.; Buist, N.; Burnette, D.; Cartwright, M.; et al. *ACS Med. Chem. Lett.* **2016**, *7*, 111–116.
5. Velazquez, F.; Chelliah, M.; Clasby, M.; Guo, Z.; Howe, J.; Miller, R.; Neelamkavil, S.; Shah, U.; Soriano, A.; Xia, Y.; et al. *ACS Med. Chem. Lett.* **2016**, *7*, 1173–1178.
6. Powell, N. A.; Kohrt, J. T.; Filipski, K. J.; Kaufman, M.; Sheehan, D.; Edmunds, J. E.; Delaney, A.; Wang, Y.; Bourbonais, F.; Lee, D.-Y.; et al. *Bioorg. Med. Chem. Lett.* **2012**, *22*, 190–193.
7. Efremov, I. V.; Vajdos, F. F.; Borzilleri, K. A.; Capetta, S.; Chen, H.; Dorff, P. H.; Dutra, J. K.; Goldstein, S. W.; Mansour, M.; McColl, A.; et al. *J. Med. Chem.* **2012**, *55*, 9069–9088.
8. Qiao, J. X.; Wang, T. C.; Ruel, R.; Thibeault, C.; Lheureux, A.; Schumacher, W. A.; Spronk, S. A.; Hiebert, S.; Bouthillier, G.; Lloyd, J.; et al. *J. Med. Chem.* **2013**, *56*, 9275–9295.
9. (a) Li, Y.; Yang, J.; Aguilar, A.; McEachern, D.; Przybranowski, S.; Liu, L.; Yang, C.; Wang, M.; Han, X.; Wang, S. *J. Med. Chem.* **2019**, *62*, 448–466. (b) Wurz, R. P.; Cee, V. J. *J. Med. Chem.* **2019**, *62*, 445–447.
10. Satoh, K.; Imura, A.; Miyadera, A.; Kanai, K.; Yukimoto, Y. *Chem. Pharm. Bull.* **1998**, *46*, 587–590.
11. Scott, R. W.; Vitale, J. P.; Matthews, K. S.; Teresk, M. G.; Formella, A.; Evans, J. W. U.S. Patent US9056860B2 (2015).
12. Chung, C. K.; Cleator, E.; Dumas, A. M.; Hicks, J. D.; Humphrey, G. R.; Maligres, P. E.; Nolting, A. F.; Rivera, N.; Ruck, R. T.; Shevlin, M. *Org. Lett.* **2016**, *18*, 1394–1397.

30 Spirooxetanes

Spirocyclic oxetanes (spirooxetanes) combine the characteristics of both oxetanes and spirocycles. As a bioisostere of dimethyl carbonyl groups, oxetane is more metabolically stable and lipophilicity neutral.[1] Since oxetane is an electron-withdrawing group, it reduces the basicity of its adjacent nitrogen atom and the subtle modulation of the basicity may lower the drug's overall lipophilicity.

Spirocyclization provides rigidification of floppy molecules. Spirooxetanes[2] may be considered bioisosteres of oxetanes with a three-dimensional (3-D) geometry. Like most 3-D structures, spirooxetanes have a trio of advantages: (a) more extensive interactions with target proteins; (b) more favorable physiochemical properties; and (c) potential novel intellectual properties (IP).

With regard to bicyclic spirooxetanes, the other ring may be fused to oxetane at either the α- or β-position of the oxygen atom. The majority of examples in the literature are the β-fused spirooxetanes.

β-fused spirooxetanes   α-fused spirooxetanes

SPIROOXETANES IN DRUG DISCOVERY

Only one spirooxetane-containing drug AZD1979 (**10**), a melanin-concentrating hormone receptor 1 (MCHr1) antagonist, has advanced to clinical trials for the treatment of diabetes (*vide infra*).

Pharmasset/Gilead's mega-blockbuster drug sofosbuvir (Sovaldi, **2**) is a phosphoramidate pro-drug of ribonucleoside PSI-6130 (**1**).[3] As a hepatitis C virus (HCV) NS5B RNA-dependent RNA polymerase (RdRp) inhibitor, sofosbuvir (**2**) has contributed significantly to the cure of HCV either as a monotherapy or as an ingredient of combination drugs.

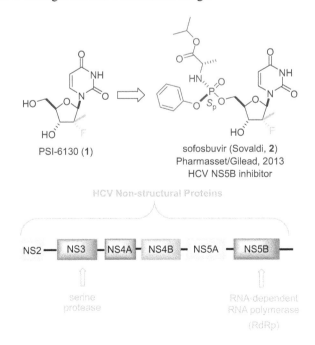

PSI-6130 (1)

sofosbuvir (Sovaldi, 2)
Pharmasset/Gilead, 2013
HCV NS5B inhibitor

HCV Non-structural Proteins

NS2 — NS3 — NS4A — NS4B — NS5A — NS5B —

serine
protease

RNA-dependent
RNA polymerase
(RdRp)

The fact that both PSI-6130 (**1**) and sofosbuvir (**2**) possess a tertiary 2′-carbon on the ribose ring suggests that sufficient space exists within the NS5B active site to accommodate an additional atom especially if it resides within a constrained ring system. Pharmasset was inspired to explore novel 2′-spirocyclic ethers and this exercise led to the discovery of 2′-spirooxetanyl-nucleoside **3**, which was moderately active *in vitro* with an HCV replicon EC_{50} value of 56.6 μM and minimal cytotoxicity with a $CC_{50} > 100$ μM. As in the case of sofosbuvir (**2**), the corresponding phosphoramidate prodrug **4** has ten-fold more potent anti-HCV activity *in vitro* with an EC_{50} value of 16.7 μM and an excellent resistance profile.[4] As a side, *in vitro* activities for both drugs **3** and **4** are measured by the active 2′-deoxy-2′-spirooxetane uridine triphosphate (TP) **6** (*vide infra*).

Great minds think the same. Janssen came up exactly with the same 2′-deoxy-2′-spirooxetane ribonucleoside **3** as a novel inhibitor of HCV NS5B polymerase.[5] Extensive prodrug exploration led to the discovery of JNJ-54257099 (**5**), a 3′-5′-cyclic phosphate ester prodrug.[6] The *in vitro* anti-HCV activities of JNJ-54257099 (**5**) are not given here because it is inactive *in vitro*. Compounds **3**, **4**, and **5** may be considered as prodrugs of 2′-deoxy-2′-spirooxetane uridine TP **6**, which is the *bona fide* actual active drug.

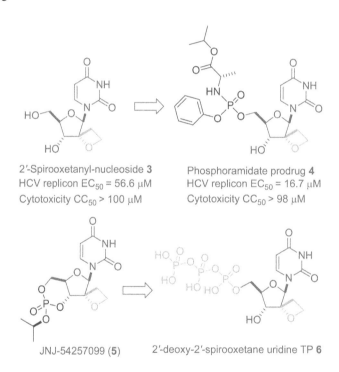

2′-Spirooxetanyl-nucleoside **3**
HCV replicon EC_{50} = 56.6 μM
Cytotoxicity CC_{50} > 100 μM

Phosphoramidate prodrug **4**
HCV replicon EC_{50} = 16.7 μM
Cytotoxicity CC_{50} > 98 μM

JNJ-54257099 (**5**)   2′-deoxy-2′-spirooxetane uridine TP **6**

Checkpoint kinase 1 (CHK1) plays a key role in the DNA damage response, facilitating cell-cycle arrest to provide sufficient time for lesion repair. This leads to the hypothesis that inhibition of ChK1 might enhance the effectiveness of DNA-damaging therapies in the treatment of cancer. Many of Genentech's 1,7-diazacarbazole ChK1 inhibitors, such as piperidine **7**, suffer from a lack of selectivity against acetylcholine esterase (AChE) and are unsuitable for development. Efforts to mitigate AChE activity led to the discovery of spirooxetane **8** as a selective, orally bioavailable ChK1 inhibitor offering excellent *in vitro* potency with significantly reduced AChE activity.[7]

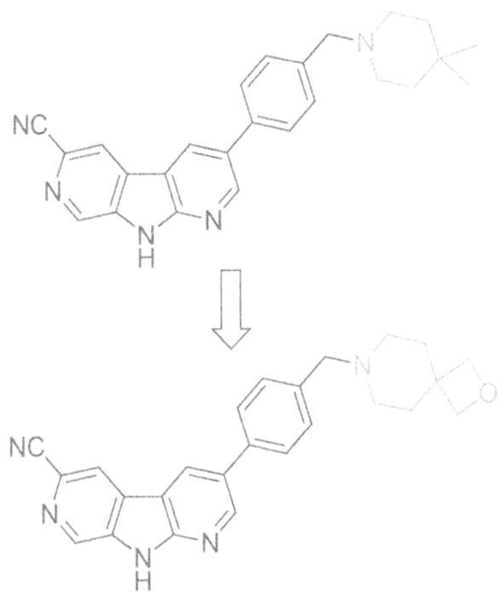

Piperidine **7**
ChK1 IC$_{50}$ = 4.4 nM
AChE IC$_{50}$ = 11 nM
Margin = 2.5-fold

Spirooxetane **8**
ChK1 IC$_{50}$ = 2.5 nM
AChE IC$_{50}$ = 2,420 nM
Margin = 968-fold

Melanin-concentrating hormone receptor 1 (MCHr1) antagonists have been explored for weight control. AstraZeneca discovered a series of oxadiazole-containing compounds as represented by morpholine **9**. Their SAR efforts culminated in clinical candidate AZD1979 (**10**) with a novel peripheral spirooxetane moiety. Its peripheral spirooxetane (2-oxa-6-azaspiro[3.3]heptane) spirocycle is an isostere of morpholine. It displayed appropriate lipophilicity for a CNS indication, showed excellent permeability with no efflux, and possessed good off-target selectivity, including human ether-a-go-go (hERG) potassium channel activity. Preclinical good-laboratory practice (GLP) toxicology and safety pharmacology studies were without findings and AZD1979 **10** was taken into clinical trials for the treatment of diabetes.[8]

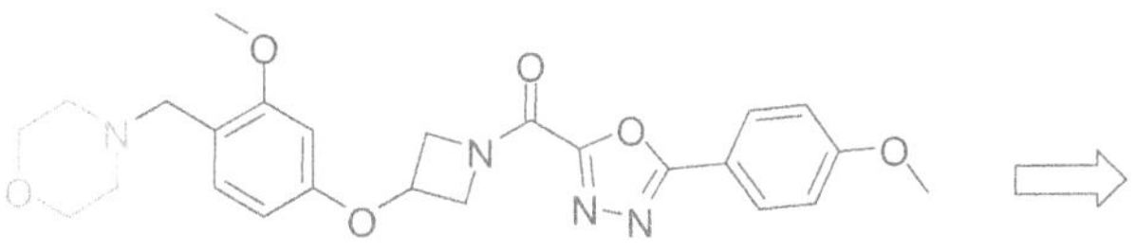

Morpholine **9**
MCH GTPγS, IC$_{50}$ = 27nM
MCH bind Mu, IC$_{50}$ = 16 nM
hERG, IC$_{50}$ = 9.8 μM

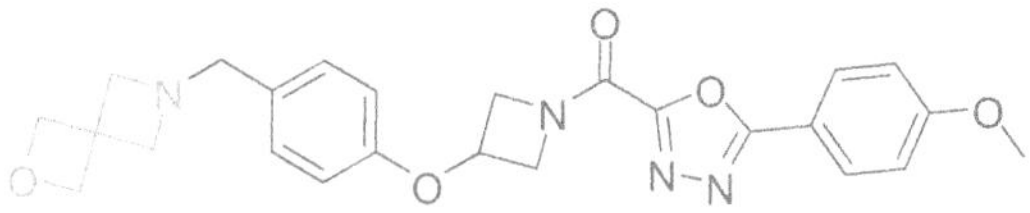

Spirooxetane AZD1979 (**10**)
MCH GTPγS, IC$_{50}$ = 27 nM
MCH bind Mu, IC$_{50}$ = 11 nM
hERG, IC$_{50}$ = 22 μM

Not all spirooxetane isosteres work like a charm.

Leucine-rich repeat kinase 2 (LRRK2) inhibitors are potentially useful in the treatment of Parkinson's disease (PD). From high throughput screen (HTS) and lead optimization, Pfizer arrived at morpholine **11** (PF-06447475) as a highly potent (LRRRK2 enzymatic assay: IC$_{50}$=3 nM), selective, brain penetrant, and *in vivo* active LRRK2 inhibitor. Attempt to replace the morpholino group with the 2-oxa-6-azaspiro[3.3]heptane isostere led to spirooxetane **12**, which was virtually inactive toward LRRK2, regrettably.[9]

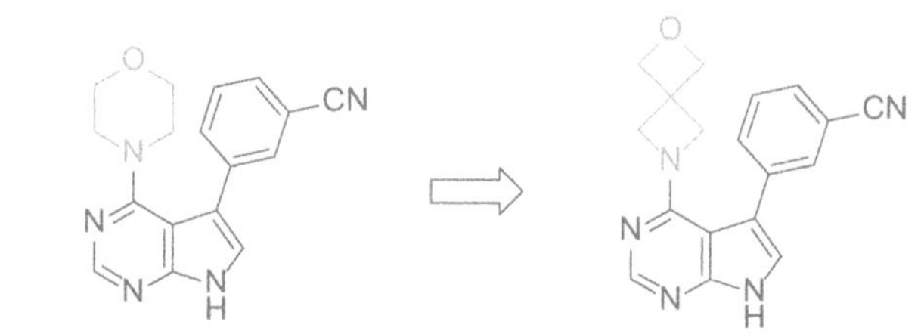

Morpholine **11**
LRRK2 IC$_{50}$ = 3 nM

Spirooxetane **12**
LRRK2 IC$_{50}$ >2,300 nM

SYNTHESIS OF SOME SPIROOXETANE-CONTAINING DRUGS

Unlike most privileged building blocks, synthetic methodology developments for making spirooxetanes[2,10–17] are ahead of utility in medicinal chemistry. In other words, many spirooxetanes have been made, yet are waiting for applications in drug discovery.

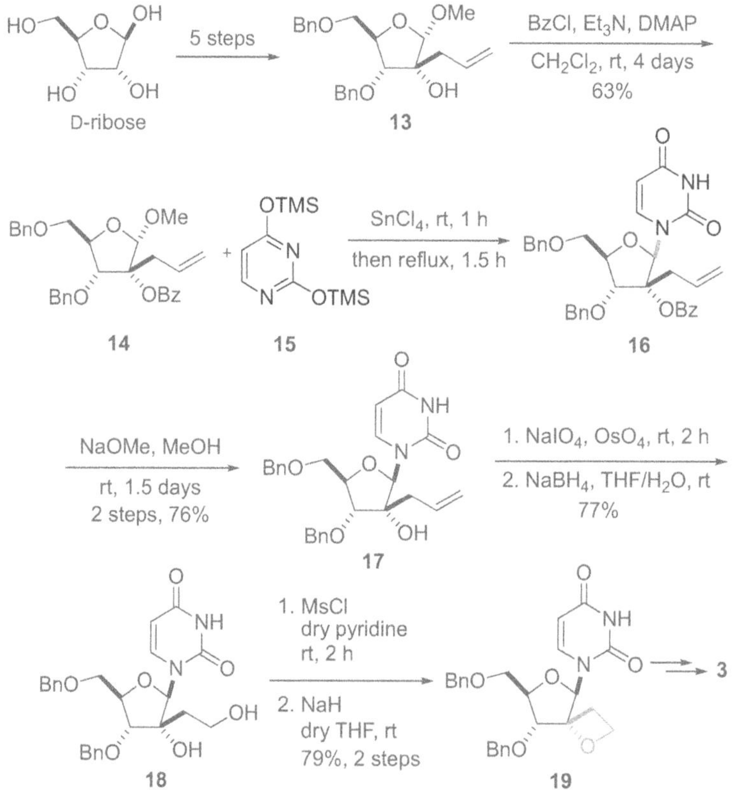

Janssen employed D-ribose as the starting material to prepare 2′-spirooxetanyl-nucleoside **3**. It took five steps to manipulate D-ribose to compound **13**, which was protected as the glycosylation substrate **14**. With the aid of SnCl$_4$, coupling between **14** and persilylated uracil **15** gave uridine derivative **16** stereoselectively. After removal of the benzoate protective group, the resulting alcohol **17** was subjected to an oxidation and then reduction sequence, giving rise to diol **18**. The primary alcohol was selectively mesylated and the resulting mesylate was subjected to a sodium hydride-mediated cyclization to offer oxetane **19**. An additional four steps of functional group transformations the delivered 2′-spirooxetanyl-nucleoside **3**.[5]

In summary, spirooxetanes combine the characteristics of both oxetanes and spirocycles. The oxygen atom on the oxetane may serve as a hydrogen bond acceptor, interacting with the target protein. The oxetane motif on a drug may also offer enhanced aqueous solubility and other physiochemical properties. Meanwhile, spirooxetanes, like most spirocycles, are bestowed with a trio of advantages: (a) more extensive interactions with target proteins; (b) more favorable physiochemical properties; and (c) potential novel intellectual properties.

REFERENCES

1. Bull, J. A.; Croft, R. A.; Davis, O. A.; Doran, R.; Morgan, K. F. *Chem. Rev.* **2016**, *116*, 12150–12233.
2. Wuitschik, G.; Rogers-Evans, M.; Buckl, A.; Bernasconi, M.; Marki, M.; Godel, T.; Fischer, H.; Wagner, B.; Parrilla, I.; Schuler, F.; et al. *Angew. Chem. Int. Ed.* **2008**, *47*, 4512–4515.
3. Sofia, M. J.; Bao, D.; Chang, W.; Du, J.; Nagarathnam, D.; Rachakonda, S.; Reddy, P. G.; Ross, B. S.; Wang, P.; Zhang, H.-R.; et al. *J. Med. Chem.* **2010**, *53*, 7202–7218.
4. Du, J.; Chun, B.-K.; Mosley, R. T.; Bansal, S.; Bao, H.; Espiritu, C.; Lam, A. M.; Murakami, E.; Niu, C.; Micolochick Steuer, H. M.; et al. *J. Med. Chem.* **2014**, *57*, 1826–1835.
5. Jonckers, T. H. M.; Vandyck, K.; Vandekerckhove, L.; Hu, L.; Tahri, A.; Van Hoof, S.; Lin, T.-I.; Vijgen, L.; Berke, J. M.; Lachau-Durand, S.; et al. *J. Med. Chem.* **2014**, *57*, 1836–1844.
6. Jonckers, T. H. M.; Tahri, A.; Vijgen, L.; Berke, J. M.; Lachau-Durand, S.; Stoops, B.; Snoeys, J.; Leclercq, L.; Tambuyzer, L.; Lin, T.-I.; et al. *J. Med. Chem.* **2016**, *59*, 5790–5798.
7. Gazzard, L.; Williams, K.; Chen, H.; Axford, L.; Blackwood, E.; Burton, B.; Chapman, K.; Crackett, P.; Drobnick, J.; Ellwood, C.; et al. *J. Med. Chem.* **2015**, *58*, 5053–5074.
8. Johansson, A.; Loefberg, C.; Antonsson, M.; von Unge, S.; Hayes, M. A.; Judkins, R.; Ploj, K.; Benthem, L.; Linden, D.; Brodin, P.; et al. *J. Med. Chem.* **2016**, *59*, 2497–2511.
9. Henderson, J. L.; Kormos, B. L.; Hayward, M. M.; Coffman, K. J.; Jasti, J.; Kurumbail, R. G.; Wager, T. T.; Verhoest, P. R.; Noell, G. S.; Chen, Y.; et al. *J. Med. Chem.* **2015**, *58*, 419–432.
10. Hamill, R.; Jones, B.; Pask, C. M.; Sridharan, V. *Tetrahedron Lett.* **2019**, *60*, 1126–1129.
11. Geary, G. C.; Nortcliffe, A.; Pearce, C. A.; Hamza, D.; Jones, G.; Moody, C. J. *Bioorg. Med. Chem.* **2018**, *26*, 791–797.
12. Nicolle, S. M.; Nortcliffe, A.; Bartrum, H. E.; Lewis, W.; Hayes, C. J.; Moody, C. J. *Chem. Eur. J.* **2017**, *23*, 13623–13627.
13. Jones, B.; Proud, M.; Sridharan, V. *Tetrahedron Lett.* **2016**, *57*, 2811–2813.
14. Beadle, J. D.; Powell, N. H.; Raubo, P.; Clarkson, G. J.; Shipman, M. *Synlett* **2016**, *27*, 169–172.
15. Brady, P. B.; Carreira, E. M. *Org. Lett.* **2015**, *17*, 3350–3353.
16. Davis, O. A.; Bull, J. A. *Synlett* **2015**, *26*, 1283–1288.
17. Carreira, E. M.; Fessard, T. C. *Chem. Rev.* **2014**, *114*, 8257–8322.

31 Tetrahydropyrans

Tetrahydropyran (THP) may be considered as a conformationally restrained ether with lower entropy. In the context of drug discovery, THP may be viewed as a bioisostere of cyclohexane with a lower lipophilicity. Having a lower lipophilicity may improve a drug's absorption, distribution, metabolism, and excretion (ADME). Furthermore, by replacing the CH_2 with an oxygen atom, THP may provide an additional point of contact with the target by offering oxygen as a hydrogen bond acceptor.

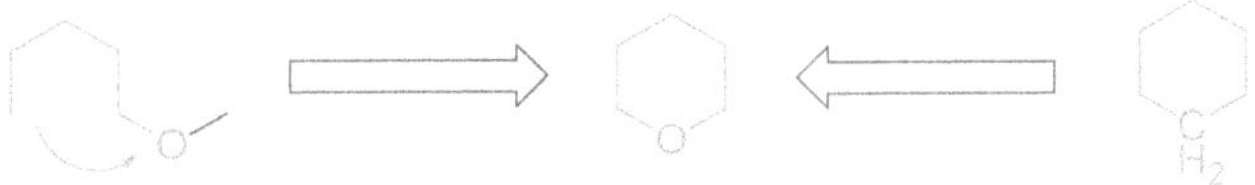

TETRAHYDROPYRAN-CONTAINING DRUGS

Some THP-containing drugs are closer to carbohydrates than to simple THPs and are not the focus of this review. An early THP-containing drug is an anticonvulsant topiramate (Topamax, **1**) as a fructopyranose O-alkyl sulfamate.[1] Similarly, neuraminidase inhibitor zanamivir (Relenza, **2**) for treating influenza infection was discovered employing terminal sialic acid, a residue from gluco-conjugates, as a starting point. It is the 4-guanadino-derivative of dehydro-2-deoxy-N-acetylneur-aminic acid (DANA), a transition-state mimetic of neuraminidase.[2]

Selective sodium-glucose cotransporter protein-2 (SGLT-2) inhibitors are one of the most recently approved drug classes for the treatment of type 2 diabetes mellitus (T2DM). All four SGLT-2 inhibitors on the market, Tanabe's canagliflozin (Invokana), BMS's dapagliflozin (Farxiga), Boehringer Ingelheim's empagliflozin (Jardiance, **3**), and Merck's ertugliflozin (Steglatro) are C-glycosides, which have improved metabolic stability over metabolically labile O-glycosides (e.g., phlorizin).[3]

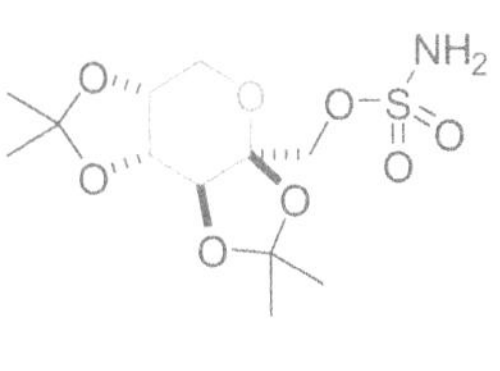

topiramate (Topamax, **1**)
Mylan, 1996
anticonvulant

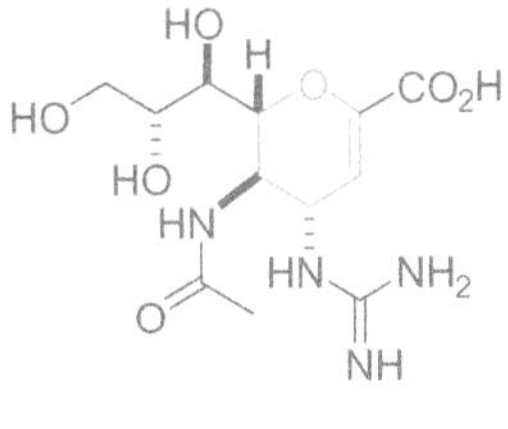

zanamivir (Relenza, **2**)
Biota/Glaxo, 1999
neuraminidase inhibitor

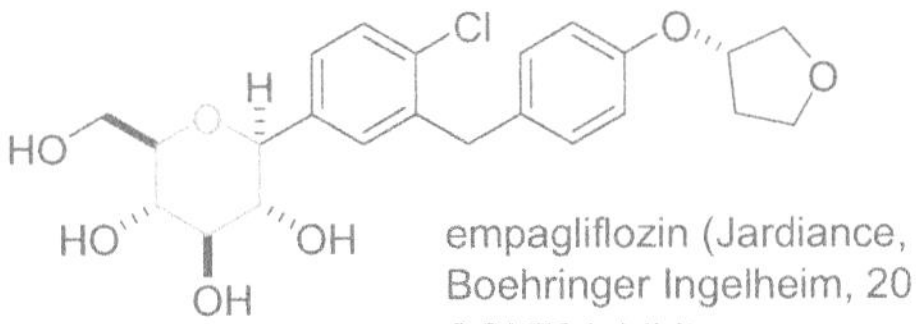

empagliflozin (Jardiance, **3**)
Boehringer Ingelheim, 2014
SGLT2 inhibitor

Here, we place emphasis on *bona fide* THP-containing drugs as exemplified by **4–9** shown below.

Lubiprostone (Amitiza, **4**), a laxative for the treatment of irritable bowel syndrome with constipation (IBS-c), contains a bicyclic THP-cyclopentanone. Derived from prostaglandin E1, lubiprostone (**4**)'s mechanism of action (MOA) is found to be a chloride channel-2 (ClC-2) opener (activator).[4a] It is not a remarkable drug except its cost, which prompted a physician to publish an article in 2017 with a title: "*$850 Per Bowel Movement?! Hard To Justify That Cost*".[4b] In contrast, few would protest the price for Eisai's eribulin (Halaven, **5**) for the treatment of metastatic breast cancer and liposarcoma. A microtubule dynamics inhibitor was discovered through a herculean effort by trimming marine natural product halichondrin B, which has seven THP rings. With three THP rings and 19 chiral centers, eribulin (**5**)'s manufacturing route entails 62 steps and even the longest linear sequence is 30 steps.[5]

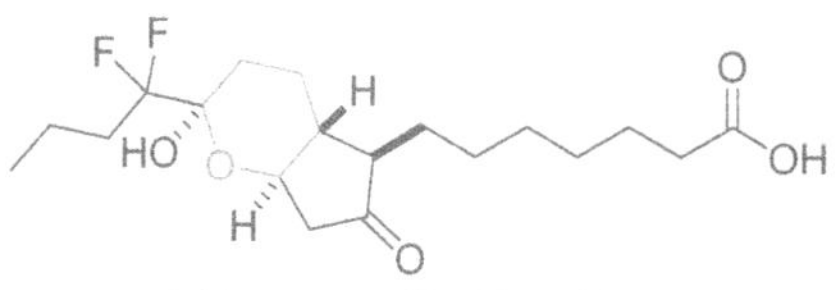

lubiprostone (Amitiza, **4**)
Mallinckrodt/Takeda, 2006
chloride channel activator

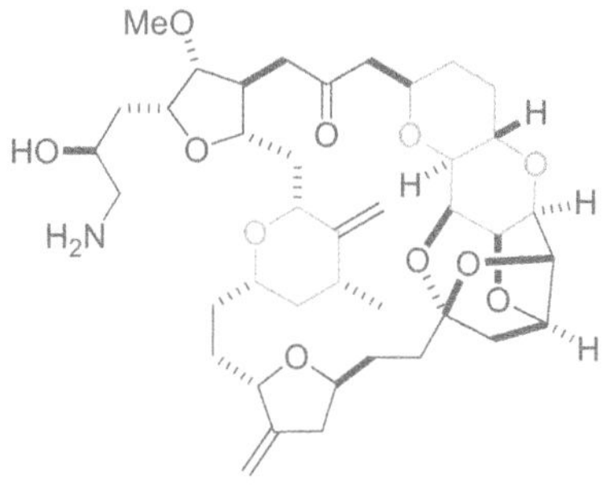

eribulin (Halaven, **5**)
Eisai, 2010
microtubule dynamics inhibitor

Merck's dipeptidyl peptidase 4 (DPP4) inhibitor omarigliptin (Marizev, **7**) is a more rigid backup drug for its initial successful drug sitagliptin (Januvia, **6**). Remarkably, omarigliptin (**7**) has such a long half-life that it may be taken once weekly whereas its progenitor sitagliptin (**6**) is given qd. Initially, when the cyclohexylamine derivative was installed to replace the linear amine on sitagliptin (**6**), the analog's selectivity against IKr ($IC_{50}=4.8$ μM) was below the desired standard ($IC_{50}>30$ μM). In addition, in the CV-dog model, the cyclohexylamine derivative was found to prolong QTc (>5% at 3 mpk). Replacement of cyclohexylamine with THP reduced the pK_a of the primary amine from 8.6 to 7.3, and the human ether-a-go-go (hERG) potassium channel activity selectivity improved accordingly ($IC_{50}=23$ μM). In addition, the THP analog was devoid of any QTc prolongation in the CV-dog model at doses up to 30 mpk iv.[6]

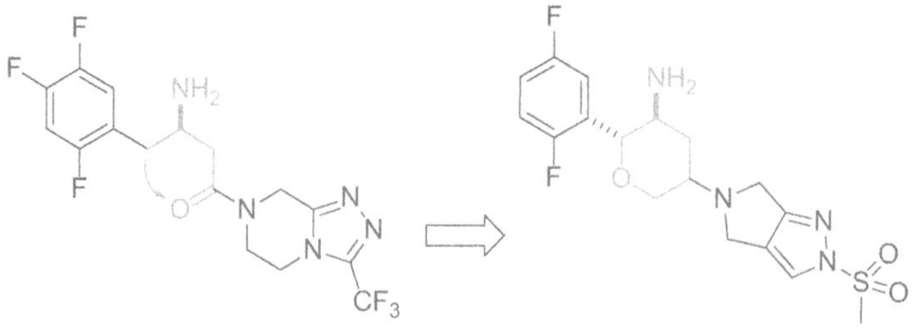

sitagliptin (Januvia, **6**)
Merck, 2006
DPP-4 inhibitor, qd

omarigliptin (Marizev, **7**)
Merck, 2015 (Japan)
DPP-4 inhibitor, once weekly

In 2018, Astellas' gilteritinib (Xospata, **8**) garnered the FDA's approval as a treatment of adult patients who have relapsed or refractory acute myeloid leukemia (AML) with an FLT3 mutation. With a popular amino-THP substituent, gilteritinib (**8**) is an AXL receptor tyrosine kinase inhibitor.[7] In addition, gilteritinib (**8**) also inhibits FLT3, ALK, LTK, and KIT kinases.

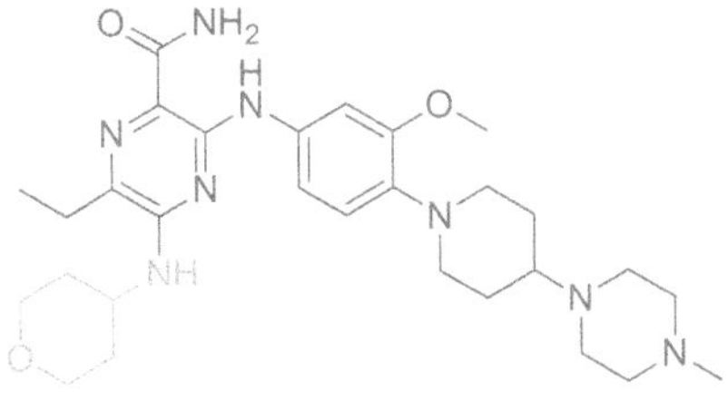

gilteritinib (Xospata, **8**)
Astellas, 2018
AXL receptor tyrosine kinase inhibitor

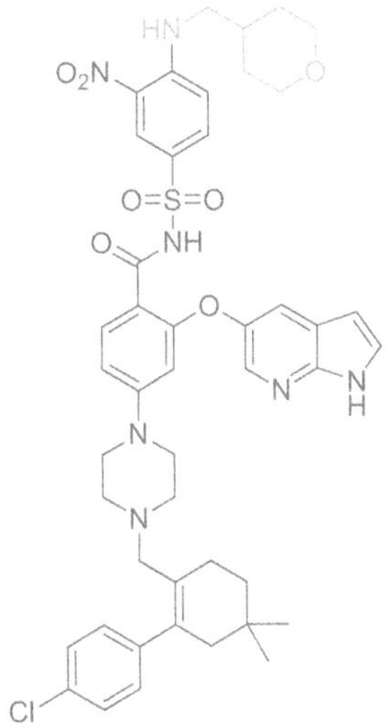

venetoclax (Venclexta, **9**)
Abbvie, 2016
Bcl-2 inhibitor

Abbvie's B-cell lymphoma 2 (Bcl-2) inhibitor venetoclax (Venclexta, **9**) is a "wonder" cancer drug for treating chronic lymphocytic leukemia (CLL) with the 17p deletion. Targeting the challenging protein-protein interactions (PPIs), it was discovered from the fragment-based drug discovery (FBDD) strategy under the guidance of the "SAR by NMR" method. Its THP tail fragment played an important role in imparting selectivity against Bcl-X_L (*vide infra*).[8]

TETRAHYDROPYRANS IN DRUG DISCOVERY

En route to the discovery of venetoclax (**9**), Abbvie prepared morpholine **10**. While morpholine **10** was a potent Bcl-2 inhibitor ($K_i = 0.2$ nM), it was not selective against Bcl-X_L ($K_i = 1.3$ nM). While inhibition of Bcl-2 offered target efficacy in leukemia and lymphoma, inhibition of Bcl-X_L led to dose-limiting thrombocytopenia, a deficiency of platelets (thrombocytes) that may increase the risk of bleeding. From ingenious reverse engineering efforts, Abbvie arrived at THP-containing compound **11**, which lost Bcl-X_L activity. Regrettably, THP **11** also had reduced Bcl-2 affinity and no cell activity. Both of the deficiencies had to be remedied by installing a 7-azaindole ether substituent to occupy the P4 hot spot. The culmination of these efforts eventually provided venetoclax (**9**), which is a potent, selective (Bcl-X_L-sparing and human platelet-sparing), and bioavailable Bcl-2 inhibitor, after an arduous and winding road of discovery.[8]

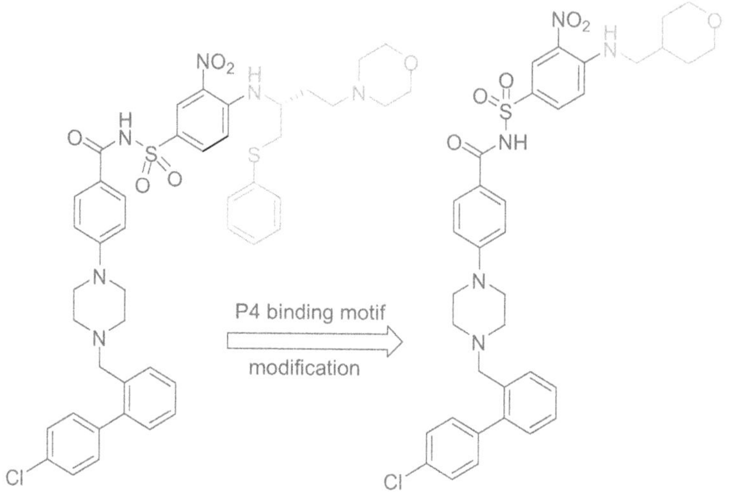

morpholine **10**
Bcl-2 K_i = 0.2 nM
Bcl-X$_L$ K_i = 1.3 nM

P4 binding motif
modification

THP **11**
Bcl-2 K_i = 18 nM
Bcl-X$_L$ K_i = >660 nM

THP rings have frequently been employed to improve a drug's ADME properties. A series of THP-containing histamine-3 (H$_3$) receptor antagonists were prepared as a treatment of allergic rhinitis. As represented by dibasic THP **12**, modulation of its partition coefficient achieved an optimal balance of blood clearance (CL = 18 mL/min/kg) and volume of distribution (V$_d$ = 94 L/kg). Remarkably, THP **12** has a half-life of 60 hours in dogs and a predicted human half-life of 250 hours.[9] Meanwhile, Actelion's THP-based inhibitor **13** of bacterial type II topoisomerases (DNA gyrase and topoisomerase IV) showed antibacterial activity against Gram-negative bacteria. These non-fluoroquinolone topoisomerase inhibitors are of great interest because they may overcome infections inflicted by multidrug-resistant (MDR) Gram-negative bacteria.[10]

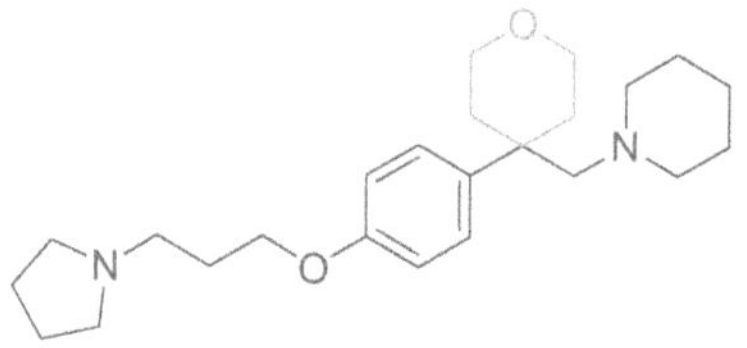

H$_3$ histamine receptor antagonist **12**

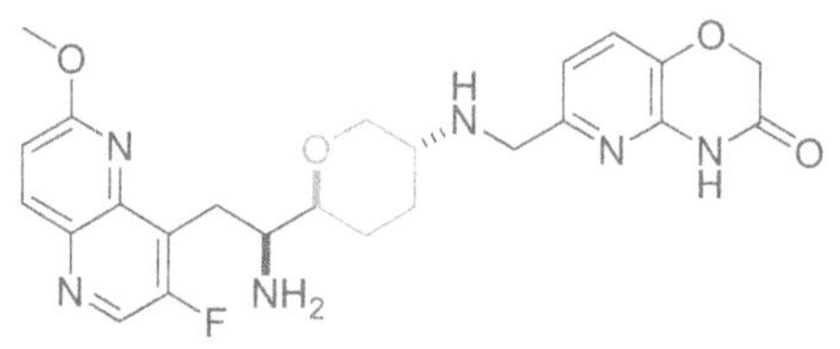

topoisomerase inhibitor **13**

Pfizer's cyclobutanol **14** is a direct activator of 5′-adenosine monophosphate-activated protein kinase (AMPK). It is potent and selective against β1-containing AMPK isoforms that allosterically activate the enzyme by binding at the "allosteric drug and metabolite" (ADaM) site at the interface of the α- and β-subunits. Its carboxylic acid, the indole N–H, and the cyclobutanol all form hydrogen bonds to protein atoms from the α1 and β1 units. As a clinical candidate for the treatment of diabetes nephropathy associated with T2DM, cyclobutanol **14** is not ideal in terms of ADME.

Namely, it undergoes rapid phase II metabolism, forming acyl glucuronide conjugate by uridine glucuronosyl-transferase (UGT) isoforms. Renal excretion of unchanged drug is observed in rats, dogs, and monkeys, and the active renal elimination process is possibly mediated by organic anion transporter (OAT) proteins expressed at the basolateral membrane of proximal tubules.[11a] As a backup drug for cyclobutanol **14**, THP-containing analog PF-06409577 (**15**) may be considered as a conformationally restrained ether, providing an opportunity to balance the lipophilicity without adding additional hydrogen bond donors. In combination with two fluorine substituents on the indole ring, PF-06409577 (**15**) offers favorable *in vitro* ADME properties including decreased CL_{int} in human hepatocytes and increased P_{app} with attenuated binding to human OAT-3 in comparison to the parent drug **14**. PF-06409577 (**15**), now in first-in-human (FIH) clinical trials, emerged from three preclinical candidates (including **14**) as a new investigational drug. Coincidentally, acyl glucuronide metabolites of both cyclobutanol **14** and PF-06409577 (**15**) are direct activators of AMPK as well.[11b]

cyclobutanol **14**
$\alpha 1\beta 1\gamma 1$-AMPK EC_{50}, 5.6 nM
rat CL_p, 23 mL/min/kg
rat CL_{renal}, 7.6 mL/min/kg

optimization of glucuronidation and renal efflux

PF-06409577 (**15**)
$\alpha 1\beta 1\gamma 1$-AMPK EC_{50}, 22 nM
rat CL_p, 1.2 mL/min/kg
rat CL_{renal}, 0.03 mL/min/kg

A THP-amine motif helped AstraZeneca to arrive at a potent, selective, and orally bioavailable inhibitor AZD0156 (**17**) of ataxia telangiectasia mutated (ATM) kinase as a potential drug to potentiate the efficacy of the approved drugs irinotecan (a DNA intercalator) and olaparib [a poly ADP ribose polymerase (PARP) inhibitor] in disease-relevant mouse models. ATM kinase is a member of the PI3K-related kinase (PIKK) family of atypical serine/threonine protein kinases (also comprising of mTOR) and plays a central role in both signaling of and the protection of cells against DNA double-strand breaks (DSB) and reactive oxygen species (ROS) that radiotherapy and a wide range of chemotherapies induce. Starting from an initial screening hit with a quinolone carboxamide scaffold, AstraZeneca arrived at ATM inhibitor **16**, which showed *in vivo* efficacy in an HT29 mouse xenograft model. But with a predicted dose of 700 mg qd for **16**, a drug with a better ADME profile and lower predicted dosage would have better chances to succeed in clinical trials. Extensive optimization, including installation of a THP-amine fragment, led to AZD0156 (**17**) with a superior profile in comparison to **16**. AZD0156 (**17**) is potent and selective against closely-related kinases such as mTOR and PI3K with superb aqueous solubility, and many other pharmacokinetic parameters. It has a predicted dose of 5 mg qd, bestowing this compound a greater chance of success in clinics.[12a] In 2018, phase I clinical trials for AZD0156 (**17**) in combination with olaparib concluded successfully. Meanwhile, a series of THP-containing 3-cinnoline carboxamides were described as highly potent, selective, and orally bioavailable ATM kinase inhibitors as well.[12b]

ATM inhibitor **16**

ATM IC_{50} = 33 nM

aq. soln. 69 μM

Pred dose = 700 mg qd

AZD0156 (17)
ATM IC_{50} = 0.58 nM
aq. soln. > 800 μM
Pred dose = 5 mg qd

THP played an important role in achieving favorable pharmacokinetic properties during Merck's efforts in optimizing a series of pyrazolyl-carboxamides as Janus kinase 1 (JAK1) selective inhibitors. THP derivative **19**, as a bioisostere of cyclohexyl derivative **18**, introduced a polar oxygen heteroatom, which offered tighter drug–enzyme binding interactions. This was reflected by a 1.4-fold increase of lipophilic ligand efficiency (LLE) although its ligand binding efficiency (LBE) value did not change much. The polarity decrease was subtle (log D 2.08 for **19** vs. log D 2.66 for **18**), but it translated to improved clearance in both rat and human and a large decrease in unbound *in vivo* rat clearance.[13] This is a good example to highlight that the cyclohexyl–THP switch may bring not only better potencies but also improved ADME properties (CL_p=plasma clearance; r; h Hept CL_{int}=rat and human hepatic intrinsic clearance).

1 atom
(C to O)
change

cyclohexyl derivative **18**
JAK1 IC_{50} = 3 nM
LBE = 0.50
LEE = 5.5
HPLC log Dp H7.0 = 2.66
rat CL_p = 484 mL/min/kg
unbound rat CL = 12,100 mL/min/kg
r:h Hept CL_{int} = 160:25 mL/min/kg

THP derivative **19**
JAK1 IC_{50} = 0.4 nM
LBE = 0.56
LEE = 7.9
2.08
rat CL_p = 20 mL/min/kg
110 mL/min/kg
70:6 mL/min/kg

A THP substituent helped AsreaZeneca to reduce the clearance of their interleukin-1 receptor-associated kinase 4 (IRAK4) inhibitors. The cyclopentyl derivative **20**, as a pyrrolopyrimidine-based IRAK4 inhibitor, had a high rate of metabolism in isolated rat hepatocytes (CL_{int}=71 μL/min/10⁶ cells). While the direct oxygen analog employing 2- or 3-tetrahydrofuran (THF) did not show significant improvement of metabolism, 4-THP derivative **21** reduced the rate of metabolism by rat hepatocytes by 5-fold. On a molecular level, the 4-THP moiety showed a lipophilic stacking interaction with Tyr262 as well as a hydrogen bond to Lys213. The optimized IRAK4 inhibitors may serve as treatment of mutant MYD^{L265P} diffuse large B-cell lymphoma (DLBCL).[14]

2 atom
(C to C–O)
change

cyclopentyl derivative **20**
IRAK4 enz IC_{50} = 6 nM
IRAK4 cell IC_{50} = 23 nM
rat heps CL_{int} = 71 μL/min/10⁶ cells

4-THP derivative **21**
IRAK4 enz IC_{50} = 6 nM
IRAK4 enz IC_{50} = 59 nM
14 μL/min/10⁶ cells

THP fragments also made an appearance in modulating receptors in addition to the examples on enzymes shown thus far. Muscarinic acetylcholine receptor (mAChR) subtype 1 (M_1) positive allosteric modulators (PAMs) hold great promises of treating Alzheimer's disease (AD) and schizophrenia. But like all potential treatments for AD, this target has encountered many failures in clinical trials. Pfizer's THP-containing M_1-selective PAM PF-06827443 (**22**) is plagued with weak agonist activity, which manifests as seizure and cholinergic adverse events.[15] Another "pure" M_1 PAM THP-containing VU6007477 (**23**) is devoid of agonist activities. Although without the cholinergic toxicity/seizure liability, it is not suitable for translation to the clinic because **23** is a P-glycoprotein (P-gp) substrate (ER = 4.5) with only moderate permeability ($P_{app} = 1.2 \times 10^{-5}$ cm/s). As a side, 7-azaindole-carboxamide **23** forms an intramolecular hydrogen bond, which helps to maintain the putative bio-activation conformation.[16]

PF-06827443 (**22**)
M_1 PAM EC_{50} = 47 nM
M_1 PAM K_i = 14 nM
M_2–M_5, EC_{50} > 10 μM

VU6007477 (**23**)
M_1 PAM EC_{50} = 230 nM
M_1 agonist EC_{50} > 10 μM
K_i = 0.28 nM, $K_{p,uu}$ = 0.32 nM

SYNTHESIS OF SOME TETRAHYDROPYRAN-CONTAINING DRUGS

The manufacturing route for an active pharmaceutical ingredient (API) is the gold standard of organic synthesis because it must consider many factors such as synthetic convergence, cost of goods (CoG), scalability, reproducibility, reaction conditions, reactors, and environmental friendliness. Merck's manufacturing route for its DDP-4 inhibitor omarigliptin (Marizev, **7**) serves as a good lesson to learn in devising a commercial production route. Ketone **24** was assembled by reaction between the Grignard reagent (2,5-difluorophenyl)magnesium chloride and the appropriate Weinreb amide. Asymmetric reduction of ketone **24** to alcohol **25** was accomplished via a dynamic kinetic resolution (DKR) asymmetric transfer hydrogenation that was facilitated by an oxo-tethered ruthenium-(II) catalyst, (R,R)-Ts-DENEB, as a highly efficient asymmetric transfer hydrogenation catalyst. Subsequent Ru-catalyzed cycloisomerization of alcohol **25** to dihydropyran **26** was carried out in the same pot without workup of the alcohol intermediate.[17]

ketone **24** → alcohol **25**

0.1 mol% (R,R)-Ts-DENEB
HCO₂H, DABCO, THF

93% yield
(94:4 dr, 99% ee)

CpRuCl(PPh₃)₂, PPh₃
N-hydroxysuccinimide

Bu₄NPF₆, NaHCO₃, DMF
86%

dihydropyran **26** → omarigliptin (Marizev, **7**)

Process Chemistry at HEC Pharma reported an alternative scalable process for the synthesis of the key intermediate of omarigliptin (**7**). Ketone **24** was reduced to alcohol **25** employing a "low-tech" reducing agent via the Meerwein–Ponndorf–Verley reaction. Exposure of alcohol **25** to iodine under basic conditions led to 5-*exo-dig* iodocyclization product tetrahydrofuran-vinyl iodide **27**, which was converted to iodoketone **28** via the addition of water and a concurrent ring-opening reaction promoted by aqueous sodium hydrogen sulfate hydrate. A simple intramolecular S_N2 displacement then gave rise to cycloetherization product tetrahydropyranone **29** as an advanced intermediate toward omarigliptin (**7**).[18]

ketone **24** $\xrightarrow[\text{CH}_2\text{Cl}_2,\ 40\ ^\circ\text{C}]{\text{Al(O}i\text{-Pr)}_3,\ i\text{-PrOH}}$ | alcohol **25** | $\xrightarrow[\text{0 }^\circ\text{C to rt}]{\text{I}_2 \quad \text{KOH, MeOH}}$

tetrahydrofuran **27**
> 99% *ee*, 72% in 2 steps

$\xrightarrow[\text{30–35 }^\circ\text{C}]{\substack{\text{NaHSO}_4\cdot\text{H}_2\text{O} \\ \text{H}_2\text{O/THF}}}$

iodo-ketone **28**

$\xrightarrow[\text{65 }^\circ\text{C}]{\text{Na}_2\text{CO}_3,\ \text{THF}}$

tetrahydropyranone **29**

$\longrightarrow$ **7**

In collaboration with Baran, BMS chemists made a heroic effort in preparing a chiral THP fragment **34** as a building block for the synthesis of the HCV NS5A inhibitor BMS-986097 (**35**). Michael's addition of imine **30** as a masked amino acid to α,β-unsaturated ester **31** gave rise to lactone **32** and its enantiomer in a 1:1 ratio. DBU-promoted epimerization and chiral supercritical fluid chromatography (SFC) separation produced lactone **33**, which was manipulated to THP **34** as a single enantiomer. THP **34** served as two tails of the symmetrical BMS-986097 (**35**).[19] For future process and manufacturing routes, an asymmetric synthesis is needed to make THP **34** without the need for epimerization and chiral SFC separations.

imine **30**   ester **30** $\xrightarrow[\text{86\%}]{\text{LiHMDS, THF}}$ lactone **32**

lactone **33**   THP **34**

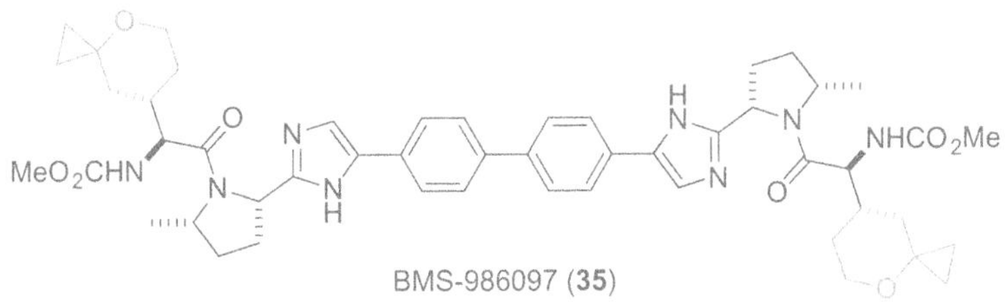

BMS-986097 (35)

Actelion's preparation of the central THP scaffold on topoisomerase inhibitor **13** involved a Sonogashira coupling of aryl bromide **36** and THP-containing terminal alkyne **37** to construct internal alkyne **38**. Hg(II)-promoted hydration under harsher acidic conditions installed ketone **39**, which could be then manipulated to deliver topoisomerase inhibitor **13**.[10,20]

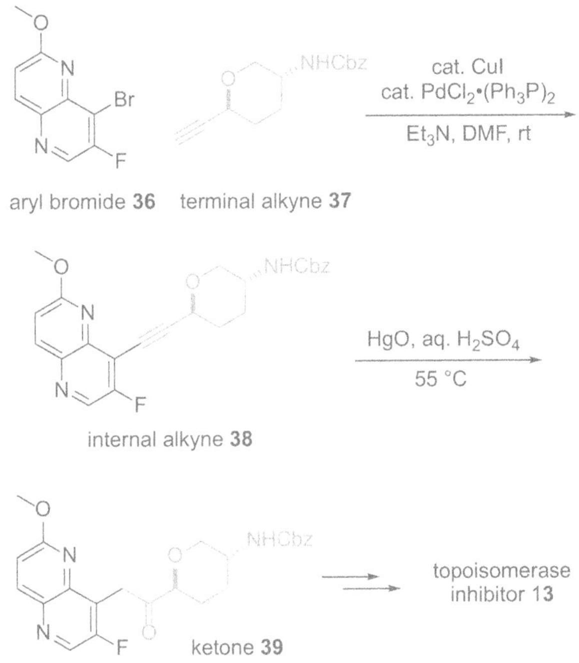

cat. CuI
cat. PdCl$_2$•(Ph$_3$P)$_2$

Et$_3$N, DMF, rt

aryl bromide **36**   terminal alkyne **37**

internal alkyne **38**

HgO, aq. H$_2$SO$_4$

55 °C

topoisomerase
inhibitor 13

ketone **39**

Thankfully, not all THP rings are so sophisticated to construct, many drug syntheses can take advantage of commercially available THP-containing building blocks. Abbvie's synthesis of the top portion **42** of venetoclax (Venclexta, **9**) entails an S$_N$Ar reaction of 4-chloro-3-nitrobenzenesulfonamide (**40**) with THP-methylamine **41** at 80°C. Replacing **40** with 4-fluoro-3-nitrobenzenesulfonamide accelerates the S$_N$Ar reaction, which may be carried out at room temperature in THF and Et$_3$N.[21]

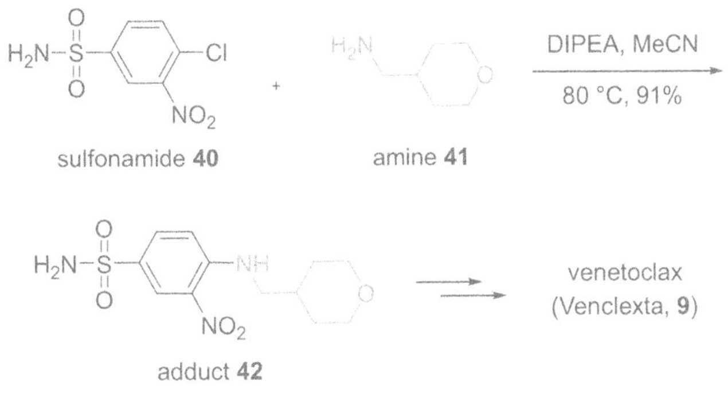

+

DIPEA, MeCN

80 °C, 91%

sulfonamide **40**   amine **41**

venetoclax
(Venclexta, **9**)

adduct **42**

Pfizer's THP-containing AMPK activator PF-06409577 (**15**) was prepared in kilogram quantities to support its FIH clinical trials. For the reduction of commercially available ketone **43** to make chiral alcohol **44**, Corey–Bakshi–Shibata reduction at −20 °C was favored over the Noyori hydrogenation because of the balance of high selectivity, yield, and control over the process. Cyclization of **44** via an intramolecular S$_N$2 reaction generated THP **45**, which was further manipulated to produce PF-06409577 (**15**).[22]

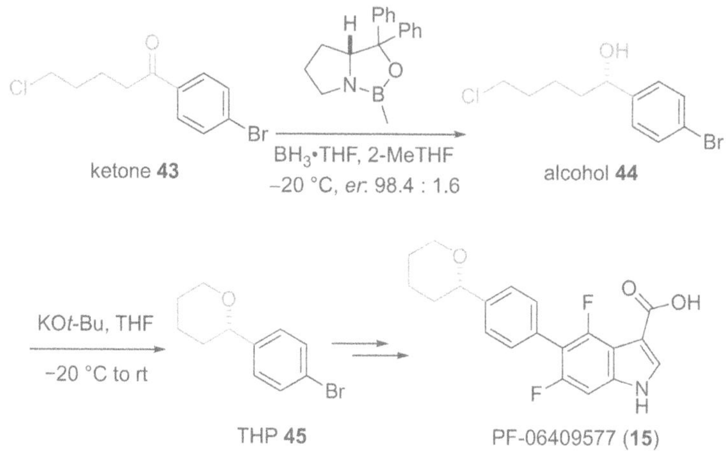

An S$_N$Ar reaction between chloroquinoline **46** and the popular THP-4-amine (**47**) was key to construct adduct **48** *en route* to the synthesis of AstraZeneca's ATM kinase inhibitor AZD0156 (**17**).[12]

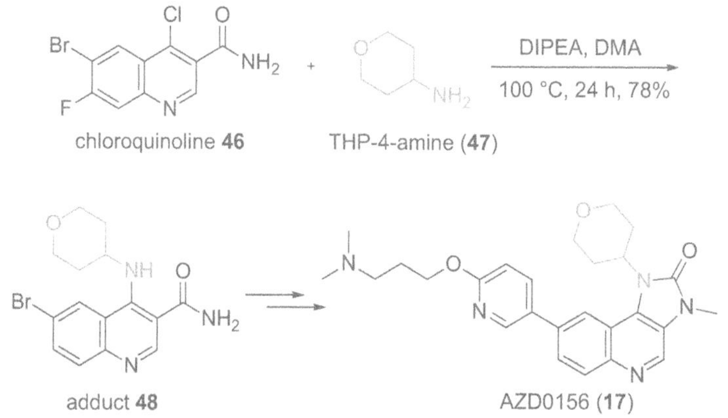

A synthesis of Merck's THP-containing selective JAK1 inhibitor **19** started with treating dihydro-2H-pyran-4(3H)-one (**49**) with TMS-CN and TMS-OTf to form the cyanohydrin, which was converted to carbonitrile **50** after treating the cyanohydrin with POCl$_3$ in pyridine. Michael's addition of pyrazole carboxamide **51**–**50** was induced by DBU to generate a mixture of products, which were separated by chiral SFC to afford enantiomerically pure **52** out of the four possible enantiomers. A Buchwald–Hartwig coupling between **52** and iodobenzene then produced the JAK1 inhibitor **19**.[13]

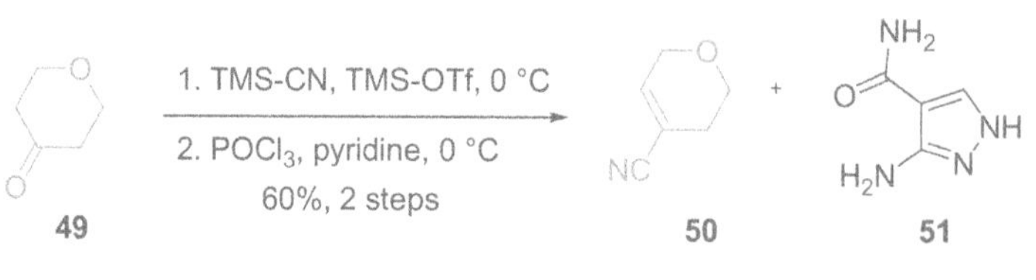

1. DBU, EtOH, 70 °C
2. Chiral SFC
   14%, 2 steps

52

Ph-I, Pd$_2$(dba)$_3$
t-Bu-XPhos, KOAc

i-PrOH, 60 °C

JAK1 inhibitor **19**

In summary, THP is a rigid form of linear ether and thus has a lower entropy. As a bioisostere of cyclohexane, THP may gain an additional point of contact with the target by offering oxygen as a hydrogen bond acceptor. In medicinal chemistry, THP substituents, with lower lipophilicity in comparison to the cyclohexyl isosteres, have been employed to modulate the pK$_a$ of drugs and improve their absorption, distribution, metabolism, and excretion (ADME).

REFERENCES

1. Maryanoff, B. E.; Nortey, S. O.; Gardocki, J. F.; Shank, R. P.; Dodgson, S. P. *J. Med. Chem.* **1987**, *30*, 880–887.
2. Smith, P. W.; Sollis, S. L.; Howes, P. D.; Cherry, P. C.; Starkey, I. D.; Cobley, K. N.; Weston, H.; Scicinski, J.; Merritt, A.; Whittington, A.; et al. *J. Med. Chem.* **1998**, *41*, 787–797.
3. da Silva, P. N.; da Conceicao, R. A.; de Castro Barbosa, M. L.; do Couto, M.R. *MedChemComm* **2018**, *8*, 1273–1281.
4. (a) Owen, R. T. *Drugs Today* **2008**, *44*, 645–652. (b) Morrow, T. *Managed Care* **2017**, *26*, 36–37.
5. Yu, M. J.; Kishi, Y. *Ann. Rep. Med. Chem.* **2011**, *46*, 227–241.
6. Biftu, T.; Sinha-Roy, R.; Chen, P.; Qian, X.; Feng, D.; Kuethe, J. T.; Scapin, G.; Gao, Y. D.; Yan, Y.; Krueger, D.; et al. *J. Med. Chem.* **2014**, *57*, 3205–3212.
7. Myers, S. H.; Brunton, V. G.; Unciti-Broceta, A. *J. Med. Chem.* **2016**, *59*, 3593–3608.
8. Souers, A. J.; Leverson, J. D.; Boghaert, E. R.; Ackler, S. L.; Catron, N. D.; Chen, J.; Dayton, B. D.; Ding, H.; Enschede, S. H.; Fairbrother, W. J.; et al. *Nat. Med.* **2013**, *19*, 202–208.
9. Hay, T.; Jones, R.; Beaumont, K.; Kemp, M. *Drug Metab. Dispos.* **2009**, *37*, 1864–1970.
10. Surivet, J.-P.; Zumbrunn, C.; Bruyere, T.; Bur, D.; Kohl, C.; Locher, H. H.; Seiler, P.; Ertel, E. A.; Hess, P.; Enderlin-Paput, M.; et al. *J. Med. Chem.* **2017**, *60*, 3776–3794.
11. (a) Edmonds, D. J.; Kung, D. W.; Kalgutkar, A. S.; Filipski, K. J.; Ebner, D. C.; Cabral, S.; Smith, A. C.; Aspnes, G. E.; Bhattacharya, S. K.; Genung, N. E.; et al. *J. Med. Chem.* **2018**, *61*, 2372–2383. (b) Ryder, T. F.; Calabrese, M. F.; Walker, G. S.; Cameron, K. O.; Reys, A. R.; Borzilleri, K. A.; Delmore, J.; Miller, R.; Kurumbail, R. G.; Ward, J.; et al. *J. Med. Chem.* **2018**, *61*, 7273–7288.
12. (a) Pike, K. G.; Barlaam, B.; Cadogan, E.; Campbell, A.; Chen, Y.; Colclough, N.; Davies, N. L.; de-Almeida, C.; Degorce, S. L.; Didelot, M.; et al. *J. Med. Chem.* **2018**, *61*, 3823–3841. (b) Barlaam, B.; Cadogan, E.; Campbell, A.; Colclough, N.; Dishington, A.; Durant, S.; Goldberg, K.; Hassall, L. A.; Hughes, G. D.; MacFaul, P. A.; et al. *ACS Med. Chem. Lett.* **2018**, *69*, 809–814.
13. Siu, T.; Brubaker, J.; Fuller, P.; Torres, L.; Zeng, H.; Close, J.; Mampreian, D. M.; Shi, F.; Liu, D.; Fradera, X.; et al. *J. Med. Chem.* **2017**, *60*, 9676–9690.
14. Scott, J. S.; Degorce, S. L.; Anjum, R.; Culshaw, J.; Davies, R. D. M.; Davies, N. L.; Dillman, K. S.; Dowling, J. E.; Drew, L.; Ferguson, A. D.; et al. *J. Med. Chem.* **2017**, *60*, 10071–10091.
15. Davoren, J. E.; Garnsey, M.; Pettersen, B.; Brodney, M. A.; Edgerton, J. R.; Fortin, J.-P.; Grimwood, S.; Harris, A. R.; Jenkinson, S.; Kenakin, T.; et al. *J. Med. Chem.* **2017**, *60*, 6649–6663.

16. Engers, J. L.; Childress, E. S.; Long, M. F.; Capstick, R. A.; Luscombe, V. B.; Cho, H. P.; Dickerson, J. W.; Rook, J. M.; Blobaum, A. L.; Niswender, C. M.; et al. *ACS Med. Chem. Lett.* **2018**, *9*, 917–922.

17. Chung, J. Y. L.; Scott, J. P.; Anderson, C.; Bishop, B.; Bremeyer, N.; Cao, Y.; Chen, Q.; Dunn, R.; Kassim, A.; Lieberman, D.; et al. *Org. Process Res. Dev.* **2015**, *19*, 1760–1768.

18. Sun, G.; Wei, M.; Luo, Z.; Liu, Y.; Chen, Z.; Wang, Z. *Org. Process Res. Dev.* **2016**, *20*, 2074–2079.

19. Mathur, A.; Wang, B.; Smith, D.; Li, J.; Pawluczyk, J.; Sun, J.-H.; Wong, M. K.; Krishnananthan, S.; Wu, D.-R.; Sun, D.; et al. *J. Org. Chem.* **2017**, *82*, 10376–10387.

20. Surivet, J.-P.; Zumbrunn, C.; Rueedi, G.; Bur, D.; Bruyere, T.; Locher, H.; Ritz, D.; Seiler, P.; Kohl, C.; Ertel, E. A.; et al. *J. Med. Chem.* **2015**, *58*, 927–942.

21. (a) Bruncko, M.; Ding, H.; Doherty, G. A.; Elmore, S. W.; Hasvold, L. A.; Hexamer, L.; Kunzer, A. R.; Song, X.; Souers, A. J.; Sullivan, G. M.; et al. U.S. Patent US8546399 (2013). (b) Hughes, D. L. *Org. Process Res. Dev.* **2016**, *20*, 2028–2042.

22. Smith, A. C.; Kung, D. W.; Shavnya, A.; Brandt, T. A.; Dent, P. D.; Genung, N. E.; Cabral, S.; Panteleev, J.; Herr, M.; Yip, K. N.; et al. *Org. Process Res. Dev.* **2018**, *22*, 681–696.

32 Trifluoromethylpyridines

A trifluoromethyl substituent can have non-covalent interactions (NCIs) with heteroatoms such as oxygen and nitrogen. Take perfluorotoluene as an example, in addition to the well-known π-hole interactions (+138 kJ/mol) afforded by the phenyl motif, the trifluoromethyl carbon atom can partake α-hole interactions (+71 kJ/mol) with the neighboring heteroatom, albeit not as strong as the π-hole interactions. The α-hole interactions, also known as *tetrel bonding*, are a testimony to the fact that the three fluorine atoms attached directly to the carbon atom are so electron-withdrawing that they make it partially positive.[1]

π-hole, X = N,O   σ-hole (tetrel)
(+138 kJ/mol)   (+71 kJ/mol)

Agios' isocitrate dehydrogenase 2 (IDH2) allosteric inhibitor enasidenib (Idhifa, **1**) complexed to the mutant IDH2 enzyme can dramatically reduce ($IC_{50} = 10–20$ nm) the production of the oncometabolite, (R)-2-hydroxyglutarate (2HG), in a cellular assay. Interestingly, it has been demonstrated that compound **2**, a "naked" analog of enasidenib without the –CF_3 groups presents a higher IC_{50} value (30 nm), suggesting that the –CF_3 group modulates the binding ability of the inhibitor, increasing the affinity to the IDH active site. Frontera and coworkers carried out docking experiments and discovered that the O⋯CF_3 distance is shorter than the sum of van der Waals radii and the directionality is adequate to establish a favorable interaction with the σ-hole (165.1°) of the CF_3. They speculated that one of two CF_3 groups present in the inhibitor interacts with an aspartate amino acid (ASP312) of the active site via the tetrel bonding.[1,2]

2
Cellular IC_{50} (2HG inh) IC_{50} = 30 nM
HLM E_h = 0.69
Solubility (pH 2/7.4) = 2/0.8 μM

enasidenib (Idhifa, **1**)
Agios/Celgene, 2017
Cellular IC_{50} (2HG inh) = 15 nM
HLM E_h = 0.16
Solubility (pH 2/7.4) = 47/23 μM

Agios identified triazine **3** as a high-throughput screening (HTS) hit. Their initial hit-to-lead chemistry led to compound **2**, the first sub-100 nM inhibitor of IDH2[R140Q]. However, despite its potency in enzymatic and cellular assays, it was highly lipophilic, resulting in solubility-limited absorption *in vivo*. To make things worse, its *in vitro* liver microsomal instability translated to high clearance *in vivo*. X-ray co-crystal structure of **2** and IDH2[R140Q] provided invaluable insight for further optimization. Installation of two trifluoromethylpyridine substituents and a 2-methyl-2-propanol led to enasidenib (**1**) with excellent potency for 2-HG inhibition, improved solubility, low clearance (0.83 L/h/kg), and good oral bioavailability (41%) *in vivo* in rats.[3]

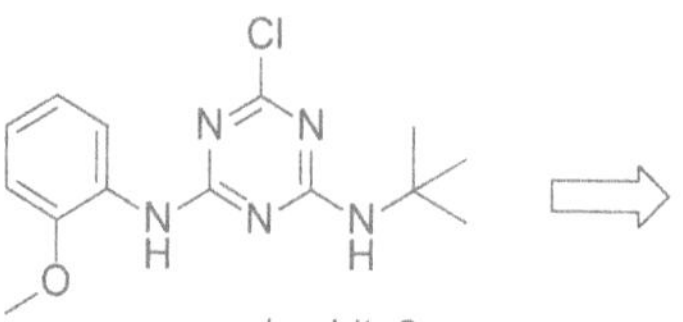

screening hit, **3**,
IDH2$_{R140Q}$, IC$_{50}$ = 1.9 μM

2		**1**
IDH2^{R140Q}, + NADPH	⟹	IDH2^{R140Q}, + NADPH
@ 16 h, IC$_{50}$ = 7 nM		@ 16 h, IC$_{50}$ = 100 nM

TRIFLUOROMETHYLPYRIDINE-CONTAINING DRUGS

In addition to enasidenib (**1**), two additional trifluoromethylpyridine-containing drugs are currently on the market. One is Upjohn's HIV protease inhibitor tipranavir (Aptivus, **4**).[4] The other is Janssen's androgen receptor antagonist apalutamide (Erleada, **5**) for treating castration-resistant prostate cancer (CRPC).[5]

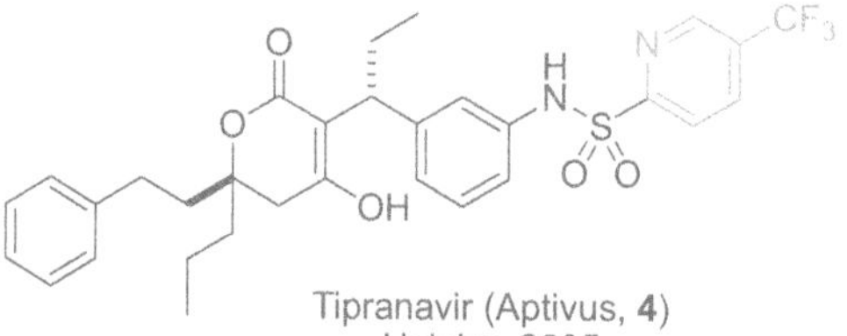

Tipranavir (Aptivus, **4**)
Upjohn, 2005
HIV protease inhibitor

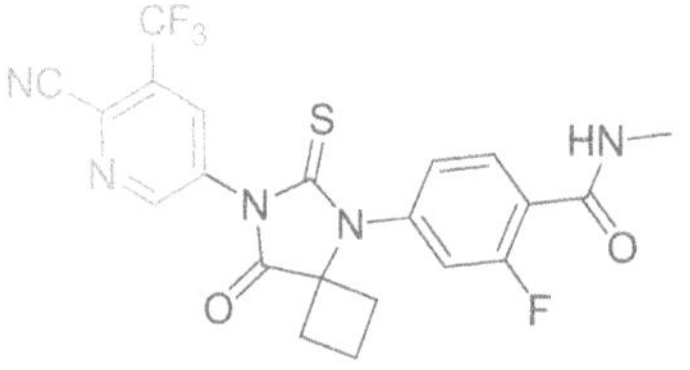

apalutamide (Erleada, **5**)
Janssen, 2018
Anti-androgen

During their structure–activity relationship (SAR) investigations, Upjohn explored three bio-isosteres for the arylsulfonamide moiety R. One was 1-methylimidazol-4-yl **6**, another was 5-cyano-2-pyridyl **7**, and the third was 5-trifluoromethyl-2-pyridyl **4**. Although compounds **7** and **4** had similar values for their K$_i$, IC$_{50}$, and IC$_{90}$, early safety studies suggested the advantages of compound **4** over compound **7**. Therefore, compound **4** was selected as a development candidate, which eventually became tipranavir (Aptivus).[4]

The structure of apalutamide (**5**), discovered by Jung's group at UCLA in the 2000s, is similar to that of enzalutamide (Xtandi, developed by Medivation and approved in 2012), also discovered by Jung. However, in murine xenograft models of metastasized-CRPC (mCRPC), apalutamide (**5**) demonstrated greater antitumor activity than enzalutamide. Furthermore, apalutamide (**5**) penetrates less effectively the BBB (blood–brain barrier) than enzalutamide, suggesting that the chance of developing seizures may be less than with enzalutamide. In the end, the fact that both Janssen and Medivation were able to secure intellectual properties for their respective AR antagonists also speaks volume of the power of the phenyl–pyridyl switch.[5]

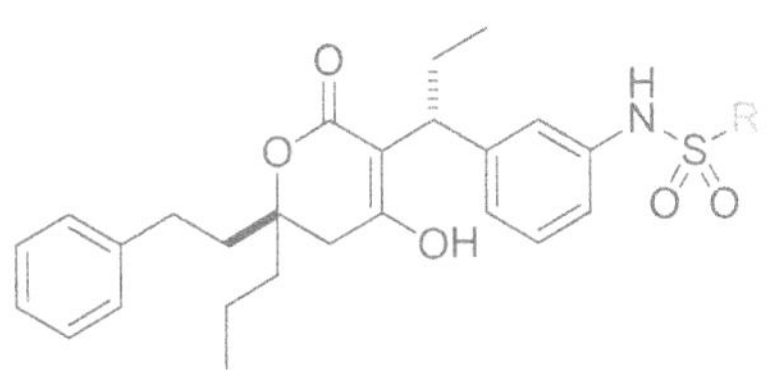

R	K_i (nM)	IC_{50} (nM)	IC_{90} (nM)
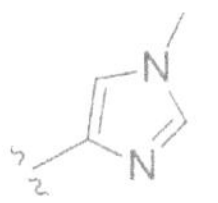	60	130	560
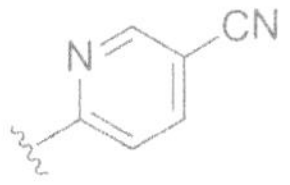	7	40	260
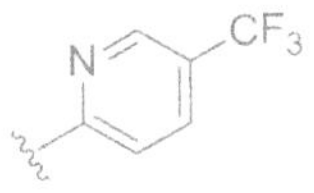	8	30	100

TRIFLUOROMETHYLPYRIDINES IN DRUG DISCOVERY

The constitutively active PI3K–AKT–mTOR signaling pathway is one of the key deregulation pathways important to cancer intervention. Phosphatidylinositol 3-kinases (PI3K) play a central role in broad cellular functions including cell growth, proliferation, differentiation, survival, and intracellular trafficking. Here RTK stands for the receptor tyrosine kinase, GFR is the acronym for growth factor receptors, and mTOR is short for a mammalian target of rapamycin. Gilead's PI3Kδ selective inhibitor idelalisib (Zydelig) was approved in 2014 and Bayer's pan-PI3K inhibitor copanlisib (Aliqopa) in 2017.

Trifluoromethylpyridine fragment has made appearances in several advanced PI3K/mTOR inhibitors.

In 2011, Novartis disclosed NVP-BKM120 (buparlisib, **8**), with a pyrimidine core structure and a 4-(trifluoromethyl)pyridin-2-amine substituent, as a potent, selective, and orally bioavailable class I PI3 kinase inhibitor for treating cancer. One of the morpholines binds to the hinge at Val882. The trifluoromethyl group here imparted superior solubility (132 μM for **8**) in comparison to the corresponding chloro-analog (45 μM) and the nitrile-analog (11 μM).[6] In 2018, Novartis sold buparlisib (**8**)'s right to China's Adlai Noryte, possibly because of some toxicity issues observed in phase III clinical trials for treating blood cancers and solid tumors. The culprit was likely due to buparlisib (**8**)'s microtubule interactions.

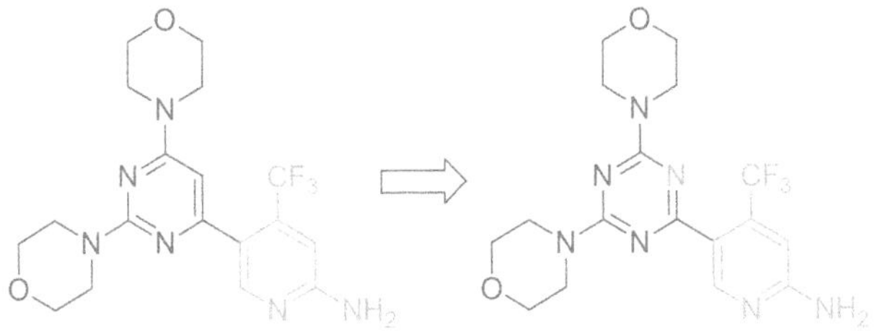

NVP-BKM120 (buparlisib, **8**)   PQR309 (bimiralisib, **9**)

PQR620 (**10**)

Wymann and colleagues at Basel sought to improve upon existing PI3K inhibitors while dialing in mTOR inhibition as well. They chose the triazine core structure aimed to (a) maximize compound solubility and bioavailability, (b) achieve blood–brain barrier (BBB) penetration, (c) avoid microtubule interactions as observed for buparlisib (**8**), and (d) introduce moderate mTOR inhibition. The fruit of their labor led to the discovery of PQR309 (bimiralisib, **9**) as a potent, brain-penetrant, orally bioavailable, pan-class I PI3K inhibitor.[7] It was in phase II clinical trials in 2017 for the treatment of advanced solid tumors and refractory lymphoma.

In another successful medicinal chemistry maneuver, Wymann et al. dialed out bimiralisib (**9**)'s PI3K activities via installation of bulkier substituted morpholines and replacing the trifluoromethyl substituent on pyridine with the corresponding difluoromethyl group. They arrived at PQR620 (**10**), which showed excellent selectivity for mTOR over PI3K and protein kinases and efficiently prevented cancer cell growth in a 66-cancer cell line panel.[8]

The trifluoromethylpyridine fragment seems to be a really fruitful substituent for PI3K inhibitors. In addition to pan-PI3K inhibitors mentioned above, Novartis discovered CDZ173 (leniolisib, **11**) as a structurally novel class of PI3Kδ selective inhibitors.[9] In 2019, it was in clinical trials for treating Sjörgren's syndrome.

CDZ173 (leniolisib, **11**)

The trifluoromethylpyridine fragment was helpful in boosting a series of anti-malarial drugs' overall profiles. Replacing the trifluoromethylphenyl fragment on compound **12** with the corresponding trifluoromethylpyridyl group led to analog **13** with improved human ether-a-go-go (hERG) potassium channel activity binding (from 3 to 20.2 μM). Furthermore, the aza-analog **13** (C_{avg}, 3.4 μM) enjoyed a 17-fold boost of plasma concentration over **12** (C_{avg}, 0.20 μM) after a single 30 mg/kg oral dose.[10]

12
hERG IC_{50} = 3 μM
$C_{avg\ 0-8h}$ = 0.20 μM

17-fold
plasma
exposure

13
hERG IC_{50} = 20.2 μM
$C_{avg\ 0-8h}$ = 0.20 μM

The phenyl–pyridine switch has done wonders in medicinal chemistry to improve a drug's *in vitro* binding affinity; *in vitro* functional affinity; *in vitro* PK/ADME profile; *in vitro* safety profile; and *in vivo* pharmacological profile.[11] The tactic made an *encore* appearance for Pfizer's phosphodiesterase 2A inhibitor (PDE2Ai) program. The PDE enzymes are well known now due to the success of sildenafil (Viagra, a PDE5 inhibitor for treating erectile dysfunction) and apremilast (Otezla, a PDE4 inhibitor for treating psoriasis). PDE2A inhibitors, on the other hand, have the potential to treat cognitive impairment associated with schizophrenia (CIAS). While one of Pfizer's PDE2Ai lead compounds pyrazolopyrimidine **14** had desired potency, selectivity, and brain penetration for a preclinical candidate, they sought to reduce the estimated human dose (180 mg/day). During SAR investigations, simply switching the trifluoromethylphenyl group on **14** to the corresponding trifluoromethylpyridyl analog indeed reduced clearance in human liver microsomes (HLMs) from 85 to 30 mL/min/kg, but the potency suffered a 4.5-fold loss. With the aid of computer-assisted drug design (CADD) and chemistry innovation to avoid using high-energy intermediates, they eventually arrived at imidazotriazine analog **15** as a potent, highly selective, and brain penetrant PDE2Ai clinical candidate.[12]

PDE2Ai **14**
PDE2A IC_{50} = 2 nM
HLM Cl_{int} = 85 mL/min/kg
MDR BA/AB = 1.4
estimated human dose = 180 mg/day
SFlogD = 2.9
LipE = 6.1

core change
H-bonding
SFlogD
Cl_{int}
LipE

PDE2Ai clinical candidate **15**
PDE2A IC_{50} = 1.6 nM
HLM Cl_{int} = 52 mL/min/kg
MDR BA/AB = 1.4
49 mg/day
SFlogD = 2.3
LipE = 6.5

Semko and colleagues at Élan discovered a novel series of sulfonamide-pyrazolopiperidines as potent and efficacious γ-secretase inhibitors. Although an early inhibitor **16** reduced brain Aβ40 levels by 25% in a wild-type FVB mouse model, its bioavailability was low (2.5%) because its ethyl group was readily oxidized and the chlorine atom on the phenyl ring was displaced by glutathione (GSH). For this particular series, the cyclopropyl group was an effective replacement for the ethyl group to achieve better stability while maintaining potency and the trifluoromethyl group helped to alleviate the glutathione conjugation issue. Therefore, trifluoromethylpyridyl analog **17** became less susceptible to glucuronidation at the NH site of the pyrazole ring but had low cellular activity since the core structure was easily oxidized to the aromatic hydroxypyridine. Thankfully, replacing the ketone with difluoromethylene led to compound **18**, which was bestowed with HLM stability (likely due to the two fluorine atoms that also lowered the susceptibility of the nearby moieties to CYP450 enzymatic oxidation) and lowered brain Aβ40 levels by 27% in a mouse model following oral administration of a dose of 1 mg/kg.[13]

16
IC_{50} (APP) = 0.3 nM
SNC cell EC_{50} = 1 nM

17
0.6 nM
2 nM

18
1.4 nM
4 nM

Novartis made a splash by revealing their allosteric SHP2 (Src homology region 2-containing protein tyrosine phosphatase) inhibitors in 2016. SHP2 is a non-receptor protein tyrosine phosphatase and scaffold protein. It is comprised of three domains: N-SH2, C-SH2, and PTP (PTP is short for PTPase, i.e., *protein tyrosine phosphatase*), where the active site resides. Since the phosphate group binding site is highly positively charged and often does not have a distinctive small molecule pocket, competitive SHP2 inhibitors mimicking phosphate are very challenging. The initial competitive SHP2 inhibitors discovered during the last two decades invariably possessed ionizable functional groups and thus had difficulty crossing cell membranes or enter the bloodstream. Novartis reported an allosteric SHP2 inhibitor SHP099, which occupies a tunnel-like binding site (a pocket formed by the confluence of the three domains) in SHP2's closed conformation. Because SHP2 is only active when adopting the open conformation, SHP099 behaves like a molecule glue that preventing the opening of SHP2. As an allosteric inhibitor, SHP099 does not need to be phosphate-like. It has appropriate affinity, cell permeability, and other properties that enable oral administration.[14] One of Novartis's allosteric SHP2 inhibitors, TNO155, went on to phase I clinical trials in 2017 and the world awaits the outcome. In 2019, they reported 3-amino-3-methylpyrimidinones such as **19** as potent, selective, and orally efficacious SHP2 inhibitors.[15]

SHP2 inhibitor **19**

SYNTHESIS OF SOME TRIFLUOROMETHYLPYRIDINE-CONTAINING DRUGS

Agios's preparation of their IDH2 allosteric inhibitor enasidenib (Idhifa, **1**) commenced with installation of the triazine core on **22** from the condensation of carbamylurea (**20**) with methyl 6-(trifluoromethyl)picolinate (**21**), followed by chlorination using POCl$_3$ to afford dichlorotriazine **22**. An S$_N$Ar displacement of the first chlorine on dichlorotriazine **22** with 4-amino-2-trifluoro-methylpyridine (**23**) gave monochlorotriazine **24**. Another S$_N$Ar displacement of the remaining chlorine with 1-amino-2-methyl-2-propanol then completed the synthesis of enasidenib (**1**).[3]

With regard to the synthesis of Upjohn's HIV protease inhibitor tipranavir (Aptivus, **4**), the major challenge was the installation of the two chiral centers on the left-hand portion **25**. Once that achieved, simply mixing aniline **25** with 5-(trifluoromethyl)pyridine-2-sulfonyl chloride (**26**) in the presence of pyridine as the base delivered the desired tipranavir (**4**) in excellent yield (81% yield includes the previous step for concurrently reducing the nitro group and a double bond).[16]

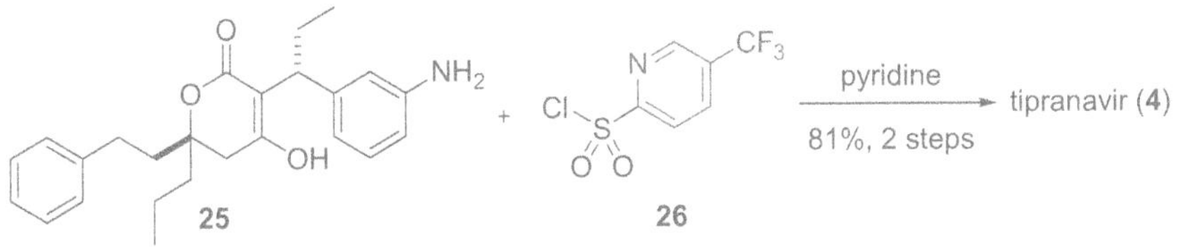

Jung's initial route on the patent was not amenable to process and manufacturing. Therefore, many process chemistry patents have been filed to improve the original route. In one case, hydrolysis of commercially available 2-chloro-3-(trifluoromethyl)pyridine (**27**) afforded pyridone **28**, which offered requisite reactivity and regioselectivity for nitration to produce 5-nitropyridone **29**. Refluxing **29** with a mixture of $POCl_3$ and PCl_5 restored the chloropyridine functionality on **30**, which was then reduced to 6-chloro-5-(trifluoromethyl)pyridin-3-amine (**31**). The choice of Raney nickel as the catalyst for hydrogenation was a wise one because a palladium-based catalyst would have caused concurrent dechlorination. After Boc protection of the amine as **32**, an S_NAr reaction took place to install 6-cyano-pyridine on **33**, which underwent an acidic deprotection to unmask the amine on **34**. Transformation of **34** to isothiocyanate **35** was accomplished by treating it with thiophosgene. Coupling between isothiocyanate **35** and aniline **36** then delivered apalutamide (**7**) after acidic hydrolysis.[5]

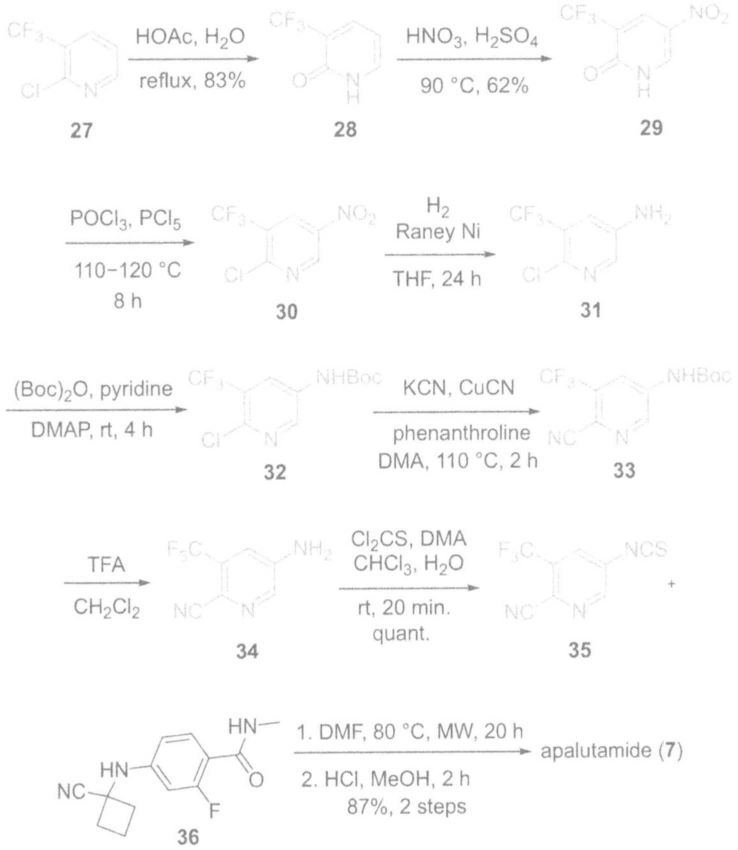

Synthesis of Basel's PI3K/mTOR inhibitor PQR309 (bimiralisib, **9**) is reasonably straight-forward. The Suzuki coupling between triazine-bromide **37** and pinacol-borane of the trifluoro-methylpyridyl fragment **38** was promoted by XPhosPdG2 as the catalyst to afford bimiralisib (**9**) in 52% yield.[7]

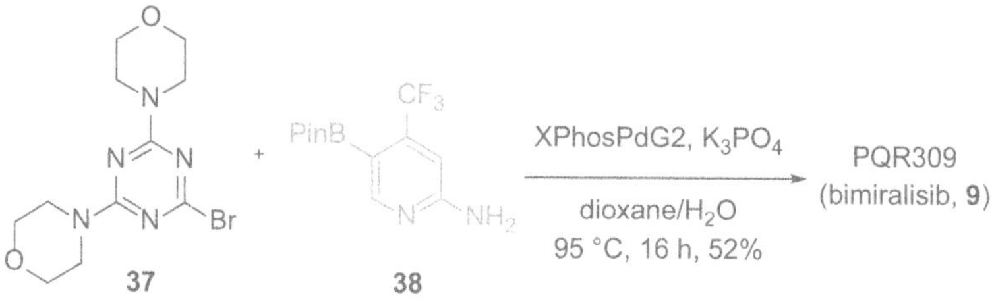

The Suzuki coupling was also key to assemble antimalarial drug **13**. The union between 3-pyridylbromide **39** and trifluoromethylpyridyl-boronic acid **40** took place with relative ease, only requiring the simple Pd(Ph$_3$P)$_2$Cl$_2$ catalyst.[10]

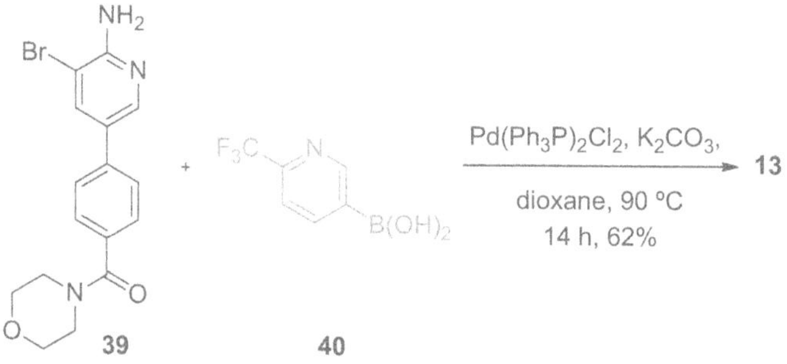

The Suzuki coupling is so ubiquitous in drug discovery, I wonder how we managed before it was discovered?! Pfizer's synthesis of their PDE2Ai clinical candidate **15** resorted to the Suzuki coupling between pyrazole bromide **41** and (5-(trifluoromethyl)pyridin-2-yl)boronic acid (**42**) to afford adduct **43**. Subsequent chlorination of **43** was followed by an S$_N$Ar replacement with azetidine to produce the API **15**.[12]

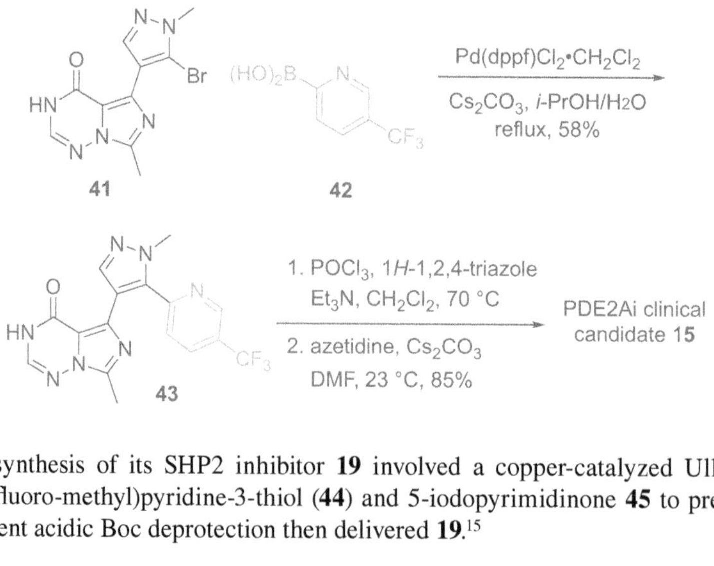

Novartis's synthesis of its SHP2 inhibitor **19** involved a copper-catalyzed Ullman coupling between 2-(trifluoro-methyl)pyridine-3-thiol (**44**) and 5-iodopyrimidinone **45** to prepare the thio-ether. Subsequent acidic Boc deprotection then delivered **19**.[15]

To conclude, the trifluoromethylpyridine fragment exists in at least three marketed drugs: Agios' IDH2 allosteric inhibitor enasidenib (Idhifa, **1**), Upjohn's HIV protease inhibitor tipranavir (**4**), and Janssen's androgen receptor antagonist apalutamide (**5**). The trifluoromethyl group may form tetrel bonding with heteroatoms on target proteins and the nitrogen atom on pyridine can serve as a hydrogen bond acceptor, establishing further binding to target proteins. As a consequence, trifluoromethylpyridine is a privileged structure in drug discovery. It may offer tighter binding to the target protein and improve upon a drug's solubility, metabolism, stability, and other drug-like properties.

REFERENCES

1. Garcia-LLinás, X.; Bauzá, A.; Seth, S. K.; Frontera, A. *J. Phys. Chem.* **2017**, *121*, 5371–5376.
2. Konteatis, Z. D.; Sui, Z. *Med. Chem. Rev.* **2018**, *53*, 525–539.
3. Yen, K.; Travins, J.; Wang, F.; David, M. D.; Artin, E.; Straley, K.; Padyana, A.; Gross, S.; DeLaBarre, B.; Tobin, E.; et al. *Cancer Disc.* **2017**, *7*, 478–493.
4. Thaisrivongs, S.; Skulnick, H. I.; Turner, S. R.; Strohbach, J. W.; Tommasi, R. A.; Johnson, P. D.; Aristoff, P. A.; Judge, T. M.; Gammill, R. B.; Morris, J. K.; et al. *J. Med. Chem.* **1996**, *39*, 4349–4353.
5. Jung, M. E.; Sawyers, C. L.; Ouk, S.; Tran, C.; Wongvipat, J. *Preparation of hydantoins as androgen receptor modulators for the treatment of prostate cancer and other androgen receptor-associated diseases.* WO2007126765 (2007).
6. Burger, M. T.; Pecchi, S.; Wagman, A.; Ni, Z.-J.; Knapp, M.; Hendrickson, T.; Atallah, G.; Pfister, K.; Zhang, Y.; Bartulis, S.; et al. *ACS Med. Chem. Lett.* **2011**, *2*, 774–779.
7. Beaufils, F.; Cmiljanovic, N.; Cmiljanovic, V.; Bohnacker, T.; Melone, A.; Marone, R.; Jackson, E.; Zhang, X.; Sele, A.; Borsari, C.; et al. *J. Med. Chem.* **2017**, *60*, 7524–7538.
8. Rageot, D.; Bohnacker, T.; Melone, A.; Langlois, J.-B.; Borsari, C.; Hillmann, P.; Sele, A. M.; Beaufils, F.; Zvelebil, M.; Hebeisen, P.; et al. *J. Med. Chem.* **2018**, *61*, 10084–10105.
9. Hoegenauer, K.; Soldermann, N.; Zecri, F.; Strang, R. S.; Graveleau, N.; Wolf, R. M.; Cooke, N. G.; Smith, A. B.; Hollingworth, G. J.; Blanz, J.; et al. *ACS Med. Chem. Lett.* **2017**, *8*, 975–980.
10. Gonzalez Cabrera, D. G.; Douelle, F.; Younis, Y.; Feng, T.-S.; Le Manach, C.; Nchinda, A. T.; Street, L. J.; Scheurer, C.; Kamber, J.; Chibale, K.; et al. *J. Med. Chem.* **2012**, *55*, 11022–11030.
11. Pennington, L. D.; Moustakas, D. T. *J. Med. Chem.* **2018**, *61*, 4386–4396.
12. Helal, C. J.; Arnold, E.; Boyden, T.; Chang, C.; Chappie, T. A.; Fisher, E.; Hajos, M.; Harms, J. F.; Hoffman, W. E.; Humphrey, J. M.; et al. *J. Med. Chem.* **2018**, *61*, 1001–1018.
13. Ye, X. M.; Konradi, A. W.; Smith, J.; Aubele, D. L.; Garofalo, A. W.; Marugg, J.; Neitzel, M. L.; Semko, C. M.; Sham, H. L.; Sun, M.; et al. *Bioorg. Med. Chem. Lett.* **2010**, *20*, 3502–3506.
14. Garcia Fortanet, J.; Chen, C. H.-T.; Chen, Y.-N. P.; Chen, Z.; Deng, Z.; Firestone, B.; Fekkes, P.; Fodor, M.; Fortin, P. D.; Fridrich, C.; et al. *J. Med. Chem.* **2016**, *59*, 7773–7782.
15. Sarver, P.; Acker, M.; Bagdanoff, J. T.; Chen, Z.; Chen, Y.-N.; Chan, H.; Firestone, B.; Fodor, M.; Fortanet, J.; Hao, H.; et al. *J. Med. Chem.* **2019**, *62*, 1793–1802.
16. Judge, T. M.; Phillips, G.; Morris, J. K.; Lovasz, K. D.; Romines, K. R.; Luke, G. P.; Tulinsky, J.; Tustin, J. M.; Chrusciel, R. A.; Dolak, L. A.; et al. *J. Am. Chem. Soc.* **1997**, *119*, 3627–3628.

Index